CHRONOSTRATIGRAPHIC UNITS

GLOBAL (EUROPEAN)														ERATHEM
CENOZOIC														
TERTIARY												QUATERNARY		SYSTEM
EOGENE						NEOGENE								
EOCENE			OLIGOCENE		MIOCENE				PLIOCENE		PLEISTOCENE	HOLO.	SERIES
MIDDLE		UPPER	LOWER	UPPER	LOWER		MIDDLE	UPPER						
LUTETIAN	BARTONIAN	PRIABONIAN	RUPELIAN	CHATTIAN	AQUITANIAN	BURDIGALIAN	LANGHIAN	SERRAVALLIAN	CHATTIAN	MESSINIAN	ZANCLEAN	PIACENZIAN			STAGE

GULF COAST		SERIES
CLAIBORNIAN	JACKSONIAN	VICKS-BURGIAN / CHICA-SAWHAYAN ; VICKS-BURGIAN / CHICASAWHAYAN	STAGE 1 / STAGE 2

577.09759 GEO
The geology of Florida /

LITHOSTRATIGRAPHIC UNITS

S.E. MISSISSIPPI, S.E. ALABAMA, W. FLORIDA PANHANDLE & OFFSHORE
Claiborne — Tallahatta — Jackson (Cockfield, Cook Mtn., Zilpha, Winona, Lisbon, Gosport, Moodys Branch) — Vicksburg (Forest Hill, Mint Spring, Marianna, Byram, Glendon, Red Bluff, Bucatunna, Bumpnose) — Yazoo — Chickasawhay — Paynes Hammock — Catahoula — Hattiesburg — Pensacola — Pascagoula — Graham Ferry — Citronelle — Alluvium

S.E. ALABAMA, CENT. FLA. PANHANDLE & S.W. GEORGIA
Claiborne — Tallahatta — Jack'n — Ocala — Lisbon — Moodys Branch — Suwannee — Chickasawhay — Suwannee — Chattahoochee — St. Marks — Alum Bluff (Shoal River, Oak Grove, Chipola, Bruce Creek) — Choctawhatchee — Intracoastal — Jackson Bluff — Citronelle — Alluvium

S. GEORGIA & E. FLA. PANHANDLE & N. FLA.
Oldsmar — Avon Park — Ocala — Suwannee — Hawthorn (Chattahoochee/St. Marks/Parachucla/Penney Farms, Torreya, Marks Head, Statenville, Altamaha/Coosahatchie) — Miccosukee — Jackson Bluff — Nashua — Cypresshead — Anastasia — Alluvium

CENTRAL & S. FLORIDA
Oldsmar — Avon Park — Ocala — Suwannee — Arcadia — Hawthorn (Arcadia, Tampa/Nocatee, Arcadia, Bone Valley, Peace River, Bone Valley, Peace River, Bone Valley/Peace River) — Tamiami — Caloosahatchee — Ft. Thompson — Anast./Miami/Key Largo — Alluvium

The Geology of Florida

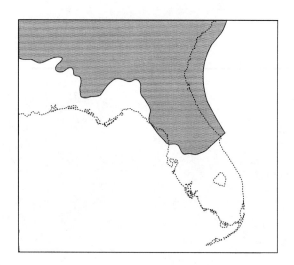
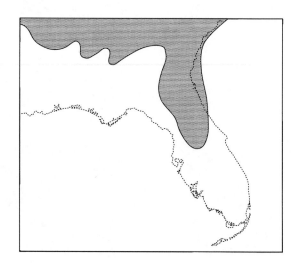

The Geology of Florida

Edited by
Anthony F. Randazzo
and
Douglas S. Jones

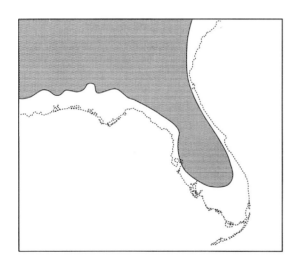

University Press of Florida

Gainesville • Tallahassee • Tampa • Boca Raton
Pensacola • Orlando • Miami • Jacksonville

Copyright 1997 by the Board of Regents of the State of Florida
Printed in the United States of America on acid-free paper
All rights reserved

12 11 10 09 08 07   8 7 6 5 4 3

LIBRARY OF CONGRESS CATALOGING-IN-PUBLICATION DATA

The geology of Florida / edited by Anthony F. Randazzo and Douglas S. Jones.
 p. cm.
Includes bibliographical references and index.
ISBN 0-8130-1496-4 (alk. paper)
1. Geology—Florida. I. Randazzo, Anthony F. II. Jones, Douglas S.
QE99.G47   1997
557.59—dc 20   96-28022

The University Press of Florida is the scholarly publishing agency for the State University System of Florida, comprising Florida A & M University, Florida Atlantic University, Florida International University, Florida State University, University of Central Florida, University of Florida, University of North Florida, University of South Florida, and University of West Florida.

University Press of Florida
15 Northwest 15th Street
Gainesville, FL 32611
http://www.upf.com

Front cover, clockwise from top: Everglades National Park, photo by Thomas M. Scott; brain coral from the Florida Keys, photo by Douglas Cook; Winter Park sinkhole, photo by Thomas M. Scott; clay lens-occurring in a massive dolomite sequence, Citrus-County, photo by Thomas M. Scott; *Ecphora bradleyae* fossil, photo courtesy of Florida Museum of Natural History; deformation in Hawthorn Group sediments, Alapaha River, photo by-Thomas M. Scott; drag-line operation for phosphate mining, Polk County, photo by Anthony-F.-Randazzo.

Back cover, top: Geologist Gene Shinn demonstrat-ing-sediment-core extraction in Florida Bay, photo by-Anthony F. Randazzo; *bottom:* sinkhole developed-in a gypstack, Polk County, photo courtesy-of Michael Graves, Richard Fountain, and-the *Lakeland Ledger.*

To William A. White, David Nicol, E. C. Pirkle, S. David Webb, Eugene A. Shinn, and Robert N. Ginsburg for their outstanding contributions to the understanding of Florida's geology.

Although we are mere sojourners on the surface of the planet, chained to a mere point in space, enduring but for a moment of time, the human mind is not only enabled to number worlds beyond the unassisted ken of mortal eye, but to trace the events of indefinite ages before the creation of our race, and is not even withheld from penetrating into the dark secrets of the ocean, or the interior of the solid globe; free, like the spirit which the poet described as animating the universe.

Charles Lyell, *Principles of Geology* (1830)

Contents

List of Figures and Plates   ix
List of Tables   xv
Preface   xvii

1. Geomorphology and
   Physiography of Florida   1
   Walter Schmidt

2. Tectonic Evolution and Geophysics
   of the Florida Basement   13
   Douglas L. Smith and Kenneth M. Lord

3. Geochemistry and Origin of
   Florida Crustal Basement Terranes   27
   Ann L. Heatherington and Paul A. Mueller

4. The Sedimentary Platform
   of Florida: Mesozoic to Cenozoic   39
   Anthony F. Randazzo

5. Miocene to Holocene
   History of Florida   57
   Thomas M. Scott

6. Hydrogeology of Florida   69
   James A. Miller

7. The Marine Invertebrate
   Fossil Record of Florida   89
   Douglas S. Jones

8. Fossil Mammals of Florida   119
   Bruce J. MacFadden

9. The Economic and
   Industrial Minerals of Florida   139
   Guerry H. McClellan and James L. Eades

10. Geology of the Florida Coast   155
    Richard A. Davis Jr.

11. Structural and Paleoceanographic
    Evolution of the Margins
    of the Florida Platform   169
    Albert C. Hine

12. Origin and Paleoceanographic
    Significance of Florida's
    Phosphorite Deposits   195
    John S. Compton

13. Environmental Geology of Florida   217
    Sam B. Upchurch and Anthony F. Randazzo

14. Geology of the Florida Keys   251
    Anthony F. Randazzo and Robert B. Halley

References   261
List of Contributors   299
Index   301

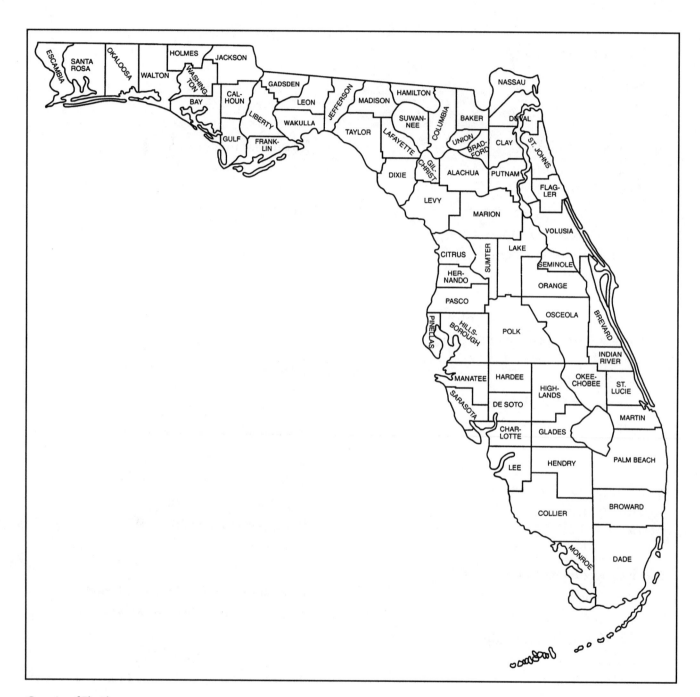

Counties of Florida

List of Figures and Plates

Figures

1. GEOMORPHOLOGY AND PHYSIOGRAPHY OF FLORIDA

1.1 Oblique view of the Florida Platform and Florida Escarpment   2
1.2 Profile across the Florida Platform showing the shoreline-shift effect of sea-level rise   2
1.3 Florida coastlines about 20,000 years ago   3
1.4 Principal physiographic provinces of the southeastern United States   3
1.5 Terraces and shorelines of Florida   3
1.6 Long Island Quadrangle   4
1.7 Juniper Creek Quadrangle   5
1.8 Crystal Lake Quadrangle   5
1.9 Distribution of closed depressions in the Lake City area   6
1.10 Cross-section of a typical Florida karst escarpment   6
1.11 Geomorphology of Florida   7
1.12 Physiographic regions of Florida   7
1.13 Postulated sequence for a Florida topographic inversion   11

2. TECTONIC EVOLUTION AND GEOPHYSICS OF THE FLORIDA BASEMENT

2.1 Geologic map of pre-Cretaceous rocks in the basement of Florida and surrounding areas   14
2.2 Map of major structural features of the West Florida Shelf and south Florida   16
2.3 Magnetic-anomaly map of the Florida Platform region   17
2.4 Gravity-anomaly map of the Florida Platform region   17
2.5 Gravity profile A-A' and modeled cross-section   18
2.6 Gravity profile B-B' and modeled cross-section   18
2.7 Structure-contour map and isometric view of the top of the basement density body   18
2.8 Theoretical depiction of the configuration of landmasses at the beginning of Triassic rifting   18
2.9 Probable locations of epicenters for earthquakes   19
2.10 Contour map of township averages of geothermal gradients in Florida and south Georgia   20
2.11 Heat-flow values in Florida   20
2.12 Possible faults in the Florida basement   22
2.13 Sequential convergence of Gondwana and Laurentia during the late Paleozoic Alleghenian Orogeny   23
2.14 Initial contact of Gondwana and Laurentia   23
2.15 Right-lateral faults and clockwise rotation of Gondwana   23
2.16 Continued motion of Gondwana and Laurentia   23
2.17 Configuration of Florida Platform portion of Gondwana–North America union after late Paleozoic closure   24
2.18 Configuration of Triassic South Georgia Rift   24

3. GEOCHEMISTRY AND ORIGIN OF FLORIDA CRUSTAL BASEMENT TERRANES

3.1 Basement components of Florida and adjacent areas   27
3.2 Sm/Nd systematics of north Florida tholeiites   29
3.3 Pb-isotope systematics of Florida basalts   29
3.4 Alkali/silica plot for southwest Florida rhyolitic rocks   29
3.5 Rb/Sr systematics of Florida basement lithologic suites   30
3.6 Locations of Avalonian terranes in eastern North America and lithotectonic belts of the southern Appalachians   34
3.7 Three postulated positions for the Tallahassee-Suwannee terrane and/or the southwest margin of Laurentia in the early Paleozoic   35

4. THE SEDIMENTARY PLATFORM OF FLORIDA

4.1 Paleogeographic and lithofacies maps of the Florida Platform, Triassic to Pleistocene   40–42
4.2 Isopach maps of Florida, Jurassic to Miocene   43–44
4.3 East-west cross-section of the Florida Platform   45

4.4 Global distribution of Cretaceous carbonate platforms 45
4.5 Example of cyclic sedimentation as depicted by carbonate rock type and skeletal/peloidal variance 46
4.6 Sea-level curve for the Cretaceous and Cenozoic 46
4.7 The Georgia Channel System and its two components, the Gulf Trough and the Suwannee Channel 46
4.8 Mesozoic and Cenozoic geochronologic and sea-level chart 47
4.9 Interpreted seismic profile across the West Florida Shelf and Escarpment 48
4.10 Foraminiferal wackestone-packstone from the Avon Park Formation 49
4.11 North-south geochronologic cross-section of the Florida peninsula 49
4.12 Cave formations in Florida Caverns State Park 51
4.13 Dolomite core material ("boulder") recovered from the "Boulder Zone" 52
4.14 Echinoid-foraminifer grainstone from the Upper Eocene Ocala Limestone 52
4.15 Dolomite displaying mimetic texture 53
4.16 Photomicrograph of medium-crystalline dolomite displaying a non-mimetic texture 54
4.17 Dolomitization models operating in the Florida Platform 54
4.18 Genetic sequence of microfabric development by heterogeneous dolomitization 55
4.19 Genetic sequence of microfabric development by homogeneous dolomitization 55
4.20 Plot of $^{87}Sr/^{86}Sr$ values of dolomites from the Florida Platform 55

5. MIOCENE TO HOLOCENE HISTORY OF FLORIDA

5.1 Structural features of Florida 58
5.2 Generalized Early Miocene sediment pattern 62
5.3 Generalized Middle Miocene sediment pattern 62
5.4 Generalized Late Miocene sediment pattern 63
5.5 Generalized Pliocene sediment pattern 63
5.6 Generalized Pleistocene sediment pattern 63
5.7 Generalized Neogene and Quaternary sediment thickness 63

6. HYDROGEOLOGY OF FLORIDA

6.1 Water budget of Florida 70
6.2 Average annual precipitation and runoff in Florida, 1951–1980 70
6.3 Major surface-water features and principal wetlands of Florida 70
6.4 Relation between water levels in the Hillsborough River at Morris Bridge and in a nearby well 70
6.5 Conditions causing a stream to gain or lose water 71
6.6 Principal aquifers exposed at or near the land surface in Florida 71
6.7 Hydrogeologic nomenclature 71
6.8 Withdrawals of freshwater by principal aquifer in Florida, 1985 72
6.9 Groundwater withdrawals in Florida, by county, 1985 73
6.10 Stratigraphic units included in the undifferentiated surficial aquifer system 74
6.11 Water movement in the surficial and Floridan aquifer systems in Indian River County 74
6.12 Withdrawals of freshwater from the surficial aquifer system by category of use, 1985 75
6.13 Stratigraphic units in the sand and gravel aquifer 75
6.14 Geohydrologic cross-section from northeastern Santa Rosa County to southwestern Escambia County 76
6.15 Withdrawals of freshwater from the sand and gravel aquifer by category of use, 1985 76
6.16 Stratigraphic units included in the Biscayne aquifer 77
6.17 Relation of the water table of the Biscayne aquifer to canals, water-conservation areas, and pumping centers 78
6.18 Movement of water between the Biscayne aquifer and canals 78
6.19 Hydrograph of well G86, completed in the Biscayne aquifer, 1942 79
6.20 Effect of canals and control structures on the position of the freshwater/saltwater interface 79
6.21 Position of the freshwater/saltwater interface in the Biscayne aquifer under natural conditions 80
6.22 Withdrawals of freshwater from the Biscayne aquifer by category of use, 1985 80
6.23 Location and approximate extent of the intermediate aquifer system 80
6.24 Stratigraphic units included in the intermediate aquifer system 81
6.25 Hydrogeologic cross-section from central Polk County to central Charlotte County 81
6.26 Potentiometric surface of the intermediate aquifer system, 1987 81

6.27 Withdrawals of freshwater from the intermediate aquifer system, by category of use, 1985   82

6.28 Extent of the Floridan aquifer system and saline water in the system   82

6.29 Stratigraphic units included in the Floridan aquifer system   83

6.30 Relation of time-stratigraphic units to subdivisions of the Floridan aquifer system   83

6.31 Thickness of the Floridan aquifer system   84

6.32 Hydrogeologic cross-section from Columbia County to Dade County   84

6.33 Thickness of the upper confining unit of the Floridan aquifer system   85

6.34 Locations of first-magnitude springs of Florida   85

6.35 Estimated transmissivity of the Upper Floridan aquifer   86

6.36 Potentiometric surface of the Upper Floridan aquifer   86

6.37 Net decline between estimated pre development potentiometric surface and observed potentiometric surface for the Upper Floridan aquifer   86

6.38 Withdrawals of freshwater from the Floridan aquifer system by category of use, 1985   87

6.39 Estimated freshwater withdrawals from the Floridan aquifer system, by county, 1990   87

6.40 Locations of injection wells in Florida, 1988   87

6.41 Hydrochemical-facies classification of waters in the Upper Floridan aquifer   88

7. THE MARINE INVERTEBRATE FOSSIL RECORD OF FLORIDA

7.1 Linguloid brachiopod in Paleozoic siltstone   90

7.2 Foraminifers from the Wood River Formation   92

7.3 Early Cretaceous benthic foraminifers   93

7.4 Late Cretaceous bivalves   93

7.5 Micro- and macrofossils from the Eocene Avon Park Formation   95

7.6 Fossil seagrass blades and epibionts from the Eocene Avon Park Formation   96

7.7 Outcrop area of the Ocala Limestone   96

7.8 Micro- and macrofossils from the Eocene Ocala Limestone   97

7.9 Boreholes of the coral-boring bivalve *Lithophaga palmerae* from the Eocene Ocala Limestone   99

7.10 Distribution of continents, oceans, and major oceanic currents during the Middle Eocene   99

7.11 Mollusks from the Florida Eocene with Tethyan affinities   100

7.12 Micro- and macrofossils from the Oligocene Suwannee and Marianna limestones   102

7.13 Miocene shallow-water marine molluscan provinces of the tropical western Atlantic   103

7.14 Macrofossils from the Tampa Member of the Arcadia Formation   104

7.15 Fossil mollusks from the Miocene Chipola Formation   106

7.16 Pliocene shallow-water marine molluscan provinces   107

7.17 Fossils from the Pliocene Jackson Bluff Formation   108

7.18 Macrofossils from the Pliocene Tamiami Formation, emphasizing the Pinecrest beds   109

7.19 Shell accumulations in a pit in Sarasota, with stratigraphic section   111

7.20 Macrofossils from the Caloosahatchee Formation   113

7.21 Modern shallow-water marine faunal provinces of the tropical western Atlantic   115

7.22 Macrofossils from the Pleistocene Bermont Formation   117

8. FOSSIL MAMMALS OF FLORIDA

8.1 A: Collecting fossil mammals from fissure deposits in the Ocala Limestone. B: Miocene and Pliocene Bone Valley phosphate deposits   120

8.2 Map of selected Florida land-mammal fossil localities and corresponding North American land-mammal ages   121

8.3 Correlation of major fossil-mammal-bearing sedimentary units and localities in Florida   121

8.4 Fossil bones at the middle Pleistocene Leisey shell pit   123

8.5 Reconstruction of *Holmesina septentrionalis* from the Pleistocene of Florida   124

8.6 Reconstruction of *Amphicyon longiramus* from the middle Miocene Thomas Farm site   126

8.7 Mandible of the procyonid *Arctonasua floridana* from the late Miocene Love site   126

8.8 *Barbourofelis* skull from the late Miocene Love site   127

8.9 Jaw of *Lynx rexroadensis* from the Bone Valley Formation   128

8.10 Evolutionary reversal in the extinct rodent *Mylagaulus* from Florida   128

8.11 Jaw and palate of the pocket gopher *Thomomys orientalis* from the Pleistocene of Florida   129

8.12 Jaw of the rhinoceros *Aphelops* from the late Miocene Love site   130

8.13 Reconstruction of *Parahippus leonensis* from the middle Miocene Thomas farm site   131

8.14 Different enamel patterns in two major groups of horses from the Bone Valley Formation   132

8.15 Dentition of the oreodont *Phenacocoelus luskensis* from the early Miocene of Florida   133

8.16 Cranial reconstruction of the ruminant artiodactyl *Kyptoceras amatorum* from the Bone Valley Formation   134

8.17 Reconstruction of cranium of the extinct pronghorn *Hexobelomeryx simpsoni* from the Bone Valley Formation   134

8.18 Reconstruction of the gomphothere proboscidean Amebelodon from the late Miocene of Florida   135

8.19 Comparison of teeth of Pleistocene proboscideans from Florida   135

8.20 Reconstructed skeleton of the extinct sirenian *Metaxytherium floridanum* from the Bone Valley Formation   136

8.21 Reconstruction of the late Eocene archaeocete whale *Basilosaurus,* and skeletons of *Basilosaurus* and *Zygorhiza*   136

8.22 Human "overkill" as a possible cause of Pleistocene megafaunal extinctions   136

9. THE ECONOMIC AND INDUSTRIAL MINERALS OF FLORIDA

9.1 Temporal variation in value of Florida mineral production   140

9.2 Florida counties producing gas and oil   140

9.3 Temporal variation in Florida gas and oil production   140

9.4 Florida counties producing phosphate   141

9.5 Temporal variation in Florida phosphate production   143

9.6 Temporal variation in Florida crushed stone production   144

9.7 Florida counties producing crushed stone   145

9.8 Florida counties producing cement   146

9.9 Temporal variation in Florida portland cement production   146

9.10 Temporal variation in Florida masonry cement production   147

9.11 Florida counties producing sand   148

9.12 Temporal variation in Florida sand and gravel production   148

9.13 Temporal variation in Florida clay mineral production   148

9.14 Florida counties producing clay   149

9.15 Temporal variation in Florida fullers earth production   150

9.16 Florida counties producing heavy mineral sands   152

9.17 Temporal variation in Florida peat production   152

9.18 Florida counties producing peat   153

10. GEOLOGY OF THE FLORIDA COAST

10.1 Florida's five major coastal sections   156

10.2 Hurricane landfalls in Florida, 1880–1980   156

10.3 Tidal range versus tidal prism for several Florida inlets   157

10.4 Generalized late Holocene sea-level curve   157

10.5 Aerial photo of the southern end of Anastasia Island   159

10.6 East coast of Florida, showing general trends in longshore sediment transport   159

10.7 Variation in beach-sand texture and mineralogy and hypothesized littoral-drift cells   160

10.8 Coast of southwest Florida, with adjacent bathymetry   160

10.9 One of the cuspate-foreland areas of the Cape Sable complex   161

10.10 Typical mangrove coast of southwest Florida   161

10.11 Idealized cross-section of the Ten Thousand Islands area   162

10.12 Islands and inlets of the west-central Florida barrier system   162

10.13 A: Anclote Key, a wave-dominated barrier. B: Caladesi Island, a mixed-energy (drumstick) barrier   163

10.14 Dunedin Pass   163

10.15 Cross-section of Anclote Key   164

10.16 Northern part of Cayo Costa Island   165

10.17 Karstic surface of Eocene limestones   165

10.18 Aerial photo of the estuary of the Crystal River   166

10.19 The panhandle coast   167

10.20 Apalachicola Delta area   167

10.21 Cape San Blas   168

11. PALEOCEANOGRAPHIC EVOLUTION OF THE MARGINS OF THE FLORIDA PLATFORM

11.1 Bathymetric map of the modern Florida Platform   169

11.2 Florida-Bahamas-Grand Banks Jurassic gigaplatform compared with modern Great Barrier Reef of Australia   170

11.3 Structure contour map of the mid-Jurassic surface   170

11.4 North-south geologic cross-section through peninsular Florida 171

11.5 Structure-contour map of the top of the Paleocene, showing sedimentary facies and current tracks 171

11.6 Cross-section from the Florida Keys to Great Bahama Bank 173

11.7 Seaward limit of subsurface Paleocene reef trend, downward flexures or folds beneath the inner shelf, and fracture traces mapped on land 173

11.8 A: Seismic line across the East Florida Shelf. B: Seismic-reflection profile across the Paleogene platform margin north of Cape Canaveral 174

11.9 Cross-section of the West Florida Platform 174

11.10 Albian paleogeographic map of the southern Florida Platform 175

11.11 Interpretation of a seismic line across the southwest margin of the Florida Platform 176

11.12 Seismic line across the Early Cretaceous margin, west Florida 176

11.13 Along-slope seismic-reflection profile, showing a submarine slide 177

11.14 Schematic summary of the carbonate-ramp slope of west-central Florida 179

11.15 Sedimentologic data from Exxon borehole 48 on the West Florida Slope 180

11.16 Interpretation of seismic line south of the Florida Keys 181

11.17 Seismic line across the West Florida Escarpment 182

11.18 Seabeam bathymetric chart of the Florida Canyon area 182

11.19 West-to-east cross-section of the Florida-Bahamas Platform 183

11.20 Schematic cross-section oriented west to east across south Florida, illustrating subsurface fluid flow within the Florida Platform 183

11.21 Diagram illustrating postulated groundwater flow within the West Florida Platform 184

11.22 Structure-contour map of subsurface unconformity R-3 in the Charlotte Harbor area 185

11.23 Structure-contour map of seismic basement beneath lower Tampa Bay 185

11.24 Sediment-facies map of the West Florida Shelf 187

11.25 Major controls on sedimentation on the West Florida Platform during the past 5.4 million years 188

11.26 Seismic line across the southern boundary of the Florida Platform 189

11.27 Seismic line off the western Florida Keys, illustrating outlier reefs 190

11.28 Bathymetry off Cape Canaveral, showing a cape-retreat massif 191

11.29 Map of relict reefs off the southeast Florida coast 192

11.30 Cartoon of lithoherms in the Straits of Florida 193

12. ORIGIN AND PALEOCEANOGRAPHIC SIGNIFICANCE OF FLORIDA'S PHOSPHORITE DEPOSITS

12.1 Estimated global abundance of P and the number of phosphate deposits throughout the Phanerozoic 196

12.2 Paleogeographic and paleoceanographic reconstruction of the Miocene continental margin of the southeastern U.S. 196

12.3 Postulated relationships among organic-C burial on the Florida Platform, the $\delta^{13}C$ record, and eustatic sea level 197

12.4 Stratigraphic correlations of the southeastern U.S. coastal plain 199

12.5 Miocene structural features and depositional pattern of the Florida Platform 199

12.6 Phosphorite from the Babcock Deep core 201

12.7 Interior of a phosphorite peloid from the Babcock Deep core 201

12.8 Diagenetic minerals from the Babcock Deep core 202

12.9 Variations in C- and S-isotopic compositions of pore-water bicarbonate and sulfate with depth of burial 203

12.10 C- and O-isotope compositions of francolite carbonate from the Hawthorn Group 204

12.11 Schematic of early burial diagenesis and reworking of sediment on the Florida Shelf during the Miocene 205

12.12 Age versus $^{87}Sr/^{86}Sr$ ratios of planktic foraminifers from DSDP sites 206

12.13 $^{87}Sr/^{86}Sr$ ratios of peloidal phosphorite grains, benthic foraminifers, shell fragments, dolomite, and shark teeth versus depth in the Babcock Deep core 209

12.14 $^{87}Sr/^{86}Sr$ age of benthic foraminifers and peloidal phosphorite grains versus depth in the Babcock Deep core 209

12.15 Comparison of eustatic sea-level fluctuations with the stratigraphy and gamma-ray log of the Babcock Deep core 211

12.16 Comparison of the $^{87}Sr/^{86}Sr$ ratios of phosphorite from the Babcock Deep core to the Sr- and C-isotopic compositions of benthic foraminifers from DSDP site 588 212

12.17 Comparison of the gamma-ray log of the Babcock Deep core and the percentage of phosphorite in bulk samples 212
12.18 Cenozoic record of $\delta^{18}O$, $\delta^{13}C$, and sea level 215

13. ENVIRONMENTAL GEOLOGY OF FLORIDA

13.1 Historic seismic events in Florida 218
13.2 Storm surge and waves breaking on the coast 219
13.3 Flood-prone areas in the Tampa–St. Petersburg area 220
13.4 Radar image of Hurricane Andrew 221
13.5 Cross-sections illustrating development of a collapse sinkhole 224
13.6 Damage caused by a cover-collapse sinkhole 225
13.7 Drilling rig that triggered a cover-collapse sinkhole 225
13.8 Cross-sections illustrating development of a solution sinkhole 225
13.9 Cross-section of an alluvial sinkhole 226
13.10 Cypress "dome" located over an alluvial sinkhole 226
13.11 Cross-sections showing stages of failure of an alluvial sinkhole 227
13.12 Cone-penetrometer-derived cross-section of an alluvial sinkhole 228
13.13 Winter Park sinkhole 229
13.14 Predicted sinkhole types in Florida 230
13.15 Sinkhole probability map of Florida 230
13.16 Examples of sinkholes used for waste disposal 231
13.17 Pattern of reported damage from sinkholes, swelling clays, and oxidizing organic materials in a Dunedin subdivision 233
13.18 Abundance of selected chemical constituents in waters of the Floridan aquifer system 238
13.19 Groundwater compositions for various aquifers in Lee County 239
13.20 Three-step process of maturation of water quality in the Floridan aquifer system 240
13.21 Gated spillway used to regulate water discharge from a flood-control canal 240
13.22 Phosphorite mining processes 241
13.23 Example of three plumes developed from a septic tank 244
13.24 Florida superfund sites, 1986 245
13.25 Decay products of ^{238}U 248

14. GEOLOGY OF THE FLORIDA KEYS

14.1 The Florida Keys in relation to water movements and the principal area of carbonate-sediment accumulations 252
14.2 Geologic map of Pleistocene limestones of the Florida Keys 252
14.3 Locations of selected test borings on the Florida Keys, and their logs 254
14.4 Panel diagram of carbonate facies of Pleistocene limestones of the Florida Keys 255
14.5 Traditional zones of an ecologic reef 256
14.6 Reconstruction of geologic conditions during the Pleistocene, showing the distribution of patch reefs and reef tract 257
14.7 Laminated crusts representing subaerial exposure and unconformities in the Florida Keys 257
14.8 Curve showing sea-level changes during the past 15,000 years 258
14.9 Future sea-level changes for Key West 258

Plates (following page 142)

1. Geologic map of outcrop and shallow subcrop rocks of Florida
2. Oolitic packstone from the Avon Park Formation
3. Upper Eocene Ocala Limestone exposed by quarry operations
4. The Anastasia Formation
5. Satellite image of the southwest Florida coast, showing the Ten Thousand Islands and adjacent areas
6. Redfish Pass, separating North Captiva and Captiva islands in Lee County
7. Oblique aerial photo of Bunces Pass
8. Cayo Costa Island
9. Nutrient-rich waters in Alaska Sink, Hillsborough County
10. Spur-and-groove development at Sand Key
11. Spur-and-groove development at Looe Key
12. A living reef community, including *Acropora palmata*

List of Tables

1.1 Primary, secondary, and tertiary division of Florida's landforms   10

2.1 Seismic events in Florida   19

4.1 Compositional ranges of dolomites from Florida   54

4.2 $^{87}Sr/^{86}Sr$ compositions of dolomite crystals from the Avon Park Formation   55

13.1 Modified Mercalli earthquake intensity scale   218

13.2 Saffir-Simpson hurricane scale   219

13.3 Criteria for differentiation of sinkhole and expansive clay damage   233

13.4 Summary of the chemistry of precipitation from selected sites   234

13.5 Common minerals in Florida and their dissolved weathering products   236

13.6 Common minerals in Florida aquifer systems and confining beds   236

13.7 Proportions of groundwater samples exceeding contaminant guidelines   243

13.8 Synthetic organic compounds found in Florida groundwater   243

14.1 Coral species of the Florida Keys and Dry Tortugas   253

Preface

WHETHER VIEWED BY AN ASTRONAUT FROM the window of the Space Shuttle or by a schoolchild on a classroom globe, few natural regions are immediately recognizable as Florida. Projecting several hundred kilometers from the southeastern corner of North America into the blue waters that gave it birth, the Florida Peninsula is surrounded by ocean yielding the longest coastline of any state except Alaska. Across millions of years the dynamics of this intimate relationship between land and sea have sculpted a unique geologic entity whose history is recorded within the rocks and strata buried below its topographically low surface. Between the covers of this book we attempt to "play back the recording" of Florida's rock record, to unravel its geologic history so that all may share the secrets of its past. For only in so doing may we accurately appreciate its present and knowledgeably anticipate its future.

Florida has enjoyed a rich, colorful history of geologic investigation, having attracted the attention of geologists for almost two centuries. Many, if not most, of these studies are cited within the chapters that follow. Yet, the first comprehensive attempt at an overview of the geology of Florida was not published until the early part of this century, appearing in the Second Annual Report of the Florida Geological Survey, authored by Matson and Clapp (1909). Updated volumes were published by the Survey approximately every two decades thereafter (Cooke and Mossom 1929; Cooke 1945; Puri and Vernon 1964).

It has been more than thirty years since Puri and Vernon last summarized the geology of Florida. The intervening years have witnessed a revolution in earth sciences, centered around the concept of a mobile Earth (seafloor spreading and plate-tectonics theory), which radically altered our understanding of fundamental geologic processes and changed irrevocably the way we view our dynamic planet. Spinoffs from this cardinal paradigm shift have permeated the earth sciences, leading to major advances in those fields of study with particular relevance to Florida, including marine geology, Cenozoic climatology, sea-level history, paleontology, and many others. As a consequence, our basic understanding of Florida's geologic evolution has been changed forever.

This book is designed to reflect the current state of knowledge of the geology of Florida, integrating older descriptions with the revolutionary perspectives developed over the past decades. As the first geology of Florida not to be published by the Florida Geological Survey, it differs from earlier volumes in several respects. First, it is organized into chapters reflecting a broad array of geologic topics related to Florida, and it is not a chronologic tracing of Florida's rock units through time. Second, it is intended to appeal to a diverse audience, from students and lay persons to scholars, while also being technically detailed enough for the geoscience specialist. Perhaps most significant, this book is an edited volume. The chapters are authored by leading authorities drawn from academia, state and federal geological surveys, and private industry. As editors, we feel that this assembly of expertise brings unprecedented breadth to the discussions of Florida's geology that follow.

We hope that the organization of chapters will introduce readers to the special landscapes of the Florida Peninsula and Panhandle (chapter 1) and provide a history of their origin and development (chapters 2 through 5). Succeeding chapters address topics ranging from hydrogeology (chapter 6), to paleontology (chapters 7 and 8), economic geology (chapters 9 and 12), coastal systems (chapter 10) and marine geology (chapter 11). The expanding field of environmental geology is discussed in relation to Florida in chapter 13, followed by a final chapter (14) concerned with the unique geology and significance of the Florida Keys.

Our goal is to provide readers with a liberally illustrated, up-to-date reference on the geology of Florida that both summarizes the current state of knowledge and interprets Florida's geologic history in light of fundamental principles. Beyond this immediate goal of communication, we hope that this book demonstrates the significance of basic geologic research and the importance of its application to issues facing a society that places increasing demands upon its physical world. With an adequate understanding of the dynamic machinery of geologic processes and an appreciation of the boundary conditions within which these processes have operated throughout geologic time, we, as a culture, will move closer to a harmony and balance with our natural environment.

But the process does not end here, just as this book is not a final product but a progress report. As scientific inquiry proceeds, an expanding supply of higher-resolution geologic information is essential for our public and private decision makers if we are to maximize our quality of life while simultaneously assuring the survival of Florida's delicate ecosystems and the management of its natural resources. We trust that as geologic concerns increasingly find their way into the daily lives of Floridians, this book will be used not just as a resource for students, scholars, and geoscience professionals, but by persons from all walks of life who are impacted by earth processes, or who merely desire to enrich their understanding of this special place called Florida.

* * *

This book was conceived nearly five years ago, following a series of conversations between the editors, who lamented the lack of an up-to-date geology of Florida that incorporated modern concepts of fundamental geologic processes. Its final publication, then, represents the realization of a long-awaited goal, attained only after much hard work, and involving the help of many people along the way. Paramount among these are the authors of each chapter, whose separate contributions define the strength and uniqueness of this volume. Special thanks are owed to David Batt and the Florida Phosphate Council, who were early supporters of this effort and made a significant financial contribution toward the production of this book.

Each chapter of this book was reviewed by the editors, and by two internal referees selected from the other chapter authors; for their help with these peer reviews, we thank each contributor who participated in this process. External referees also reviewed each chapter. For their knowledgeable opinions and comments we thank the following: Richard Buffler, University of Texas, Austin; David Dallmeyer, University of Georgia; Pamela Hallock-Muller, University of South Florida, St. Petersburg; Richard Hulbert, Georgia Southern University; Stanley Locker, University of South Florida, St. Petersburg; Donald McNeill, University of Miami; Henry T. Mullins, Syracuse University; David Nicol, University of Florida; C. K. Paull, University of North Carolina, Chapel Hill; E. C. Pirkle, University of Florida; Roger Portell, Florida Museum of Natural History; Stanley Riggs, East Carolina University; Eric Rosencrantz, University of Texas, Austin; John Sharp, University of Texas, Austin; E. A. Shinn, U.S. Geological Survey, St. Petersburg; William Tanner, Florida State University; and Paul A. Thayer, University of North Carolina, Wilmington.

At the University Press of Florida in Gainesville, we thank Walda Metcalf, associate director and editor-in-chief, and her staff, for invaluable assistance during each phase of production. Their ideas, enthusiasm, and special expertise are evident in a greatly enhanced final product.

For assorted contributions to one or more chapters we thank the following people who provided data, maps, photographs, opinions, references, illustrations, interpretations, editorial advice, word-processing skills, and/or ideas essential to the completion of this book: Paul Huddlestun, Georgia Geologic Survey; Jon Jee, Clemson University; Ed Lane, Joel Duncan, and Frank Rupert, Florida Geological Survey; William Wisner, Florida Department of Transportation; John Knaub, John Wollinka, Hallie Smith, Kelly Barber, Linda Parsons, Kirsten Hale, Ray Thomas, and David Hodell, University of Florida; Gary Morgan, Roger Portell, Craig Oyen, and Kevin Schindler, Florida Museum of Natural History; Eugene Shinn and Wendy Danchuck, U.S. Geological Survey; Jon Bryan, Okaloosa-Walton Community College; Chris Robinson, St. Johns County GIS Division; Stephen Snyder, North Carolina State University; Dave Mallinson, University of South Florida, St. Petersburg; and David Butler, South Florida Water Management District.

We gratefully acknowledge the assistance of the Florida Phosphate Council for a publications grant that made possible many of the graphics and illustrations appearing in this book. Of course, the production of each chapter was facilitated not only by the considerable efforts of the authors, but with the cooperation and resources of their respective institutions. For this indirect support we thank the following: Florida Geological Survey; U.S. Geological Survey; University of Florida (Department of Geology and Florida Museum of Natural History); University of South Florida (Department of Geology, Tampa, and Department of Marine Science, St. Petersburg).

A final acknowledgment of our appreciation is given to Lynne Randazzo and Sheila Jones. Their support and encouragement sustained our efforts to complete this work.

All royalties realized from the sale of this book will be used for scholarships for geology students at the University of Florida (Department of Geology) and the University of South Florida (Department of Geology and Department of Marine Science).

Geomorphology and Physiography of Florida

Walter Schmidt

The landforms we observe throughout our lifetimes seem to change little—they seem static. To the untrained eye, the mountains, shorelines, valleys, and plains appear the same year after year. Geologists know, however, that these features do change, and change continuously. The surface of the Earth is a dynamic system. The landforms we see today are but temporary forms in a long sequence of crustal modifications that began nearly 5 billion years ago with the origin of our planet. The same physical processes and laws that we observe in operation today have operated throughout geologic time, although at varying intensities and rates. By understanding these processes and their relationships to one another, we can imagine how the land may have looked in the distant past, or predict how it might look in the future.

Landforms appear limitless in variety, but geologists have grouped them using various schemes based on the processes that formed them. The dominant factors affecting landforms are the local geologic structure and rock type, and the type and rate of erosion or aggradation. Each geologic structure, through the associated geomorphic processes acting on it, develops its own characteristic group of landforms. In addition, varying rates of weathering and erosion will produce a decipherable sequence of landforms. The landforms of Florida are an expression of its geologic history.

The branch of science devoted to the study of landforms is variously known as geomorphology or physiography, and these two terms are often used interchangeably; in recent decades, however, they have come to have different meanings, each with its own usefulness to geologists.

Physiography is the older of the two. It was introduced into the science of geography in the late 1860s for the description of natural phenomena. It later came to mean "a description of the surface features of the Earth."

Still later, physiography became a branch of physical geography that dealt with the description and origin of landforms. Modern usage tends to assign the descriptive aspect of the study of landforms to physiography.

Geomorphology is a term which has replaced physiography in describing the study of the origin of landforms. It encompasses the classification, description, nature, origin, and development of landforms (Bates and Jackson 1980). Because of this more comprehensive, genetic definition, geomorphology is now considered a branch of geology rather than geography.

Physiography, then, generally refers to the descriptive side of the study of the Earth's surface, and geomorphology refers to the interpretive side.

Florida's Dominant Geomorphic Processes and the Resulting Landforms

Florida's landforms show the dominant effect of marine forces in shaping the land surface. Whenever the sea covered the Florida Platform, the shallow marine currents and their associated erosion and deposition shaped the shallow seabed, leaving subsequent erosional forces to modify this geometry. Ancient seas have left behind extensive flat plains that were their shallow floors, and scarps where old coastlines were cut into the uplands. Many present-day rivers and streams follow paths of old lagoons or near-shore swales.

Florida's ridges and highlands also owe their origin to shallow-water marine environments; they were left high when bordering currents eroded away sediments, or they were produced by windblown sand that formed coastal dunes and beach ridges. These features continue to be modified by the constant action of erosional and depositional forces acting on the emergent part of the Florida Platform known as the state of Florida.

MARINE INFLUENCES FORMING TERRACES AND SCARPS

The Florida Platform is a product of the sea. It has been in existence for tens of millions of years, and throughout that history it has been alternately flooded by shallow seas and salt lakes or exposed as dry land. The state of Florida as we know it today occupies only about half of the larger Florida Platform (fig. 1.1).

Because of the gentle slope and general lack of surface relief on the platform, a relatively small change in sea level can have a very dramatic effect on the area covered by the sea (figs. 1.2 and 1.3). The rest of the Gulf and Atlantic coastal plains also are relatively low in elevation compared to the mountains and continental interior (fig. 1.4), but Florida is especially low, with a maximum elevation of about 104 meters; this highest point in the state is in northern Walton County, in the panhandle. All the rocks and sediments that underlie the southeastern coastal plains were deposited by eolian, fluvial, or marine processes associated with near-shore marine environments during ancient high stands of sea level.

Today, as we observe the various landforms throughout the state, we can see the dominant influence marine forces have had in shaping Florida's landscape. The coastal areas are composed of negative features such as estuaries and lagoons, and positive features such as barrier islands, coastal ridges, and relict spits and bars, with intervening coast-parallel valleys.

Mapping of these landforms in conjunction with general elevation-zone mapping has resulted in identification of several well defined marine terraces throughout the state (Cooke 1945; MacNeil 1949; Alt and Brooks 1965; Pirkle et al. 1970; Healy 1975). These elevation zones represent periods of deposition and sediment reworking when ancient seas occupied higher levels. The most recent attempt at statewide correlation of these "terraces" and the associated scarps (the erosional slope break between terraces) was by Healy (1975), who identified eight terrace intervals and depicted them on a state map (fig. 1.5). They are, in ascending order, the Silver Bluff Terrace, less than 1 to 3 m above mean sea level; the Pamlico Terrace, 2.5 to 7.6 m above; the Talbot Terrace, 7.6 to 12.8 m; the Penholoway Terrace, 12.8 to 21.3 m; the Wicomico Terrace, 21.3 to 30.4 m; the Sunderland (or Okefenokee) Terrace, 30.4 to 51.8 m; the Coharie Terrace, 51.3 to 65.5 m; and the Hazlehurst Terrace

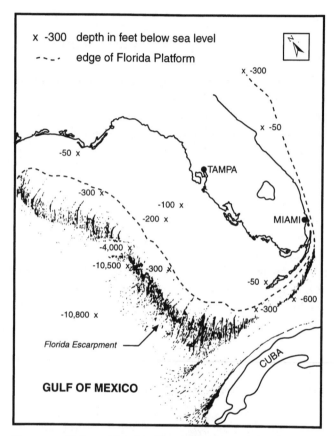

Figure 1.1. Oblique view of the Florida Platform and Florida Escarpment, showing the islands of the Florida Keys fringing its southern rim (Lane 1986, fig. 2.).

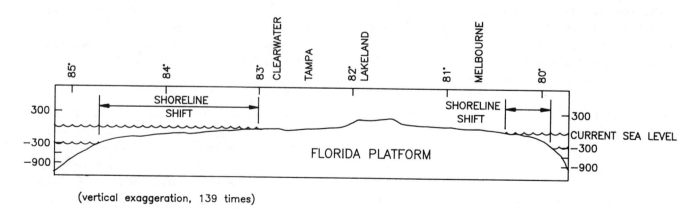

Figure 1.2. Profile across the Florida Platform along latitude 28°N, showing the shoreline-shift effect of sea-level rise over the last 20,000 years (modified from Cooke 1939).

(coastwise delta plain or high Pliocene terrace of various authors), 65.5 to 97.5 m.

Numerous methods have been used to delineate these terraces, each having its own advantage and contribution to the overall picture of developing landforms. Originally, topographic maps were simply correlated, and elevation zones or seemingly related features were identified and defined. Physiographic or geomorphic maps were also used to group the areas with common elevations. Stratigraphic, lithologic, and granulometric evidence has also been used to distinguish the marine terraces from deltaic, fluvial, or terrestrial sediments. Some researchers have used fossils to try to date the various surfaces, some have mapped soil types or drainage patterns, and some have described the botanical assemblage associated with a particular zone. Superimposed on these variables is the realization that even this evidence undergoes constant change: erosion, both surface and subsurface, reconfigures and modifies the natural evidence left from original deposition when ancient seas occupied the area.

COASTAL REGIONS AND ASSOCIATED LANDFORMS

The coastal zone—the transition from terrestrial uplands to marine environments—is a complex, dynamic system. Basically, two processes cause the system to respond in any of several ways. Sea-level changes or land-level changes in any combination can change the rate of erosion or deposition, or change the system from erosional to depositional.

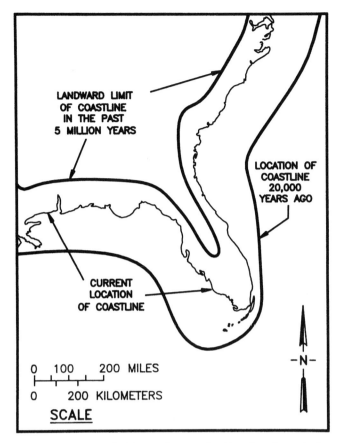

Figure 1.3. Ancient Florida coastlines: sea level has risen dramatically since the end of the last ice age, about 20,000 years ago (modified from Williams et al. 1990).

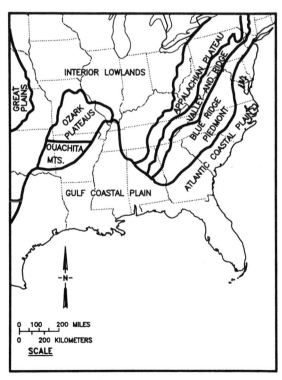

Figure 1.4. Principal physiographic provinces of the southeastern United States.

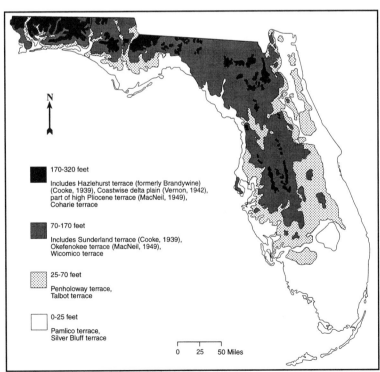

Figure 1.5. Terraces and shorelines of Florida (modified from Healy 1975).

Common topographic features created by marine erosion include wave-cut benches, sea cliffs, and rocky headlands flanked by coves or bays. Marine depositional features include beaches, bars, barrier beaches, and barrier chains. Still other types of coasts have coastal wetlands (including bayous), tidal flats, coastal marshes, and coral reefs, which in Florida include living reefs and dead reef-rock formations.

The classification of coasts is not a simple matter, and different classification systems have been devised, depending on the types of information included in the assessment; however, Florida's coastal features can be summarized as follows (see also King 1972; Tanner 1960a, 1960b): The majority of the east coast and the middle west coast are marine depositional coasts dominated by barrier beaches, barrier islands and spits, and overwash fans. The south Florida coast, south of Cape Romano around to Florida Bay, includes salt marshes with mud flats, marsh grasses, and mangroves. The Florida Keys, of course, include modern and ancient exposed coral reefs. The Big Bend Coast, from Pasco County north to the Ochlockonee River, is also a salt swamp and marsh, and a drowned karst topography, with abundant near-shore oyster reefs. From the Ochlockonee River west to Port St. Joe, along the Apalachicola River delta, the coast is dominated by barrier islands. Farther west, to the Alabama state line, drowned estuaries are bordered by barrier beaches and spits. R. A. Davis (chapter 10) provides a more complete discussion of Florida's modern coasts.

DRAINAGE SYSTEMS

After new land is formed, either by structural uplift, aggradational depositional systems, or drop in sea level, water moves across the land toward lower levels, eroding the existing surface as brooks and rivers continually dissect the land. These streams form valleys that are directly related to the drainage basin and to such parameters as amount of flow, gradient, and bedrock or sediment type. The valleys eventually coalesce, forming integrated groups called drainage systems.

By observing the drainage patterns evident on a topographic map or an aerial photo of an area, a geologist can gain a general understanding of the local subsurface geology and geologic history. Distinctive drainage patterns result from subsurface domes, folds, faults, or joint lines. Geomorphic processes related to coastal marine systems also leave their telltale patterns. A young drainage system differs markedly from a mature, or old, system. Various examples of Florida's drainage systems are discussed below.

The Florida Everglades is a unique area. A flat plane with negligible relief slopes gently to the south-southwest, from an elevation of about 4.6 m south of Lake Okeechobee to sea level about 120 km away. Shallow depressions formed by carbonate dissolution have created wetlands now filled in with cypress trees, forming cypress domes. The surrounding flat grasslands have been called the "River of Grass" by Douglas (1947). This sluggish, sheet-flow drainage system forms distinctive linear patterns where the cypress domes are elongated in the direction of flow (fig. 1.6).

The peninsula as a whole is dominated by north-to-south orientation of rivers. This reflects the near-shore marine environment which contributed to the basic landform construction. Relict beach ridges, constructed during past sea-level high stands, are separated by swales previously occupied by shallow lagoons. When sea level dropped, these lagoons became valleys, and streams eroded the sands and clays, creating several coast-parallel river systems seen today as the St. Johns, Oklawaha, Kissimmee, and Withlacoochee rivers.

Many of the streams of the clayey sand hills of the Northern Highlands in the northern Florida Peninsula and Panhandle display a typical dendritic drainage pattern. This reflects the lack of subsurface structural control on most streams, and in general a uniform resistance to erosion by most of the Neogene and Quaternary clayey sands. A dendritic drainage pattern is character-

Figure 1.6. Long Island Quadrangle, Dade County, Florida (7.5 minute orthophotomap).

ized by irregular, treelike branching of tributary streams, generally at angles less than 90°.

An anomalous area occurs on the west side of the Apalachicola River in Calhoun County, where an apparent trellis pattern exists (fig. 1.7). A trellis pattern is a subparallel system of streams, with primary tributary streams typically at right angles to the main stream; this pattern generally is associated with drainage following the strike of rock formations or a series of parallel faults or joints. In this area, however, other influences apparently cause the pattern: As the Apalachicola delta prograded southward, successive barrier islands and their associated landward lagoons became stranded high and dry. There is an analogous barrier system on the east side of the river (approaching symmetry?), suggesting similar environments on both sides of the prograding delta. Fourmile Creek and Juniper Creek in Calhoun County (fig. 1.7) occupy remnant lagoon floors. Their tributary streams, developed at right angles, are short and straight, reflecting the homogeneous nature of the sands and the steep flanks of the old barrier islands.

Internal drainage is another interesting pattern common in Florida. Where the limestones are close to the ground surface and the water table is low, there is typically little surface runoff. Most surface water in these areas drains through sinkholes into the limestone bedrock, or seeps into the bedrock through highly permeable surface sands. This type of drainage is typical of karst topography, which is characterized by sinkholes, caves, natural bridges, and springs, and generally lacks well developed rivers. These features occur because the underlying limestone dissolves more easily than most rocks, which leads to increased porosity and voids in the bedrock. Figure 1.8 is an aerial photo of part of the Crystal Lake Quadrangle, in Bay County, showing a transition from karstic internal drainage in the northern and central parts of the area to typical stream development in the southern part.

Figure 1.7. Juniper Creek Quadrangle, Calhoun County, Florida (7.5 minute orthophotomap).

Figure 1.8. Crystal Lake Quadrangle, Florida (7.5 minute orthophotomap).

KARST GEOMORPHOLOGY:
THE PATTERN OF SINKHOLES

The karst geomorphology of Florida has been summarized by White (1958, 1970) and Lane (1986). Typical landforms include small sinkholes on nearly planar karst platforms in the Coastal Lowlands; rolling hills and star-shaped sinkholes on the ridges, such as Brooksville Ridge and Lake Wales Ridge in north-central Florida; and isolated collapse sinkholes in buried-karst and/or newly developing karstic uplands.

Figure 1.9 illustrates the pattern of closed depressions and a geologic cross-section in the Lake City area of north-central Florida. Note that the pattern of depressions is different in the three physiographic domains. In the upland domain, the Hawthorn Group is intact, and few sinkholes occur. A number of the sinkholes in the Highlands province are deep, steep-walled, rock- or cover-collapse sinkholes, which form as the dolostones in

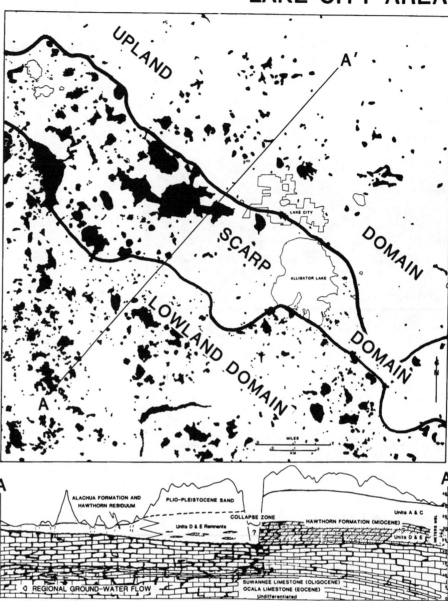

Figure 1.9. Distribution of closed depressions in the Lake City area (Columbia County). The region is subdivided into domains based on sinkhole size and density (courtesy of S. B. Upchurch).

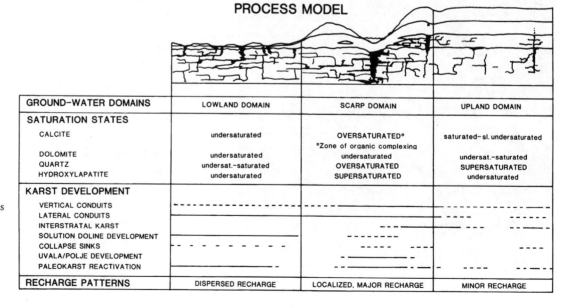

Figure 1.10. Cross-section of a typical Florida karst escarpment, showing the karst morphologies and groundwater chemistries associated with escarpment domains (courtesy of S. B. Upchurch).

the lower part of the Hawthorn Group are "sapped" by interstratal karst development. Devil's Millhopper State Geological Site and Little Salt Springs are examples of sinkholes that have formed on the outer, erosional edge of the Hawthorn Group. In the scarp domain (which is a subtle karst escarpment called the Cody Escarpment in north Florida, and may not be a topographic escarpment at all elsewhere), large, complex sinkholes (uvalas, poljes, and cockpits) and large sinkhole lakes form. These are large because thick remnants of the Hawthorn Group remain as a mantle, and the size of the funnel-shaped part of a sinkhole is a function of cover thickness; thus, large depressions develop over relatively small conduits. The lowland domain contains many small sinkholes, sinking streams, and other features in a bare to very thinly mantled karst plain. The sinks are small because the overlying sands have been largely removed, and only the rock conduits remain.

Figure 1.10 summarizes the groundwater chemistry required for scarp retreat, and the resulting karst landforms (Upchurch and Lawrence 1984). This process continues today, but apparently it was most intense in the past, during times of lower stands of sea level.

RIDGES AND HIGHLANDS

As discussed above, ancient high stands of sea level caused both erosion of the uplands and deposition and relief reduction in the shallow-water, near-shore zones. Highlands were eroded both by stream dissection and by reworking of sediments along the coast. The highlands of the northern peninsula and panhandle represent the dissected sedimentary remains of Neogene fluvial, deltaic, and shallow-water marine systems. These sediments were eroded from the southeastern coastal plain and southern Appalachians as the siliciclastic coastal-plain wedge prograded southward toward the sea. When these prograding

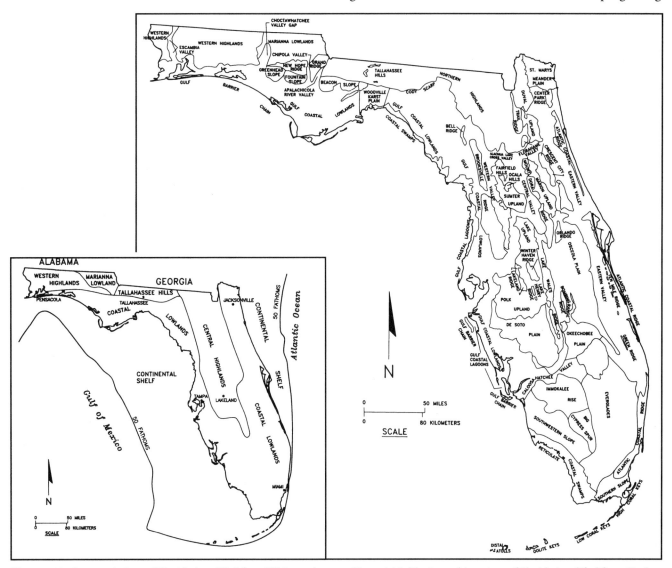

Figure 1.11. Geomorphology of Florida (modified from White et al. 1964 and White 1970).

Figure 1.12. Physiographic regions of Florida (modified from Cooke 1939).

sediments eventually filled the Gulf trough and spilled onto the carbonate platform, the siliciclastics were transported farther southward (Schmidt 1984). This siliciclastic invasion into the clear, carbonate-producing shallow waters, covered the limestone and formed the spine of clayey sands on the peninsula.

Subsequent sea-level fluctuations and associated nearshore, coast-parallel currents reworked and reshaped these deposits, leaving the elongate system of upland ridges we see today: Lake Wales Ridge, Trail Ridge, Mount Dora Ridge, and Brooksville Ridge are most prominent (fig. 1.11). In chapter 5, T. M. Scott presents a thorough discussion of platform development during the Neogene.

WETLANDS

Wetlands have become a very controversial topic in recent times, not only with respect to scientific study, but also with respect to environmental protection, regulation, and jurisdiction. Wetlands represent a complex, dynamic system, and considerations of their extent and preservation often impact zoning decisions by land-use planners. A better understanding of the origins of wetlands, then, is clearly desirable to aid in these decisions, and at the foundation of this understanding are the subsurface geology and geologic history of an area.

Wetlands are generally defined by three basic elements—hydrology, vegetation, and soils. Each of these parameters responds in different time frames to changes in the local water level, which by and large is directly related to seasonal precipitation rates (man's ability to modify or manipulate local surface waters and groundwater notwithstanding). It is important to note that most regulatory agencies in Florida distinguish wetlands and submerged lands: the latter are always covered by marine or freshwater (offshore marine areas, lakes, etc.), whereas wetlands are frequently inundated by water (lake-edge marshes, tidal zone, etc.).

Without an appreciation of the geologic makeup of an area, a complete understanding of wetland dynamics is impossible. Florida has numerous wetland types formed in various ways. The basic generic types are coastal marshes; estuaries; river and stream flood plains; low-relief plains; areas underlain by near-surface water tables; and groundwater potentiometric highs, or areas fed by artesian springs.

Regulatory agencies must understand that maps of wetlands represent the areal extent of the wetland at only one instant in time; wetland limits constantly change, and have a life cycle, just as lakes do. Earth science can offer a better appreciation of the long stretches of time involved in the formation and evolution of these landforms.

Historical Literature Summary

The authors who have addressed geomorphic nomenclature or described topographic features in Florida are too numerous to list here; instead, some of the more significant contributions are summarized. In many cases these publications contain references to studies which may be of local interest.

Most of the geomorphic subdivisions were recognized and described by early workers in the field, such as Shaler (1890), Matson and Clapp (1909), Matson and Sanford (1913), Harper (1914, 1921), Cooke and Mossom (1929), Fenneman (1938), Cooke (1939, 1945), and Davis (1943).

These pioneers did a herculean job of traversing the state, describing the exposed sediments, and compiling their descriptions of the local and regional topography. Matson and Clapp (1909) were first to publish a topographic and geologic map of Florida; in their text, they discussed the topography and drainage, and specifically addressed shorelines, coastal regions, lake regions, and north and west Florida. In the same volume, Sanford (1909) described the topography and geology of southern Florida. Matson and Sanford (1913) described four constituent sections of the central peninsula: the "Flatwoods," or "Hammock Lands," along the Gulf Coast; the "Lime Sink Region" inland from the Gulf Coast; the "Lake Region" along the axis of the peninsula; and the "East Florida Flatwoods" along the Atlantic Coast. Fenneman (1938) later provided brief descriptions of these areas.

Harper (1914, 1921) and Davis (1943), both botanists working with their geologic counterparts at the Florida Geological Survey, compiled valuable notes during their field work, providing a better understanding of the complex relationship between native vegetation, topography, drainage, and soils. Botanical assemblages were found to be valuable parameters in geologic and topographic mapping in the low-relief coastal plains; extensive species lists with associated soil types and physiographic conditions were compiled, and these relationships were documented in many photographs.

Cooke (1939) discussed at length the physical processes responsible for observed landforms. He described the past marine high stands of sea level and how they shaped the terrain, and how running water, dissolution, and waves and winds have eroded the land. His primary physiographic divisions of the Floridan Plateau (Cooke 1939, 1945) are still generally cited as the basis for statewide divisions (fig. 1.12).

Vernon (1951) subdivided the physiography of Florida into four basic divisions based on origin. He divided the Highlands into a "Delta Plain Highlands" and a "Tertiary Highlands," and the lowlands into the

"Terraced Coastal Lowlands" and the "River Valley Lowlands." He further stated that these divisions "can be subdivided to any degree and local names can be applied."

Among the more modern and comprehensive works addressing the state's geomorphology that have resulted in maps of the entire state are those of White (1958), White et al. (1964), White (1970), and Brooks (1981). White (1958) named five highlands for the cities located on them: Lake Wales Ridge, Winter Haven Ridge, Lakeland Ridge, Brooksville Ridge, and Orlando Ridge. He also described the processes that contributed to the formation of many land features throughout the peninsula. His discussion included river and valley formation, beach-ridge and dune formation, lake types, and a general review of the karst processes evident on the Floridan Platform.

White et al. (1964) proposed a hierarchical (primary, secondary, and tertiary) division of Florida's landforms (table 1.1); they included a 1:1,000,000 physiographic map of Florida in the form of three figures (their figs. 1.5, 1.6, and 1.7). However, the most often cited publication on Florida peninsular geomorphology is probably White (1970); although his map is identical to that published by Puri and Vernon (1964), White described in detail the secondary and tertiary subdivisions, which were merely named by White et al. (1964).

Brooks (1981) published a map at 1:500,000 titled "Physiographic Divisions," with an accompanying text. To delineate the geomorphic features, Brooks considered rock type, soil type, geologic structure, geomorphic processes, and relief; other criteria used to define minor divisions included degree of dissection, relationship to the water table, and the soil-forming processes.

Interesting Observations and Interpretations

Florida has numerous examples of both classic and uncommon geomorphic processes and features that reflect its geologic history and geographic location: stream capture, meander patterns, offset river courses, disappearing water bodies, and topographic lineaments and inversion are well illustrated in the geomorphology of Florida.

STREAM CAPTURE: CHATTAHOOCHEE, FLINT, AND APALACHICOLA RIVERS

Hendry and Yon (1958), through analysis of topography and drainage basins, and using data from shallow auger holes, postulated capture of the Chattahoochee and Flint rivers by the Apalachicola River. Their figure 4 diagrammatically illustrates the postulated stages of capture. The data call for an ancient Chattahoochee River valley farther west than its present course in Jackson County, and the Flint flowing into this system farther west than the present junction at the city of Chattahoochee. The ancient Apalachicola was a tributary to this old Chattahoochee River. As the Apalachicola cut farther headward into the highlands, it eventually captured the Flint River, and the Chattahoochee was diverted eastward by the enlarged, inverted stream (the beheaded part of the Flint River). The old valley of the ancestral Chattahoochee is now occupied by the Chipola River.

TECTONIC DEFORMATION INFERRED FROM TERRACE MAPPING OR SURFACE LINEAMENTS

Several authors have compiled data on regional terraces or surface lineaments, and proposed various tectonic solutions to account for the observed patterns. For example, Tanner (1965) and Bond et al. (1981) discussed a fault-control origin for Lake Okeechobee in south Florida; they suggested that the lake occupies a down-faulted depression, based on geomorphologic evidence and a postulated vertical offset observed in wells northeast of the lake. Various surface features in south Florida (primarily lineaments) were ascribed to subsurface geologic structures.

Hoyt (1969) studied Plio-Pleistocene sediments in north Florida to detect tectonic displacement. His correlations of previously mapped ancient coastlines showed that these features, described as essentially horizontal, were in fact warped. He suggested uplift in the area around Ocala.

Winker and Howard (1977) correlated three shoreline sequences along the southern Atlantic Coastal Plain. The geomorphic features they mapped included shoreline scarps, beach ridges, and fluvial terraces extending into the northern peninsula of Florida. They used a combination of published and direct geomorphic evidence to suggest that the shoreline sequences have been deformed, and concluded that warping continued through Pleistocene time.

Otvos (1981) reported that tectonism, reflected in structural lineaments, was probably responsible for coastal escarpments formed in the Florida Panhandle. He reported that lineaments appeared in the configuration of escarpments, aligned streams, shoreline directions, and other features. He believed that vertical movements caused erosion of Pleistocene and Pliocene sediments.

PLATFORM REBOUND FROM CARBONATE LOSS

Opdyke et al. (1984) plotted elevations of Pleistocene marine fossils and concluded that Florida had been uplifted epeirogenically during the Pleistocene. They further offered a mechanism for the uplift: They estimated

Table 1.1. Primary, secondary, and tertiary division of Florida's landforms

Atlantic Coastal Lowlands	Intermediate Coastal Lowlands	Gulf Coastal Lowlands	Central Highlands	Northern Highlands	Marianna Lowlands
The Eastern Valley	Okeechobee Plain	Reticulate coastal swamps and Ten Thousand Islands	Marion Upland	Trail Ridge	Holmes Valley
Peoria Hill	Caloosahatchee Incline		Mount Dora Ridge	Florahome Valley	Escarpment
Palatka Hill	Everglades	Gulf coastal lagoons and barrier chains	Central Valley	Tallahassee Hills	
San Mateo Hill	Caloosahatchee Valley		Lake Wales Ridge	Cody Escarpment	
Ten Mile Ridge	Immokalee Rise	Gulf coastal estuaries	Inter-Ridge Valley	New Hope Ridge	
Green Ridge	Big Cypress Spur	Coastal swamps and drowned coastal karst	Fairfield Hills	Washington Co. outliers	
St. Mary's meander plain	Southwestern Slope		Martel Hill	High Hill	
Roses Bluff	Southern Slope	Aeolian features	Ocala Hill	Orange Hill	
Yulee Hill	Florida Bay mangrove islands	Cape Sable	Cotton Plant Ridge	Rock Hill	
Evergreen Hill		Desoto Plain	Sumter Upland	Oak Hill	
St. Johns River offset	Keys	Polk Upland	Lake Upland	Falling Water Hill	
Wekiva Plain	High coral Keys	Bell Ridge	Rock Ridge Hills	Western Highlands	
Atlantic Coastal Ridge, lagoons and barrier chain	Low coral Keys	Greenhead Slope	Winter Haven Ridge	Drowned valley lakes	
	Oolite Keys	The Deadenings	Gordonville Ridge		
	Distal atolls	Fountain Slope	Lake Henry Ridge		
Silver Bluff Scarp		Beacon Slope	Lakeland Ridge		
Duval Upland		Wakulla Hills	Western Valley		
Teasdale Hill		Lake Munson Hills	Tsala Apopka Plain		
Crescent Ridge		Calico Hill	Brooksville Ridge		
Deland Ridge		Pea Ridge	Alachua Lake		
Geneva Hill		Interlevee Lowlands	Cross Valley		
Osceola Plain		Tates Hell Swamp	Lake Harris Cross Valley		
Orlando Ridge		Pickett Bay	Dunnellon Gap		
Bombing Range Ridge		Relict spits and bars	High Springs Gap		
		Terraces of the panhandle	Kenwood Gap		
		Braided channel of Wacissa River	Zephyr Hills Gap		

Source: White et al. 1964.

the sum of dissolved solids discharging from Florida springs and calculated a loss equivalent to 1 m of surficial limestone every 38,000 years. They postulated that this loss caused isostatic uplift of at least 36 m during the Quaternary, consistent with observed elevations of marine terraces; however, fossiliferous Quaternary sediments have not been identified above 20 m in elevation in peninsular Florida, although they have been reported at 49 m in southeasternmost Georgia, behind Trail Ridge (Pirkle and Czel 1983).

DISAPPEARING LAKES AND RIVERS

Many authors have described the numerous lakes in Florida that occasionally drain through lake-bottom sinkholes. One of the first was Gunter (1934), who described the major role of limestone dissolution in the origin of many lakes. He explained how limestones underlying these lakes are slowly dissolved by circulating groundwater, forming solution channels and, finally, sinkholes. Continued circulation of these slightly acidic groundwaters progressively lowers an area, and ultimately forms a lake basin. As a rule, the "disappearance" of most lakes seems to occur after prolonged periods of dry weather, when the water table is lowered, decreasing buoyant support to cavern roofs and causing collapses that drain the remaining water from lake low spots. Classic examples of this phenomenon are lakes Jackson and Iamonia in Leon County, Lake Miccosukee in Jefferson County, Paynes Prairie in Alachua County, and the Alapaha River in Hamilton County.

A TOPOGRAPHIC INVERSION

White (1970) and Knapp (1977) described an interesting phenomenon in the northern Brooksville Ridge area. They noted that the core of the ridge is incised into the surrounding limestone plain, and is composed primarily of Hawthorn Group siliciclastics deposited in a tidal channel or marine lagoon. Because of the relative insolubility of these clays and quartz sands, they were not prone to the dissolution that profoundly affected the contiguous limestones: the siliciclastics have remained topographically high, while the surrounding limestone plain has been lowered by dissolution (fig. 1.13).

MEANDER PATTERN OF
THE SUWANNEE RIVER

The Suwannee River is entrenched in a shallow solution valley along most of its length. Outcrops of limestone are common from White Springs, in Hamilton County, downstream to the Gulf of Mexico. The river traverses sandy limestones and dolostones of the Miocene Hawthorn Group, the Oligocene Suwannee Limestone, and the Upper Eocene Ocala Limestone.

Vernon (1951) postulated that regional joint patterns control the courses of many Florida rivers, including the Suwannee. He suggested that the joint patterns represent fracturing along stress lines associated with the Ocala Platform. Two regional patterns were noted— one roughly NW-SE, and a subordinate NE-SW trend. Colton (1978) measured these linear patterns along the Suwannee and reported their averages to be N 65° W and N 39° E. He further noted that the direction of flow for 92% of the total length of the river (cutting the Suwannee Limestone) could be attributed to these joint patterns.

Vernon (1951) noted that the Live Oak area in Suwannee County was structurally high. He suggested that the crest of the Ocala uplift separates into two folds near the northern boundary of Levy County, one trending north and passing east of Live Oak, and the other continuing the northwest trend of the main fold. Vernon believed that the courses of the Suwannee and Santa Fe rivers were influenced by the dip of bedrock along the flanks of these two folds. In this same context, Colton (1978) recognized five domelike features on his structure-contour map of the top of the Suwannee Limestone. These features likely contributed to the drainage pattern of the Suwannee River in the vicinity of Hamilton and Suwannee counties; however, more recent stratigraphic interpretation by the Florida Geological Survey indicates that these features merely represent a typical, undulatory karstic surface.

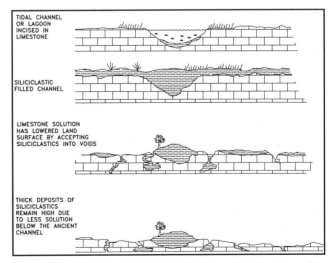

Figure 1.13. Schematic sequence postulated for a Florida topographic inversion.

Brooks (1966) compiled information on supposed faults in the northern peninsula area. He postulated a fault trending NW-SE that separates the karst limestone plain developed over the Ocala Platform from the "Okefenokee high flatlands." Brooks stated that "it is claimed by some that the abrupt change in the course of the Suwannee River at White Springs to the northwest is in relation to a fault."

Pliocene and Pleistocene landforms resulting from sea-level fluctuations clearly impacted the river's course as we see it today. Okefenokee Swamp is the headwaters of the Suwannee and St. Mary's rivers. The swamp lies in an old strath of an eastward-flowing Pliocene river that graded to a sea-level stand at about 36.5 m (Brooks 1966). The northern extension of Trail Ridge subsequently cut off the eastward course of this ancestral river, which was in turn captured by a stream (the present Suwannee River) flowing westward toward the Gulf of Mexico. Brooks postulated that this may have occurred during the early Pleistocene.

THE OFFSET COURSE OF THE NORTH-FLOWING ST. JOHNS RIVER

Most rivers and streams in the Northern Hemisphere tend to flow toward the equator; that is, level, homogeneous, structureless, massive sediments would not bias or force drainage away from that trend. The Earth's centrifugal force seems to account for this, and, indeed, north-flowing rivers are rare north of the equator, the most famous being the Nile in North Africa.

The St. Johns River in northeast Florida is also an exception. The river comprises three segments. The southern segment includes the headwaters in southern Brevard County and continues north approximately to the latitude of Sanford. The middle segment of the river lies between Sanford and Palatka. At Palatka, the northern segment jogs roughly east and then north to Jacksonville, where it makes its final jog east to the Atlantic.

Whereas some geologists believe that the St. Johns River originated in a lagoon developed in Pamlico time (late Pleistocene), the more recent consensus is that the river developed on a regressional or progradational beach-ridge plain (White 1970; Pirkle 1971). The course of the river may have been determined in part by the location of swales between the relict Pleistocene beach ridges. The middle segment of the St. Johns has been referred to as the "St. Johns River offset" by White and Pirkle, and it may have had a history different from that of the northern and southern segments. The offset segment was probably formed earlier, during the late Tertiary to early Pleistocene.

A number of coast-parallel topographic features, such as relict lagoons and swales between beach ridges, would have controlled the flow direction of surface water. This water then would have moved into the underlying limestone and sandy shell beds and dissolved them, forming a series of fairly straight valleys. It has also been suggested that some type of faulting has occurred, bringing the underlying limestones closer to the surface in the area of the offset (Wyrick 1960; Barraclough 1962; Leve 1966; Lichtler et al. 1968; Duncan et al. 1994). Factors believed to contribute to the establishment of this older offset segment include faulting and associated fracturing, solution of carbonate sediments, and resulting enhanced artesian discharge. At some time when sea level was low, the offset part of the St. Johns captured the present headwaters of the river east of Sanford. When sea level rose, the offset part of the St. Johns became an estuary, and when sea level subsequently fell, the St. Johns became the integrated river system we now see (White 1970).

Summary

At first glance, Florida's geomorphology seems mundane and unchallenging compared to areas where there is greater surface relief in the form of mountains, canyons, and other spectacular landforms, which suggest a history of land changes greater and more varied than Florida's. But, as briefly described in this chapter, Florida's geomorphology often reflects a subtle and complex history of geologic processes.

The Florida Platform has been described as a child of the sea. This is reflected in the thousands of meters of rocks deposited in marine environments. It is also evident from the near-shore sand and shell deposits which have been laid down and reworked numerous times before assuming their present resting place. Marine and coastal processes have been the dominant factors shaping and modifying the submerged parts of the platform during most of the Cenozoic. The most recently emerged part of the platform, which we call the State of Florida, is now exposed to the winds and rains of our atmosphere, whereby riverine deposits and valleys, eroded highlands, karst landforms, windblown dunes, and other terrestrial physiographic features are created. Florida's long history near sea level has provided unique opportunities for geologists to compile information on sea-level changes, shallow-water marine life and environmental evolution, extinctions of fossil plants and animals, and ocean-circulation patterns of past epochs.

2

Tectonic Evolution and Geophysics of the Florida Basement

Douglas L. Smith and Kenneth M. Lord

SHROUDED WITH RELATIVELY YOUNG sedimentary rocks and devoid of current tectonic activity, the crustal basement of the Florida Platform harbors the clues to unravel its tectonic history. The prominent peninsular shape of the plateau, and of the exposed eastern half that constitutes the state, is not geographically unique, but is sufficiently rare to justify plausible reconstructions of geologic events as causal mechanisms. Evidence from the Florida basement, including its peninsular shape, can contribute to viable explanations of its evolution and placement.

The term *basement* usually describes some underlying crystalline complex that is overlain unconformably by sedimentary strata. By implication, the basement can extend to the base of the crust and may have experienced intense deformation. As a relative term, however, it refers to those rocks under any pronounced unconformity, below which the rocks are poorly known.

Descriptions of borehole cuttings and interpretations of their correlations yielded early definitions of the Florida basement as the pre-Mesozoic subsurface (Applin 1951; Milton and Grasty 1969) and, more specifically, the sub-Zuni surface (Barnett 1975). The Florida basement can be recognized as those rocks underlying the coastal plain sequence of Cretaceous through Holocene sedimentary rocks and the post-rift sediments that begin with the Jurassic Louann and Werner evaporites. It includes not only igneous and metamorphic rocks, but also the Triassic redbed rocks filling tensional basins and nonconformable early Paleozoic sedimentary rocks.

Banks (1978) described the basement as the surface of the crystalline rock complex underlying Florida. This surface is beneath the early Paleozoic sandstones of the Tippecanoe sequence. Subsequent reviews of the Florida basement by Smith (1982), Klitgord et al. (1984), Dallmeyer et al. (1987), and McClain and Karr (1989) have retained Barnett's definition, which is adopted here.

The nature of the Florida basement and models of its tectonic history can only be deduced from assessments of direct evidence and interpretations of geophysical data. The direct evidence (beyond the plateau shape) consists exclusively of rock samples from the few boreholes that have penetrated the basement. The geologic identity of borehole samples permits spatial delineations of certain basement rock types, but the paucity of boreholes and their poor distribution limits mapping efforts and frustrates reconstructions of history. Accordingly, geophysical data, which are presumed to be representative of physical properties characterizing the basement complex, augment the direct evidence.

Interpretations of geophysical data are constrained by boundary parameters, but are flawed by non-uniqueness. Nevertheless, they serve to link lithologies to models, complement geochemical analyses (see chapter 3), and allow the preparation of histories of tectonic evolution consistent with presently accepted accounts of major tectonic events. This chapter introduces and summarizes our knowledge of geophysical evidence of the basement: the anomaly patterns of the gravity and magnetic fields, the distribution of known earthquakes and seismic properties of the peninsula, and the geothermal nature of the basement.

The other chapters of this book eloquently characterize the geology of and geologic processes related to the surficial rocks of Florida. Although the thick, progradational wedge of coastal plain sedimentary rocks represents a quasi-continuous depositional history, intriguing accounts of internal sedimentation changes, hydrology, faunal distributions, and karstification also provide a rich portrayal of the geology of the peninsula.

The well-documented geologic history of the Florida surficial rocks contributes little information about the basement. One can neither extrapolate the events represented by surface evidence past a certain time nor associate any surficial conditions to those tectonic activities that formed the basement. Accordingly, our only bases for formulating not only a tectonic history of the basement, but also the present configuration of rock types and structural features, are the sparse borehole samples and inferences from geophysical data.

Basement Studies

Although Campbell (1939a, 1939b) reported pre-Cretaceous rocks in assessments of the petroleum potential of Florida, Applin (1951) presented the first systematic accounting of basement rocks in Florida. Relying on borehole cuttings and cores, Applin distinguished and described the pre-Mesozoic sedimentary rocks, extrusive and intrusive igneous rocks, and metamorphic rocks that compose the basement. Contributions of new data, analyses of existing samples, or updated tabulations of the basement-penetrating boreholes were presented by Carroll (1963), Berden (1964), Bass (1969), Milton and Grasty (1969), Milton (1972), Barnett (1975), and Smith (1982). Subsequent reports of rock types (Smith 1983; Mueller and Porch 1983; Dallmeyer et al. 1987; Arthur 1988; McClain and Karr 1989; Heatherington and Mueller 1991; and others) basically discuss conclusions regarding previously described samples, and do not introduce new samples.

Rock Types and Ages

Early Paleozoic sedimentary rocks have been penetrated by more than sixty boreholes. These basement rocks, distributed across northern Florida (fig. 2.1), have been described as quartzitic sandstones and dark shales that are horizontally bedded and unmetamorphosed (Applin 1951). Their top lies 0.8 to 2.3 km below the surface. Based on paleontologic analysis, Berden (1964) assigned an Early Ordovician age to the basal unit of quartzitic sandstones interbedded with shales.

Applin (1951) described samples from two boreholes that penetrated more than 600 m of white to light tan, micaceous, feldspathic, quartzitic sandstone. Based on petrographic analyses, Carroll (1963) suggested a granitic or metamorphic origin for the sandstones with several cycles of sedimentation. However, some samples yielded fragments of euhedral crystals, and others appear to contain recrystallized quartz indicative of hydrothermal alteration.

Middle Ordovician to late Silurian (and perhaps early Devonian) shales with sand lenses overlie the sandstones and thicken to the north and west. At least one borehole penetrated more than 300 m of black shale (Applin 1951). An inspection of samples from a depth of approximately 800 m in Alachua County (Opdyke et al. 1987) recognized highly micaceous, undeformed shales showing low-grade metamorphism, in a complex dominated by fine-grained siltstones and quartzites.

Many borehole samples have yielded either macrofossils or microfossils. The diverse fossils include graptolites, brachiopods, trilobites, crinoids, mollusks, conodonts, palynomorphs, and chitinozoans (Berden 1964; Cramer 1973; Pojeta et al. 1976). No single well has revealed more than one faunal zone, and few wells have penetrated more than one lithology. The faunas are dissimilar to North American assemblages, and can be correlated with assemblages in northwestern Africa and South America (Cramer 1971; Pojeta et al. 1976; see also chapter 7).

Arthur (1988) has summarized the descriptions of samples from thirty-nine boreholes in Florida that encountered diabase and basalt within basement rocks. They are all described as tholeiite, are assigned to the eastern North American tholeiite suite by Chowns and Williams (1983), and range in age from 129 ± 60 to 244 ± 10 million years. Depths to the tholeiites range from 1.1 to 1.3 km in Columbia, Madison, and Levy counties

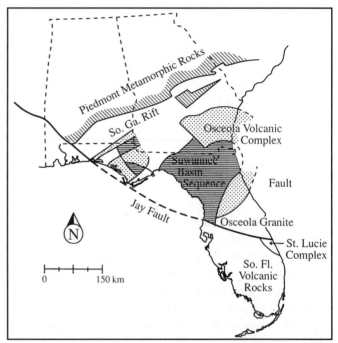

Figure 2.1. Geologic map of pre-Cretaceous rocks in the basement of Florida and surrounding areas (adapted from Dallmeyer et al. 1987 and Thomas et al. 1989). The Suwannee Basin sequence of undeformed Ordovician to Devonian sedimentary rocks is shown by horizontal lines. The Osceola complex includes the Osceola Granite, the St. Lucie metamorphic rock sequence, and the Osceola volcanic complex (stippled). Where present in the panhandle, the Osceola volcanic complex includes undelineated granitic subcrops. The South Florida volcanic rocks are Mesozoic. The South Georgia Rift area is occupied by Mesozoic sedimentary rocks. The boundary of Piedmont metamorphic rocks is indicated by diagonal lines.

to 3.9 to 4.9 km in Highlands, Lee, Okaloosa, and St. Lucie counties. Barnett (1975) believed that the tholeiites overlie Triassic rhyolite of undefined areal extent.

In addition to the Mesozoic tholeiitic rocks, sequences of variably deformed rhyolitic and calc-alkaline andesitic rocks have been penetrated. These volcanic rocks, underlying the Silurian and Ordovician sedimentary rocks in northern Florida, have Cambro-Ordovician ages and are geochemically and petrologically similar to those found along modern convergent (ocean/continent) plate boundaries (Bass 1969; Mueller and Porch 1983). Thomas et al. (1989) identified similar rocks in the southwestern extreme of the peninsula. The "southern" suite of rhyolitic rocks (as deep as 5.7 km in Collier County) appears, however, to have Mesozoic ages, and displays geochemical patterns characteristic of continental rifting environments with a mantle plume (Mueller and Porch 1983; Heatherington and Mueller 1991).

Samples from boreholes penetrating granite and diorite have been described by Applin (1951), Bass (1969), Barnett (1975), and Dallmeyer et al. (1987). Various descriptions identify the samples as granodiorite, alaskite granite, quartz monzonite, and granite; ages ranging from approximately 527 to 535 million years, consistent with those presented by Bass (1969), have been measured by Dallmeyer et al. (1987).

The distribution of subsurface granitic rocks permits identification of a triangular batholith centered approximately 1.7 to 2.4 km beneath Osceola County (fig. 2.1). Banks (1978) correlated this feature to Avalonian rocks in New England and named it the "Avalon complex"; however, in present usage the term "Osceola complex" is preferred for the granitic batholith and associated early Paleozoic volcanic rocks (Chowns and Williams 1983).

South of the known extent of the Osceola Granite in St. Lucie County, a 3.9 km borehole penetrated an amphibolite-grade metamorphic sequence. Bass (1969) interpreted detrital granitic and rhyolitic clasts in the basal sedimentary layer immediately overlying the metamorphic sequence as evidence of a nearby source and, based on K/Ar and Rb/Sr dates, assigned an age of approximately 530 million years to a causal metamorphosing event.

Configuration of Units

The Osceola complex is often defined to include the known subsurface extent of the Osceola Granite, the St. Lucie metamorphic complex immediately to the southeast, and the calc-alkaline rhyolitic assemblage that overlies the granite to the north (fig. 2.1). Thomas et al. (1989) suggested that the volcanic rock portion of the Osceola complex extends east from the Florida peninsula and then curves north and west into southern Georgia. Several drill holes in southeastern Georgia have bottomed in felsic tuffs and rhyolite (Chowns and Williams 1983).

The northern extent of the Osceola complex is uncertain. Geophysical data suggest that it may continue northward beneath the sequence of early Paleozoic sedimentary rocks and into northern Georgia (Wicker and Smith 1978; Jarrett et al. 1984). Also, several undated rhyolitic samples have been obtained from the south-central part of the Florida Platform in the South Florida Basin; it has been suggested that these samples are either from horsts of this same felsic terrane (Thomas et al. 1989) or from a separate, southerly Mesozoic bimodal suite (Barnett 1975; Mueller and Porch 1983). Felsic igneous rocks also have been encountered in the Florida panhandle and are supposed to be related to the Osceola complex (Arden 1974; Barnett 1975; Smith 1983; Thomas et al. 1989).

Overlying the Osceola complex and extending north from it is the Suwannee Basin sequence of massive early Ordovician quartzitic sandstone and Ordovician to Devonian intercalated sandstones and shales. This sequence extends into southern Georgia (fig. 2.1) and westward at least to the Florida panhandle, as well as to the Middle Ground Arch in the north-central part of the platform (Ball et al. 1988; Dobson and Buffler 1991).

Although the Suwannee Basin sequence has never been penetrated more than about 700 m, gravity modeling suggests that the western part has a maximum thickness of approximately 2.5 km (Wicker and Smith 1978; Banks 1978). Seismic profiling (Arden 1974; Jarrett et al. 1984; Nelson et al. 1985a) suggests equivalent or greater thicknesses. The western part of the basin was block-faulted and intruded by diabase, presumably during Triassic rifting, and subsequently overlain by redbeds of the Triassic Eagle Mills Formation.

Underlying the north-central part of the plateau is a 95 km wide, northeast-trending, complex graben system extending from the Florida panhandle to eastern Georgia. Variously described as the South Georgia Rift (Daniels et al. 1983; Klitgord et al. 1984; McBride 1991), the Southwest Georgia Embayment (Barnett 1975; Miller 1982), the Apalachicola Embayment (Schmidt 1984), and the Tallahassee Graben (Smith 1983), this feature is a Triassic extensional graben or abortive rift with as much as 1,800 m of arkosic sandstones and mafic igneous rocks intruded by diabase sills. Faulting within and adjacent to the graben allowed individual blocks to independently respond to stresses and seek separate elevations, yielding post-erosion subsurface exposures of different basement lithologies (Suwannee Basin and Osceola complex) (Smith 1983).

Paleo-Depositional Patterns

Although various petrographic descriptions of core samples from the Suwannee Basin sedimentary rocks suggest different sediment-transport distances and/or environments of deposition, no detailed reconstruction of depositional patterns has been completed. Based on descriptions and interpretations of cores (Applin 1951; Carroll 1963; Smith 1982; Chowns and Williams 1983; Opdyke et al. 1987), the Ordovician and Silurian (and perhaps lower Devonian) strata thicken and appear to grade upward into deeper-water shales to the northwest. These trends suggest early Paleozoic deposition from a southeastern source, perhaps the Osceola complex, toward a northwestern, deep-water marine environment.

Reflection Seismology

Seismic reflection studies have contributed extensive evidence toward understanding the nature of subsurface features in Florida. From a seismic reflection survey in the panhandle, Arden (1974) distinguished faulting and gentle folds within the felsic volcanic sequence. Although Arden contended that these features are related to Triassic tectonic activity, to which he attributes faulting within the Tallahassee Graben, it is possible that the reflections represent primary Paleozoic features.

Ball et al. (1988) interpreted seismic reflection profiles from offshore the panhandle area as indicative of a northeast-dipping master fault, with approximately 1.4 km of throw, that marks one boundary of the graben system. Subsequent reflection interpretations (Dobson and Buffler 1991) suggest a continuity of the graben system southwestward to a truncation by the Jay Fault.

Southwest of the Jay Fault, a broad, positive sea-floor feature, the Middle Ground Arch, has been recognized in reflection profiles (Martin 1978; Ball et al. 1988; Winker and Buffler 1988; Dobson 1990). This is a Jurassic feature that separates the Apalachicola Basin from the Tampa Basin (fig. 2.2).

Within the Tampa Basin, a Jurassic depression at the western edge of the Florida Platform, seismic reflection profiles exhibit dipping reflection surfaces that suggest an underlying Paleozoic synclinal structure (Ball et al. 1988; Dobson 1990). Other studies have contributed to resolutions of features within the South Florida Basin and near the western escarpment of the Florida Platform.

Extensive onshore reflection profiling was conducted in northern Florida by the Consortium for Continental Reflection Profiling (COCORP), as part of a program to identify a Paleozoic suture zone in southern Georgia (Nelson et al. 1985a; Nelson et al. 1985b). COCORP profiles above the Paleozoic sedimentary sequence were interpreted as indicative of approximately 6 km of layered reflections representing the Suwannee Basin sequence and basal felsic volcanic rocks. Nelson et al. (1985a) recognized only flat or gently dipping reflectors, and suggested that the Paleozoic sequence had experienced only minor deformation.

Patterns of Gravity and Magnetic Anomalies

The general configurations of gravity and magnetic anomalies are similar. Because the contributions of overlying sedimentary rocks to either the gravity or the magnetic field are considered minor, the anomaly patterns are indicative of compositional or structural variations within the basement.

An early representation of magnetic field values in Florida was presented by Lee et al. (1945) as a series of east-west traverses across the peninsula. These and other data were incorporated by King (1959) into a regional magnetic anomaly map (fig. 2.3). Subsequent data from Gough (1967), and U.S. Geological Survey and Naval Oceanographic Office data summarized by Zeitz (1982)

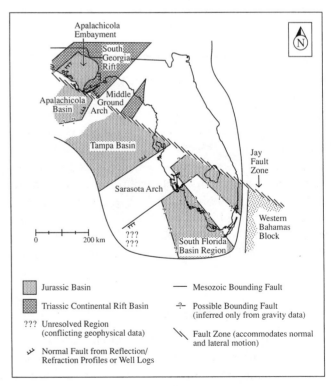

Figure 2.2. Major structural features of the West Florida Shelf and south Florida.

and Klitgord et al. (1984), provide additional detail to the dominant trends.

The magnetic anomalies exhibit two dominant trends: a well-defined northeasterly trend in the northern half of the peninsula, and a northwesterly trend in the southern half. Although the anomaly pattern in northern Florida parallels that of the southern Appalachians, it extends northward only to southernmost Georgia, where individual, prominent, positive anomalies and a sweeping east-west negative anomaly, the Brunswick Magnetic Anomaly, disrupt the pattern. Magnetic anomaly trends in southern Florida appear to be continuous with those to the southeast in the Bahamas Platform.

Early gravity data for Florida yielded Bouguer anomaly patterns very similar to those described above (Lyons 1950; Oglesby and Ball 1971; Chaki and Oglesby 1972; Oglesby et al. 1973). These and other data were incorporated into regional maps that included Florida (Krivoy and Pyle 1972; Society of Exploration Geophysicists 1982; Klitgord et al. 1984; Simpson et al. 1987); these maps exhibit a Bouguer-anomaly range of +42 to –40 mgal (fig. 2.4). The major positive anomalies are in southern and central peninsular Florida (coincident with positive magnetic anomalies). The largest negative gravity anomaly is in Jackson County, near (but not coincident with) positive magnetic anomalies.

Recognizing the probable influence of crystalline basement rocks on the gravity and magnetic fields of Florida, most interpreters of local anomalies (Gough 1967; Wolansky and Spangler 1975; Smith and Taylor 1977; Bennett and Smith 1979) have identified varying basement depths or compositions as the causes of the anomalies. The marked contrast in orientations of both gravity and magnetic anomalies between southern Florida and northern Florida is often described as indicative of a major compositional change in the underlying crust.

Wicker and Smith (1978) modeled the gravity field as indicative of a boundary between oceanic and continental crust in central Florida. Using known and estimated density values and thicknesses of sedimentary rock sequences in peninsular Florida, they inverted the gravity field to create a three-dimensional basement reconstruction (figs. 2.5, 2.6, and 2.7) that required an oceanic crustal density of 3.0 g/cm^3 for the south Florida basement. Basement-rock density values of 2.7 g/cm^3 more

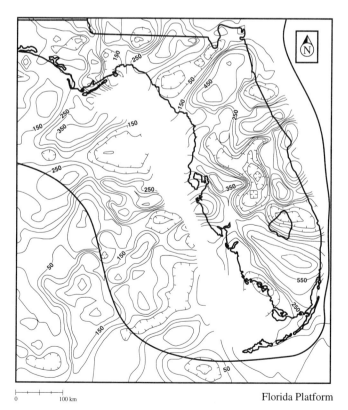

Figure 2.3. Magnetic-anomaly map of the Florida Platform region. Data from sources cited in text.

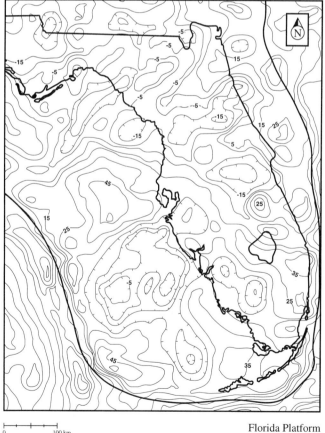

Figure 2.4. Gravity-anomaly map of the Florida Platform region. Data from sources cited in text.

representative of continental crust values, yielded the best approximation of gravity values in northern Florida.

The disparity of both gravity anomaly and magnetic anomaly patterns between northern Florida and southern Florida has been cited as evidence for a transform plate boundary through Florida during the Jurassic (Klitgord et al. 1984). The postulated feature is an extension of the Bahamas Fracture Zone (described as the Jay Fault earlier in this chapter), and may have formed the southern edge of the North American plate during the Jurassic (fig. 2.8). Detailed magnetic studies in Polk County (Smith and Graves 1986) have yielded anomalies interpreted as consistent with a basement granitic feature truncated by a major northwest-trending fault.

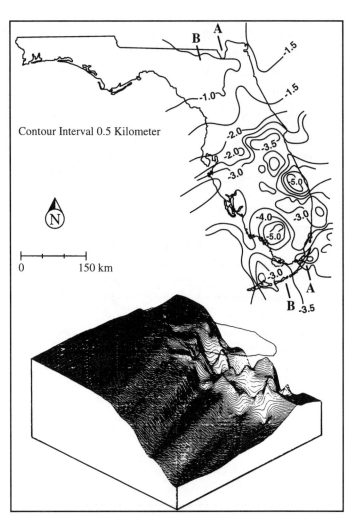

Figure 2.7. Structure-contour map (km below sea level) and isometric view of the top of the basement density body (from Wicker and Smith 1978). Vertical exaggeration 80:1 on isometric view. Profiles A-A' and B-B' shown as figures 2.5 and 2.6.

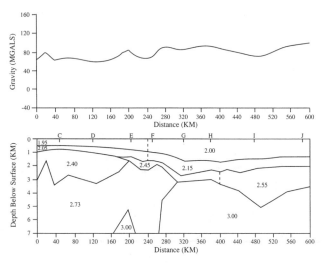

Figure 2.5. Gravity profile (on adjusted scale) and modeled cross-section for profile A-A' (Wicker and Smith 1978). Density values in g/cm³.

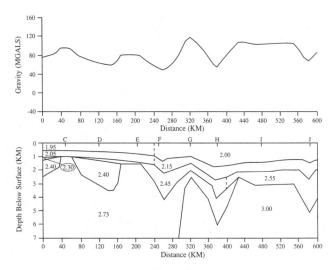

Figure 2.6. Gravity profile (on adjusted scale) and modeled cross section for profile B-B' (Wicker and Smith 1978). Density values in g/cm³.

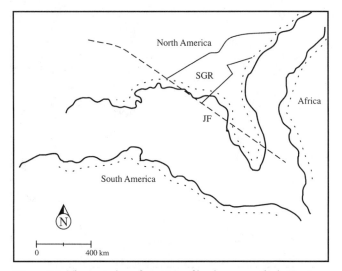

Figure 2.8. Theoretical configuration of landmasses at the beginning of Triassic rifting. Present extent of the South Georgia Rift is labeled SGR. The dashed line shows the Jay Fault-Bahamas Fracture Zone (JF) and the northwest continuation as the Pickens-Gilbertown Fault.

Seismicity

The Florida Platform is characterized by a unique seismologic stability that has yielded very few confirmable earthquakes. Although many historical events have been reported as earthquakes in Florida, and some descriptions conclusively suggest actual earthquakes, no damaging events are known to have occurred within the state.

An early compilation (Campbell 1943) of reported Florida earthquakes cited fifteen dates and estimated intensity values based on newspaper accounts, weather-bureau records, and private accounts or files; however, most of the reported events are associated with vague locations, uncertain or conflicting times, and obvious confusion with publicized events beyond Florida. An expanded report of Florida earthquakes (Mott 1983) listed thirty-three events beginning in 1727, but conceded that many may be events that occurred outside the state. In the early nineteenth century, the population of Florida was less than 35,000 and was concentrated in the St. Augustine and Apalachicola River areas. Accordingly, the initial geographic pattern of reported events was limited to those areas, and gradually expanded to mimic the distribution of the growing population.

Reagor et al. (1987) included Mott's and other data in a comprehensive review of earthquake reports and a seismicity map of Florida. Estimated intensities were assigned to each reported event, and estimated magnitudes (3.5 and 2.9) were assigned to the most recent events (Merritt Island in 1973 and Daytona in 1975, respectively). However, many of the reports can be attributed to blasting, military activities, and other non-seismic phenomena. A critical review of all seismic data for Florida (Smith and Randazzo 1989) identified only six events from 1879 to 1975 that could be accepted as possible earthquakes. Subsequent reviews have reduced the number of plausible events to five (table 2.1; fig. 2.9).

Although the 1973 and the 1975 events each were felt throughout a large local area, they were recorded only by seismograph stations outside the state. Consequently, a single component seismograph station (GAI) was installed in Gainesville in 1977 (Smith 1978), and no reportable events in Florida have been recorded since that time. Beginning in 1989, the Gainesville station was upgraded with three-dimensional short-period and long-period digital instruments, and remote stations were installed in the Everglades, Sarasota County, Wakulla Springs, and Waycross, Georgia.

ORIGIN OF FLORIDA EARTHQUAKES

Seismic activity in any area can be attributed to stress accumulation and release, usually related to displacement on active faults, or to isostatic imbalances requiring adjustment. Correspondingly, low levels of seismicity are an indicator of tectonic stability. Lithospheric plate margins and well-studied intraplate flexural features (e.g. New Madrid fault zone) are too far removed from the Florida Platform to be a source of crustal stress. Indeed, there is no evidence for active faulting or deformation in Florida during the Holocene Epoch, and probably for most of the Neogene. However, Long (1974) has suggested that Florida earthquakes may occur along extensions of crustal block edges where irregularities of crustal structure and differential vertical uplift can amplify stresses.

Seismicity maps for the U.S. (e.g. Stover 1986) and for the southeastern U.S. (Bollinger 1973; SEUSSN Contributors 1992) demonstrate a seismic quiescence for Florida and suggest an abrupt decrease of seismicity south of a proposed suture zone linking Appalachian basement with a more stable Gondwana basement (Lord and Smith 1991). Only three of the five events listed in

Table 2.1. Seismic events in Florida attributed to a tectonic origin

Date	Location
13 January 1879	Uncertain; felt throughout north Florida and South Georgia
14 November 1935	Palatka
22 December 1945	Offshore Miami
27 October 1973	Merritt Island
4 December 1975	Daytona

Note: Exact locations shown in fig. 2.9.

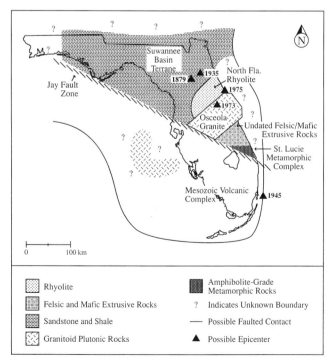

Figure 2.9. Probable epicenters for earthquake events listed in table 2.1 and their relationships to major subsurface features.

table 2.1 (1935, 1973, 1975) are well located. The 1945 event has credible newspaper documentation, but a poor epicenter determination; it coincided with a recording at a seismic station in Alabama, but no clear association could be established (Mott 1983). Early newspaper accounts of the 1879 event suggest its authenticity, but an exact epicenter cannot be identified, and the event could have originated within a wide area in northern peninsular Florida or southern Georgia.

Figure 2.9 locates probable epicenters of the five identified events, and shows their relationships to basement features. None of the events appear to be definitively associated with the Jay Fault or with the Tallahassee Graben boundaries, and those features must be regarded as substantially stable. Continued subtle tectonic adjustments within the Osceola complex are plausible, and are the probable sources of the few Florida earthquakes.

Geothermal State of the Basement

Early summaries of subsurface thermal conditions in Florida were based on interpretations of uncorrected bottom-hole temperatures recorded in boreholes during or shortly after drilling activity. Using bottom-hole temperatures from approximately 300 boreholes, Griffin et al. (1969) and Reel and Griffin (1971) developed a geothermal gradient map of Florida with the purpose of defining areas potentially favorable for oil and gas maturation. Their map (fig. 2.10) identified subsurface temperature gradients that range from approximately 27° C/km in the western panhandle to 13.5° C/km in the southern peninsula, and even lower in the Keys.

Temperature values from oil well test holes also were used by Kohout (1967), Henry and Kohout (1972), and Kohout et al. (1977) to demonstrate a negative thermal gradient in the lower part of the Floridan aquifer in southern Florida. They proposed a "normal" geothermal gradient of approximately 20° C/km below the aquifer system, but described a lateral extraction of heat by an influx of cold seawater circulating into the aquifer system.

Using bottom-hole temperatures and estimates of thermal conductivity, Reel (1970) and Griffin et al. (1977) calculated heat flow values for Franklin County and Palm Beach County. These estimates were similar to values computed from measured temperatures in equilibrated boreholes, and to laboratory determinations of thermal conductivity presented by King and Simmons (1972), Fuller (1976), and Smith and Fuller (1977). Ten new heat flow values, mostly from the western panhandle, were published by Smith et al. (1981a), Smith et al. (1981b), and Smith and Dees (1982). Figure 2.11 is a summary of all the heat flow values determined for Florida.

In general, heat flow in Florida is relatively low, ranging from 64 mW/m² in the panhandle to 20 mW/m² or less near the northeast coast and in the north-central highlands. Most of the higher values were recorded in the western panhandle, which is consistent with the observation (Smith and Dees 1982) that heat flow along the Gulf Coastal Plain increases from east to west. High thermal gradients (31 to 34° C/km) were detected in limited segments of boreholes in the Pensacola area. The highest actual temperatures (approximately 35°

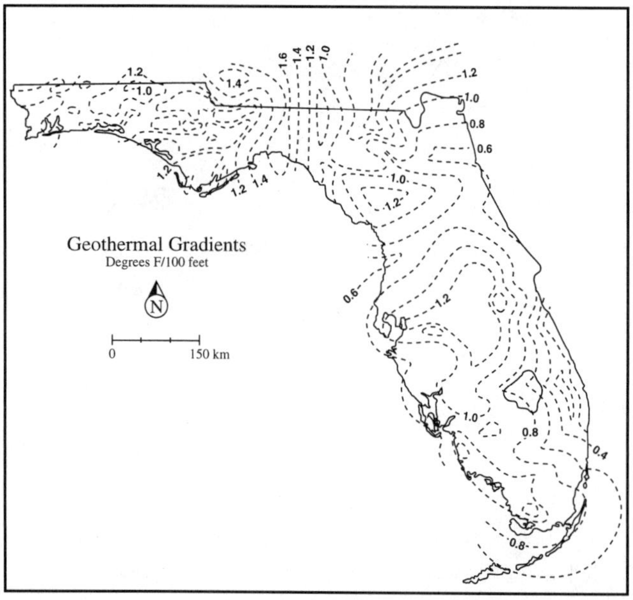

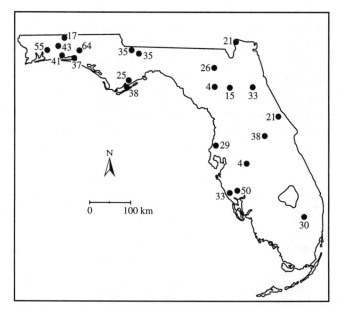

Figure 2.10. Contour map of township averages of geothermal gradients in Florida and south Georgia (Reel and Griffin 1971).

Figure 2.11. Heat-flow values (in mW/m²) determined from measurements and estimates in Florida. Data from sources cited in text.

C at 500 m depth at Fort Walton Beach) are suggestive of possible low-grade geothermal heat sources.

HYDROLOGIC INFLUENCES

Directly measured temperatures in thermally stable boreholes in Florida reveal significant variations in geothermal gradients among separate segments of individual boreholes, as well as variations among boreholes (Smith and Fuller 1977). The resulting diversity of calculated values of conductive heat flow suggests an influence on the upward migration of heat at relatively shallow depths. Heat flow calculations from measurements at greater depths (below the aquifers) appear to be more uniform (approximately 40–44 mW/m^2), and may be more representative of the heat flow from the Florida basement.

Because of the extensive aquifers and other permeable rock layers in Florida, possible disruption of conducted heat by convective overturn or lateral migration of groundwater must be recognized. The simplest pattern of temperature disturbance by circulating groundwater is based on the concept of a horizontal aquifer with high permeability and fluid circulation inserted within a medium of heat-conducting solid rock. Within the aquifer, conduction will be insignificant compared to convective heat transport, and the aquifer will essentially behave as a heat sink at its base and a heat source at its top, because temperature equilibrium will be established by internal circulation. A very low (ideally zero) temperature gradient may be established within the aquifer, while anomalously high gradients may extend immediately above and below the convecting layer.

Groundwater temperatures measured in some producing wells in southern Florida reveal anomalous relationships between temperature and depth that can be attributed to upwelling of warmer water along fracture systems (Sproul 1977). Osmond and Cowart (1977) have demonstrated upwelling of deeper waters in southwestern Florida through interpretations of uranium-series isotopic anomalies.

Thermal effects from the aquifer systems in Florida are obvious, as demonstrated by temperature/depth profiles (Smith and Fuller 1977; Smith and Dees 1982). Thus, the credibility of calculated gradients and heat flow values must be tempered by an awareness of uncertainties about heat transport within sedimentary sections. Accordingly, the anomalously low heat flow values of figure 2.11 (4–20 mW/m^2) probably are more representative of the surficial thermal regime at their respective locations than of the steady-state heat flow from the basement.

THERMAL MODELS

Long and Lowell (1973) cited inhomogeneous heat production within the crustal rocks of northern peninsular Florida and adjacent ocean basins as a causal factor in differential thermal subsidence of the Florida continental margin. Their model, later modified (Lowell and Long 1977), showed that thermal contraction could be responsible for more than 200 m of elevation differences at the peninsular margin. Although heat flow values should vary horizontally in response to subsurface heat-generation variations (decreasing both to the south and in adjacent ocean basins), the authors attribute low heat flow in Florida, in part, to insulating effects of recent sedimentation.

Regional Tectonic History

The pre-Tertiary configuration of the continental block(s) underlying the Florida Platform and the adjacent Bahamas Platform is based primarily from limited sample information from deep drill holes. Understanding the evolution of these blocks is crucial for developing a detailed tectonic history for the Gulf of Mexico and the southeastern U.S., particularly since an allochthonous origin (with respect to North America) for the continental material of the Florida Platform basement has been widely accepted.

GONDWANAN ORIGIN

A Gondwanan provenance for the Florida Platform block during the early Paleozoic has been documented with continental reconstructions (Bullard et al. 1965; Wilson 1966; Dietz and Holden 1970; LePichon and Fox 1971; Pindell and Dewey 1982; Klitgord et al. 1983; Roussel and Liger 1983; Van Siclen 1984; Pindell 1985; Venkatakrishnan and Culver 1988), with paleontologic analyses from Florida and Africa (Cramer 1971, 1973; Pojeta et al. 1976), with comparisons of isotopic ages (Dallmeyer 1987, 1988; Dallmeyer et al. 1987), with paleomagnetic data (Van der Voo et al. 1976; Opdyke et al. 1987), and with stratigraphic correlations between Florida and West Africa (Smith 1982; Chowns and Williams 1983). As a result, it is reasonably well established that the Florida Platform block was originally a part of the West African continental margin near Senegal, and was only rifted from that margin during the Triassic breakup of Pangaea.

Specific units in West Africa can be correlated with their apparent counterparts underlying the Florida Platform. These units provide the sole record of the pre-Mesozoic tectonic history of the Florida Platform. The extensive rhyolitic terrane of northern Florida and southern Georgia has been correlated with rhyolites throughout the western margin of Africa (Dillon and Sougy 1974; Dallmeyer 1987; Dallmeyer and Villeneuve 1987). Geochemical analyses of the Florida suite suggest deposition in a convergent plate margin setting, an ocean/continent subduction environment being the most likely setting (Mueller and Porch 1983).

The St. Lucie metamorphic complex has been correlated with metamorphic rocks of the central Rokelide Orogen in Guinea (Chowns and Williams 1983; Dallmeyer and Villeneuve 1987). In addition to similarities in petrology and degree of metamorphism, both of these units appear to have been affected by an extensive thermotectonic event that occurred along the northwest margin of Gondwana about 550 million years ago (Dallmeyer and Villeneuve 1987).

There is evidence to support the correlation of the Osceola granitoid complex of east-central Florida with the Coya Granite of Senegal, exposed in the northern Rokelide orogen. Both are post-tectonic, early Paleozoic plutons (about 530 million years old) with similar petrographic and petrologic characteristics (Dallmeyer et al. 1987). These similarities, as well as numerous continental reconstructions, suggest that during the early Paleozoic the Osceola and Coya granites likely formed as part of the series of granitoids emplaced along Gondwana's northwest margin during the aforementioned thermotectonic event (about 550 million years ago), recognized throughout West Africa (Dillon and Sougy 1974; Dallmeyer et al. 1987).

The Suwannee Basin complex of north-central Florida was deposited in a middle Paleozoic sedimentary basin. This basin was filled first with quartz sands, and subsequently with graptolite-bearing black shales, which are indicative of a restricted marine environment. Based largely on stratigraphic and paleontologic similarities, the Suwannee Basin has been correlated by a number of workers with the Bove Basin of Guinea (Cramer 1971, 1973; Pojeta et al. 1976; Smith 1982; Chowns and Williams 1983; Venkatakrishnan and Culver 1988). Subsequently, this correlation has been validated by a paleomagnetically determined Early Ordovician paleolatitude of 49° S (Opdyke et al. 1987) and a comparison of $^{40}Ar/^{39}Ar$ plateau ages for detrital muscovite (about 505 million years) from both basins (Dallmeyer 1987). It is likely that these deposits represent the disjointed remnants of a single, extensive, restricted basin that existed along the northwest margin of Gondwana during the middle Paleozoic.

PLATE CONVERGENCE

Hatcher (1989) has summarized evidence for the assemblage of microplates outboard of the North American continent to form the forward edge of the North American landmass which locked with the Afro-South American landmass. In the area of the present southern Appalachians, accretion of the Avalon-Carolina composite terrane preceded continental closure. The part of the Gondwanan landmass now outboard of the Avalon-Carolina terrane (the Florida Platform block north of the Jay Fault, including the Osceola complex and Suwannee Basin) is the Suwannee terrane.

Smith et al. (1992) have postulated that these terranes were juxtaposed close to their present-day relative positions in the Florida Platform basement during the late Paleozoic closure of the Iapetus Ocean. In this scenario (figs. 2.12 through 2.18), fragments of the Gondwanan Rokelide metamorphic and volcanic complexes were moved laterally into relative positions along strike-slip faults, such as the Jay Fault, that formed in response to the continental collision between Laurentia and Gondwana. Specifically, development of right-lateral faults is a possible response to the Gondwanan landmass closing with, then rotating clockwise around, the Alabama promontory. Basement fragments southwest of

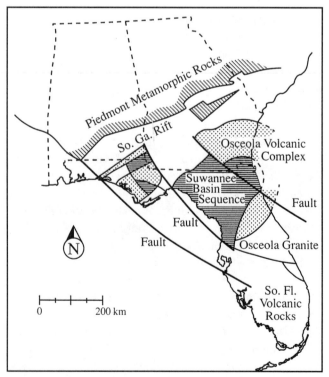

Figure 2.12. Possible faults in the Florida basement. Features as described for figure 2.1.

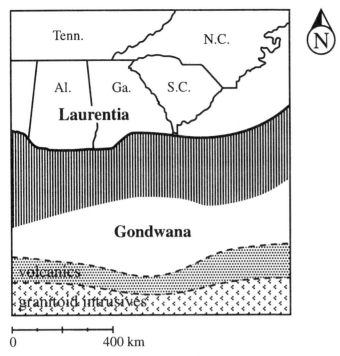

Figure 2.13. Schematic representation of the sequential convergence of Gondwana and Laurentia during the late Paleozoic Alleghenian Orogeny. The leading edge of Gondwana consists of layered early Paleozoic sedimentary rocks (Suwannee Basin), a felsic volcanic fringe, and a granitic core (Osceola complex). Sequence continued in following figures.

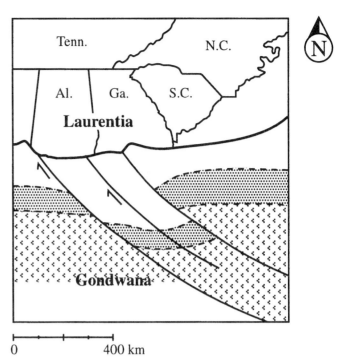

Figure 2.15. Right-lateral faults accommodate clockwise rotation of Gondwana around the promontory and up to the Ouachita closure.

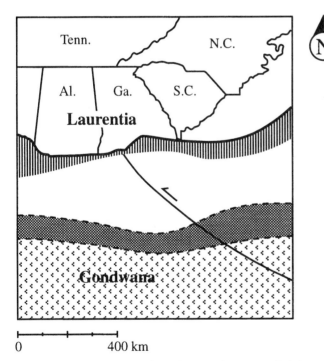

Figure 2.14. Initial contact of Gondwana and Laurentia. Relative motion to accommodate Alabama Promontory.

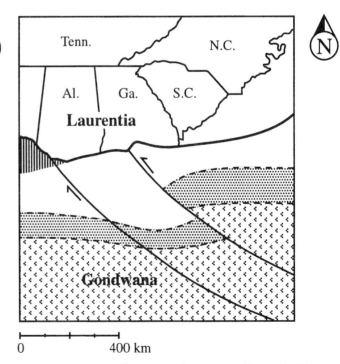

Figure 2.16. Continued motion of Gondwana and Laurentia. Many accommodating faults, including the precursors of the Jay Fault and Pickens-Gilbertown fault system.

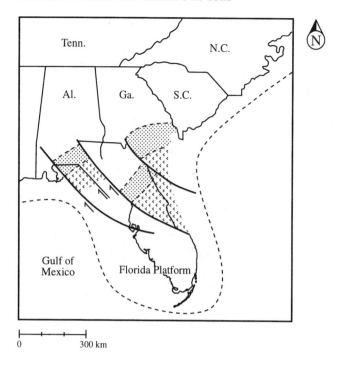 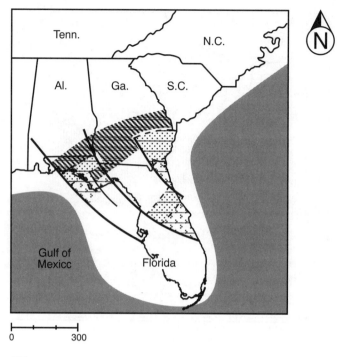

Figure 2.17. Postulated configuration of Florida Platform part of the Gondwana–North America union after late Paleozoic closure. Numerous right-lateral faults have displaced the leading edge of Gondwana to the Florida panhandle area, and an initial left-lateral fault has emplaced the Osceola complex into southeast Georgia. Ocean areas for reference only.

Figure 2.18. Configuration of the Triassic South Georgia Rift as controlled by basement block pattern shown in figure 2.17. Right-lateral faults shown for West Florida are reactivated to accommodate left-lateral motion.

the Jay Fault and associated faults could then have been displaced northwest toward the Ouachita Orogen. The process may have reduced stress on the Florida Platform, and limited deformation in the Suwannee Basin sequence to the isolated thrust folds at the western edge of the platform (D. D. Arden, pers. comm. 1992), in addition to suturing these Gondwanan fragments to North America during the amalgamation of Pangaea.

TRIASSIC-JURASSIC RIFTING

Most late Paleozoic to early Mesozoic reconstructions of Pangaea place the continental basements of the Florida Platform and the Bahamas Platform in the reentrant at the junctions between the juxtaposed North American, South American, and African plates (fig. 2.8). Triassic rifting that created the Atlantic Ocean probably was initiated with rift propagation from hot spots, including one near the southern tip of the Florida Platform, as suggested by isotopic signatures of volcanic rocks in the south Florida basin (Mueller and Porch 1983; Heatherington and Mueller 1991). An interior rift system, including many of the present-day Triassic basins of the Atlantic margin, presumably developed simultaneously, and facilitated rifting of North America from Gondwana along the traces of the Paleozoic margins.

A large graben, the South Georgia Rift, formed across southern Georgia, nearly along the trace of the Alleghenian suture between Florida and North America. Shifting of the rift geometry caused the South Georgia Rift to become an aulacogen, and, as rifting there ceased, the basements of the Florida Platform, the Yucatan, and the Bahamas Platform were left appended to southeastern North America (Mullins and Lynts 1977; Pindell and Dewey 1982; Smith 1982; Burke et al. 1984; Van Siclen 1984).

In the early Jurassic, the Gulf of Mexico opened, as North America separated from the South American section of Gondwana, and the Yucatan Peninsula started to rotate away from North America. The Gulf existed as a restricted basin from about 165 to 150 million years ago, resulting in extensive evaporite deposition (Pindell and Dewey 1982; Salvador 1987). The mechanisms of relative motion between the two continents during the Jurassic are not well constrained; however, it is likely that a left-lateral strike-slip movement along a series of northwest-southeast fracture zones served to accommodate this relative motion (Klitgord et al. 1984). It has been suggested that the system of the Bahamas Fracture Zone and Jay Fault was the most important of these transforms, because it formed the southern edge of North America during the Jurassic and connected spreading centers in the Atlantic and Gulf of Mexico through left-lateral motion (Pindell and Dewey 1982; Klitgord et al. 1984; Van Siclen 1984; Pindell 1985).

During this separation of North and South America and rotation of the Yucatan, not only did the South Georgia Rift form, but a series of basement horsts and grabens also formed around the periphery of the Gulf of Mexico basin (Pindell and Dewey 1982; Burke et al. 1984; Klitgord et al. 1984; Buffler and Sawyer 1985; Pindell 1985; Winker and Buffler 1988). Some of the horsts, such as the Wiggens Arch underlying southern Alabama and Mississippi, are considered to be residual blocks of South American origin (Smith et al. 1981b; Pindell and Dewey 1982; Pindell 1985; Van Siclen 1990); however, the series of large horsts and grabens that form the basement of the western Florida Platform are either the result of differential crustal attenuation during the opening of the Gulf (Pindell 1985; Ball et al. 1988; Winker and Buffler 1988; Dobson and Buffler 1991), or they represent stranded horsts of North American crust, which may have been broken off during the separation of North and South America (Klitgord et al. 1984). During this period, mafic volcanic and hyperabyssal rocks were emplaced in the southwest Florida grabens (Barnett 1975; Martin 1978; Ball et al. 1988).

HOLOCENE STABILITY

In the time since the formation of these basement highs and lows, the Florida Platform has been tectonically quiescent. A period of tectonism in the southeastern Gulf of Mexico associated with the collision of Cuba and the Bahamas during the early Paleogene has been documented; however, there is no evidence for associated tectonic activity on the Florida Platform (Pindell 1985). Rather, the undisturbed Upper Cretaceous and Tertiary strata on the plateau document an extended, nearly continuous period of shallow-water marine carbonate deposition, occurring at a rate approximately equal to the relatively rapid rate of regional subsidence, which possibly has been interrupted by short periods of epeirogenic uplift.

EPEIROGENIC UPLIFT

The distribution throughout Florida of Miocene and younger marine sedimentary rocks at the surface demonstrates an emergence of the Florida Platform from a submarine environment during the Neogene. However, no evidence exists to suggest tectonic deformation or orogenic activity of any nature throughout the Cenozoic. Popularization and common adoption of the term "Ocala Uplift" for an area of exposed Eocene limestones has led to a concept of tectonic uplift, if not folding, since the Eocene. Careful stratigraphic analysis (Winston 1976) indicates that no uplift or folding has occurred, and that the exposure pattern can be attributed to regional tilting.

Epeirogenic uplift and regional elevation of Plio-Pleistocene beach ridges in northern Florida have been attributed to density changes within limestone formations experiencing Pleistocene and Holocene karstification (Opdyke et al. 1984). Estimates of dissolved limestone based on current erosion rates suggest an isostatic uplift of approximately 50 m in response to density changes.

Summary

The evidence and interpretations presented here portray a basement of varied components and complex history. The striking peninsular shape of the Florida basement seemingly can be explained only by a unique pattern of Triassic rifting which was constrained by preexisting suture-zone weaknesses, a hot spot, and individual crustal blocks. The southern part of the peninsula may even be attributable to a fortuitous alignment of basements by lateral motion along the Jay Fault.

Discontinuities in the anomaly patterns of the gravity and magnetic fields suggest abrupt basement contrasts at mid-peninsula, and a unique seismic quiescence suggests a lithosphere stress field distinct from that north of the late Paleozoic suture zone. Our understanding of the Florida basement suggests that it somehow was part of a leading edge of continental convergence, but paradoxically escaped significant deformation or metamorphism.

Borehole evidence reveals a basement consisting of an early Cambrian granitoid batholith, associated felsic volcanic rocks, and an overlying sequence of early Paleozoic sedimentary rocks, all of which can be assigned a Gondwanan origin. Superimposed on these Paleozoic features are a series of Mesozoic grabens and one or more Jurassic transform faults (which may have had a late Paleozoic origin). Subsequently, from the Cretaceous until the present, a thick sequence of undisturbed carbonates and evaporites has been almost continuously deposited over these older features.

A number of lines of evidence from a variety of basement studies demonstrate that the Florida Platform has experienced a complex tectonic history. The Paleozoic features record a long period of convergence that culminated in the suturing of North America to Gondwana; however, the relatively pristine condition of the Suwannee Basin terrane demonstrates that this suturing was not accompanied by significant deformation on the Gondwanan margin. The Mesozoic extensional features record the Triassic to Jurassic rifting of Pangaea and the openings of the central North Atlantic and Gulf of Mexico. Finally, the thick sedimentary sequences of the Cretaceous to the present record a more recent history of regional tectonic quiescence accompanied by relatively rapid subsidence.

3

Geochemistry and Origin of Florida Crustal Basement Terranes

Ann L. Heatherington and Paul A. Mueller

Sedimentary rocks of the coastal plain of Florida and adjacent areas of Alabama and Georgia rest unconformably on pre-Cretaceous basement. The basic configuration is one of thick sequences of Cretaceous and younger carbonate rocks with a minor clastic sedimentary component overlying igneous, and to a lesser extent, sedimentary and metamorphic lithologies of late Proterozoic to early Mesozoic age (Applin 1951; Bass 1969; Milton and Grasty 1969; Milton 1972; Barnett 1975; Lloyd 1985; Chowns and Williams 1983). Samples of this basement have been recovered from more than 140 petroleum exploration test wells. Drill-hole samples and seismic data have revealed that the depth to basement is greatest in the southern half of the peninsula—as much as 3.9 km, compared with 0.8 to 1.6 km in the northern peninsula. As a consequence, drill holes penetrating the pre-Cretaceous surface are most numerous in northern Florida, southern Georgia, and southern Alabama.

The youngest igneous basement rocks are basalts of Triassic to Jurassic age, recovered from sites in southwestern Florida, as well as the northern peninsula and panhandle. In southwestern Florida, the basalts are accompanied by high-potassium rhyolites and dacites, whereas in the north they are associated with sedimentary rift sequences containing redbeds. The next oldest lithologies are early Paleozoic clastic sedimentary rocks, often referred to as "Suwannee Basin" lithologies, recovered from wells in north-central Florida and the panhandle. Adjacent to these, in northeastern Florida, is a suite of intermediate to silicic volcanic rocks of probable early Paleozoic or late Proterozoic age. The basement in east-central Florida consists of the "Osceola Granite" (Chowns and Williams 1983), a late Proterozoic dioritic to granitic plutonic complex. The southeasternmost samples are associated with high-grade metamorphic rocks called the "St. Lucie metamorphic complex" by Thomas et al. (1989). The general locations of these lithologic associations are shown in figure 3.1.

Because the depth of penetration of test wells into basement is limited, and field relations cannot be directly observed, the nature of the contacts (erosional, faults, etc.) among the various subcrop lithologic provinces is poorly known. In addition, the only available samples are drill-hole cores and cuttings, and alteration and limited sample quantity have been consistent impediments to

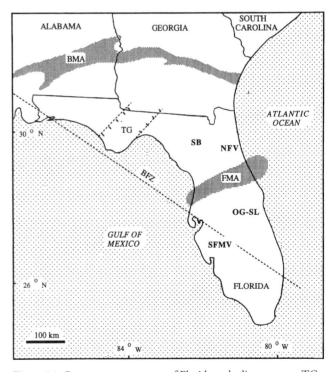

Figure 3.1. Basement components of Florida and adjacent areas. TG: Tallahassee Graben; BMA: Brunswick Magnetic Anomaly; FMA: Florida Magnetic Anomaly (Hall 1990); BFZ: Bahamas Fracture Zone; OG-SL: General location of the Osceola Granite and St. Lucie metamorphic complex; NFV: North Florida calc-alkaline volcanic suite; SFMV: General location of the South Florida Mesozoic volcanic province.

their investigation. However, recent advances in geochemical techniques have provided the opportunity to extract useful information from these small samples. This chapter will review the geochemical and geochronologic characteristics of the Florida basement rocks, with an emphasis on recently obtained isotopic and geochemical data, and their tectonic implications. All cited well numbers are those of the Florida Geological Survey (Lloyd 1985). For well locations, detailed well logs, and petrographic descriptions of Florida basement rocks, the reader is referred to Lloyd (1985), Applin (1951), Bass (1969), Milton and Grasty (1969), Milton (1972), Barnett (1975), and Chowns and Williams (1983).

Mesozoic Rocks

Rocks of apparent Mesozoic age dominate the basement subcrop in southwestern Florida and in parts of northern peninsular Florida and the panhandle. Two distinctly different, but typical rift-related suites are represented. Redbeds and tholeiitic basalts occur in the north, and a bimodal suite of alkaline to tholeiitic basalts and high-K rhyolites occurs in the southwest. It is generally accepted that the basalts are genetically related to the breakup of Pangaea (Chowns and Williams 1983; Arthur 1988; Cummins et al. 1992), but it is uncertain whether the two provinces were originally formed in their present close proximity. Also unclear are the relationships of the Florida Mesozoic basalts (1) with other Mesozoic basalts of eastern North America that have been related to the breakup of Pangaea and (2) with Mesozoic basalts of West Africa and northeastern South America, which may be correlative sequences on opposite sides of Mesozoic rifted continental margins. These relationships are critical to understanding the fit of the Florida basement in the Pangaean puzzle.

NORTHERN SEQUENCE

Mesozoic tholeiitic basalts, dacitic ash-fall tuffs, and a rift-basin sedimentary sequence occur in a northeast-trending basin interpreted as a Triassic graben (Barnett 1975), the Tallahassee graben of Smith (1983), in the eastern Florida panhandle (fig. 3.1). These rocks are locally overlain by Jurassic evaporites (Barnett 1975). In the southwestern parts of the graben, drill holes have penetrated granitic horsts of probable Paleozoic age (Cambrian to Permian; Barnett 1975; Smith 1983). Tholeiitic basalt flows and diabases are also common outside the Tallahassee Graben, in northern peninsular Florida, the western panhandle, southern Georgia (the Georgia Basin of Chowns and Williams 1983), and southern Alabama (Neathery and Thomas 1975; Guthrie and Raymond 1992). The north Florida basalts have been studied by Milton and Grasty (1969), Arthur (1988), and Heatherington and Mueller (1990). Arthur (1988) reported concentrations of ten major and ten trace elements for basaltic rocks from twelve locations in northern Florida and two locations in southern Florida. All samples are tholeiitic, and defined no systematic variation of composition with geographic location. These Florida tholeiites are compositionally similar to many Mesozoic mafic rocks of eastern North America, particularly diabase dikes from Georgia and incompatible element depleted tholeiitic basalts of the northeastern U.S. In contrast, the north Florida tholeiites show few similarities to Mesozoic basalts from Liberia and Morocco, or incompatible element enriched basalts of the northeastern U.S. (Arthur 1988).

Pb-, Nd-, and Sr-isotopic data for the north Florida basaltic rocks were reported by Heatherington and Mueller (1990). Whole-rock Rb/Sr data for the north Florida suite lie near a reference isochron corresponding to 240 million years, consistent with a link to the Mesozoic breakup of Pangaea (Heatherington and Mueller 1990). The data also indicate a source with a moderately enriched isotopic composition (initial $^{87}Sr/^{86}Sr$ = 0.7062 to 0.7093; initial ϵ_{Nd} = −4.7 to +1.2), similar to the source proposed by Pegram (1990) for the generally coeval eastern North American diabase suite. Pegram (1990) also reported data for two suites of tholeiitic basalts, those within and those outside Mesozoic rift basins in the exotic Carolina terrane of the southern Piedmont; initial $^{143}Nd/^{144}Nd$ ratios of the two suites (corrected to 190 million years) produced nearly parallel arrays of about 900 million years when plotted on Sm/Nd isochron diagrams (fig. 3.2), suggesting that they were derived from a source approximately 900 million years older than the basalts themselves. Initial $^{143}Nd/^{144}Nd$ ratios for the north Florida samples lie between the fields for basinal and extra-basinal basalts, and produce an array suggesting an age of about 800 million years. This array is subparallel to the arrays reported for the Carolina terrane basalts and suggests that both suites may have been extracted from sources of similar ages and compositions, or were involved in similar mixing scenarios. In either case, the coincidence of these data suggests at least a limited commonality of source and/or process in their petrogenesis.

With the exception of samples from Levy County (well no. 2012, Humble Oil and Refining Co. C. E. Robinson no. 7), the north Florida tholeiites form linear trends on common-Pb diagrams. They are generally slightly less enriched in ^{207}Pb and ^{208}Pb for a given ^{206}Pb compared with diabases from the Carolina terrane (data from Pegram 1990), again suggesting similar, though not identical, source components for the two suites (fig. 3.3). The Levy County samples are relatively enriched in ^{207}Pb and ^{208}Pb, indicating involvement of an older, enriched component, either continental crust or ancient, enriched mantle (Heatherington and Mueller 1990).

SOUTHWESTERN SEQUENCE

The geochemistry of the Southwest Florida Mesozoic Volcanic suite (SFMV) has been investigated by Mueller and Porch (1983), Arthur (1988), and Heatherington and Mueller (1991). The rhyolitic rocks of the SFMV have K_2O contents of 5.6% to 6.0% and are classified as alkalic by the criteria of Miyashiro (1978), and as high-K rhyolites by the criteria of Jakes and White (1972) (fig. 3.4). High-K and alkalic rhyolites are most commonly found in continental-rift environments and as late-stage eruptives in magmatic arcs. Although some investigators have suggested that the Southwest Florida rhyolites are correlative with the late Proterozoic rhyolites of northern Florida (Thomas et al. 1989), there are several reasons why we do not think this is justified. First, the major-element chemistry of the two silicic suites is quite different. The Southwest Florida rhyolites are much richer in K_2O and total alkalis compared with the north Florida rocks (fig. 3.4). Second, on Rb/Sr diagrams the two suites define entirely different arrays with very different slopes (fig. 3.5). The basalts and rhyolites of the SFMV define a single array from which a date of 238 ± 20 million years (1σ) can be calculated. This date is consistent with a tectonic origin associated with the breakup of Pangaea and the opening of the North Atlantic, but is somewhat older than whole-rock K/Ar and $^{40}Ar/^{39}Ar$ dates of 180 to 190 million years reported by Milton and Grasty (1969) and Mueller and Porch (1983). The Rb/Sr, K/Ar, and $^{40}Ar/^{39}Ar$ dates may not necessarily be in conflict, because the Rb/Sr dates may be related to the

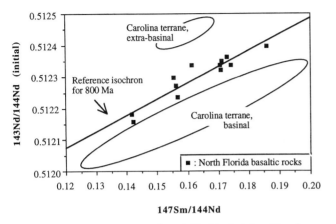

Figure 3.2. Sm/Nd systematics of north Florida tholeiites (data from Heatherington and Mueller 1990) compared with Sm/Nd systematics of Carolina terrane tholeiites (data from Pegram 1990). $^{143}Nd/^{144}Nd$ ratios are corrected to 190 million years, the approximate age of the rocks.

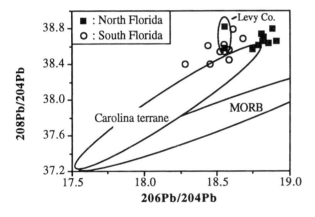

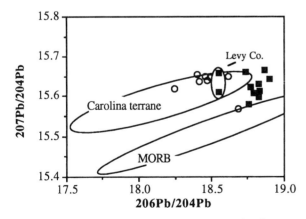

Figure 3.3. Pb-isotope systematics of Florida basalts (data from Heatherington and Mueller 1990 and 1991) compared with Pb systematics of Carolina terrane tholeiites (data from Pegram 1990). The field for mid-ocean-ridge basalts (MORB) is shown for reference (data from Basaltic Volcanism Study Project 1981).

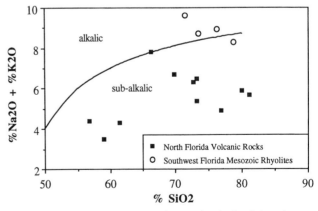

Figure 3.4. Alkali-silica plot for southwest Florida rhyolitic rocks (Hillsborough County) and north Florida Paleozoic-Proterozoic volcanic rocks (Putnam, Flagler, and Marion counties). Data from Mueller and Porch (1983). The dividing line between alkalic and sub-alkalic magmas is from Miyashiro (1978).

source region of these rocks—that is, to a time of isotopic homogenization related to pre-melting metamorphism—whereas the K/Ar and ^{40}Ar/^{39}Ar dates represent cooling ages of the rocks.

The basalts of southwestern Florida vary in composition from tholeiitic to alkalic (Mueller and Porch 1983; Arthur 1988; Heatherington and Mueller 1990, 1991). The alkali basalts recovered from well no. 966 (Humble Oil and Refining Co. C. C. Carlton Estate no. 1) in Highlands County are the only alkali basalts reported from the Florida basement (Mueller and Porch 1983; Heatherington and Mueller 1991). They have many characteristics to suggest that they underwent little modification (i.e. fractional crystallization and/or assimilation) after derivation from a mantle source: high MgO (8.9 to 9.2%), high Na_2O/K_2O (8.0 to 8.7%), and high TiO_2 (3.4 to 3.6%). The high TiO_2 values are comparable to those found in ocean-island basalts (Mueller and Porch 1983). Ratios of high-field-strength elements (HFSE) to large-ion lithophile elements (LILE) are also high, and compatible with a plume-related origin (Heatherington and Mueller 1991). Isotopic data confirm that the magmas do not represent uncontaminated, depleted mantle. Initial Sr isotopic ratios (^{87}Sr/^{86}Sr = 0.7060 at 190 million years) indicate derivation, at least in part, from an aged, enriched source. While more depleted than bulk Earth, initial ϵ_{Nd} values (+1.5 to +0.3) are lower than would be expected from a depleted mantle (i.e. MORB) source. It is most probable that these basalts are products of a mantle plume that may have assimilated continental lithosphere to some degree during ascent. In contrast, basaltic rocks from other locations (e.g. Hardee County well no. 1655 (Humble Oil and Refining B. T. Keen no. 1); and Highlands County well no. 3578 (Continental Oil Co. C. C. Carlton et al. no. 1) are tholeiitic to transitional, with higher silica contents, much lower concentrations of TiO_2 (1.3 to 2.0%) and MgO (3.8 to 4.9%), and lower ratios of Na_2O/K_2O (1.5 to 2.8) (data from Arthur 1988; and Mueller and Porch 1983). In general, these elemental abundances are less suggestive of a mantle-plume-derived magma, as are the Sr and Nd isotopic data (initial ^{87}Sr/^{86}Sr = 0.7072 to 0.7075, initial ϵ_{Nd} = −2.5 to −3.1). In contrast with the geochemical data for well no. 966, these geochemical data more strongly indicate derivation from, or significant interaction with, an enriched source that is most likely either crust or aged lithospheric mantle. Direct derivation of these rocks from the same combination of sources as the alkali basalts (well no. 966) without contamination is precluded by the isotopic data (Heatherington and Mueller 1991).

Compared to north Florida basaltic rocks, the SFMV basalts have higher ^{207}Pb/^{204}Pb and ^{208}Pb/^{204}Pb ratios for a given ^{206}Pb/^{204}Pb ratio (fig. 3.3), indicating involvement of an ancient, ^{207}Pb- and ^{208}Pb-enriched component in their genesis. Elevated ^{207}Pb/^{204}Pb and ^{208}Pb/^{204}Pb ratios in certain mantle-plume sources are well documented in the literature, and are often attributed to the presence of ancient, recycled lithosphere within the mantle (Hart 1984, 1988; Dupre and Allegre 1983; Allegre and Turcotte 1985). Alternatively, these elevated ratios may be a result of crustal contamination of

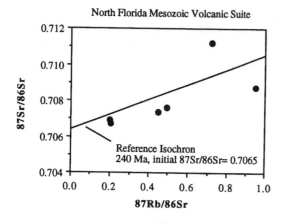

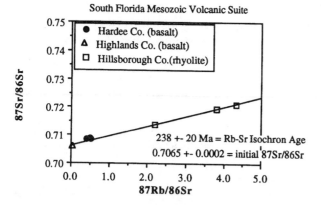

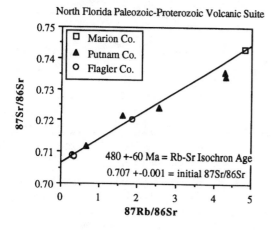

Figure 3.5. Rb/Sr systematics of Florida basement lithologic suites.

a depleted mantle source; however, this latter possibility is unlikely, because mass-balance calculations show that an unrealistically large amount of crust must be assimilated before Pb-isotope ratios are affected, because of the relatively high concentrations of Pb in the SFMV basalts (7 to 8 ppm).

It has been suggested (Heatherington and Mueller 1991; Mullins and Lynts 1977; Smith 1983; see chapter 2) that southwest Florida is a likely location for a Mesozoic triple junction or hot spot related to the breakup of Pangaea. In particular, the geochemistry of the basalts from Highlands County well no. 966 suggests involvement of a mantle-plume component unique in the Mesozoic rock record of Florida. The tholeiitic basalts of northern Florida may be related to the same extensional tectonic regime, but their major-element compositions reflect equilibration at a shallower depth, from a different (non-plume) source.

Paleozoic Rocks

Although calc-alkaline volcanic rocks subcropping in northeast Florida were thought at one time to be Paleozoic, data now suggest that they actually may be of late Proterozoic age (Dallmeyer 1989a; Heatherington et al. 1992). Consequently, the only unequivocal Paleozoic rocks within the Florida basement are clastic sedimentary units of the subsurface Suwannee Basin (name from King 1961) of northern Florida. The Suwannee Basin sedimentary rocks consist predominantly of sandstones, siltstones, and shales, which have in some cases undergone low-grade metamorphism that produced phyllitic textures (Opdyke et al. 1987). Similar Paleozoic siliciclastic sedimentary units in southeastern Alabama, southern Georgia, and the western panhandle of Florida have been interpreted as a continuation of the Suwannee terrane (Bridge and Berden 1952; Neathery and Thomas 1975; Chowns and Williams 1983). The age of sedimentation is constrained loosely by paleontologic evidence. Small inarticulate brachiopods *(Lingulepis floridaensis)* and *Skolithos* burrows indicate an Early Ordovician age for the oldest recovered Suwannee Basin samples (Pojeta et al. 1976; Howell and Richards 1949). The uppermost parts of the sequence are black shales containing Silurian to Devonian fossils (Applin 1951; Pojeta et al. 1976). The succession appears to be a prograding sequence, with Early Ordovician near-shore sandstones passing into offshore shales of the Middle or Late Ordovician (Chowns and Williams 1983). The lower boundary for the age of recovered Suwannee Basin sediments is constrained by an $^{40}Ar/^{39}Ar$ date of 504 million years for detrital muscovite from well no. 1904 (Sun Oil Co. H. T. Parker no. 1), in Marion County (Dallmeyer 1987).

The ages and sedimentologic character of these rocks commonly have been used to support the suggestion that the Florida basement evolved as a Gondwanan or peri-Gondwanan terrane that accreted to North America during the Alleghenian Orogeny (Chowns and Williams 1983; Dallmeyer 1987, 1989a, 1989b; Rankin et al. 1989; Thomas et al. 1989). Several investigators have proposed West African source terranes for the Suwannee Basin sediments, based on tectonic reconstructions that place the Florida peninsula adjacent to present-day West Africa during the early Paleozoic (Chowns and Williams 1983; Dallmeyer 1987). The Bové Basin of Senegal and Guinea has been suggested as a correlative to the Suwannee Basin by Chowns and Williams (1983) and Dallmeyer (1987). Chowns and Williams (1983) cited general similarities in the stratigraphic successions in the two regions (e.g. quartz sandstones overlain by black to dark-gray shales).

Geochronologic data from detrital minerals within Suwannee Basin and Bové Basin sandstones have been used to test this proposed correlation. Dallmeyer (1987) determined that muscovite concentrates from the Suwannee Basin and the Bové Basin yield similar $^{40}Ar/^{39}Ar$ dates of 504 ± 2 and 500 to 510 million years, respectively, showing that the source terranes for the two basins contain components with similar cooling ages. U/Pb geochronologic data from multi-grain splits of detrital zircons from Suwannee Basin sediments from two different wells have yielded Pb/Pb ages of 1,650 to 1,800 million years (Opdyke et al. 1987: Alachua County well no. 11447, Chevron-Container Corp. no. 1), and at least 1,850 million years (Odom and Brown 1976: Hernando County well no. 994, Ohio Oil Co. Hernasco Corp. no. 1). Because the zircons are detrital, each multi-grain sample split may be a mixture of zircons of different ages; thus, the calculated ages of 1,650 to 1,800 and 1,850 million years may represent averages of older and younger zircons. Recently, U/Pb analyses of forty single zircon grains from well no. 11447 in Alachua County using the SHRIMP ion microprobe at the Australian National University have revealed that this is the case. Two principal populations were identified from these analyses: (1) 515 to 637 million years (x_{20} = 574 million years; $^{206}Pb/^{238}U$ ages), and (2) 1,967 to 2,282 million years (x_{12} = 2,130 million years; $^{207}Pb/^{206}Pb$ ages). Other identified age components are Archean (about 2680 million years), Early Proterozoic (2,463 million years), and Middle Proterozoic (1,750 million years) (Mueller et al. 1994). The two dominant age groups correspond to the Pan-African (550–650 million years) and

Birimian (Eburnian, about 2,300–1,900 million years) orogenic cycles of West Africa, and to the Brasiliano and Trans-Amazonian cycles of northern South America (Cahen and Snelling 1984; Brito Neves and Cordani 1991; Caby 1989; Bernasconi 1987). Worldwide, co-occurrences of rocks of these ages are rare; therefore, these data strongly support a Gondwanan provenance for the Suwannee Basin sediments. West Africa, specifically, is most often mentioned as a source for these sediments; however, given the close proximity of northeastern South America and West Africa during the early Paleozoic (Bond et al. 1984a; and others), it may not be possible or reasonable to make a distinction between the two as potential sediment sources with currently available data.

Paleozoic-Proterozoic Calc-Alkaline Volcanic Suite

A suite of volcanic rocks spanning the compositional range from basaltic andesites to rhyolites subcrops in northeastern and north-central Florida. The volcanic rocks have calc-alkaline geochemical characteristics according to the criteria of Miyashiro (1978) (fig. 3.4), and Mueller and Porch (1983) and Heatherington et al. (1988, 1989) suggested that they were generated at an ocean/continent convergent margin. Trace-element data show that the basaltic andesites are depleted in high field strength elements (HFSE) such as Nb and Ta with respect to large ion lithophile elements (LILE), a geochemical feature typical of island arc basalts (Heatherington et al. 1988, 1989); however, the ages of these rocks are not well constrained. ^{40}Ar/^{39}Ar measurements for rocks from Putnam and Flagler counties yielded complex gas-release patterns with poorly defined plateaus corresponding approximately to 410 to 420 million years (Mueller and Porch 1983). Whole-rock Rb/Sr data from several intervals in three wells suggest a composite isochron corresponding to an age of 480 ± 60 million years (Heatherington et al. 1988, 1989) (fig. 3.5). These are presently the only dates available for these rocks, but they must be viewed only as lower limits, due to the complexity of the ^{40}Ar/^{39}Ar data and the possibility that the whole-rock Rb/Sr data may represent, in part, a mixing array. Furthermore, both the ^{40}Ar/^{39}Ar and Rb/Sr systems are easily reset in these types of rocks by low-grade thermal events. Dallmeyer (1989a, 1991) used general lithologic comparisons to propose that the north Florida volcanic suite is directly correlative with Late Proterozoic calc-alkaline volcanic rocks of the Niokolo-Koba Group in Senegal, West Africa, which, in turn, are proposed coeval with granites dated by ^{40}Ar/^{39}Ar and Rb/Sr at 650 to 700 million years (Bassot and Caen-Vachette 1983; Dallmeyer et al. 1987). Dallmeyer (1989a and pers. comm.) therefore suggests assigning the Florida rocks an age of about 650 to 700 million years, based on this inferred correlation. Striking similarities between major- and trace-element abundances and Nd-, Sr-, and Pb-isotope systematics of the Florida and Senegalese suites (Heatherington et al. 1992) support the correlation and, therefore, the suggestion that the measured ^{40}Ar/^{39}Ar and Rb/Sr dates for the Florida rocks may be too young. Any more accurate age for the north Florida volcanic suite must come from a method that is more resistant to low-temperature resetting, such as U/Pb zircon geochronology.

Osceola Granite and St. Lucie Metamorphic Complex

The Osceola Granite and St. Lucie metamorphic complex form a sector of the Florida basement that is generally thought to be bounded on the north by a northeast-trending magnetic high (the Florida Magnetic Anomaly; Hall 1990) and on the south by a proposed extension of the Bahamas Fracture Zone (BFZ), as inferred from offsets in magnetic and gravity contours (Klitgord et al. 1984) (fig. 3.1). Wells penetrating the Osceola Granite are located in Lake, Orange, St. Lucie, and Osceola counties. The name "Osceola Granite" is something of a misnomer, because compositions of the unit range from diorite to granodiorite. Texturally, the rocks are phaneritic and undeformed. Several dates have been published for the unit; these include ^{40}Ar/^{39}Ar dates of 527 and 535 million years reported by Dallmeyer et al. (1987) for biotite samples from two different wells (well no. 11341, Atlantic Richfield Bronson no. 2, Osceola County; and well no. 3673, Warren Petroleum Co. George Terry no. 1, Orange County), and a Rb/Sr date of about 530 million years reported by Bass (1969) for feldspar fractions from well no. 1014 (Humble Oil and Refining Co., Ray Carroll no. 1, Osceola County).

Nd-isotopic data obtained for the Osceola Granite are intriguing. Depleted-mantle Nd model ages are derived by using the measured ^{143}Nd/^{144}Nd and ^{147}Sm/^{144}Nd ratios of a rock to extrapolate the growth curve for radiogenic Nd back in time until it intersects a model growth curve for depleted (sub-lithospheric) mantle (DePaolo 1988). The calculated age of intersection is an estimate of the time that has passed since Nd in the rock or its source was extracted from depleted mantle. If contamination or mixing with a different reservoir occurs, the resultant model age will represent an average of the times of extraction of Nd from model depleted mantle for the source and contaminating reservoirs. Depleted-mantle model ages (T_{DM}) are distinctly different for whole-rock samples from well no. 11342 (Atlantic Richfield Bronson no. 1; 1,500 million years) and well no. 1014 (Humble Oil and Refining Co. Ray Carroll no. 1; 3,360 million years), both located in Osceola County. These results suggest that the Late Proterozoic to Early Cambrian Osceola Granite was derived from two or more older sources, with different ages, at least one of which was Archean. Reconnaissance single-grain ion-probe (SHRIMP) analyses of zircons from the Osceola Granite (well no. 1014) corroborate both the $^{40}Ar/^{39}Ar$ cooling age of approximately 530 million years determined by Dallmeyer et al. (1987) and the Archean component suggested by the Sm/Nd analyses (Mueller et al. 1994). Several grains produced $^{206}Pb/^{238}U$ dates of about 550 to 600 million years, consistent with the $^{40}Ar/^{39}Ar$ date of 530 million years as a cooling age; other grains appeared to be inherited from preexisting lithosphere, yielding dates of about 3.0 billion years.

The St. Lucie metamorphic complex has been penetrated by wells in St. Lucie and Martin counties in southeastern Florida. The recovered rock types are predominantly amphibolites, accompanied by schist and layers or veins of quartz diorite (Bass 1969). Radiometric dates for the St. Lucie rocks have been reported by Bass (1969) and Dallmeyer (1988). Bass reported K/Ar dates of 503 and 470 million years for hornblende from amphibolite in well no. 4323 in St. Lucie County (Amerada no. 2 Cowles Magazines Inc.). Dallmeyer (1988) reported a more reliable $^{40}Ar/^{39}Ar$ date of 513 ± 9 million years for a hornblende concentrate from amphibolite recovered from well no. 13082, also in St. Lucie County (Peacock Fruit and Cattle Corp. no. 1). On the basis of this date, Dallmeyer (1988) has suggested that the St. Lucie amphibolite is correlative with amphibolites from the northern Rokelide Orogen (Pan-African II) in Sierra Leone, West Africa, which have similar cooling ages.

Tectonic Affinities and Interrelationships of Florida Lithospheric Provinces: Where Do the Pieces Fit?

Considerable progress has been made in recent years toward understanding the Paleozoic and later tectonic evolution of the eastern margin of North America. This work has led to our present view of the Appalachians as a collage of accreted terranes associated with the Paleozoic collision of North America, Europe, and Africa (Williams and Hatcher 1982; Rowley 1981; Zen 1983; Keppie 1989; Gromet 1989; Horton et al. 1989; Rankin et al. 1989; and others). Detailed geochronologic and geochemical studies have played key roles in the identification of these terranes and their origins (Dallmeyer 1988, 1989a, 1989b; Rast and Skehan 1981; Black 1980; Bland and Blackburn 1980; Ayuso 1986; Zartman 1988; and others). The lithologic and faunal associations of early Paleozoic to late Proterozoic sedimentary sequences of many terranes are dissimilar to those of adjacent parts of North America, but instead resemble those of some European and African terranes. In the northern Appalachians, these terranes have generally been designated as "Avalonian" on the basis of their resemblance to lithologic assemblages of the Avalon peninsula, Newfoundland (Rodgers 1972). They are characterized by late Precambrian to Cambrian felsic igneous rocks and meta-igneous rocks overlain by late Proterozoic to early Paleozoic sedimentary and metasedimentary rocks (Rankin et al. 1989; Keppie et al. 1991).

In the southern Appalachians, the exotic Carolina terrane exhibits some similarities to the Avalonian terranes located to the north, and has on occasion been grouped with them as a possible Avalonian terrane itself (Rankin et al. 1989). Avalonian-type terranes (including the Carolina terrane) have been identified over nearly the entire north-south extent of the Appalachian Orogen, from Newfoundland to Georgia, with the exception of gaps at the latitudes of southern Maine, New York, and New Jersey (Rankin et al. 1989). Paleomagnetic data for the Carolina and Avalonian terranes suggest Gondwanan latitudes during the early Paleozoic, and it is generally accepted that these terranes are of Gondwanan or peri-Gondwanan affinity, and were accreted to Laurentia during the assembly of Pangaea. However, their precise origins are uncertain, as are the relationships of individual terrane fragments to each other.

The pre-Cretaceous rocks of Florida and adjacent parts of Alabama and Georgia are also generally thought to be exotic to North America (Chowns and Williams 1983; Dallmeyer 1989a, 1989b; Rankin et al. 1989; Thomas et al. 1989; and others). They have been termed the "Tallahassee-Suwannee terrane" by Williams and Hatcher (1982, 1983). The concept that the Tallahassee-Suwannee terrane is non-Laurentian originated from limited paleontologic and paleomagnetic data and the undeformed nature of the Paleozoic sedimentary package (Wilson 1966; Cramer 1973; Pojeta et al. 1976; Opdyke et al. 1987), and is supported by geochronologic and geochemical evidence (Dallmeyer 1987, 1988, 1989a; Dallmeyer et al. 1987; Heatherington et al. 1992; Mueller et al. 1994). Although not usually classified as "Avalonian" in the literature, the lithostratigraphic succession of the Tallahassee-Suwannee terrane has much in common with the Carolina and Avalonian terranes. Like these other exotic terranes, the Tallahassee-Suwannee

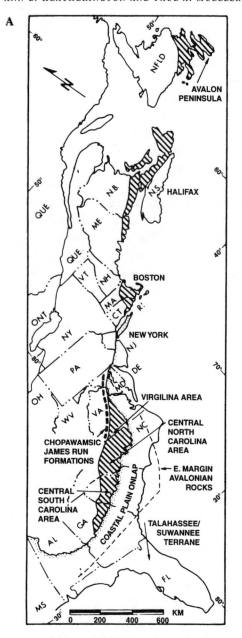

terrane is characterized by Late Proterozoic to Paleozoic (500–600 million years) calc-alkaline, felsic volcanic rocks and undeformed granitoids, overlain by Ordovician to Silurian clastic sedimentary rocks. Figure 3.6 (from Rankin et al. 1989) shows the distribution of Avalonian-Carolina rocks in North America and their present geographic relationship to the Tallahassee-Suwannee terrane, and also the boundaries of lithotectonic provinces in the southern Appalachians. The Kiokee, Belair, Charlotte, and Carolina slate belts all have been classified as parts of the exotic Carolina terrane (Secor et al. 1986).

The prevailing view is that the Tallahassee-Suwannee terrane was accreted to North America during the late Paleozoic Alleghenian Orogeny (Dallmeyer 1989a, 1989b; Rankin et al. 1989; Thomas et al. 1989), after accretion of the more-inboard Carolina terrane, which, according to paleomagnetic evidence, probably occurred during the Ordovician (Vick et al. 1987; Rankin et al. 1989). As we have discussed, Dallmeyer et al. (1987), Dallmeyer (1987, 1988, 1989a, 1991), and Chowns and Williams (1983) have proposed very specific correlations of the Tallahassee-Suwannee terrane with the West African craton, based on $^{40}Ar/^{39}Ar$ cooling ages and general lithologic associations. In contrast, Van der Voo (1988) placed all of the Avalonian terranes and the Tallahassee-Suwannee terrane in an offshore, non-cratonic, peri-Gondwanan location during the Ordovician (fig. 3.7a). This assignment was based on paleomagnetic data and marine affinities of Paleozoic sedimentary rocks of the Avalonian Delaware Piedmont (Brown and Van der Voo 1983).

Although the boundaries between most of the Avalonian-Carolina terranes and pre-Alleghenian North

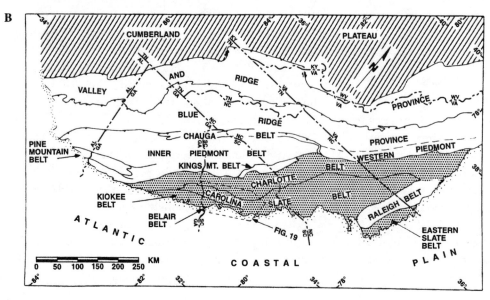

Figure 3.6. A: Locations of Avalonian terranes in eastern North America. B: Lithotectonic belts of the southern Appalachians. The Carolina Slate Belt, Kiokee Belt, and Belair Belt were classified as parts of the Carolina terrane by Secor et al. (1986). Maps are from Rankin et al. (1989).

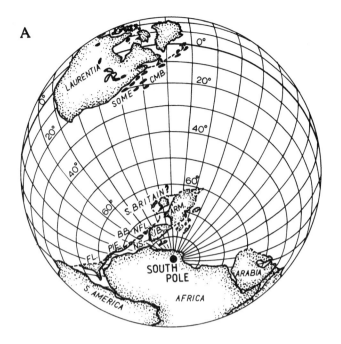

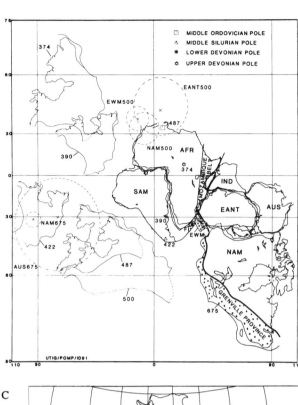

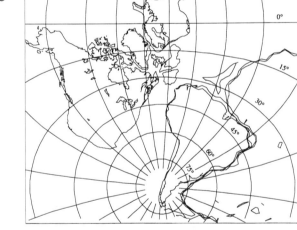

Figure 3.7. Three postulated positions for the Tallahassee-Suwannee terrane and/or the southwest margin of Laurentia in the Early Paleozoic, as indicated primarily by paleomagnetic data. A: Van der Voo (1988) for the Middle to Late Ordovician. B: Dalziel (1991) for ages shown on map. C: Channell et al. (1992) for the Siluro-Devonian.

America are relatively well defined tectonic zones or faults, this is not the case for the boundary (or boundaries) of the buried Floridan basement terrane(s). It has been postulated that the suture between the Tallahassee-Suwannee terrane and previously accreted terranes of North America is buried beneath sedimentary rocks of the coastal plain, coinciding with the Brunswick Magnetic Anomaly (BMA) (Daniels et al. 1983; McBride and Nelson 1988, 1990; Chowns and Williams 1983; Nelson et al. 1985a). The BMA is a major magnetic lineation that trends approximately east-west across southern Georgia and Alabama and consists of a northern, negative component and a southern, positive component (fig. 3.1). The positive and negative components diverge and become indistinct in southwestern Alabama where the pre-Cretaceous basement rocks pass under the northeastern limit of the Louann Salt (Jurassic). The negative branch, sometimes termed the "Altamaha anomaly" (Higgins and Zietz 1983), is approximately 100 km north of the positive branch at the point where the anomaly disappears (fig. 3.1).

A currently popular view is that the anomaly is a manifestation of an Alleghenian suture between the Tallahassee-Suwannee terrane and preexisting North American terranes (Daniels et al. 1983; McBride and Nelson 1988; Chowns and Williams 1983), and that the paired positive and negative branches are the signature of a south-dipping lithospheric slab (McBride and Nelson 1988). COCORP seismic profiles indicate a south-dipping reflector coinciding approximately with the trace of the BMA (Nelson et al. 1985a; McBride and Nelson 1988). Alternatively, Tauvers and Muehlberger (1987) suggested that the Brunswick anomaly represents the limit of North American lithosphere at deep levels, but that the Tallahassee-Suwannee terrane was thrust over North America, so that upper levels of the crust are allochthonous as far as 100 km north of the magnetic anomaly. McBride and Nelson (1988, 1990) agreed with this model, and further suggested that the suture was reactivated as a rift basin during Mesozoic extension, and was the site of basaltic magmatism during that time.

Some investigators disagree with the interpretation of the BMA as a suture between lithospheric terranes. Austin et al. (1990), Hutchinson et al. (1983, 1990), and Higgins and Zietz (1983) viewed the BMA as a Mesozoic rift basin, rather than a suture, the magnetic and seismic anomalies representing the edge of a seaward-dipping graben. Furthermore, Hall (1990) called attention to the northeast-trending magnetic high that crosses the Florida peninsula (the Florida Magnetic Anomaly; fig. 3.1), and suggested that a suture should be coincident with this anomaly, rather than associated with the BMA in southern Georgia and Alabama.

Regardless of whether these magnetic anomalies represent sutures, the available paleomagnetic and paleontologic evidence indicates a Gondwanan affinity for the Tallahassee-Suwannee terrane; however, specific evidence for a West African correlation, as opposed to a South American correlation, is less abundant. Paleomagnetic correlations are based on a single paleolatitude determination of 49° for a clastic sedimentary rock of uncertain age—probably Late to Middle Ordovician, but perhaps Silurian or Devonian (Opdyke et al. 1987). Although considerable importance has been attached to this paleopole, these data should be viewed with caution, because of the uncertainty of the age of the samples. According to conventional tectonic reconstructions that place the southeastern margin of Laurentia at about 20° to 30° during the Ordovician to Devonian (Van der Voo 1988; and others) (fig. 3.7a), the paleopole is clearly non-Laurentian. Because the paleolatitude of Gondwana changed dramatically from the Ordovician to the Devonian (Van der Voo 1988; and others) and the age of the sample is not well constrained, it is not possible to determine whether the Florida sample had a specifically West African latitude at the time of deposition. In addition, paleomagnetic data cannot be used to specify paleolongitude. Further complicating the issue is the recent SWEAT hypothesis of Dalziel (1991) and Moores (1991), which places the Appalachian margin of Laurentia at about 50° during Ordovician-Silurian time (fig. 3.7b). If this new tectonic reconstruction is correct, the Florida paleopole is consistent with either a Laurentian or a Gondwanan affinity. Geochronologic evidence is more compatible with a Gondwanan affinity (Dallmeyer 1987, 1988, 1989a; Dallmeyer et al. 1987; Heatherington et al. 1992; Mueller et al. 1994); however, distinctions between West African and other geographically proximal Gondwanan origins (e.g. northeastern South America) are not discernible from the available data.

Tectonic reconstructions involving the Florida Platform, or Tallahassee-Suwannee terrane, are made more problematic by the possibility that the terrane may actually be an amalgamation of several terranes. The absence of surface outcrop makes it difficult to trace lithostratigraphic units over long distances and obscures possible suture zones within the Florida Platform. Consequently, geophysical evidence (i.e. discontinuities within gravity and magnetic patterns) has been the basis for postulated tectonic boundaries that may bisect the Florida peninsula. The primary potential lithospheric boundary (see chapter 2) has been referred to as an extension of the Jay Fault of Alabama and the Florida panhandle, or described as an extension of the Bahamas Fracture Zone (Klitgord et al. 1984) (fig. 3.1). Smith and Lord (chapter 2) have proposed that right-lateral strike-slip motion occurred along this feature during the assembly of Pangaea.

Geochemical evidence, however, does not support this feature as a major translational boundary. For example, Nd model ages for rhyolites (rocks most likely to be representative of crustal sources) range from 1,000 to 1,350 million years for northern Florida, and from 1,200 to 1,350 million years for southwestern Florida (Heatherington et al. 1988). The similarity in these ages suggests that rhyolites from north and south of this feature were derived from a source (or sources) of essentially the same age and Nd isotopic composition; therefore, these data suggest that lithosphere north and south of this feature may have been geographically proximal since the Proterozoic. If the boundary is a fault, it may have a strong normal component, as evidenced by the sediment cover, which in southern Florida is more than twice as thick as in northern Florida. In addition, the extensional nature of Mesozoic tectonomagmatic activity is consistent with normal faulting (Heatherington and Mueller 1991). Regardless of whether this geophysically evident feature ultimately proves to be a boundary between lithospheric blocks of fundamentally different character, it represents only one of the major lithologic transitions within the Florida basement that must be investigated before a clear understanding of the lithospheric heritage of the Florida peninsula can be realized.

Summary

Five distinct lithostratigraphic associations are evident in the pre-Cretaceous basement of Florida. Mesozoic units are located in northern and southwestern Florida. The northern zone is characterized by tholeiitic basalts and diabases geochemically similar to tholeiites of eastern North America associated with the breakup of Pangaea. The southwestern zone is characterized by alkalic rhyolites and alkalic to transitional basalts, and also is an extension-related suite of rocks. Isotopic and geochemical data suggest that southwestern Florida was in close proximity to a Mesozoic hot spot and/or triple junction. The early Paleozoic Suwannee Basin subcrops in northern Florida and consists of sandstones overlain by black to gray shales. A calc-alkaline igneous association, probably arc-related, is located east and south of the Suwannee Basin; although the age of the suite is uncertain, available data indicate that it is probably Late Proterozoic or Early Paleozoic. The Osceola Granite makes up the basement subcrop in east-central Florida. These rocks have Late Proterozoic to Early Cambrian cooling ages, and display evidence of partial derivation from an Archean source. The St. Lucie metamorphic complex lies generally south of the Osceola Granite and helps separate it from the Mesozoic volcanic province of southwestern Florida.

Paleomagnetic and paleontologic evidence suggests that the Florida basement terranes originated at Gondwanan latitudes—or possibly Laurentian latitudes if the SWEAT hypothesis of Dalziel (1991) and Moores (1991) is considered—and were accreted to North America during the Paleozoic. Geochemical and geochronologic data have provided support for proposed correlations of Florida basement terranes with West Africa and northeastern South America, but also show possible links to Avalonian terranes in eastern North America. The origin of Avalonia itself is obscure, but it is generally thought to have been generated as a magmatic arc (or set of arcs) on the margin of Gondwana during the Late Proterozoic (Murphy and Nance 1989; Nance et al. 1991). Much of Avalonia may have originated as an island arc on oceanic basement, but recent investigations suggest continental basement in the source of some Avalonian rocks (Keppie et al. 1991). Zircon geochronology indicates a significant component of Archean material in the source of Paleozoic and Late Proterozoic Florida volcanic rocks (Mueller et al. 1994), suggesting that the Florida rocks formed on cratonic Gondwanan basement. Based on known geochronologic and lithostratigraphic similarities, we suggest that the Florida basement, and perhaps other Avalonian terranes, may represent fragments of a single, large, Late Proterozoic orogenic belt that extended from the Gondwana continent to adjacent oceanic lithosphere.

4

The Sedimentary Platform of Florida: Mesozoic to Cenozoic

Anthony F. Randazzo

THE SEDIMENTARY ROCK RECORD OF Florida (inside front cover) provides significant insight into geologic environments and tectonic settings of the past 500 million years. To achieve a comprehensive understanding of its history, this record should be considered within the context of a rifting continent, which was subsequently flooded to create a tectonically stable platform where marine sedimentation occurred. The eastern Gulf of Mexico, the Caribbean, and the Bahamian area also represent an integral part of Florida's geologic history. The entire region and sedimentary record were strongly influenced by tectonic-plate development during the Phanerozoic.

Early Depositional History

The precise age and derivation of clastic sedimentary rocks of the Paleozoic and early Mesozoic are poorly understood (see chapters 2 and 3). Extensive Mesozoic deposition of clastic sediments began with a rifting phase during the late Triassic to early Jurassic (fig. 4.1a). Subsequent carbonate sedimentation was constrained by a latitudinally controlled climatic belt and tectonism in the region. Passive-margin depositional environments were in place thereafter, and tectonically stable conditions prevailed during the late Cretaceous and throughout the Cenozoic.

Carbonate, evaporite, and siliciclastic sediments began to accumulate on the Florida Platform with the development of the Gulf of Mexico basin of deposition, probably during the middle Jurassic (Winston 1992). Paleogeographic reconstructions and facies patterns are displayed chronologically in figure 4.1. Isopach maps for Jurassic-Miocene rocks are shown in figure 4.2. Within the eastern Gulf, the Apalachicola Basin and Tampa Embayment were two principal depocenters for thick evaporite sedimentation (Louann Salt sequence). Evaporites eventually gave way to deposition of clastic fluvial, eolian, and marine sediments (Norphlet Sand) in the northern parts of the platform, and, with a rise in sea level, carbonate buildups (Smackover Limestone) (Wu et al. 1990a, 1990b). The Smackover carbonate unit, which grades up-dip into clastic sequences, displays shoaling-upward cycles of sedimentation, with ramp progradation westward and southwestward from the peninsula and panhandle. Oolitic sandbars, shallow- to deep-water shelf bioclastic deposits, and reefs of corals, sponges, and stromatolites characterize this unit. Alternating clastic fluvial and deltaic deposition and carbonate sedimentation (Haynesville-Cotton Valley sequences) reflect sea-level variation throughout the middle to late Jurassic and early Cretaceous. Basement configuration and faulting affected the distribution, thickness, and character of the late Triassic to Jurassic sedimentary sequences (Miller 1982; Salvador 1987; Dobson 1990). The Peninsular Arch (see chapter 2) was a positive area and a natural divide for sedimentation between the eastern and western parts of the peninsula (Miller 1986).

Depositional sequences of the Cretaceous and Cenozoic formed in a relatively stable tectonic setting and display a general west-to-east and north-to-south gradation of clastic to carbonate sedimentation (Salvador 1991b). Evaporite formation became less prevalent, gradually decreasing during the late Cretaceous and early Cenozoic, the time interval which saw the greatest accumulation of sedimentary rocks in Florida. The Cretaceous also marked the separation of the Florida sedimentary regime from the Bahamas (see chapter 11).

Florida Sedimentary Platform: Cretaceous to Paleogene

The relatively stable passive-margin tectonic setting of that time produced a broad, shallow-water marine platform that experienced little tilting or disturbance. Steep submarine escarpments now bound the gently dipping Florida Platform (Paull and Neumann 1987) (fig. 4.3).

Figure 4.1 (a-s). Paleogeographic and lithofacies maps of the Florida Platform, Triassic to Pleistocene (Florida Geological Survey; Murray 1961; Chen 1965; Poag 1985; Miller 1986; Salvador 1987, 1991b; Winston 1991, 1992, 1993a, 1993b; McFarlan and Menes 1991; Sohl et al. 1991; Gallaway et al. 1991).

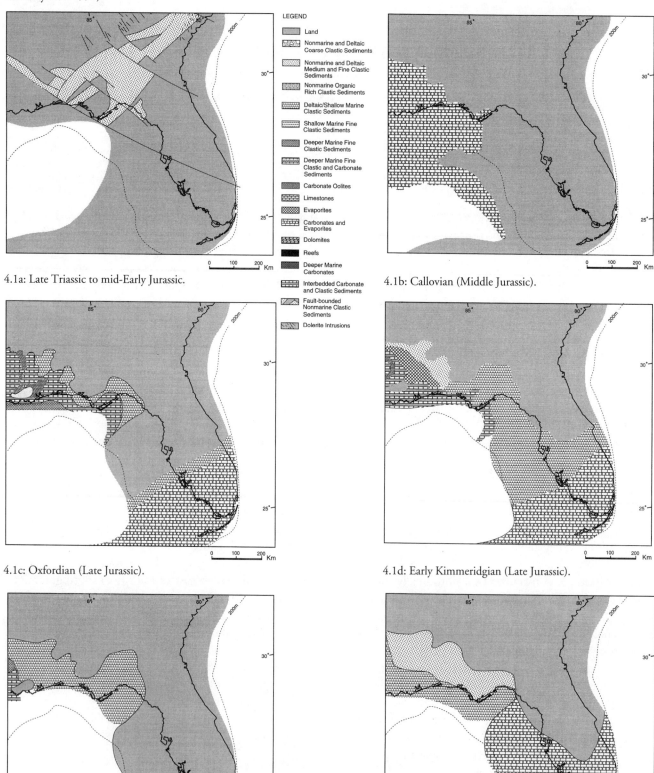

4.1a: Late Triassic to mid-Early Jurassic.

4.1b: Callovian (Middle Jurassic).

4.1c: Oxfordian (Late Jurassic).

4.1d: Early Kimmeridgian (Late Jurassic).

4.1e: Late Kimmeridgian (Late Jurassic).

4.1f: Tithonian (Late Jurassic).

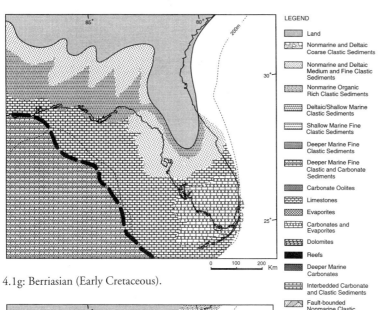

4.1g: Berriasian (Early Cretaceous).

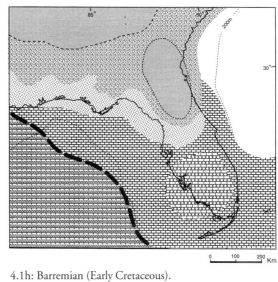

4.1h: Barremian (Early Cretaceous).

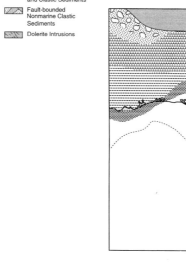

4.1i: Albian (Early Cretaceous).

4.1j: Turonian (Late Cretaceous).

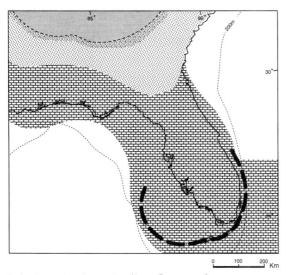

4.1k: Coniacian-Santonian (Late Cretaceous).

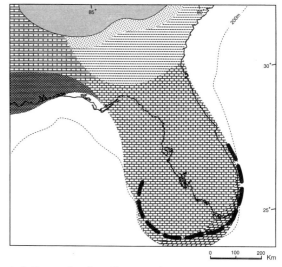

4.1l: Campanian (Late Cretaceous).

Figure 4.1 (continued)

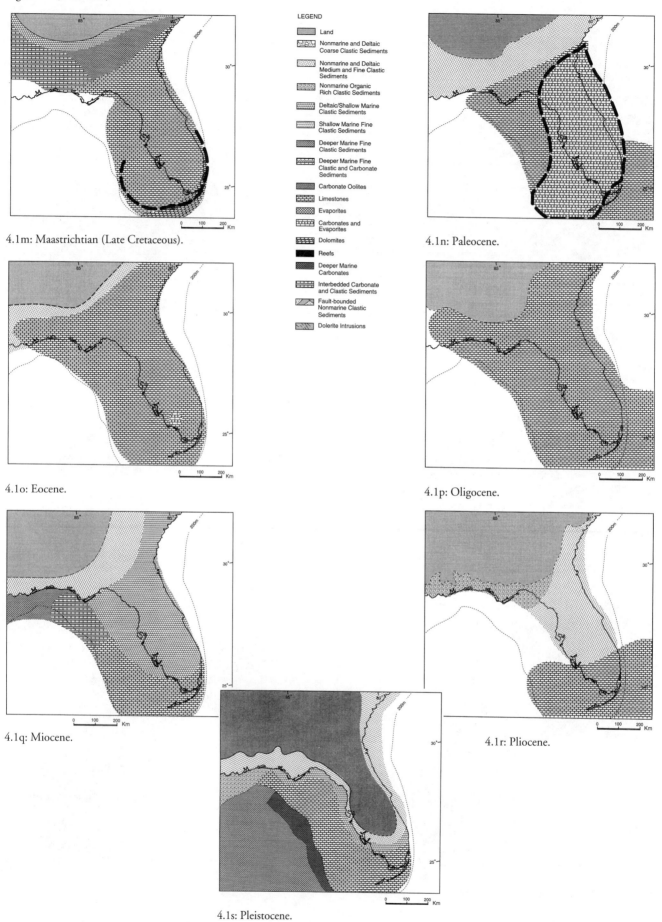

4.1m: Maastrichtian (Late Cretaceous).
4.1n: Paleocene.
4.1o: Eocene.
4.1p: Oligocene.
4.1q: Miocene.
4.1r: Pliocene.
4.1s: Pleistocene.

Figure 4.2 (a-i). Isopach maps of Florida, Jurassic to Miocene (Miller 1986 and pers. comm.).

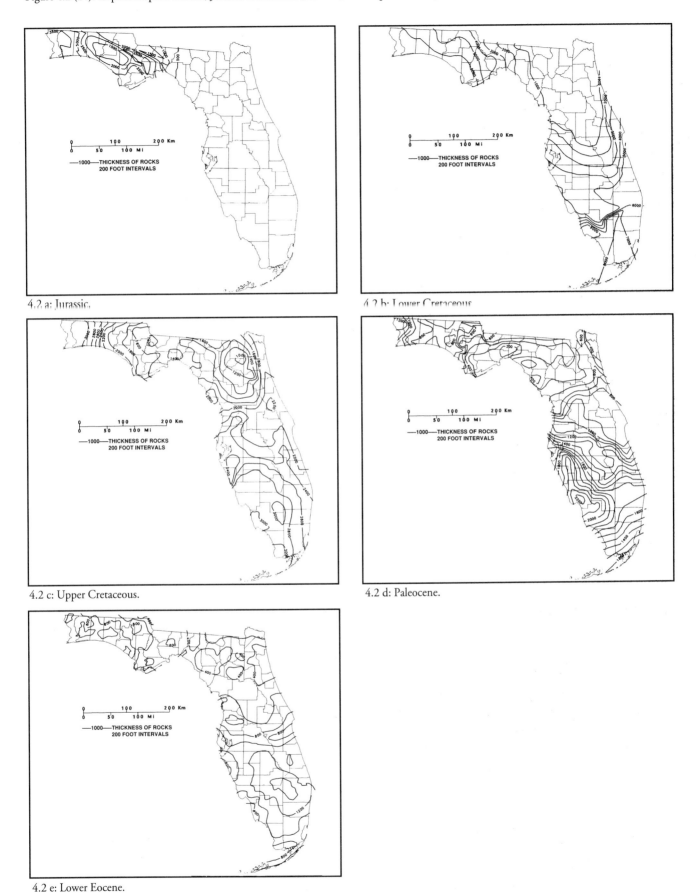

4.2 a: Jurassic.

4.2 b: Lower Cretaceous.

4.2 c: Upper Cretaceous.

4.2 d: Paleocene.

4.2 e: Lower Eocene.

Figure 4.2 (continued)

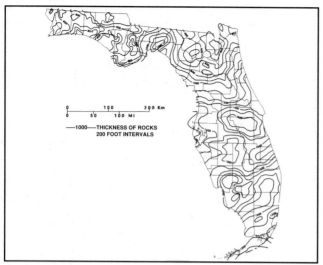

4.2 f: Middle Eocene.

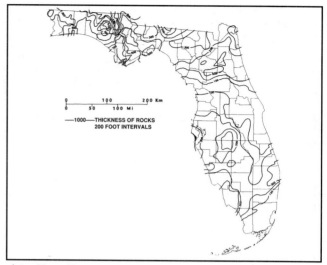

4.2 g: Upper Eocene.

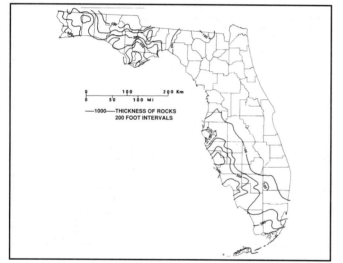

4.2 h: Oligocene.

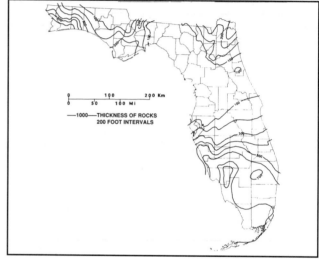

4.2 i: Miocene.

The sedimentary sequences consist of nearly flat-lying marine rocks stacked some 7 km thick and terminating at these boundary escarpments.

Depositional Dynamics

Thorough treatments of the principles of carbonate-rock buildups are presented in several texts (Wilson 1975; Scoffin 1987; Tucker and Wright 1990; Schlager 1992; Sims et al. 1993). Carbonate and evaporite sedimentation is controlled by climatic settings conducive to high marine organic productivity and evaporation. Today these settings are in warmer, low-latitude areas (tropical and subtropical belts) where ocean waters are saturated with carbonate and sulfate compounds. Physiographic configurations also influence the intrabasinal accumulations of these sediments. Carbonate platforms are often referred to as "carbonate factories," but, as is the case for the Florida Platform, they represent integrated carbonate-evaporite-siliciclastic facies systems. These systems respond to low-stand, transgressive, and high-stand sea-level conditions (Handford and Loucks 1993). The types of platform margins (e.g. reef-rimmed; unrimmed, but bounded by steep faulted or erosional escarpments; gently sloping ramps), influenced or created by dynamic sea-level variations, are a principal component of carbonate-platform classification. The Florida Platform was located at or near the mid-Triassic equator and migrated northward (along with North America) during the Jurassic to the Oligocene. This platform evolved from a rimmed shelf in the Jurassic and Cretaceous to a carbonate ramp in the Paleogene (Corso 1987; Corso et al. 1989; Winston 1991). The global paleogeographic reconstruction for the Cretaceous (fig. 4.4) displays the position of the Florida Platform and others existing at that time.

Episodes of cyclic sedimentation (fig. 4.5) are documented locally for the Florida Platform (Randazzo 1972;

Figure 4.2 (a-i). Isopach maps of Florida, Jurassic to Miocene (Miller 1986 and pers. comm.).

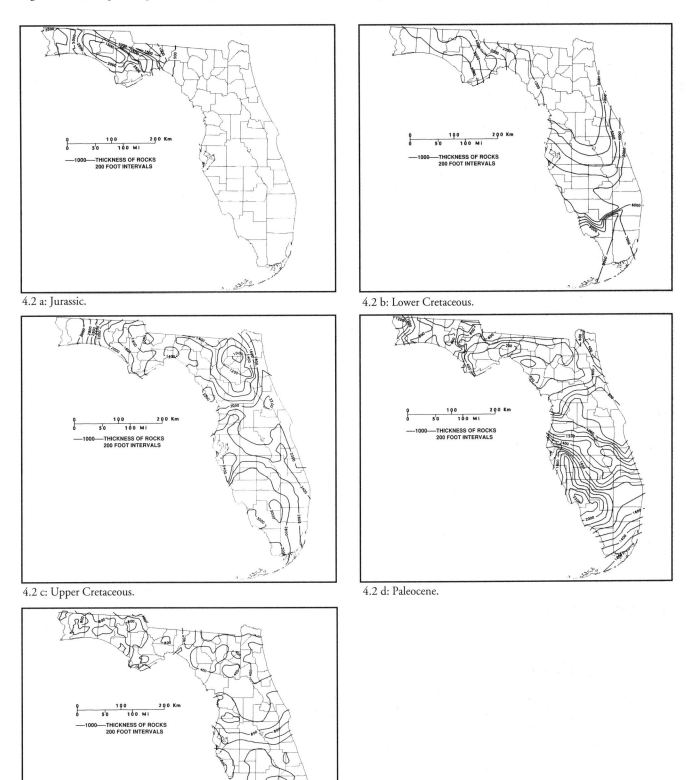

4.2 a: Jurassic.

4.2 b: Lower Cretaceous.

4.2 c: Upper Cretaceous.

4.2 d: Paleocene.

4.2 e: Lower Eocene.

Figure 4.2 (continued)

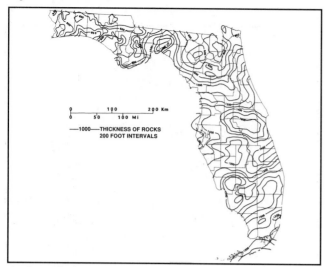

4.2 f: Middle Eocene.

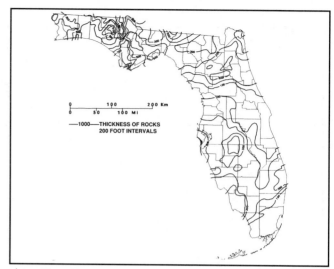

4.2 g: Upper Eocene.

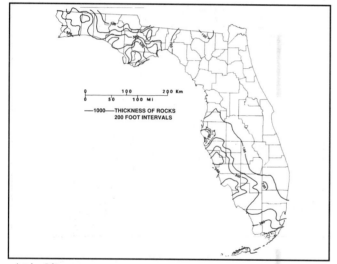

4.2 h: Oligocene.

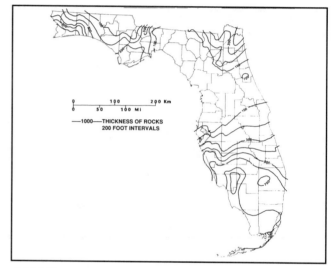

4.2 i: Miocene.

The sedimentary sequences consist of nearly flat-lying marine rocks stacked some 7 km thick and terminating at these boundary escarpments.

Depositional Dynamics

Thorough treatments of the principles of carbonate-rock buildups are presented in several texts (Wilson 1975; Scoffin 1987; Tucker and Wright 1990; Schlager 1992; Sims et al. 1993). Carbonate and evaporite sedimentation is controlled by climatic settings conducive to high marine organic productivity and evaporation. Today these settings are in warmer, low-latitude areas (tropical and subtropical belts) where ocean waters are saturated with carbonate and sulfate compounds. Physiographic configurations also influence the intrabasinal accumulations of these sediments. Carbonate platforms are often referred to as "carbonate factories," but, as is the case for the Florida Platform, they represent integrated carbonate-evaporite-siliciclastic facies systems. These systems respond to low-stand, transgressive, and high-stand sea-level conditions (Handford and Loucks 1993). The types of platform margins (e.g. reef-rimmed; unrimmed, but bounded by steep faulted or erosional escarpments; gently sloping ramps), influenced or created by dynamic sea-level variations, are a principal component of carbonate-platform classification. The Florida Platform was located at or near the mid-Triassic equator and migrated northward (along with North America) during the Jurassic to the Oligocene. This platform evolved from a rimmed shelf in the Jurassic and Cretaceous to a carbonate ramp in the Paleogene (Corso 1987; Corso et al. 1989; Winston 1991). The global paleogeographic reconstruction for the Cretaceous (fig. 4.4) displays the position of the Florida Platform and others existing at that time.

Episodes of cyclic sedimentation (fig. 4.5) are documented locally for the Florida Platform (Randazzo 1972;

Randazzo and Zachos 1984). These repetitive cycles involve shallowing-upward alternations of carbonate and evaporite sedimentation. They are likely related to eustatic controls (glacial-ice volumes) driven by the Earth's orbital perturbation (Milankovitch cycles) (Tucker and Wright 1990; Fischer and Bottjer 1991). Such astronomical cycles as precession of the equinoxes, obliquity of the ecliptic, and eccentricity affect insolation and the extent of polar icecaps. Lag-time complications, tectonic effects, and organic carbonate productivity would all influence the resulting sediment record. Autocyclic sedimentation, brought on by responses of organic carbonate productivity to sea level induced sediment-supply changes can also produce shallowing-upward cycles.

Sea level was higher during the Cretaceous and most of the Paleogene than at present (fig. 4.6). Fluctuations resulted in incursions of clastic sediments from the north and west during low stands, which became more frequent during the late Eocene and Oligocene.

A negative surface feature in the panhandle marked the boundary between carbonate deposition to the southeast and clastic deposition to the northwest. This feature was an interface between two very different sedimentologic regimes which shifted through time. It was either the site of slower sediment accumulations, or higher-energy currents which prevented thick sediment deposits from forming. Huddlestun (1993) has proposed the comprehensive name "Georgia Channel System" for

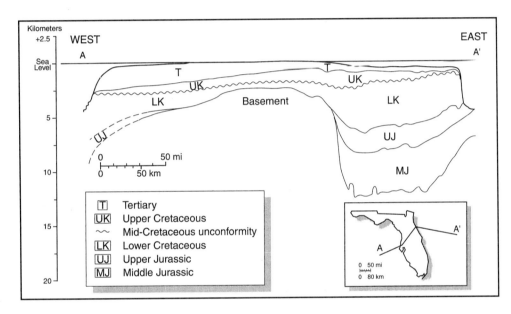

Figure 4.3. East-west cross-section of the Florida Platform, West Florida Escarpment to the Blake Plateau (Miller 1986; Randazzo and Cook 1987; Klitgord et al. 1988; Salvador 1991a).

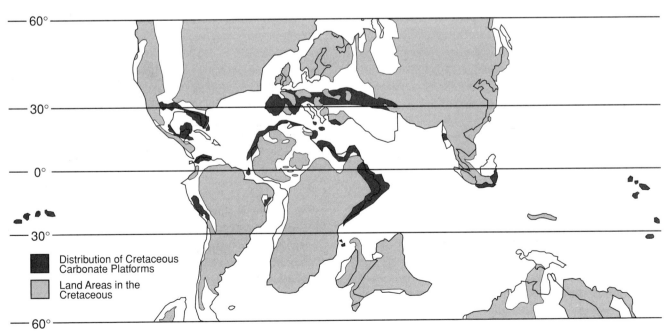

Figure 4.4. Global distribution of Cretaceous carbonate platforms (modified from Simo et al. 1993).

this feature, consisting of two distinct, but related sedimentologic systems (figure 4.7). The older system is known as the Suwannee Strait, or Channel (Dall and Harris 1892; Hull 1962; Gelbaum 1978; Miller 1986); this channel existed from the Late Cretaceous to the Middle Eocene and was associated with two embayments, the Tallahassee Embayment and the Southeast Georgia Embayment. The younger system is the Gulf Trough (Herrick and Vorhis 1963; Gelbaum and Howell 1982), which existed from the Middle Eocene to the Middle Miocene. The Gulf Trough is believed to have only a westerly embayment (Chatahoochee Embayment). The Tallahassee and Chatahoochee embayments are referred to collectively as the Apalachicola Embayment (Huddlestun 1993).

With the southward influx of clastic sediments, carbonate sedimentation was suppressed (Walker et al. 1983; McKinney 1984). During high stands of sea level, the channel migrated northward, and the area for carbonate sedimentation increased accordingly. Destruction of the Suwannee Channel during the late Oligocene is attributed to increased clastic deposition, caused by a drop in sea level, and structural adjustments in southern Florida accompanied by a diversion of the Gulf Stream (McKinney 1984). The supply and influence of clastic sediments were at least partly controlled by rejuvenation of the Appalachian Mountains during the late Paleogene.

The evidence for high sea levels in the Eocene rocks of Florida is not entirely consistent with the global sea-level curve of Haq et al. (1987a, 1987b, 1988) (fig. 4.8). Field and laboratory evidence points to higher sea-level stands for the late Eocene and parts of the Oligocene than are indicated by the first-order cycles of Haq et al. (1988).

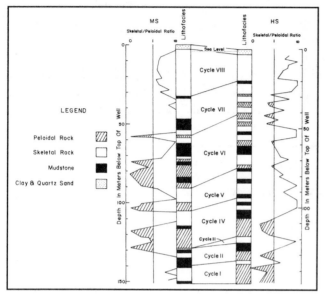

Figure 4.5. Example of cyclic sedimentation as indicated by carbonate rock type and skeletal/peloidal variance (Randazzo and Zachos 1984).

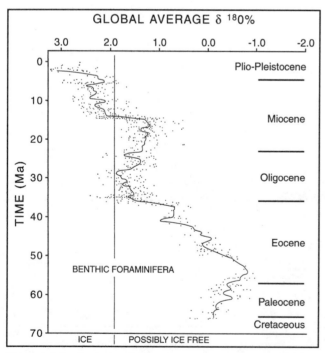

Figure 4.6. Sea-level curve for the Cretaceous and Cenozoic, based on $\delta^{18}O$ compositions of benthic foraminifers (after Miller and Fairbanks 1985; Williams 1988).

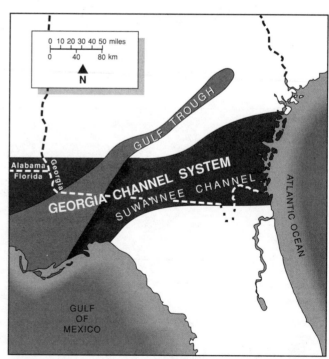

Figure 4.7. The Georgia Channel System and its two components, the Gulf Trough and the Suwannee Channel (Huddlestun 1993).

The rock-derived sea-level interpretations may represent a shorter subcycle that can not be precisely correlated with the curve of Haq et al. Other factors that could have affected the Florida area include changes in the geoid (configuration of the Earth's gravitational field), rapid sea-floor spreading, and local intraplate stresses (Cloetingh 1988). Other factors may be the fundamental differences in the use of the $\delta^{18}O$ record with onlap events (Vail et al. 1977; Williams 1988) (fig. 4.8) and the inaccurate application of the influence of tectonism and metamorphic CO_2 degassings during the Eocene (Berner et al. 1983; Raymo and Ruddiman 1992; Caldeira et al. 1993; Kerrick and Caldeira 1994).

Evaporitic sediments are most prevalent in the Mesozoic part of the Florida section. Thick, massive deposits of gypsum and anhydrite are believed to be the result of a silled basin with restricted water circulation (Handford and Loucks 1993). Great thicknesses of salt accumulated during the Jurassic and Cretaceous (figs. 4.1 and 4.2). Later (Paleocene and Eocene) evaporitic sedimentation produced thinner beds and nodules over smaller areas, associated with sabkha-derived carbonate sediments. These deposits reflect drastic changes, such as a shallower basin configuration, a drier climate, and cyclic sea-level variations.

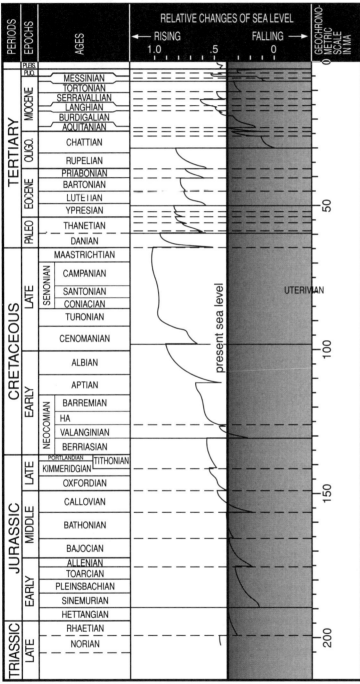

Figure 4.8. Late Mesozoic and Cenozoic geochronologic and sea-level chart based on magnetostratigraphy, biostratigraphy, and sequence chronostratigraphy (after Haq et al. 1988).

Sequence Stratigraphy

Sequence stratigraphy (Vail 1987) involves the geometric configurations of distinctly related sedimentary deposits bounded by unconformities (Van Wagoner et al. 1988). These geometries are routinely recognized from seismic profiles (fig. 4.9) and coordinated with drilling data to produce a chronostratigraphic sequence of sedimentologic events (Loucks and Sarg 1993). These sequence tracts are responsive to sea-level changes that interrupt deposition and often partially destroy the sedimentary record by erosion. Eustacy (climatically controlled) and tectonism (rates of sea-floor spreading, subsidence, and sediment supply) are the fundamental driving mechanisms for sequence development. Drowning events and the resulting transgressive-systems tracts and high stand systems tracts can bring about an onlap or backstepping distribution of carbonate sediments and platform evaporite deposition. Low-stand conditions can result in an increased rate of siliciclastic sediment dispersal onto the platform, suppressing carbonate sedimentation (see chapter 11). If subaerial exposure occurs, karstification and erosional sequence-boundary surfaces develop.

The Jurassic to Holocene sequence record is well documented for the eastern Gulf of Mexico (Angstadt et al. 1985; Winker and Buffler 1988; Corso et al. 1989; Jee 1993). Generally, there is an absence of prograding clinoforms, suggesting a very low relief carbonate platform upon which carbonate and evaporite buildups occurred. Conformable, relatively flat-lying sequences appear to be the rule for this time interval in the eastern Gulf.

Of special note are two carbonate sequences from the early to late Cretaceous, recognized on seismic profiles by the application of sequence stratigraphy (fig. 4.9). Buffler et al. (1980) proposed the term "Mid-Cretaceous Unconformity" (MCU) for the seismic reflection separating these two carbonate sequences. This feature has not been recognized everywhere (Faust 1990), and has been renamed the "Mid-Cretaceous Sequence Boundary" (MCSB). The MCSB represents a major change in depositional regimes: a global fall of sea level, followed by a dramatic rise in sea level, significant drowning of the platform, and deposition of deeper-water carbonate sediments during the Late Cretaceous. Limited drilling and seismic and sequence-stratigraphic evidence document the characteristic geometric configuration and lithologies resulting from these Cretaceous sea-level fluctuations and responding carbonate sedimentation (Buffler et al. 1980; Schlager 1981, 1989; Addy and Buffler 1984; Faust 1986, 1990; Corso et al. 1989; Gardulski et al. 1991; Jee 1993).

The controversial Cretaceous/Tertiary (Mesozoic/Cenozoic) boundary is present in the Florida Platform and notably indistinct in seismic profiles (Jee 1993). Subsurface sampling of upper Cretaceous and Paleocene sequences reveals the same general distribution of clastic to carbonate sedimentation from the panhandle to the peninsula (Miller 1986). The Cretaceous/Tertiary boundary is better known in Alabama, where it is exposed at the surface (Jones et al. 1987; Donovan et al. 1988; Baum and Vail 1988; King and Skotnicki 1990; Mancini and Tew 1990, 1991, 1992).

Peninsula and panhandle lithostratigraphic correlations document transgressive and regressive marine sequences with diminished evaporite deposition by the early Eocene, an acme of carbonate sedimentation in the middle to late Eocene and Oligocene, and predominant clastic sedimentation from the Miocene to Holocene (fig. 4.1n-s). Sequence stratigraphy has been employed in equivalent Paleogene sequences of the Gulf and Atlantic coastal plains (Loutit et al. 1988; Baum and Vail 1988), where sea-level variations have been interpreted from onlap sequences and condensed sections (abnormally thin deposits). An interesting alternative explanation to sea-level falls and the development of unconformities is a sea-level rise, creating deeper water, a starved basin, and a condensed section. Both of these sea-level trends may be represented by the rock record in the eastern Gulf of Mexico and Florida for this time frame.

Facies represented by the carbonate sequences include deep- and shallow-water subtidal, intertidal, and supratidal deposits. Faunal and floral assemblages (Cheetham 1963; Randazzo and Saroop 1976; Randazzo et al. 1990; Ivany et al. 1990; see chapter 7) substantiate the interpreted water depths of these deposits. The carbonate rock types representing these environmental facies range from mudstones to grainstones (Dunham 1962). Most of these

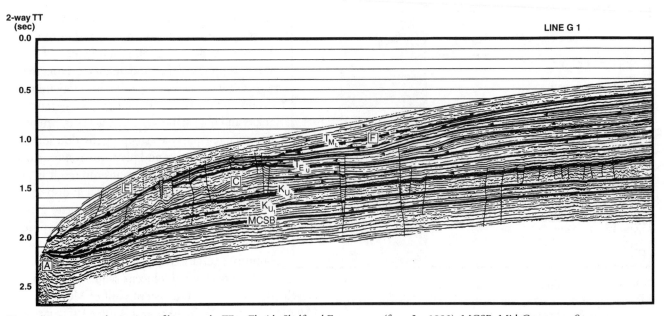

Figure 4.9. Interpreted seismic profile across the West Florida Shelf and Escarpment (from Jee 1993). MCSB: Mid-Cretaceous Sequence Boundary. K_{U1}: Top of Cenomanian-Santonian subunit. K_{U2}: top of Campanian-Maastrichtian subunit. T_{EU}: Top of Upper Eocene. T_{ML}: Top of Lower Middle Miocene. A: "Seismic reef." C: Disrupted, contorted reflections. E: Youngest seismic unit. F: Lower Miocene and younger strata.

are bioclastic, reflecting significant biologic productivity in normal marine environments (fig. 4.10). Of special note are the carbonate boundstones (organically bound reef rocks). Platform-margin reefs of the Cretaceous comprised a community of invertebrate marine organisms, most notably rudists, corals, algae, and stromatoporoids (Yurewicz et al. 1993). Paleogene reefs are more difficult to document because of a lack of fossil preservation, most likely resulting from the evolutionary development of reef organisms. Rudists, a principal component of Cretaceous reefs, became extinct at the end of the Mesozoic (see chapter 7); scleractinian corals replaced them as a principal reef-builder in the Cenozoic. The more highly permeable colonial-scleractinian coral reefs, composed of relatively unstable aragonite, would have been susceptible to dissolution and replacement by calcite and dolomite, and their apparent absence from the Paleogene is attributed to diagenetic alteration (Randazzo 1987). Similarly, the scarcity of Paleogene oolitic facies (mostly aragonite in modern carbonate environments) is believed to be a result of diagenesis. Duncan et al. (1994) were first to report oolitic carbonate rock from the Avon Park Formation in the area of Melbourne, Florida (plate 2).

Stratigraphic Units

The stratigraphy of Florida is mired in contradictory and confusing nomenclature because of the inconsistent application of the North American Stratigraphic Code and the limitations of the Code with respect to Florida's rock record (Applin and Applin 1944; Cooke 1945; Puri and Vernon 1964; Randazzo 1976; Miller 1986). The Code relies upon definition of lithostratigraphic units for formation names. This requires recognition of distinct, field-recognized lithologies. Unfortunately the carbonate rocks of Florida do not lend themselves to easy lithostratigraphic definition; thus, many workers have relied upon key fossils in naming stratigraphic units. The Code recognizes this practice as a biostratigraphic exercise defining zones, rather than formations. A further problem is reliance upon subsurface identification of most of the sedimentary units because of their lack of exposure. Figure 4.11 displays a generalized north-south cross-section of geochronologic units.

The U.S. Geological Survey's nomenclature has consistently utilized the Code to establish formation names, and its usage is endorsed here for the Cretaceous to Oligocene. Miocene stratigraphic nomenclature is based on the work of Scott (1983, 1988b) and advocated by the Florida Geological Survey (see the stratigraphic correlation chart on the inside front cover of this book). The Correlation of Stratigraphic Units of North America (COSUNA) chart for the Gulf Coast region (Braunstein

Figure 4.10. Photomicrograph of a foraminiferal wackestone-packstone from the Avon Park Formation (Middle Eocene). Numerous benthic foraminifers are cemented by dark micrite. Bar scale = 1 mm.

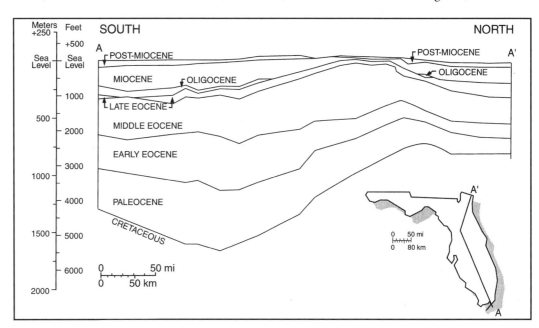

Figure 4.11. North-south geochronologic cross-section of the Florida peninsula (after Miller 1986).

et al. 1988) provides a regional summary of the stratigraphic units in Florida. Because of the striking contrast between clastic rocks of the panhandle and carbonate rocks of the peninsula, a number of stratigraphic names designate units of the same age, but different lithology and geography. Winston (1991) presented a series of facies maps and a correlation chart for the Jurassic through the Paleocene. Further, he offered an unconventional regional analysis of the Paleogene section of the panhandle and peninsula (Winston 1993a, 1993b).

PENINSULAR FLORIDA

The Cretaceous rocks of the peninsula are deeply buried and incompletely known. In the southern part of the peninsula, the Lower Cretaceous consists of more than 3,000 m of carbonate rocks, with at least two thick evaporite units (Punta Gorda Anhydrite of the Glades Group, and Panther Camp Formation of the Naples Bay Group). Because of the limited stratigraphic control, units of the Lower Cretaceous have been lumped together as the Marquesas Supergroup. Most notable among the carbonate units of this supergroup is the Sunniland Limestone (Aptian of the Ocean Reef Group), the principal oil-producing formation in peninsular Florida. The carbonate formations are generally dolomitic limestones, with varying amounts of interbedded evaporites; they represent peritidal and subtidal shelf environments of deposition, reflecting small- and large-scale sea-level fluctuations. The carbonate-dominated Marqueses Supergroup is thinner (600–2,000 m) in northern peninsular Florida, where its lower part has more interbedded clastics and fewer evaporites.

The MCSB and related sea-level fall is recorded by the Atkinson Formation (Cenomanian and possibly Turonian), a mixed carbonate-clastic unit. Carbonate deposition predominated during the rest of the late Cretaceous and is represented by the Pine Key Formation, which contains at least two extensive dolomite sequences in southern Florida (Card Sound Dolomite and Rebecca Shoal Dolomite). Winston (1978, 1989) postulated that the Rebecca Shoal Dolomite represents a fringing reef that existed from the late Cretaceous to the early Eocene (fig. 4.1n).

The Cedar Keys Formation (Paleocene and Lower Eocene) is pervasively dolomitized and consists of peritidal carbonate rocks with interbedded and intergranular evaporites (anhydrite and gypsum). The Oldsmar Formation (Middle Eocene) is less pervasively dolomitized and is represented by shallow-water shelf deposits and peritidal carbonates and evaporites. The environments of deposition represented by these units reflect increasingly higher sea levels from the Paleocene to the Eocene. Peritidal evaporite and carbonate sedimentation was gradually replaced by more normal marine carbonate deposition.

The Avon Park Formation (Middle Eocene) is the oldest stratigraphic unit exposed in Florida (plate 1). It is unconformable with the underlying Oldsmar Formation, suggesting an episode of subaerial exposure and erosion (sea-level fall). The boundary (seen only in drill cores) is generally recognized by the contrast between older, porous, foraminiferal grainstones and packstones and younger dolomitic wackestones-mudstones. The Avon Park is a carbonate mud dominated peritidal sequence, pervasively dolomitized in places, and undolomitized in others, and it contains some intergranular and interbedded evaporites in its lower part (Randazzo and Saroop 1976; Randazzo and Zachos 1984; Randazzo et al. 1990). Fossils are mostly benthic forms showing limited faunal diversity. Seagrass beds are well preserved at certain horizons.

The Upper Eocene Ocala Limestone consists of two units. The lower unit is partially dolomitized and comprises both restricted- and open-marine carbonate lithologies. No evaporite minerals are reported, but peritidal conditions are partly responsible for the carbonate deposits of this lower unit. The boundary between the Ocala Limestone and the Avon Park Formation is unconformable in some places and conformable in others. The Avon Park unit often is thin-bedded, with algal laminations, and finer-grained in its upper part, contrasting with the interbedded skeletal-rich packstones-wackestones and mudstones of the lower Ocala.

The upper Ocala is generally composed of white to gray foraminiferal and molluscan packstones and grainstones, with lesser amounts of wackestones and mudstones. Faunal diversity is high. Open-marine, shallow-water, and middle-shelf deposition prevailed during the late Eocene.

Oligocene carbonate deposition is represented by the Suwannee Limestone, another generally open-marine carbonate unit dominated by packstones and grainstones. The boundary between the Ocala and Suwannee limestones is often difficult to identify because of their similar lithologic appearances. The Suwannee Limestone gradually becomes sandier (quartz sand) toward the top, and it has several dolomitized sections; this gradual, but continuous change in the lithologic character of the Suwannee Limestone reflects the return of lower sea levels and the influx of clastic sediments from the north.

PANHANDLE FLORIDA

The Lower Cretaceous rocks of the panhandle consist of a series of undifferentiated sandstones, shales, limestones, and evaporites (more than 2,000 m thick), which unconformably overlie the Jurassic to possibly Lower Cretaceous Cotton Valley Group. The MCSB separates these rocks from the Upper Cretaceous Tuscaloosa Group, a quartz sand rich clastic unit time-equivalent to the mixed carbonate-clastic Atkinson Formation of the peninsula. The upper part of the Tuscaloosa is finer-grained in the eastern panhandle and reflects marine, deltaic, and possibly fluvial environments of deposition, respectively, from east to west; it is unconformable with the sandy and conglomeratic Eutaw Formation of the western panhandle and unnamed carbonate and mixed carbonate-clastic sequences of the central panhandle. Carbonate deposition progressed westward through the panhandle during the latest Cretaceous, as represented by the Selma Group. This carbonate unit is noted for its chalk lithology, but it is dolomitic and interbedded with calcareous clay.

The Paleocene rocks of the panhandle are unconformable with the Cretaceous sequences. The Wilcox and Midway groups, principally thick (more than 500 m), mixed carbonate-clastic units with a more clastic lithofacies in the west, are overlain by more carbonate rich sediments (still mixed with clastics) of the Claiborne Group (Lower and Middle Eocene).

Late Eocene and Oligocene sediments of the panhandle are characterized by an extension of carbonate systems from the peninsula to the panhandle, represented by the Ocala Limestone and Suwannee Limestone in both areas. The early Oligocene is represented by another incursion of clastic sedimentation in the western panhandle (Vicksburg Group), but carbonate deposition (Suwannee Limestone) resumed at least in the central and eastern panhandle until the late Oligocene. The panhandle stratigraphic units reflect sea-level variations during the Cretaceous and Paleogene, and the responding change in positions of carbonate and clastic deposition.

Post-Depositional Processes

The carbonate-evaporite sequences of Florida are extremely vulnerable to post-depositional changes in composition and texture (i.e. diagenesis). Diagenetic processes include inversion, recrystallization, replacement, dissolution, and cementation (Folk 1965). Most of the original marine carbonate sediments were aragonite, a metastable form of $CaCO_3$. Time and exposure to nonmarine conditions result in the inversion of aragonite to calcite, or the aragonite may dissolve. Calcite commonly recrystallizes, producing different sedimentary textures, or it too may dissolve. Dolomite, $CaMg(CO_3)_2$, and silica, SiO_2, are the two most abundant minerals replacing primary aragonite, calcite, and evaporite minerals. Calcite itself often replaces older carbonate-evaporite phases through cementation or pore filling.

DISSOLUTION

The dissolution of primary mineral phases greatly affects the porosity and permeability of the sediments and the composition of the dissolving waters. Generally this phenomenon occurs in a freshwater phreatic system in which groundwater has filled open spaces. Because groundwater is weakly acidic, a chemical reaction occurs that causes solid carbonate and evaporite minerals to dissolve.

Dissolution of Florida's carbonate rocks has produced significant changes in rock fabric, accompanied by development of many different types of pore spaces. Moldic, vug, and interparticle pore types are most common. Extensive dissolution can create caves and caverns (fig. 4.12), the underlying cause of sinkholes and karst landscapes (see chapters 1 and 13). Secondary porosity is also extremely important in oil and gas accumulation.

Figure 4.12. Cave formations in Florida Caverns State Park (photo courtesy of E. Lane, Florida Geological Survey).

The major sea-level low stand that occurred during the late Oligocene and Miocene in response to a global cooling trend had a profound influence on marine carbonate rocks of Florida. Lowered sea levels subjected these rocks to nonmarine phreatic and vadose hydrologic conditions. Surface-water runoff physically and chemically eroded much of the early Oligocene rocks from positive areas in the panhandle and peninsula. Extensive groundwater dissolution created cavern systems in Eocene and older rocks—precursors to the karst terrain that later developed (plate 3). Neogene marine transgressions resulted in filling and burial of many of these cavernous areas by sediment. In places, these infilling sediments include concentrations of marine and nonmarine vertebrate fossils, and today these filled cavities provide an invaluable source of specimens reflecting the diversity of life that existed during the past 10 million years (see chapter 8).

One of the most dramatic and enigmatic examples of dissolution is Florida's "Boulder Zone" (Vernon 1970; Puri and Winston 1974; Lane 1986; F. W. Meyer 1989). This deeply buried cavernous zone, approximately 500 to 2,500 m below the land surface, has extremely high groundwater transmissivity (2.8×10^5 m^2/day); it is called the Boulder Zone because of the phenomenon experienced by drillers ("like drilling through a pile of boulders"). When a large cavity (or carbonate rock weakened by extensive dissolution) is encountered during drilling, the roof and walls of the cavity often collapse or spall, forming rock fragments that move about the drill stem, producing manmade boulders (fig. 4.13). Drilling in the Boulder Zone usually results in a loss of fluid circulation, and drilling tools commonly are damaged or lost because of collapse of the drill hole. Rock-sample recoveries are poor, and the samples represent survivors of the dissolution process and drilling fragmentation.

The Boulder Zone is developed in the Paleocene and Lower Eocene section of southern Florida, in the Cedar Keys Formation and Oldsmar Limestone; these units principally consist of limestone, dolomite, and varying quantities of evaporites. Various hypotheses for the origin of the Boulder Zone have been proposed (Vernon 1951, 1970; Kohout 1965; Hanshaw et al. 1971; Puri and Winston 1974); these hypotheses invoke groundwater geochemical reactions involving saltwater/freshwater mixing that causes dissolution of limestone and partial replacement by dolomite (see discussion of dolomitization, below). Geothermal convection of dolomitizing fluids is postulated as one mechanism for the reaction process. The cavities of the Boulder Zone may thus represent a stage in the dolomitization process.

A hypothesis presented here involves a more holistic approach, and an extension of the hypotheses proposed to explain erosion of Florida's marine escarpments (see chapter 11). Tectonic history and the sedimentary facies in which the Boulder Zone formed played important roles. The carbonate and evaporite deposits of the Cretaceous and lower Tertiary were subjected to tectonic and structural forces occurring in the Caribbean and Gulf of Mexico. Resulting fractures and faults (Safko and Hickey 1992) provided conduits for future migration of diagenetic fluids. Nonmarine and marine groundwaters would have repeatedly flushed this sedimentary package in response to sea-level fluctuations (fig. 4.8). The thick gypsum and anhydrite deposits in this part of the stratigraphic section could have provided the source material for the natural production of sulfuric acid, a potent and efficient agent for massive dissolution of carbonate rocks

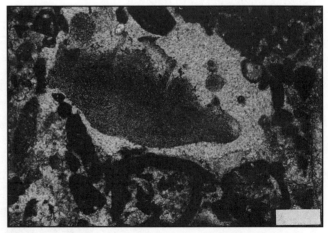

Figure 4.13. Dolomite core material ("boulder") recovered from the "Boulder Zone." The specimen has been fragmented, striated, and rounded by a rotating drill bit during the drilling operation (photo courtesy of E. Lane, Florida Geological Survey).

Figure 4.14. Photomicrograph of an echinoid-foraminifer grainstone from the Upper Eocene Ocala Limestone. At the center is a large echinoid plate with a sparry-calcite overgrowth that has filled pore spaces during rock cementation. Bar scale = 1 mm.

(sulfuric-acid dissolution of Permian limestones has been hypothesized for the creation of Carlsbad Caverns in New Mexico) (Davis 1980; Egemeier 1988; Hill 1990). Geothermal convection could have enhanced the effectiveness of this process in southern Florida. With some 15 to 20 million years of global sea-level fluctuations, periodic migrations of groundwater into fractured carbonate and evaporite sequences could have produced the Boulder Zone. Groundwaters of varying compositions and oxidation states, perhaps in concert with bacterial chemosynthetic productivity and fluid-hydrocarbon migrations from Cretaceous sequences, could have caused evaporite and carbonate dissolution (Hill 1990; Paull et al. 1991a; Bischoff et al. 1994). Cowart et al. (1978) reported high concentrations of H_2S in waters of the Boulder Zone, and suggested that this might have resulted from activity of sulfate-reducing bacteria. H_2S-rich waters ascending through fractures could have mixed with more oxygen laden groundwater to produce sulfuric acid, causing extensive dissolution of limestone. Dolomitization in a phreatic mixing zone environment could have been promoted as well. The absence of a boulder zone in northern Florida can be attributed to the thinner early Tertiary evaporite sequence and the lesser tectonic fracturing in that area.

LITHIFICATION

Lithification of carbonate sediments is generally achieved by recrystallization and cementation. Many small crystals fuse and suture themselves into fewer larger crystals in response to burial load pressures or crystal-lattice energy differences (Folk 1965). This recrystallization can occur in a gradational or site-specific way, but, with time, results in the transformation of relatively unconsolidated sediments into hard, rigid rocks.

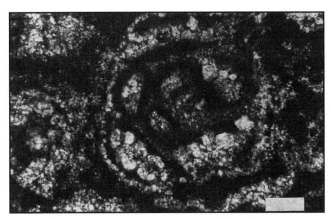

Figure 4.15. Photomicrograph of a fine- to medium-crystalline dolomite displaying a mimetic texture. The original calcite microfabric has been preserved by the dolomite replacement process. The internal architecture of the foraminiferal test in the center of view has been preserved. Bar scale = 1 mm.

A definitive order of grain recrystallization is recognized for the Paleogene rocks of Florida. Mollusk shells, composed predominantly of aragonite, undergo the most intense recrystallization, and are represented by mosaics of pseudospar. Porcelaneous foraminifer tests (originally high-Mg calcite) are recrystallized to microspar, but hyaline foraminifers and bryozoans show little or no evidence of recrystallization. This order of recrystallization is a reflection of relative mineral stability. Matrix diagenesis is generally aggrading in nature and results in mosaics of irregular, tightly interlocking crystals.

Cementation involves chemical precipitation of a mineral phase (most commonly calcite, dolomite, or quartz) in a pore space. The crystal growth of the cement may be projecting into a pore space or as a coating on a grain (fig. 4.14). The resulting reduction of porosity and anchoring of a grain network produces lithification. Numerous cement types are represented in the carbonate rocks of Florida; these include meniscus, fibrous, equant or blocky, and dogtooth forms. These types have been attributed to particular chemical environments (Folk 1974; Lahann 1978) associated with marine and nonmarine hydrologic conditions (Longman 1980; Moore 1989). The complexity of carbonate cement types and their distribution is a result of Florida's carbonate rocks having been subjected to alternating marine and nonmarine chemical environments and the superimposing of one diagenetic fabric upon another.

DOLOMITIZATION

The occurrence of dolomite as a pore-filling cement and replacement mineral is recognized in Florida and elsewhere (Randazzo and Hickey 1978; Randazzo and Bloom 1985; Randazzo and Cook 1987; Cander 1994). Dolomite can faithfully preserve original depositional fabrics, or completely obliterate them (figs. 4.15 and 4.16). Dolomite formation affects rock porosity and permeability. Explaining the origin of this mineral is difficult, and numerous models for its formation have been proposed (Tucker and Wright 1990). The fascination with dolomite stems from its unique chemical characteristics, forms, and occurrences, as well as its economic significance: more than half of the world's petroleum is found in carbonate rocks, and most of these carbonate rocks are dolomites.

Dolomite is a dynamic and ubiquitous mineral phase that forms in a number of diagenetic settings, but its origins are problematical. As a replacement product, it can be penecontemporaneous with its host, or millions of years younger. As an original precipitate, it can evolve into other forms having various physical and chemical characteristics.

Figure 4.16. Photomicrograph of medium-crystalline dolomite displaying a non-mimetic texture. The original calcite microfabric has been completely obliterated by the dolomite replacement process. Dark areas are pore spaces into which replacement dolomite has crystallized. Bar scale = 30 μm.

Because dolomite has not been precipitated in a laboratory at surface temperatures from natural waters, explaining the origin of dolomite has been a particular challenge to researchers. Dolomite formation requires a concentration of Mg^{2+} ions, a dynamic transport mechanism to maintain this supply, and time to allow the slow kinetics of its crystallization to occur. The origin of Florida's dolomite is compatible with proposed models involving evaporative pumping, seepage reflux, groundwater mixing zones, and seawater (fig. 4.17).

The dolomite of Florida has numerous variations in crystal size, form, structure and spatial position, stoichiometry, and trace-element and isotopic compositions (table 4.1). These variations are a function of the time and mode of dolomite formation. The diverse dolomitic microfabrics have been classified descriptively as equigranular (unimodal) or inequigranular (multimodal) (Randazzo and Zachos 1984). Two genetic trends of microfabric development have been recognized: heterogeneous dolomitization and homogeneous dolomitization (figs. 4.18 and 4.19). Because dolomite is a mineral phase found in both modern and ancient sediments, many models invoke uniformitarianism. The Florida

Table 4.1. Summary of compositional ranges of dolomites from Florida

Dolomite host age	Mol% $MgCO_3$	Sr (ppm)	Na (ppm)	$\delta^{18}O°/00$ PDB	$\delta^{13}C°/00$ PDB
Eocene	39–50	142 – <1,200	334–3,678	+0.27 – +4.18	−0.60 – +3.10

Sources: Randazzo et al. 1983; Randazzo and Bloom 1985; Randazzo and Cook 1987; Cander 1994; Eppler, pers. comm. 1994.

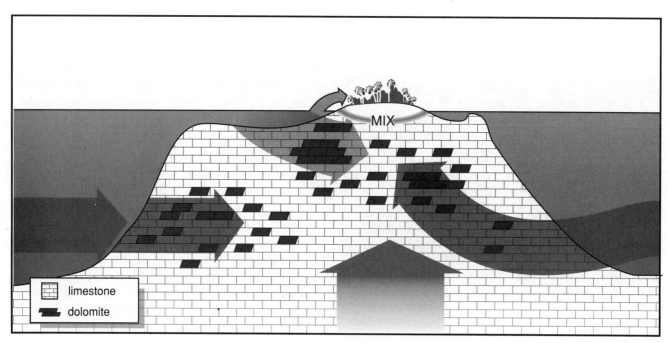

Figure 4.17. Dolomitization models operating in the Florida Platform (after Tucker and Wright 1990).

Keys were one of the first areas in which modern dolomite was discovered (Shinn 1964). There, its low abundance, poor crystal development, and impurities are in stark contrast to the very abundant, well crystallized, stoichiometric and near-stoichiometric dolomite found in older parts of Florida's rock record.

Determining the time of dolomitization of Florida's carbonate rocks may be possible by analysis of Sr-isotope ratios ($^{87}Sr/^{86}Sr$). Measuring this ratio for single dolomite crystals, and comparing these values to the Sr-isotope composition of seawater for the Neogene and latest Paleogene (fig. 4.20) (Koepnick et al. 1985; DePaolo 1986; Hodell et al. 1991), has produced encouraging results. Oligocene and Miocene ratios have been measured for dolomite crystals in Eocene host rocks (table 4.2). The presence of "younger" dolomite in these older host rocks

Table 4.2. $^{87}Sr/^{86}Sr$ compositions of dolomite crystals from the Avon Park Formation, west-central Florida

$^{87}Sr/^{86}Sr$	"Age"*	Reference
0.708369	Miocene	Cander 1991
0.708196	Oligocene	Cander 1991
0.708184	Oligocene	Eppler, pers. comm.
0.708183	Oligocene	Cander 1991
0.708124	Oligocene	Cander 1991
0.708084	Oligocene	Eppler, pers. comm.
0.708048	Oligocene	Cander 1991
0.708045	Oligocene	Eppler, pers. comm.
0.708039	Oligocene	Cander 1991
0.708013	Oligocene	Cander 1991
0.708008	Oligocene	Cander 1991
0.707806	Oligocene	Eppler, pers. comm.
0.707804	Oligocene	Eppler, pers. comm.

*Based on seawater curve of Koepnick et al. 1985.

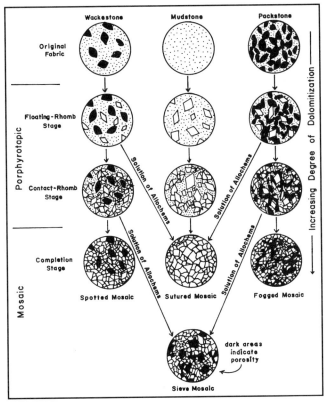

Figure 4.18. Genetic sequence of microfabric development by heterogeneous dolomitization (Randazzo and Zachos 1984).

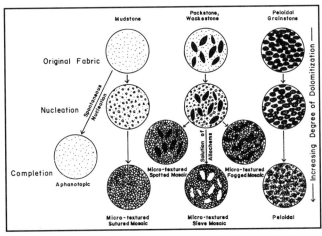

Figure 4.19. Genetic sequence of microfabric development by homogeneous dolomitization (Randazzo and Zachos 1984).

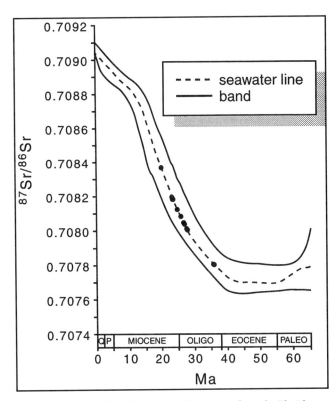

Figure 4.20. Plot of $^{87}Sr/^{86}Sr$ values of dolomites from the Florida Platform (see table 4.2). The dashed line is the published estimate of seawater ratios of $^{87}Sr/^{86}Sr$ versus time (after Koepnick et al. 1985). The upper and lower solid lines define a band that encloses 99% of the data used to establish the seawater line.

confirms a post-depositional origin for the dolomite and supports the hypothesis of formation in different hydrologic regimes (e.g. marine phreatic, coastal mixing zone) (Cander 1991, 1994; S. Eppler, pers. comm.).

Summary

Climatic variations represented by Florida's rocks for the past 75 million years reflect global conditions. By the mid-Cretaceous the world was experiencing a climatic warming trend or "greenhouse" environment. Sea level was higher, latitudinal temperature gradients were less steep, and polar icecaps were greatly diminished. Extensive carbonate sedimentation occurred in Florida during the late Cretaceous, Paleocene, Eocene, and early Oligocene. There is global evidence of episodes of cooling—short time intervals when climatic interruptions occurred (Berger 1982). Oxygen-isotopic compositions of sediments at the Eocene/Oligocene boundary suggest relatively cooler temperatures (McGowan 1990), but an even cooler climate or "icehouse" environment prevailed during the late Oligocene and Miocene (fig. 4.6). These cooling periods were marked by inundations of clastic sediments on the peninsula and/or erosion of some of the previously deposited carbonate rocks (fig. 4.1a-s).

Climatic variations and the associated sea-level changes deduced from oxygen- and carbon-isotopic studies are not always compatible with paleontologic and lithologic data. Rapid plate movements are also recognized as a factor in displacing water from ocean basins and causing sea level to rise. The Florida rock record reflects action of the Caribbean plate, and the global warm climate of Cretaceous-Paleogene times. There was a pronounced and continuous decrease in evaporite sedimentation, variable shallow- and deeper-water carbonate sedimentation, and an increased influx of terrigenous sediments during the late Oligocene to Miocene that may have been tectonically as well as eustatically induced (fig. 4.1g-q).

The marine conditions responsible for carbonate sedimentation would produce skeletal, muddy, oolitic, and reef facies, but reefs are noticeably lacking in the Florida Paleogene record; this apparent absence is most likely related to diagenetic changes involving fabric obliteration by aragonite dissolution and dolomitization.

Florida's sedimentary rocks record drowning and exposure events which explain the nature and distribution of carbonate, evaporite, and siliciclastic sedimentation chronologically and geographically. The development of the Florida Platform was controlled by organic evolutionary patterns, climatic and tectonic events, and continental linkages. High-resolution sequence-stratigraphic studies, integrated with detailed petrographic and geochemical analyses, will facilitate application and modification of existing carbonate-evaporite-siliciclastic sedimentation models. These studies will also allow for more precise and accurate comparisons of the Florida Platform to other platforms on a regional and global scale.

5

Miocene to Holocene History of Florida

Thomas M. Scott

THE FLORIDA PLATFORM EXTENDS southward from the North American continent, separating the Atlantic Ocean from the Gulf of Mexico. The exposed part of the platform, the state of Florida, is offset to the east of the platform's axis, which more or less coincides with the present-day west coast of the peninsula. This leaves much of the platform covered with water and nearly inaccessible, so very little is known about the Neogene to Holocene sediment distribution in those areas (see chapter 11).

Virtually the entire Florida Platform is covered by a blanket of sediments ranging in age from Miocene to Holocene. Older sediments most often are exposed only along rivers and streams, and in sinkholes that cut through the younger sediments. Most of these younger sediments were deposited under marine conditions, but this sediment package also contains fluvial, freshwater, and eolian deposits. The upper Cenozoic sediments range from less than a meter thick in parts of the west-central peninsula and the north-central panhandle to more than 300 m thick in southern Florida. These sediments contain a wide assortment of minerals, including some that form under unusual conditions (e.g. phosphate, palygorskite, sepiolite). The Miocene to Holocene sediments include sands, clays, shell material, heavy minerals, and phosphate resources (see chapter 9).

Hydrostratigraphically, the Miocene sediments form the confining unit (intermediate confining unit) of the Floridan aquifer system (Miller 1986). The intermediate aquifer system is developed in the Miocene-Pliocene sediments in many areas of the state (Scott et al. 1991). The Pliocene-Pleistocene sediments constitute the shallow aquifer system. The intermediate aquifer system and the shallow aquifer system are important water sources in areas of the state where waters of the Floridan aquifer system are not potable.

This chapter discusses the deposition and subsequent modification of the Miocene to Holocene sediments of Florida.

Depositional History

Deposition of carbonate sediments was prevalent on the Florida Platform after it separated from the African Plate during the Jurassic. Periodically, siliciclastic sediments became prevalent on the northern and western parts of the platform, in response to sea-level fluctuations and sediment supply. These cycles of shifting sedimentation have intermittently covered parts of the platform with siliciclastics, followed by reestablishment of carbonate deposition as siliciclastic influx diminished.

During the first 35 million years of the Cenozoic, carbonate sediments predominated on the Florida Platform, forming the thick sequence of limestones, dolostones, and evaporites that constitute the Floridan aquifer system (see chapters 4 and 6). Approximately 25 million years ago, near the beginning of the Miocene, siliciclastic sediments began to be transported onto the platform in sufficient quantities to effectively suppress carbonate deposition (Walker et al. 1983). By the mid-Pliocene, siliciclastic sediments virtually covered the entire platform, leaving only isolated areas of significant carbonate deposition. During the Late Pliocene, the siliciclastic sediment supply began to diminish, and carbonate sedimentation was reestablished in southernmost peninsular Florida. The transition from carbonate deposition to siliciclastic deposition on the Florida Platform is well documented in the Neogene to Holocene strata, and is cited as an example of a carbonate-siliciclastic transition (Evans and Hine 1991). However, the cause of the influx of siliciclastics is still debated.

Before the end of the Paleogene, the Georgia Channel System (Huddlestun 1993; including features called the "Gulf Trough," or the "Suwannee Straits") acted as a barrier to siliciclastic sediment transport onto the Florida Platform. The Suwannee current occupied the channel system until the Late Oligocene. Following the Late Oligocene sea-level low stand, the current did not reoccupy the channel system, and siliciclastic sediments began to encroach onto the Florida Platform.

The source of the siliciclastic sediments is a question. Was the influx of siliciclastics due to a climatic change that allowed significantly more erosion (Poag 1982)? Or a result of a broad uplift that rejuvenated the Appalachians (Schlee et al. 1988; Stuckey 1965) and renewed southward transport of sediments onto the Florida Platform (Scott 1988b)? The answer is still unclear, and this question remains a topic of much discussion among geologists.

Structural Framework

The Florida Platform is generally considered to be part of a relatively stable tectonic setting on the passive margin of the North American plate, but even in this setting, features suggesting folding, faulting, uplift, and subsidence have been recognized (Vernon 1951; Chen 1965; Meisburger and Field 1976; Missimer and Gardner 1976; Winker and Howard 1977; Bond et al. 1981; Scott 1983; Opdyke et al. 1984). These features, although minor in comparison to those recognized in the Appalachians and other areas, indicate that the Florida Platform has endured and responded to tectonic forces during the Neogene to Holocene.

The Neogene to Holocene sediments cover Paleogene carbonate sediments. In the northern two-thirds of the peninsula and much of the panhandle, the Paleogene carbonates are eroded and karstified, providing for an unconformable contact with the overlying Neogene sediments. Sedimentation in southern peninsular Florida may have been more or less continuous across the Paleogene/Neogene boundary. Sediments that were once considered to be Miocene are now recognized as straddling the boundary (Scott 1988b; J. M. Covington, pers. comm. 1992; Wingard et al. 1993).

The structural template on the Paleogene surface consisted of a number of highs and basins. The highs include the Ocala Platform (or "uplift"), Chattahoochee "Arch," Sanford High, Saint Johns Platform, and Brevard Platform. The basins include the Gulf of Mexico Basin, the Georgia Channel System (comprising the Apalachicola Embayment, Gulf Trough, and Southeast Georgia Embayment, including the Jacksonville Basin), the Osceola Low, and the Okeechobee Basin. The Peninsular Arch, a Mesozoic to early Cenozoic feature often miscorrelated with the Ocala Platform, lies east of and parallel with the Ocala Platform (fig. 5.1).

Many of the structural features that had affected deposition during the earlier Cenozoic continued to do so in the later Cenozoic. The Gulf of Mexico Basin, Apalachicola Embayment, and Southeast Georgia Embayment continued to be major depocenters during the early Neogene, and they contain thick sequences of Miocene sediments. These basins appear to have been filled during the late Neogene, and their Pliocene sediment sequences are much thinner.

The influence of the South Florida Embayment, a major Mesozoic and early Cenozoic depocenter that is easily recognized on isopach and facies maps, was altered in the Neogene. The South Florida Embayment is easily recognized on maps of the Paleogene sediments (Miller 1986); however, it had less influence on deposition during the Oligocene than during the Paleocene or Eocene. The distribution of Miocene sediment indicates a shift from the South Florida Embayment with a south-southeast dipping axis (Scott 1988b); this shift appears to have been in response to downwarping and tilting of the southern peninsular area.

The major high that affected deposition during the early Cenozoic was the Peninsular Arch, which appears to have had an effect on sedimentation into the Oligocene, based on maps by Miller (1986). The Peninsular arch did not affect deposition of the Neogene to Holocene sediments. The major high recognized on the top of the Paleogene carbonates is the Ocala Platform. Although there is continuing argument about the origin of the Ocala Platform, it exerted an influence on Neogene deposition. Miocene sediments thin onto the Ocala Platform, but were deposited across it (Scott 1988b). Post-Middle

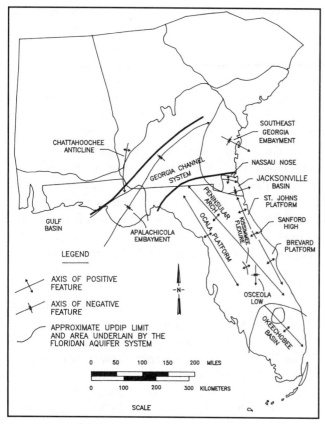

Figure 5.1. Structural features of Florida.

Miocene erosion removed sediments of the Hawthorn Group from the crest of the Ocala Platform, exposing the Eocene carbonates (plate 1). Subsequently, an often thin siliciclastic-sediment section was deposited on the exposed carbonates. The term "Pleistocene undifferentiated sediments (or sands)" is often used to describe these sediments, although there is no evidence that they are Pleistocene only. This section consists of residual clays and clayey sands, and eolian sands deposited during the mid-Miocene to Holocene.

The Sanford High, in the east-central part of the peninsula, is a positive feature on the Paleogene surface, and the Ocala Limestone and Hawthorn Group have been eroded from its crest. Post-Hawthorn sediments cover this high, so that it has no surface expression. The St. Johns and Brevard platforms are subsurface extensions of the Sanford High.

In the panhandle, Paleogene carbonates are exposed on the Chattahoochee "arch." Miller (1986) considered the Chattahoochee "arch" an erosional feature, rather than a true structure. Neogene sediments covered at least part of the "arch," but they have been removed by erosion.

Faults in the Cenozoic sediments of Florida—postulated in many areas of the state by a wide variety of authors—have long been a source of controversy. Some geologists have found it difficult to accept many of the faults, based on the evidence at hand (Scott 1976). Many of the postulated faults in the state have been identified as offsets in the top of the Ocala Limestone, a karstified, unconformable surface that may have 50 m or more of relief. It is very difficult to identify faulting in the extremely heterogeneous Neogene sediments, especially with incomplete cores or rock cuttings, but some investigators have suggested the existence of faults affecting the Neogene section of Florida (Sproul et al. 1972; Bond et al. 1981; Scott 1983; and others).

Historical Perspective and Stratigraphic Problems

Dall and Harris (1892, 85) believed that Florida presents the "most complete succession of Tertiary and post-Tertiary fossil-bearing strata of any part of the United States." They referred to Florida as "a sort of standard, with which it may be convenient to compare the Cenozoic beds of other parts of the coast." Since the late 1800s, our understanding of the geology of the Florida Platform has evolved tremendously. Geologists now attempt to understand the geology of Florida in the light of global sea-level fluctuations, sequence stratigraphy, and broad-scale biostratigraphic zonations.

Cooke (1945, 109) subdivided the Miocene into three parts, "each of which was ushered in by an expansion of the sea upon the land." The Early Miocene was represented by the Tampa Limestone statewide. The Middle Miocene comprised the Shoal River and Chipola formations in the panhandle, and the Hawthorn Formation in the peninsula. The Late Miocene was represented by the Duplin Marl. Cooke also recognized a number of formations in the Pliocene section, including the Citronelle, Caloosahatchee, and Tamiami formations. He also recognized seven or eight shoreline "terrace formations" related to the Pleistocene interglacial episodes, as well as the Miami Oolite, Key Largo Limestone, Anastasia Formation, and Fort Thompson Formation.

The evolution of geologic thought continued through the ensuing two decades, as illustrated by Puri and Vernon (1964). The Miocene was subdivided into three stages: the Tampa Stage (Lower Miocene), Alum Bluff Stage (Middle Miocene), and Choctawhatchee Stage (Upper Miocene). The Tampa Stage contained the Chattahoochee and St. Marks formations. The Alum Bluff Stage comprised the Chipola, Oak Grove, Shoal River, and Hawthorn formations. The Choctawhatchee Stage included the Jackson Bluff and Tamiami formations, and other, less distinct units. In the peninsula, these authors recognized a "continental facies" that extended from a Miocene "delta" in southern Georgia and northern Florida.

Puri and Vernon (1964) believed that there were no marine Pliocene units in Florida. They placed the Caloosahatchee Formation in the Pleistocene, along with the Fort Thompson Formation, Key Largo Limestone, Miami Oolite, and Anastasia Formation.

Geologists investigating the Miocene through Holocene sediments in Florida encounter several important problems: (1) the use of biostratigraphic means to identify "formations"; (2) the limited number of good exposures, due to the generally low relief of the state; (3) complex sediment-facies patterns, and phosphorite sedimentation; (4) diagenetic alteration of sediments that obliterated chronostratigraphically important fossils used to place the lithostratigraphic units in the proper frame of reference; and (5) lack of pelagic micro- and nannofossils that could be utilized for correlation.

Lithostratigraphic Framework

In order to discuss and comprehend the geologic history of the Neogene and Quaternary in Florida, the lithostratigraphic framework first must be understood. Although the lithologic variability of these sediments can be extreme, the generalized lithostratigraphy is relatively straightforward. Because the state is large—nearly 650 km east to west, and 800 km north to south—there are many differences among lithostratigraphic units from area to area. Continental influences were more dramatic in the panhandle, whereas marine influences prevailed in the peninsular area.

Distinct sequences of lithostratigraphic units are recognized in different areas: the western panhandle, the central panhandle (including the Apalachicola Embayment), the eastern panhandle and northern peninsula, and the central and southern peninsula (inside front cover). Many rock units in these sequences are bounded above and/or below by unconformities caused by nondeposition or erosion, and these unconformities account for much of the time comprehended by the entire sequence. The lithostratigraphic units may grade laterally into each other from area to area. The lithostratigraphic units recognized here generally follow the COSUNA (Correlation of Stratigraphic Units of North America) Gulf Coast chart (Braunstein et al. 1988); however, some modifications have been made as the result of more recent information, and in an effort to generalize and limit the number of unit names.

The following discussion summarizes the lithostratigraphy of the several areas, and covers the last 25 million years in the geologic development of Florida.

WESTERN PANHANDLE

The basal Miocene sediments of this area are carbonates containing varying percentages of siliciclastics. These variably fossiliferous carbonates have been called the "Tampa Limestone," "St. Marks Formation," and "Chattahoochee Formation." Overlying this section are an unnamed, Chipola-equivalent siliciclastic limestone and the fossiliferous Bruce Creek Limestone.

During the early Middle Miocene, sedimentation shifted from predominantly carbonate to siliciclastic deposition with subordinate carbonates. The Pensacola Clay was deposited during the Middle to Late Miocene. It is overlain by Pliocene coarse clastics, followed by the Plio-Pleistocene siliciclastic Citronelle Formation, and Pleistocene-Holocene alluvium. These units may be sporadically fossiliferous.

CENTRAL PANHANDLE

As in the western panhandle, siliciclastic-bearing carbonates occur in the Lower Miocene, represented here by the Chattahoochee and St. Marks formations. Carbonate deposition was replaced as the dominant sediment early in the Middle Miocene. The Middle and Upper Miocene Alum Bluff Group, consisting of variably fossiliferous siliciclastics and carbonates, includes the Chipola, Oak Grove Sand, Shoal River, and Choctawhatchee formations. The Alum Bluff Group, Bruce Creek Limestone, and lower parts of the mixed siliciclastic-carbonate Intracoastal Formation overlie the Lower Miocene carbonates.

The Pliocene Jackson Bluff Formation, an often highly fossiliferous siliciclastic unit, and the upper Intracoastal Formation overlie the Alum Bluff Group. Overlying the Jackson Bluff Formation and interfingering with the Intracoastal Formation is the siliciclastic Citronelle Formation. The Citronelle Formation is overlain by siliciclastic Pleistocene-Holocene alluvium.

EASTERN PANHANDLE AND NORTHERN PENINSULA

The siliciclastic-bearing carbonates of the Chattahoochee and St. Marks formations are the oldest Neogene units in the eastern panhandle. They are overlain by the Torreya Formation of the Hawthorn Group. The basal Torreya is carbonate, grading into a siliciclastic upper part. The Hawthorn Group is overlain by the Jackson Bluff Formation, which underlies the siliciclastic Miccosukee Formation. Undifferentiated Pleistocene-Holocene siliciclastic deposits cap the Neogene sequence.

In the northern peninsula, the Hawthorn Group includes most of the Miocene section. The St. Marks Formation is present only in limited areas in the northwestern part of the region. The Hawthorn Group comprises (in ascending order) the Penney Farms, Marks Head, and Coosawhatchie formations, and, in a limited area, the Statenville Formation. Carbonate sediments are abundant in the Penney Farms and Marks Head formations, interbedded with siliciclastics. Siliciclastic sediments predominate in the Coosawhatchie Formation and younger units. Phosphate is virtually ubiquitous in the Hawthorn Group sediments of the peninsular area.

The Hawthorn Group underlies the siliciclastic Cypresshead Formation and the variably fossiliferous, siliciclastic Nashua Formation, which are overlain by undifferentiated, variably fossiliferous, siliciclastic Pleistocene-Holocene deposits. The Anastasia Formation consists of widely varying percentages of whole shells and shell fragments mixed with quartz sand, representing ancient beach deposits along the east coast of the state.

CENTRAL AND SOUTHERN FLORIDA

The entire Miocene section and part of the lower Pliocene section of central and southern Florida is represented by sediments of the Hawthorn Group, which comprises (in ascending order) the Arcadia Formation (with the Tampa and Nocatee members) and the Peace River Formation (with the Bone Valley Member and the

Wabasso beds). Siliciclastic-bearing carbonates predominate in the Arcadia Formation, except in the Nocatee Member, where siliciclastics predominate. Siliciclastics are prevalent in the Peace River Formation. Phosphate in varying amounts is generally present in the Hawthorn Group sediments, except in the Tampa Member, where it often is absent.

The section overlying the Hawthorn Group is a complex of often highly fossiliferous siliciclastic sediments and siliciclastic-bearing carbonates. Pliocene sediments are included in the Tamiami Formation (with its component members), Ochopee Limestone, Pinecrest Sand, and Buckingham Limestone, and the lower part of the carbonate and siliciclastic Caloosahatchee Formation. The Cypresshead Formation extends into central and southern Florida from the north.

The Caloosahatchee Formation overlies the Tamiami Formation, and in turn is overlain by the Pleistocene "Bermont Formation" (informal name, but widely used). The Bermont is overlain by the Fort Thompson Formation. Both the Bermont and the Fort Thompson are composed of fossiliferous siliciclastics interbedded with carbonates. A sequence of upper Pleistocene carbonate rocks in part overlies and in part interfingers with the Fort Thompson Formation in southeastern Florida; these are the Miami and Key Largo Limestones and the Anastasia Formation. The youngest sediments in the southern peninsula are undifferentiated Pleistocene-Holocene deposits.

Miocene to Holocene Development of the Florida Platform

The Neogene siliciclastic sediments encroached upon a Paleogene surface that had been exposed to erosion and underwent karstification during the late Paleogene. The oldest unit underlying this surface is the Avon Park Formation on the crest of the Ocala Platform in Levy County; other, isolated occurrences are found in central peninsular Florida. Across much of the northern half of the peninsula, the Ocala Limestone comprises the Paleogene surface; elsewhere in the state, the Suwannee Limestone and other Oligocene carbonates form this surface. The area where the Suwannee is missing has been referred to as "Orange Island" by Vaughan (1910). The dissolution features formed in these Paleogene carbonates provided the template for later collapse and further karstification.

Cenozoic sea levels were higher during the Paleogene than during the ensuing Neogene to Quaternary. The Late Oligocene regression recognized by Haq et al. (1987b) was responsible for major changes in the depositional history and development of the Florida Platform. The Georgia Channel System provided an effective barrier to siliciclastic-sediment transport onto the Florida Platform, allowing the platform to persist as a subtropical carbonate bank. The depositional environments of the Florida Platform and the Bahamas Platform were very similar until the Georgia Channel System stopped intercepting the siliciclastic sediments, and since the latest Oligocene the depositional environments of the two platforms have been significantly different. The Bahamas Platform persisted as a carbonate bank, while the Florida Platform was increasingly dominated by siliciclastics.

The influx of siliciclastic sediments onto the Florida Platform was a result of a number of interacting factors. Before the Late Oligocene, the volume of siliciclastic sediments shed from the Appalachians was very low, as indicated by carbonate deposition that extended well into South Carolina. It is postulated that during the latest Oligocene the Appalachians were broadly uplifted, rejuvenating the sediment source and dramatically increasing the input of siliciclastic sediment to the marine environment (Stuckey 1965; Schlee et al. 1988; Scott 1988b). A dramatic lowering of sea level during the Late Oligocene forced the Suwannee current to abandon the Georgia Channel System (Huddlestun 1993). The increased sediment supply essentially filled the channel and blocked the current from reoccupying it when sea level rose again during the earliest Miocene.

The encroachment of siliciclastic sediments onto the former carbonate platform occurred gradually (figs. 5.2–5.6). It was not until the Late Miocene to Pliocene that siliciclastic sediments dominated the Florida Platform, with only subordinate amounts of carbonate deposition. Siliciclastics replaced carbonates on the northern part of the Florida Platform first. By the Middle Miocene, Alum Bluff Group and Hawthorn Group siliciclastic sediments were deposited over the panhandle and the northern half of the peninsula. Carbonate deposition on the southern part of the platform appears to have persisted into the late Middle Miocene to earliest Pliocene. Deposition of the Arcadia Formation (Hawthorn Group) in southernmost peninsular Florida may have continued into the Late Miocene to earliest Pliocene. Arcadia carbonate deposition ceased in response to the increase influx of siliciclastic sediments.

It should be noted that thin beds of siliciclastic sediments occur throughout the Lower Miocene sediments nearly platform-wide. These beds are often very clayey in the earliest Miocene deposits, and contain more coarse materials upward in the section.

How early in the Neogene did sediments first cover the entire platform? Vernon (1951) stated that the Ocala Platform was not covered by younger sediments, and that the Paleogene surface has been exposed for some 30 million years. Scott (1981) postulated that the sediments of at least part of the Hawthorn Group completely covered the Ocala Platform. It appears that by the mid-Middle Miocene the carbonates on the Ocala Platform were covered by siliciclastic sediments. These sediments were very clayey, varying from clay to clayey sands, and deposition of these sediments on the platform limited infiltration of water, dramatically reducing the further development of dissolution features. As these sediments were eroded from the higher parts of the platform, the rate of dissolution increased, creating the karst landscape recognized today across much of Florida.

Neogene-Quaternary sediments vary from thick sequences in the central and western panhandle and the northeastern and southern peninsula to thin sequences on the Ocala Platform and Chattahoochee "anticline" (fig. 5.7). Although Miocene sediments were deposited across the Ocala Platform, and possibly the Chattahoochee "anticline," erosion has removed them, exposing the underlying Paleogene carbonates.

Regressions of the sea during the Neogene and Quaternary periodically exposed vast expanses of the Florida Platform. During these episodes of subaerial exposure, dissolution of the underlying carbonate sediments continued. Weathering and reworking of the marine sediments filled karst features and fluvial features with terrestrial sediments. These terrestrial deposits include numerous sand-dune fields scattered across the state (White 1970).

MIOCENE

Sea-level fluctuations during the Miocene and consequent changes in the position of the Gulf Stream were responsible for the formation of a phosphorus-rich mineral suite and cyclic depositional sequences (Riggs 1984). The idealized lithologic sequence of the depositional cycles begins with dominant siliciclastic sediments in an early transgressive phase. Late in the transgressive phase, carbonate sediments become dominant. Riggs (1984) hypothesized that the initial phosphogenesis occurred early in the transgressive phase on the outer shelf, and spread to a maximum across the shelf midway during the transgression. It appears that during later phases of sea-level fluctuation and subsequent transgressions and regressions phosphate was reworked into younger sediments (Scott 1988b; Compton et al. 1993; see also

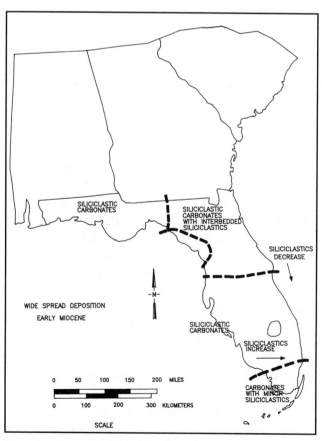

Figure 5.2. Generalized Early Miocene sediment pattern.

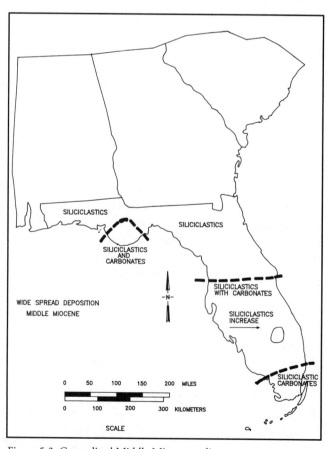

Figure 5.3. Generalized Middle Miocene sediment pattern.

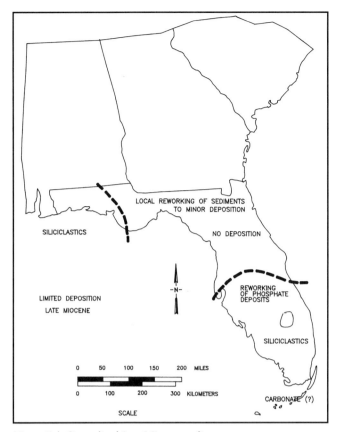

Figure 5.4. Generalized Late Miocene sediment pattern.

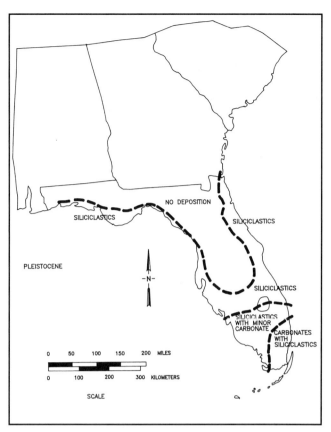

Figure 5.6. Generalized Pleistocene sediment pattern.

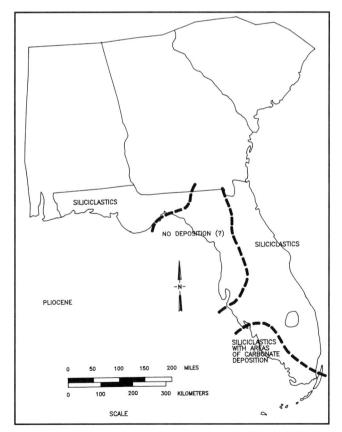

Figure 5.5. Generalized Pliocene sediment pattern.

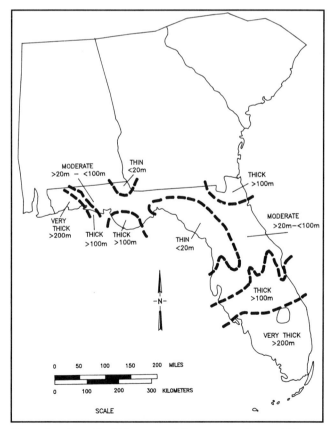

Figure 5.7. Generalized Neogene and Quaternary sediment thickness.

chapter 12). Cyclic sedimentation in the Hawthorn Group sediments in southwestern Florida has also been discussed by Missimer and Banks (1982).

The large and frequent sea-level fluctuations of the Miocene spread sediments over the Florida Platform, but left an intermittent stratigraphic record at best. Although biostratigraphic data defining the ages of stratigraphic units are still very sparse, new data from fossils, Sr isotopes, and magnetostratigraphy have helped to reform our concept of Miocene to Holocene sediment distribution (Jones 1992; Compton et al. 1993; Mallinson et al. 1994; Wingard et al. 1994; Weedman et al. 1995; Brewster-Wingard et al. 1996).

The Miocene sediments on the exposed part of the Florida Platform consist of highly variable admixtures of clay, silt, sand, and carbonate. Much of the carbonate sediment is dolostone to dolomitic limestone showing the effects of diagenesis. Hawthorn Group sediments have been more strongly affected by dolomitization in the northern two-thirds of the peninsula, whereas limestone is the dominant carbonate phase in southern Florida. Diagenesis destroyed most fossils in the sediments, rendering biostratigraphic interpretations nearly impossible.

Diagenesis has not affected the sediments in the panhandle as strongly. Sediments of the Alum Bluff Group in the central panhandle contain a diverse mollusk assemblage that paleontologists have analyzed for many years (Gardner 1926–50; Puri 1953; and others). The best exposures of Miocene sediments in the state occur in this area at Alum Bluff and along a number of small creeks. The subsurface units in the panhandle often contain diverse micro- and macrofossil assemblages (Schmidt and Wiggs-Clark 1980; Huddlestun 1984; Schmidt 1984).

During the Miocene, the basins on the Florida Platform filled significantly. In the Okeechobee and Jacksonville basins, for example, the base of the Hawthorn Group lies 175 to 300 m below sea level. The top of the Hawthorn Group lies at 35 to 50 m below sea level in the Okeechobee basin, and around 15 m below sea level in the Jacksonville basin. Although there is some debate over the ages of the lower and upper boundaries of the Hawthorn Group in various areas of the state, most of the filling of the basins appears to have occurred during the Miocene. Data from the Jacksonville basin (Mallinson et al., in press) indicate that in Florida Geological Survey core hole W-14619 (Carter no. 1) 100 m of 135 m of Hawthorn Group sediments is Miocene. Thick sequences of Miocene sediments also occur in the Apalachicola Embayment and the Gulf of Mexico Basin, overlain by a thinner Plio-Pleistocene sequence (Schmidt 1984).

Based on lithostratigraphic, biostratigraphic, and Sr-isotope research, an interesting picture of the distribution and preservation of Miocene sediments on the various parts of the platform is emerging. In the western panhandle (Walton County), nannofossil data indicate that Upper Miocene sediments are missing. Along the axis of the peninsula in southwestern Florida, Sr-isotope and nannofossil investigations suggest that Middle and Upper Miocene sediments may be absent (T. M. Missimer, pers. comm. 1992; J. M. Covington, pers. comm. 1992). On the southeastern part of the platform, nannofossil biostratigraphy suggests that Upper (and possibly Middle) Miocene sediments are present (J. M. Covington, pers. comm. 1992). It must be noted that fossil preservation in the Hawthorn Group is often poor, and that suggested gaps in the depositional record may be due to the paucity of diagnostic fossils (Scott 1988b).

Miocene sediments over much of the Florida Platform contain phosphorite in amounts ranging from a trace to more than 50%. Riggs (1986) stated that phosphogenic episodes occurred during the transition between deposition of two major lithologic groups, such as the carbonate to siliciclastic transition on the Florida Platform. Associated with phosphogensis was the formation of palygorskite and sepiolite (Mg-rich clays), along with dolomite (see chapter 12). Phosphate-bearing sediments were eroded from the positive features, and redeposition of the phosphorites was widespread in sediments of the Hawthorn Group, forming admixtures with dolomite, quartz sand, and clays. The nearly ubiquitous quartz sand admixed with the phosphorite grains was cited by Scott (1988b) and Compton et al. (1993) as evidence of reworking. Further reworking of the Hawthorn Group sediments has widely distributed phosphorite and Mg-rich clays in post-Miocene deposits.

PLIOCENE

Our knowledge of the distribution of Pliocene sediments is hindered by the paucity of age-diagnostic fossils in many areas. Although Pliocene strata occur widely in Florida (fig. 5.5), the sediments are not often naturally exposed, and frequently occur in areas of low relief. Much of our understanding of the Pliocene stratigraphy is derived from fossiliferous exposures and cores in two areas: the southwestern peninsula and central panhandle. In the southwestern peninsula, Pliocene sediments are occasionally exposed in shallow pits and canals. Pliocene units in the central panhandle are best known from a few "classical" exposures, including Alum Bluff and Jackson

Bluff. Both areas have yielded a number of high-quality cores that have provided significant insight into the biostratigraphy and lithostratigraphy.

Following the dramatic lowering of sea level during the Late Miocene, the sea rose during the Pliocene to levels that should have inundated much of the platform. The present distribution of Pliocene sediments, with minor modification by erosion, reflects those sea levels (fig. 5.5). Pliocene sediments in Florida are often highly fossiliferous, containing massive molluscan faunas. These fossils have attracted the attention of paleontologists since the 1800s (Dall and Harris 1892; and others), and they continue to be a focus of investigations (Allmon 1992; Ketcher 1992; Zullo 1992; and others). Sequence stratigraphy of the marine Pliocene beds high on the Florida Platform can be tied to molluscan zonations (Zullo and Harris 1992).

The patterns of Pliocene sediment distribution reflect the dominance of the siliciclastic sediment source to the north. Continuing the trends initiated in the Miocene, siliciclastics dominated depositional environments over much of the Florida Platform. Carbonate sediments remained a dominant component of the sediments in southernmost Florida during much of the Pliocene. This Late Pliocene siliciclastic-sediment dominance appears to represent the maximum influx and farthest southward advance of the siliciclastics. Siliciclastic sediments of presumed Pliocene age underlie the Florida Keys, forming the foundation for the Pleistocene carbonate bank and reefs.

The Okeechobee Basin persisted as a depocenter during the Pliocene, as did the Apalachicola Embayment. However, the Jacksonville Basin (Southeast Georgia Embayment) appears to have been essentially filled, and it contains only a thin Pliocene section. Pliocene sediments are more than 100 m thick in the Apalachicola Embayment, and as much as 30 m thick in the Okeechobee Basin.

In the panhandle, most of the Pliocene sediments lack fossils. The sediments form parts of a large delta complex that prograded southward across the coastal plain. These sediments—"coarse clastics" once thought to be Miocene, and the Citronelle and Miccosukee formations—underlie the Northern and Western Highlands (see chapter 1). In the Gulf Trough and Apalachicola Embayment, sediments of the Jackson Bluff Formation and the upper part of the Intracoastal Formation are significantly fossiliferous. These panhandle formations thin onto the Chattahoochee "arch" and toward the Ocala Platform, and are absent from the crests of the highs.

The Pliocene sediments of the peninsula were deposited in a marine environment. Sediments of the Cypresshead Formation (Huddlestun 1988; Scott 1988a) were deposited in a near-shore environment and occasionally reveal molds of mollusks long since removed by dissolution. Down-dip in northern Florida, the Cypresshead Formation grades into the fossiliferous Nashua Formation, and in southern Florida it appears to grade into part of the Caloosahatchee Formation. The Cypresshead Formation is the Citronelle-Miccosukee time-equivalent in the peninsula, where it contains very fine- to coarse-grained siliciclastic sediments including discoid quartz pebbles. Cypresshead sediments were eroded during later Neogene transgressions of the sea, and were spread southward along the peninsula. The resulting deposits appear similar to the Cypresshead, and can be difficult to distinguish from the original deposit.

DuBar (1991) provided maps showing the distribution of Plio-Pleistocene units in peninsular Florida. Although his stratigraphic nomenclature has not been accepted, the distribution patterns of the groups provide useful information.

During the Pliocene, much of southern Florida south of present-day Lake Okeechobee was a broad lagoon. Petuch (1992) envisioned the lagoon as being nearly encircled by barrier islands and reefs. The lagoon was a warm, shallow-water, protected environment where marine organisms thrived. Petuch has identified a number of endemic mollusk species from this area, several of which attained unusually large proportions due to the very hospitable conditions, but numerous duracrusts attest to many episodes of subaerial exposure during minor sea-level fluctuations.

The Peace River Formation (upper Hawthorn Group) is in part Pliocene, based on nannofossil data compiled by Covington (1992), and based on vertebrate fossils (Webb and Crissinger 1983). The Wabasso beds of Huddlestun (1988) also date from the Early Pliocene. Lithologically, these sediments appear nearly identical to the mid-Miocene Peace River Formation. Further study of the biostratigraphy of the Mio-Pliocene section in southern Florida will provide better understanding of the stratigraphic sequence of the Hawthorn Group.

The Tamiami Formation is present over much of southern Florida and is an important hydrostratigraphic unit. Tamiami-equivalent sediments extend northward along the east coast of the state into Georgia, where they are identified as the Raysor Formation, or Raysor-equivalent sediments (Huddlestun 1988). The Tamiami Formation is a complex unit comprising sand, clay, carbonate, and reef facies (Meeder 1987; Missimer 1992), all of which may contain variable, but limited amounts of phosphate. The formation contains more carbonate in southwestern Florida, and grades northward and eastward into siliciclastic beds.

The Tamiami Formation is unconformably overlain by the Caloosahatchee Formation in southern Florida. In northern Florida, the Tamiami-equivalent unit is overlain by the Nashua Formation.

The Caloosahatchee Formation of southern peninsular Florida is an often highly fossiliferous carbonate and siliciclastic unit. It has been extensively studied in the southwestern peninsula by DuBar (1958b, 1962), Lyons (1992), and others. As with the Tamiami Formation, the Caloosahatchee grades northward and eastward into siliciclastics. To the north, it becomes the siliciclastic Nashua Formation. Southeastward, it appears to become more carbonate-rich under parts of Dade and Monroe counties. Both the Nashua and Caloosahatchee formations appear to straddle the Pliocene/Pleistocene boundary (R. W. Portell, pers. comm. 1993).

PLEISTOCENE

Pleistocene seas did not inundate the entire Florida Platform. Low-lying parts of the platform, including most of southern Florida, were covered by shallow marine waters during the transgressions. The highlands, for the most part, remained exposed throughout the Pleistocene. Fluvial erosion and karstification extensively sculpted the uplands, producing areas of high relief (more than 70 m). Some of the most aesthetically pleasing, oldest, and continuously exposed landscapes in the state are found in the highlands of Brooksville Ridge, the Lake Upland, and the Tallahassee Hills.

It was formerly thought that Pleistocene sea levels were more than 70 m higher than the present level, and terrace deposits were correlated with each stillstand (Puri and Vernon 1964). It now appears that Pleistocene sea levels were not more than 20 m above the present level, and not all the terraces are Pleistocene (Colquhoun et al. 1968). Also, the Pleistocene sediments in Florida once were included in "terrace formations" deposited in response to widely fluctuating sea levels; these terrace formations (Cooke 1930) are no longer considered proper stratigraphic units, and they have been abandoned by most stratigraphers. Due to the paucity of distinguishing features other than elevation, the sediments forming the terraces are often referred to as "undifferentiated sands."

Pleistocene sediments in the state are predominantly siliciclastics (fig. 5.6). Erosion of the Mio-Pliocene siliciclastic sediments provided much of the material for these deposits. Most of the Pleistocene sediments in the panhandle are sparsely fossiliferous to unfossiliferous, and the siliciclastic sediments that predominate in the Pleistocene across much of the peninsula are variably fossiliferous. Fossils are more common and the assemblages more diverse in southern Florida south of Lake Okeechobee. The carbonate content of the sediments increases southward in this area, and the sediments gradually grade into the carbonates of southernmost Florida.

Pleistocene sediments on the platform show significant variability in the preserved thicknesses. DuBar (1991) suggested that the Pleistocene sediments generally do not exceed 9 m in thickness; however, investigations in southeastern Florida have revealed thicknesses exceeding 60 m (Perkins 1977), and Pleistocene sediments along a trend paralleling the edge of the continental shelf in southern Florida may exceed 70 m in thickness (Perkins 1977). Pleistocene sediments in both the panhandle and peninsula are generally undifferentiated, because they are unfossiliferous siliciclastics with a paucity of distinctive characteristics upon which to base lithostratigraphic units.

The recognized siliciclastic to mixed siliciclastic-carbonate stratigraphic units in the Pleistocene include the Nashua Formation, the upper part of the Caloosahatchee Formation, the Bermont Formation, and the Fort Thompson Formation. These units are variably fossiliferous (Matson and Clapp 1909; Sellards 1919; Parker and Cooke 1944; Brooks 1968). The Caloosahatchee and Bermont formations, in particular, contain well preserved, diverse fossil assemblages. The Bermont Formation (informal) is distinguished from the Caloosahatchee Formation solely by differences in molluscan faunas (DuBar 1974).

The development of barrier islands and beaches during the Pleistocene can be seen in the stratigraphic record of the peninsula. Deposits of coquina—lithified accumulations of shells deposited on a developing beach—occur well inland from the present coastline. Along the present coastline and inland, from near Jacksonville to West Palm Beach, lithified beach deposits of the Anastasia Formation are sporadically exposed. The Anastasia Formation forms the Atlantic Coastal Ridge.

The Anastasia Formation is a multicyclic deposit formed during several transgressions of the sea. Perkins (1977) recognized at least two disconformities within the formation. Osmond et al. (1970) measured two different ages for the Anastasia Formation, suggesting two episodes of accumulation. Based on fossil assemblages, coquina deposits well inland from the present coast were deposited early in the Pleistocene (E. J. Petuch, pers. comm. 1992). The Anastasia Formation grades into the Miami Limestone in southern Palm Beach County.

Southern Florida was periodically covered by shallow, warm marine waters inhabited by an abundance of organisms. Minor fluctuations in sea level exposed large areas, creating numerous disconformities in the section. The frequency of the fluctuations is evident from the numerous subaerially formed duracrusts and from the freshwater and brackish-water sediment sequences present in these units. Perkins (1977) recognized five regional disconformities in an investigation of Pleistocene sediment distribution in southern Florida; his study also revealed the influence of siliciclastic sediments reworked from older sediments exposed to the north and west.

Carbonate deposition resumed on the southern part of the Florida Platform during the Pleistocene. Perkins (1977) traced the evolution of the southern platform from an area dominated by quartz sands mixed with carbonate sediments during the earliest Pleistocene to a carbonate-dominated environment during the later Pleistocene. The later Pleistocene carbonates include the Miami Limestone and the Key Largo Limestone. The Miami Limestone consists of two facies: an oolitic limestone that underlies the Atlantic Coastal Ridge south of Boca Raton, and a bryozoan-rich limestone that underlies the Everglades and Florida Bay (Hoffmeister 1974; Perkins 1977). The paleo-reef trend in the Florida Keys is preserved in the Key Largo Limestone (see chapter 14).

HOLOCENE

A thin band of Holocene sediments forms the present coastline of the state. These deposits are beach, dune, marsh, and lagoon sediments that developed in response to the latest rise in sea level. The Holocene deposits are siliciclastic, carbonate, and organic sediments. Siliciclastic sediments dominated the Holocene depositional environments of the panhandle and much of the peninsula. Carbonate sediments become important in southernmost Florida, where siliciclastic sediment influx was not sufficient to mask the carbonate deposition (Enos 1977). The transition from predominantly carbonate to siliciclastic sediments south of Boca Raton has been documented by Duane and Meisburger (1969a, 1969b) and Meisburger and Field (1976; see also chapter 10).

Summary

Carbonate sedimentation dominated deposition on the Florida Platform since the mid-Mesozoic. Periodically, pulses of siliciclastic sediments encroached from the north and spread over parts of the platform, temporarily interrupting production and deposition of carbonate sediments. A major episode of siliciclastic sedimentation began during the late Paleogene, when siliciclastic sediments shed from the Appalachians, the Piedmont, and inner Coastal Plain flooded into the carbonate-producing environments of southeastern North America. The influx of siliciclastics suppressed carbonate deposition in northern Florida and, finally, by the middle to late Pliocene, the entire platform. This trend reversed during the Quaternary, and carbonate sedimentation resumed in southern Florida. The Florida Neogene and Quaternary sections provide an excellent example of siliciclastic and carbonate interaction and transition.

Miocene to Holocene sediments cover virtually the entire Florida Platform, leaving only limited exposures of older sediments. This sedimentary blanket ranges from very thin (less than a meter) to thicknessess exceeding 300 m. Contained within this sediment package are valuable natural resources, including limestone, phosphorite, heavy minerals, clay, and shell deposits, which are mined extensively across the state. These sediments include hydrostratigraphic units of the intermediate and surficial aquifer systems, which provide potable groundwater in many areas of the state. The intermediate confining unit provides a competent barrier to the downward migration of pollutants into the regionally extensive Floridan aquifer system.

Sea-level fluctuations played a primary role in the distribution of Neogene and Quaternary sediments on the Florida Platform. Following a major regression during the Late Oligocene, when deposition appears to have been restricted to southern Florida, sea level rose during the Early and Middle Miocene, and reached a maximum during the mid-Middle Miocene. During this transgressive phase, the entire Florida Platform experienced marine conditions. Sediments deposited on the crest of the Ocala Platform during the Miocene were subsequently removed by erosion, leaving very little record of these events. Sea level fell during the Late Miocene, exposing much of the platform.

The sea once again covered much of the platform during the Early Pliocene. Late Pliocene sea level was significantly lower, exposing much of the platform to subaerial conditions. Although limited in their distribution, the mollusk-rich Late Pliocene sediments have significant faunal diversity.

Pleistocene sea level did not rise higher than approximately 20 m above the present level, as shown by the distribution of Pleistocene sediments in Florida.

Following the latest Pleistocene regression, sea level has risen during the Holocene to its present position. Holocene sediments form the present coastline of the state, and represent beach, dune, marsh, and lagoon environments.

6

Hydrogeology of Florida

James A. Miller

ABUNDANT SUPPLIES OF FRESH SURFACE water and groundwater in Florida constitute one of the state's most important natural resources. Like many other natural resources, these waters are not evenly distributed, and are subject to damage or partial depletion as a result of human activities. Unlike most resources, however, the volume of freshwater available for withdrawal and use varies with time. Because freshwater in Florida is naturally replenished by precipitation, the volumes of water, and water levels in wells, lakes, and streams, can recover to pre-withdrawal conditions under favorable circumstances. The purpose of this chapter is to describe the geology and hydrology of the aquifers that contain Florida's groundwater resources, and to discuss some of the changes in hydrologic conditions that have occurred over time.

The source of all freshwater in Florida is precipitation. Most of the water either stored in or moving through surface water bodies and groundwater reservoirs in Florida originates from precipitation falling directly on the land surface within the state's boundaries; however, some of the water originates as precipitation in Alabama and Georgia and moves southward into Florida either in surface streams or as groundwater inflow (fig. 6.1). Average annual precipitation in mainland Florida from 1951 to 1980 ranged from more than 64 inches in the panhandle to less than 48 inches near Tampa Bay (fig. 6.2a).

Most of the precipitation is returned to the atmosphere by evapotranspiration. Rates of evapotranspiration are high in Florida because of the relatively high mean annual temperature, and because water is at or near the land surface in most places. Some of the precipitation moves directly to streams, rivers, and lakes as direct runoff; however, some of it infiltrates through the soil to shallow aquifers, where it is either discharged to surface streams as base flow or percolates downward to recharge deeper aquifers. The sum of direct runoff and base flow is total runoff. The average annual total runoff for Florida from 1951 to 1980 is shown in figure 6.2b. Except for southern Florida, where evapotranspiration rates are extremely high, the distribution of runoff is similar to that of precipitation during the same period—that is, runoff is greatest where precipitation is greatest. Throughout Florida, precipitation is greater than runoff. A substantial part of the precipitation neither is evapotranspired nor runs off, and is thus available for aquifer recharge.

Florida is characterized largely by high rainfall and generally low relief (see chapter 1). Across large parts of the state, highly permeable soil or rock is present at or near the land surface. Drainage density is low, except in the panhandle (fig. 6.3), but surface-water features include extensive wetlands (Hampson 1984), in addition to more than 7,700 lakes. The largest wetland is the vast system of freshwater marshes known as the Everglades, which extends over more than 1.5 million acres in southern Florida. The Everglades are characterized by sheetflow that originates directly from precipitation, whereas some widespread wetlands elsewhere in Florida, such as the coastal areas of Taylor, Dixie, and Levy counties, represent groundwater discharge areas. Lake Okeechobee, Florida's largest lake (fig. 6.3), covers about 436,000 acres, and is the second-largest freshwater lake in the conterminous United States that is wholly within one state. Before development, water that accumulated in Lake Okeechobee and spilled over its southern shore during periods of high rainfall provided much of the sheetflow that covered the Everglades. The lake is the principal reservoir that supplies water to an extensive network of canals and three large water-conservation areas in southern Florida. Many of the lakes in central and western peninsular Florida occupy basins formed by sinkholes that result from dissolution of part of the

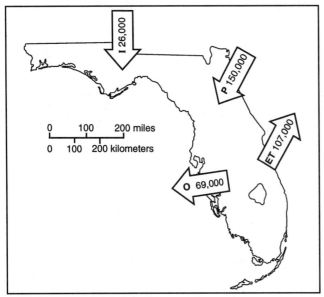

Figure 6.1. Water budget of Florida, in million gallons per day. ET: Evapotranspiration. I: Surface- and groundwater inflow. 0: Surface- and groundwater outflow, including consumptive use. P: Precipitation (Fernald and Patton 1984).

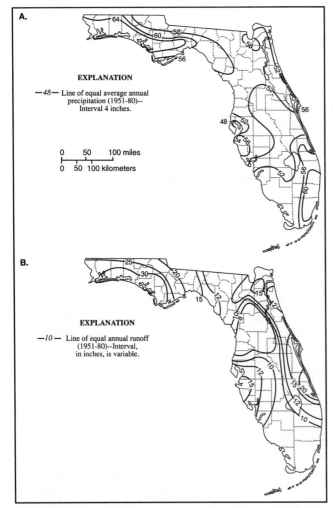

Figure 6.2. Average annual precipitation and runoff in Florida, 1951–80. A: Precipitation (U.S. Geological Survey 1986b). B: Runoff (Gebert et al. 1987).

limestone bedrock. Water levels in some of these lakes fluctuate directly with variations in groundwater levels in the limestone.

Water levels in many Florida streams are related to aquifer water levels. The water level, or stage, of the Hillsborough River at the Morris Bridge gage site northeast of Tampa is shown in figure 6.4. The figure also shows the water level in a well located 500 feet from the gage and completed in a limestone aquifer. The rise and fall in the river stage is generally paralleled by rise and fall in aquifer water levels, and both vary in response to changes in precipitation. Most of the time, the water level in the aquifer is higher than that in the river; under such conditions, the river is a gaining stream (condition

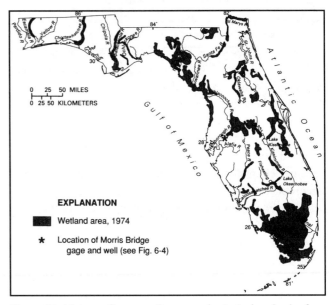

Figure 6.3. Major surface-water features and principal wetlands of Florida (Fernald and Patton 1984).

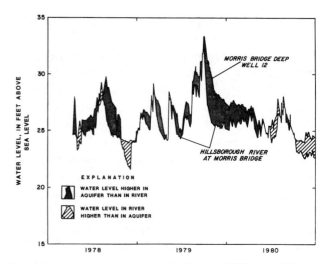

Figure 6.4. Relation between water levels in the Hillsborough River at Morris Bridge and in a nearby well (Wolansky and Thompson 1987).

A in fig. 6.5), and the groundwater contribution to the streamflow is called base flow (or "fair weather flow"). During periods of low precipitation, base flow may account for most of the water in the stream. The Hillsborough River at Morris Bridge receives base flow most of the time, as represented by the shaded areas between the hydrographs in figure 6.4. During periods of high precipitation, the river stage may rise above the water level in the aquifer, and water moves from the stream into the aquifer (hachured areas in fig. 6.4, and condition B in fig. 6.5); the stream becomes a losing stream. In places where the water level in an aquifer is always lower than the bottom of a stream channel (condition C in fig. 6.5), the stream is a losing stream that provides recharge to the aquifer as long as there is sufficient precipitation to provide stream flow.

There are five principal aquifers, or aquifer systems, in Florida. Four of these crop out at the land surface or are covered by a thin veneer of soil and/or weathered rock; a fifth, the intermediate aquifer system of southwestern Florida, is completely buried by shallower aquifers or confining units. An aquifer system consists of two or more aquifers with sufficient hydraulic connection that the groundwater flow systems of the aquifers function similarly and a change in conditions in one aquifer affects the other aquifer(s).

Figure 6.6 shows the extent of the three shallowest aquifers in Florida: the undifferentiated part of the surficial aquifer system, and its two named parts, the sand and gravel aquifer and the Biscayne aquifer. In contrast, only a small part of the deepest aquifer system, the Floridan, crops out. The Floridan aquifer system extends in the subsurface throughout the state; it also underlies large areas in Georgia, and smaller areas in Alabama and South Carolina (Miller 1986). The clayey confining unit that overlies the Floridan aquifer system also is exposed at the land surface (fig. 6.6).

The terminology used in this report for aquifers and confining units conforms to that used in the *Ground Water Atlas of the United States* (Miller 1990), and is not everywhere the same as that recommended by the Southeastern Geological Society for use in Florida (Florida Geological Survey 1986). Comparison of the two naming schemes is shown in figure 6.7.

The undifferentiated part of the surficial aquifer system, the sand and gravel aquifer, and the Biscayne

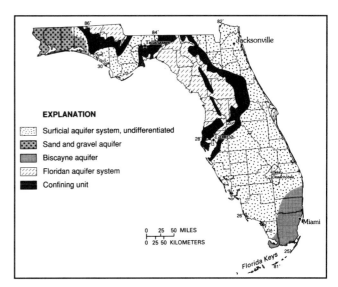

Figure 6.6. Principal aquifers exposed at or near the land surface in Florida. One principal aquifer, the intermediate aquifer system, does not crop out (Miller 1990).

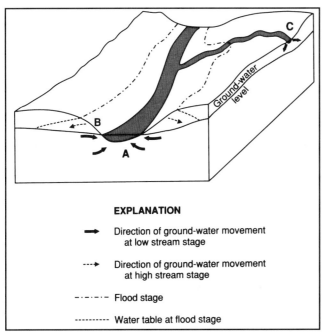

Figure 6.5. A: Diagram of a gaining stream under low flow conditions. B: The same stream becoming a losing stream in flood stage. C: A losing stream.

Southeastern Geological Society *in* Florida Bureau of Geology Special Publication 28	This report	
Surficial aquifer system	Surficial aquifer system, undifferentiated Sand and gravel aquifer Biscayne aquifer	
Intermediate aquifer system or intermediate confining unit	Upper confining unit of the Floridan aquifer system	Unnamed confining unit
		Intermediate aquifer system
Floridan aquifer system	Floridan aquifer system	Upper Floridan aquifer
		Middle confining unit
		Lower Floridan aquifer
Sub-Floridan confining unit	Lower confining unit of the Floridan aquifer system	

Figure 6.7. Comparison of hydrogeologic nomenclature proposed by the Southeastern Geological Society with that used in this chapter.

aquifer consist of strata that are mostly equivalent in age. These three water-yielding units are in lateral hydraulic connection, but differ in lithology and permeability. The undifferentiated surficial aquifer system consists mostly of relatively thin, unconsolidated sand, with some beds of shell and limestone, and yields only small volumes of water to wells. The sand and gravel aquifer consists mostly of complexly interbedded lenses of gravel and coarse sand, with some clay beds, and yields moderate volumes of water. The Biscayne aquifer is mostly limestone and sandy limestone, and yields large volumes of water. A thick, clayey confining unit underlies these aquifers in most places and hydraulically separates them from deeper aquifers. The intermediate aquifer system, so named for its position above the Floridan and below the surficial aquifer systems, consists of sand interbedded with limestone or dolomite. The limestone beds may yield large volumes of water, but the sands yield only small volumes. A clayey confining unit underlies the intermediate aquifer system and hydraulically separates it from the deeper Floridan aquifer system. The Floridan consists of a thick sequence of carbonate rocks and is the most prolific aquifer system in Florida. The Floridan supplies many of Florida's cities, and is the source of much of the water withdrawn for irrigation and mining needs.

The water-yielding properties of the aquifers in Florida are directly related to the geology of the beds that compose the aquifers. Texture, degree of sorting, and the amount of disseminated or interbedded clay and silt are the major factors that influence permeability of aquifers in siliciclastic sediments (Miller 1988). In general, fluvial deposits are the most permeable; deltaic deposits are of intermediate permeability (but may be highly permeable locally); and marine deposits are the least permeable. The original composition and texture of siliciclastic rocks in the Coastal Plain usually are preserved, unless the rocks have been deeply buried. The siliciclastic rocks included in Florida's aquifers have been buried only to shallow depths.

The original texture and porosity of carbonate rocks, like those of siliciclastic sediments, are determined to an extent by the energy conditions that existed where the rocks were deposited. For example, grainstones may form in tidal channels, swash areas, and other high-energy environments, whereas lime mudstones are more likely to be deposited in deeper water; however, unlike siliciclastic rocks, the original texture and permeability of carbonate rocks may be considerably altered after deposition. The processes of dissolution and dolomitization greatly affect the permeability of carbonate rocks. Dissolution increases the permeability of limestone by enlarging the original pore space or fractures that formed later. In extreme cases, large caverns are developed, resulting in a hydraulic conductivity for the rock so large that it is meaningless to determine. Dolomitization of lime mudstone can create a porous, loosely interlocked mosaic of dolomite crystals, replacing what was originally an almost impermeable sediment; however, overgrowths of dolomite or secondary calcite may partially or completely fill the pore space that originally existed in a grainstone or coquina, thereby decreasing permeability.

Evaporite minerals, common in parts of the Tertiary sequence in Florida, are subject to dissolution by circulating fresh groundwater, but the dissolved constituents may be transported and redeposited elsewhere. Where present, evaporite minerals cause the rocks to be almost impermeable.

Groundwater is the primary source of freshwater for all uses in Florida (U.S. Geological Survey 1990). During 1985, groundwater withdrawals provided about 65% of the total freshwater used in the state, and accounted for almost 90% of the water used for public supply (U.S. Geological Survey 1990). The Floridan aquifer system was the source of more than 60% of the total groundwater withdrawn in Florida during 1985 (fig. 6.8); the Biscayne aquifer provided about 20% of the water pumped during that same period. Most of the water withdrawn from the Biscayne was used for public supply in the rapidly growing urban areas of eastern Dade, Broward, and Palm Beach counties.

The surficial aquifer system was the third-largest source of groundwater withdrawn in Florida during 1985 (U.S. Geological Survey 1990). Because of its relatively low permeability, the surficial aquifer system does not provide large volumes of water to wells; however, the aquifer system is widespread, and accordingly is used extensively for domestic and small-community supplies. The intermediate aquifer system, fourth in importance in terms of the volume of water withdrawn, is nevertheless an important source of supply where water in the underlying Floridan aquifer system is brackish or saline. The sand and gravel aquifer provided 100 million

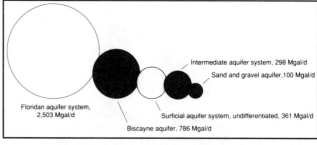

Figure 6.8. Withdrawals of freshwater (in Mgal/day) by principal aquifers in Florida, 1985 (U.S. Geological Survey 1990).

gallons of water per day (Mgal/d) during 1985, principally in the Pensacola area, where this aquifer is the primary water source (U.S. Geological Survey 1990).

Florida is the largest user of groundwater of any state east of the Mississippi River (U.S. Geological Survey 1990), but withdrawals are not evenly distributed. Groundwater withdrawals during 1985 were greatest in Dade, Broward, and Hillsborough counties (fig. 6.9) because of public-supply use, and in Polk County because of the combination of public-supply, agricultural (primarily citrus irrigation), and industrial (phosphate mining) uses. Sparsely populated counties, primarily in the Big Bend area and the panhandle, used the least amount of groundwater during 1985 (U.S. Geological Survey 1990).

The natural, or background, quality of the water in the shallow parts of all the principal aquifers in Florida generally makes it suitable for most uses. The chemical quality of natural groundwater primarily is affected by the mineralogy of aquifer materials and the length of time the water is in contact with aquifer minerals; accordingly, water deep within an aquifer generally is more mineralized than that in the shallow parts of the aquifer. Except for water in the siliciclastic aquifers, groundwater in Florida is hard to very hard. It is either a calcium-bicarbonate type or a calcium-magnesium-bicarbonate type, because of the partial dissolution of calcite and dolomite in the aquifers. The sand and gravel aquifer and part of the surficial aquifer system consist mostly of quartz, which is relatively insoluble; therefore, water in these aquifers is soft and not highly mineralized.

Excessive iron concentrations are present locally in water from the surficial aquifer system, the sand and gravel aquifer, and the Biscayne aquifer. In places, the Floridan aquifer system includes gypsum and/or anhydrite, which contribute large concentrations of calcium sulfate to the groundwater. All of the aquifers contain saltwater in places, either because they (1) are connected to the ocean, the Gulf, or other saltwater bodies, or (2) extend to depths where they contain unflushed saltwater. Therefore, intrusion of saltwater is a possibility, especially if an aquifer is intensely pumped near a freshwater/saltwater interface, or where the aquifer is vertically separated from a saltwater bearing aquifer by a leaky confining unit.

Undifferentiated Surficial Aquifer System

The undifferentiated surficial aquifer system in Florida (fig. 6.6) includes all aquifers that (1) are predominantly unconsolidated sand, (2) are at the land surface and contain water mostly under unconfined conditions, and (3) are not part of the Biscayne aquifer or the sand and gravel aquifer. Although the Biscayne and the sand and gravel aquifers are lateral equivalents of, or interfinger with, the surficial aquifer system, these aquifers are mapped separately because they are highly permeable and supply large municipalities. In contrast, the surficial aquifer system principally is used for domestic, commercial, or small-community supplies.

Beds of unconsolidated sand, shelly sand, and shell compose most of the surficial aquifer system. Limestone beds are locally important, high-yielding parts of the aquifer system, particularly in the area immediately north and west of the Biscayne aquifer in southern Florida. Clay beds locally create confined conditions within the surficial aquifer system, and in places these clay beds are thick and continuous enough to divide the system into discrete aquifers. For the most part, the aquifer system consists of complexly interbedded fine and coarse clastic sediments that are less than 50 feet thick; however, the aquifer system is as much as 150 feet thick in part of St. Johns County, 250 feet thick in parts of Martin and Palm Beach counties, and 400 feet thick in parts of Indian River and St. Lucie counties (Healy 1982).

Late Miocene to Holocene sediments compose the surficial aquifer system (fig. 6.10). The various formations shown in figure 6.10 are not all present at any single location; rather, in most places the aquifer system comprises only three or four of the formations (see chapter 5 for a detailed description of Miocene and younger rocks). The stratigraphic nomenclature used in figure 6.10 and other correlation charts in this chapter is that of the Florida Geological Survey, and does not

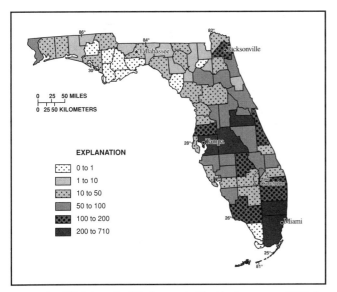

Figure 6.9. Withdrawals of freshwater in Florida, by county, 1985 (in Mgal/day) (U.S. Geological Survey files).

necessarily correspond with that used by the U.S. Geological Survey. Many of the formations shown in figure 6.10, particularly those of Pleistocene age, laterally interfinger with each other. Some of the listed units, such as the Caloosahatchee Formation, provide only small volumes of water. Limestone beds of the Fort Thompson and Tamiami formations in Collier and Hendry counties are the most productive parts of the surficial aquifer system, especially where dissolution of part of the limestone has created large openings in the rock. Moderate yields are obtained in places where the combined thickness of the Pamlico Sand and younger, unnamed sand deposits exceeds 40 feet (Healy 1982). Overall, the surficial aquifer system yields only small volumes of water.

Water in the surficial aquifer system is under unconfined, water-table conditions, except where clay beds in the aquifer system create local confined or semiconfined conditions. Precipitation falling directly on the aquifer system is the source of most recharge. Most of the water that enters the aquifer system moves along short flow paths and discharges as base flow to streams. Part of the water leaks downward across clayey confining beds and

Series	Stratigraphic and hydrologic units		Lithology
Holocene	Undifferentiated alluvium and terrace deposits	Surficial aquifer system, undifferentiated	Sand with local shell beds
Pleistocene[1]	Pamlico Sand		Fine to medium sand
	Miami Limestone		Oolitic limestone
	Fort Thompson Formation		Interbedded sand, shell, and limestone
	Anastasia Formation		Sandy limestone and marl
	Caloosahatchee Formation		Marl with minor sand and silt
Pliocene	Tamiami Formation		Marl with beds of fossiliferous limestone
Miocene	Hawthorn Group	Peace River Formation	Phosphatic sand and clay
		Arcadia Formation	Sand and limestone

[1](Stratigraphic units are equivalent in part. Order does not necessarily reflect relative age)

Figure 6.10. Stratigraphic units included in the undifferentiated surficial aquifer system (Healy 1982; Scott 1988b).

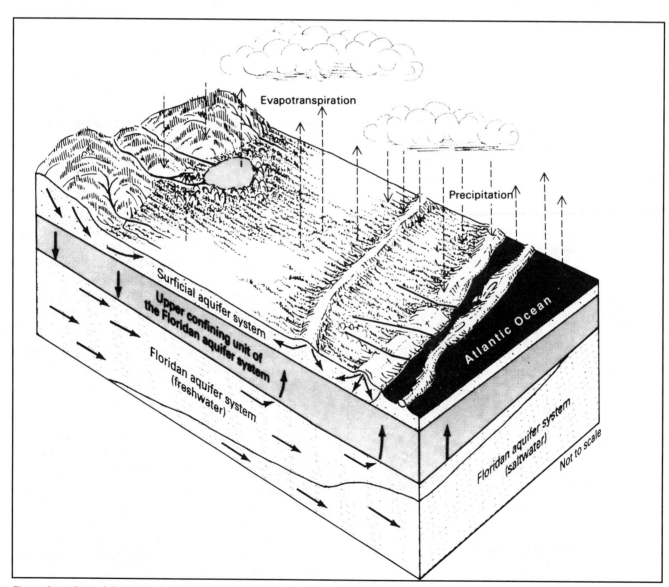

Figure 6.11. General direction of water movement in the surficial and Floridan aquifer systems in Indian River County (Crain et al. 1975).

recharges the intermediate aquifer and the Floridan aquifer system where the hydraulic heads in those aquifers are lower than the water table of the surficial aquifer system (fig. 6.11). Where the head relationship is reversed, leakage can occur from the deeper aquifers to the surficial aquifer system.

The permeability and transmissivity of the surficial aquifer system are extremely variable because of the complex interbedding of aquifer sediments. Most reported values of transmissivity range from 1,000 to 10,000 feet squared per day (ft^2/d), but values as large as 50,000 ft^2/d are reported where the aquifer system is thick and permeable. Well yields generally are less than 50 gallons per minute (gal/min), but Healy (1982) reported yields of 1,000 gal/min from a thick sand section in Indian River County.

The surficial aquifer system provided about 361 Mgal/d of water for all uses during 1985 (fig. 6.12). Domestic and commercial use and public supply accounted for most of the withdrawals, about 157 and 154 Mgal/d, respectively, being pumped for these use categories. Practically all remaining withdrawals were for agricultural use, about 46 Mgal/d being pumped for this purpose. About 4 Mgal/d was withdrawn for industrial and mining use.

Sand and Gravel Aquifer

The sand and gravel aquifer extends over all or part of the four westernmost counties in the panhandle of Florida (fig. 6.6). This aquifer is the principal source of water supply in Escambia and Santa Rosa counties, including the city of Pensacola. Much of the water in the aquifer is under unconfined conditions, but beds and lenses of clay interspersed with the coarser-grained sediments form confining beds that create local artesian conditions. Regional movement of the water in the aquifer is coastward.

Miocene and younger rocks compose the sand and gravel aquifer (fig. 6.13). In most places, the aquifer can be divided into two high-permeability zones, separated by a sand and clay unit that is less permeable (Barr et al.

1981). The upper zone, called the "surficial zone," consists mostly of fine- to medium-grained sand with gravel beds and lenses, all part of the Citronelle Formation of Pilocene age, combined with younger terrace and alluvial deposits. The surficial zone is recharged directly by precipitation, and most of the water entering the zone moves along short flow paths to small streams, where it discharges as base flow. Some of the water percolates downward through the middle, less-permeable zone of the aquifer and recharges the lower, high-permeability zone. Clay beds and hardpan layers of iron oxide locally create perched water tables or artesian conditions in the surficial zone. Accordingly, the permeability of the zone and the yields of wells completed in it are extremely variable.

The lower of the two high-permeability zones is called the "main producing zone." This lower zone is the one from which most water is withdrawn in Escambia and Santa Rosa counties. The main producing zone consists of gravel and coarse sand beds of Miocene age (fig. 6.13) and contains water under confined conditions. This zone is recharged by downward leakage from the surficial zone and discharges water to major streams, the Gulf, or bays and sounds. Wells completed in the main producing zone commonly yield more than 1,000 gal/min, and transmissivity values of 20,000 ft^2/d have been calculated for the zone in places (Hayes and Barr 1983).

The sand and gravel aquifer is more than 700 feet thick in southwestern Escambia County (fig. 6.14). The aquifer is underlain in most places by a confining unit (Trapp et al. 1977) that separates it from the underlying Floridan aquifer system; however, where this confining unit pinches out to the northeast, the Floridan is in direct hydraulic contact with the sand and gravel aquifer and receives some recharge from water percolating downward through the sand and gravel aquifer. Aquifer tests and laboratory tests of cores from the Pensacola Clay show that this confining unit has very low permeability (Hayes and Barr 1983).

Variations in hydraulic head with depth characterize

Series	Stratigraphic and hydrologic units			Lithology
Holocene and Pleistocene	Alluvium and terrace deposits		Sand and gravel aquifer	Undifferentiated silt, sand and gravel, with some clay. Surficial zone of aquifer
Pliocene	Citronelle Formation			Sand, very fine to very coarse and poorly sorted. Hardpan layers in upper part. Part of less permeable zone
Miocene	Unnamed coarse clastics	Choctawhatchee Formation		Sand, shell, and marl. Part of surficial zone
		Alum Bluff Group Shoal River Formation Chipola Formation		Sand with lenses of silt, clay, and gravel Main producing zone of aquifer
	Pensacola Clay		Confining unit	Dark to light gray sandy clay. Is basal confining unit in southern one-half of area
	St. Marks Formation		Floridan aquifer system	Limestone and dolomite—top of the Floridan aquifer system

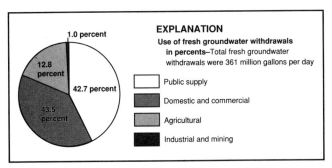

Figure 6.12. Withdrawals of freshwater from the surficial aquifer system, by category of use, 1985 (U.S. Geological Survey 1990).

Figure 6.13. Stratigraphic units in the sand and gravel aquifer (Cushman-Roison and Franks 1982).

the sand and gravel aquifer in many places, because local hardpan and clay confining units can perch water at different elevations above the regional water table, especially in the surficial zone of the aquifer. Heads generally decrease with depth in the aquifer, except near the coast, where some water leaks upward to the Gulf of Mexico, sounds, or bays. The general coastward movement of water from recharge areas at higher elevations is interrupted locally by streams and well fields.

The sediments that compose the sand and gravel aquifer are quartz rich and are not readily dissolved. Concentrations of dissolved solids in water from the aquifer are less than 50 milligrams per liter (mg/L), except near the coast and adjacent to large bays and sounds, where there is a freshwater/saltwater transition zone. Chloride concentrations greater than 1,000 mg/L are reported in water from wells located near the interface (Cushman-Roisin and Franks 1982). Dissolved-iron concentrations locally reach objectionable levels. The sand and gravel aquifer is susceptible to contamination, particularly in the surficial zone of the aquifer; for example, the surficial zone was contaminated by creosote wastes from a wood-preserving plant near Pensacola (Mattraw and Franks 1986).

Withdrawals from the sand and gravel aquifer in Florida totaled 100 Mgal/d during 1985 (fig. 6.15). Roughly equal volumes were withdrawn for public-supply use (about 44 Mgal/d) and for industrial and mining uses (about 43 Mgal/d). About 7 Mgal/d was pumped for agricultural use, and the remainder of about 6 Mgal/d was used for domestic and commercial purposes. The single largest user of water from the sand and gravel aquifer is the city of Pensacola (U.S. Geological Survey 1990).

Biscayne Aquifer

The Biscayne aquifer underlies an area of about 4,000 square miles in Broward, Dade, Monroe, and Palm Beach counties (fig. 6.6). The aquifer is named after Biscayne Bay (Parker et al. 1955), and it extends under the bay to the Atlantic Ocean. The aquifer contains saltwater under the ocean and the bay, and some of the saltwater has moved inland in response to the lowering of water levels in the freshwater parts of the aquifer by withdrawals at municipal well fields and by drainage-canal construction. The Biscayne aquifer is the only source of drinking water for about three million people who live primarily in urban areas from Homestead, in Dade County, northward to Boca Raton, in Palm Beach County (Klein and Causaras 1982). This aquifer also is a source of water that is transported by pipeline to the Florida Keys.

The Biscayne aquifer is at shallow depths everywhere and is highly permeable. Accordingly, the aquifer is in direct hydraulic connection with streams, canals, and other natural and manmade surface-water bodies. Because of this connection, the aquifer, Lake Okeechobee, three large water-conservation areas, and an extensive network of canals, control structures, and pumping stations are continually monitored and managed as an integrated hydrologic system. The agency responsible for this management is the South Florida Water Management District. Management functions include flood control, storage and subsequent release of surplus stormwater, prevention of saltwater intrusion, and maintaining groundwater levels and supplies.

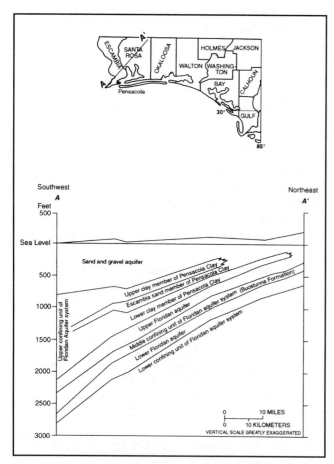

Figure 6.14. Geohydrologic cross-section from northeastern Santa Rosa County to southwestern Escambia County (Miller 1990).

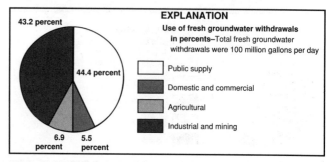

Figure 6.15. Withdrawals of freshwater from the sand and gravel aquifer, by category of use, 1985 (U.S. Geological Survey 1990).

Highly permeable limestone and less-permeable sandy limestone and sand compose the Biscayne aquifer. Most of the formations composing the aquifer are of Pleistocene age, but some Pliocene strata also are included (fig. 6.16). These formations were mostly deposited in shallow marine to marginal marine environments that shifted rapidly in response to minor changes in sea level. Accordingly, most of the formations are thin and commonly lenticular, and they interfinger with other formations (Causaras 1985, 1987); thus, the entire sequence of formations listed in figure 6.16 is not present at any one place. The most productive water-yielding formations in the aquifer are the Fort Thompson Formation, the Miami Limestone, and, locally, the Anastasia Formation. Yields are greatest where solution cavities have developed in limestone units: Klein and Causaras (1982) reported yields of 7,000 gal/min from large-diameter wells in Dade County, with only small drawdowns, and Fish (1988) calculated transmissivity values greater than 1 million ft^2/d from aquifer tests in Brevard County. The Biscayne aquifer as a whole is more sandy to the north and east, and contains more limestone, sandy limestone, and calcareous sandstone to the south and west (Klein and Hull 1978). As it thins and becomes more sandy, the aquifer grades laterally into, and interfingers with, the undifferentiated surficial aquifer system. The Biscayne is underlain by a thick sequence of low-permeability clay beds that are as much as 1,000 feet thick in places and separate the aquifer from the underlying Floridan aquifer system, which contains saltwater in southeastern Florida.

Groundwater and surface water form a united hydrologic system in southern Florida. For example, before development of land and water resources in the Everglades area, most of the 50 to 60 inches of annual rainfall moved as sheetflow to the south and southwest across the flat topography. Much of the sheetflow was returned to the atmosphere by evapotranspiration, but some ultimately discharged to Florida Bay and the Gulf of Mexico. Most of the Everglades and other low-lying areas were inundated during the wet season; however, during dry periods water covered much less of these areas, and moved through troughs and sloughs. Lake Okeechobee functioned as a retarding basin, temporarily storing the water from southward-flowing streams that drain into the lake.

An extensive network of canals was constructed during the late 1800s and early 1900s. Some of the earliest canals were dredged for navigation, but most were constructed with the intent of draining the land for agriculture and settlement (Fernald and Patton 1984). Today, much of the water that became sheetflow under natural conditions is channeled by canals into large impoundments bounded by levees (water-conservation areas), and is gradually released.

The water-conservation areas and the large canals used to control water movement in southeastern Florida are shown in figure 6.17. Only the major canals are shown in the figure, and many other canals exist. The water level of Lake Okeechobee is controlled by levees and by control structures on canals that channel water into the lake from the north and northwest and drain it to the west, east, and southeast. Natural land-surface elevations and levee heights are such that water impounded during wet periods moves from Lake Okeechobee sequentially to water-conservation areas 1, 2, and 3. Freshwater is released from conservation area 3 to partly sustain flow to Everglades National Park. A network of pumping stations allows transfer of excess stormwater from canals into the water-conservation areas, thus providing flood control. During dry periods, water stored in the conservation areas is diverted into the canal system and flows to areas where it is needed to help maintain groundwater levels in the Biscayne aquifer. The high permeability of the aquifer allows rapid movement of water from the canals into the aquifer. Control structures located near the coast on the major canals help prevent encroachment of saltwater into the canal system during dry periods, but permit excess stormwater to move coastward freely.

The generalized configuration of the water table in the Biscayne aquifer is shown by contours in figure 6.17. The map was constructed using water levels measured during the dry season in May 1978. The general movement of water is seaward, as shown by the arrows. Water-table contours are not shown in the water-conservation areas, because in these impoundments the water table has no slope. In upstream areas, water moves from the aquifer toward natural surface drainage and canals, a situation shown schematically in figure 6.18a. Downstream, the movement is reversed, and the aquifer receives recharge by drainage from the canals (fig. 6.18b). Major well fields

Series	Stratigraphic and hydrologic units[1]		Lithology and water-yielding characteristics	Thickness (feet)
Holocene	Organic soils	Confining unit	Peat and muck; water has high color content. Almost impermeable. Lake Flirt is shelly, calcareous mud.	0–18
	Lake Flirt Marl[2]			
Pleistocene[1]	Pamlico Sand	Biscayne aquifer	Quartz sand; water high in iron. Small yields to domestic wells	0–40
	Miami Limestone		Sandy, oolitic limestone. Large yields	0–40
	Fort Thompson Formation		Alternating marine shell beds and freshwater limestone. Generally high permeability. Large yields	0–150
	Anastasia Formation		Coquina, sand, sandy limestone, marl. Moderate to large yields	0–120
	Key Largo Limestone		Coralline reef rock. Large yields	0–60
	Caloosahatchee Formation		Sand, shell, silt, and marl. Moderate yields	0–25
Pliocene	Tamiami Formation[3]	Confining unit	Limestone, clay, and marl. Occasional moderate yields in upper few feet. Remainder forms upper part of basal confining unit.	25–220

[1]Stratigraphic units are equivalent in part. Order does not necessarily reflect relative age.
[2]Low-permeability material is of local extent.
[3]Part of undifferentiated surficial aquifer where permeable.

Figure 6.16. Stratigraphic units included in the Biscayne aquifer (Klein and Causaras 1982).

are represented by the closed depression contours near Miami and Fort Lauderdale (fig. 6.17). Intensive pumping has lowered the water table near the well fields and has reversed the natural seaward flow direction of water in some places. The water-table contours are closely spaced north of Fort Lauderdale; there, the Biscayne aquifer is more sandy and less permeable than in Dade County and southern Broward County, where the aquifer mostly consists of highly permeable limestone.

Fluctuations in the water table of the Biscayne aquifer may range from 2 to 8 feet per year, depending primarily on location and on variations in precipitation and pumpage. Pumpage is generally greater during dry periods. During prolonged droughts, the water table in the aquifer may decline below sea level throughout large areas near the coast or near well fields (Schroeder et al. 1958). Such declines in coastal areas can result in encroachment of saltwater and contamination of the aquifer. In places where the aquifer is exposed at the land surface or is covered only by a thin veneer of soil, water levels in the aquifer respond rapidly to rainfall (fig. 6.19). As shown in figure 6.19a, the water level in a well north of Miami rose more than 4 feet within a few hours following 11 inches of rain on April 16, 1942 (fig. 6.19b). Rainfall of more than 6 inches on April 17 produced an additional water-level rise of more than 1.5 feet, again within only a few hours following the intense rainfall. Such rapid response can occur only in a highly permeable, unconfined aquifer with a shallow water table.

Under natural or predevelopment conditions, there was a balance between freshwater and saltwater in the coastward parts of the Biscayne aquifer. The predevelopment interface position was similar to that shown in figure 6.20a (Klein et al. 1975). Drainage canals (fig. 6.20b) lower levels of freshwater, and thus allow saltwater to rise to a higher elevation in the aquifer. Canals dug in tidewater areas also allow saltwater to migrate inland by flowing directly up the canal during high tide, or during periods of low rainfall when the coastward movement of freshwater in the canals is minimal. Levels of freshwater may be further lowered by intensive pumping from wells located near canals (fig. 6.20c), possibly resulting in saltwater contamination at the wells. Damlike control structures constructed along coastward reaches of canals artificially raise levels of freshwater in the canals and in the adjacent aquifer when the structures are closed, thus preventing saltwater encroachment (fig. 6.20d). In some places, saltwater encroachment in the Biscayne aquifer has been reversed following emplacement of control structures (Hughes 1979).

Under natural conditions, the theoretical configuration of the freshwater/saltwater interface is generally like a landward-sloping wedge-face, the freshwater body

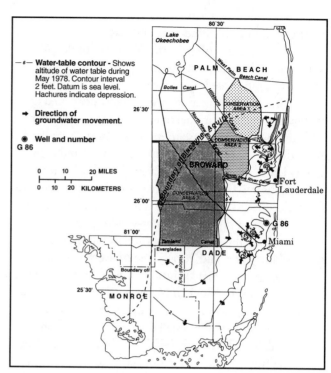

Figure 6.17. Relation of the water table of the Biscayne aquifer to canals, water-conservation areas, and pumping centers (Klein and Causaras 1982).

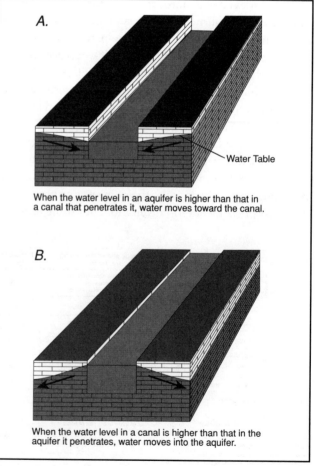

Figure 6.18. Movement of water between the Biscayne aquifer and canals. A: Upstream areas. B: Downstream areas (Klein et al. 1975).

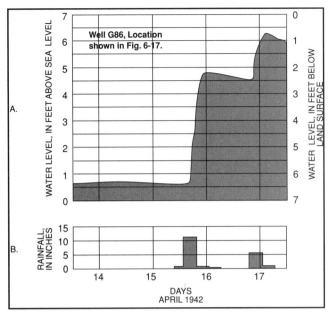

Figure 6.19. A: Hydrograph of well G86, completed in the Biscayne aquifer, mid-April 1942. B: Rainfall during same period (Klein and Hull 1978).

being thinnest near the coast, and thickening inland. The maximum inland extent of saltwater is thus in the deeper parts of the aquifer. The exact position of the interface changes as recharge to and discharge from the aquifer vary. These variations are responsible for convective flow in the saltwater wedge (Kohout 1964) and account for the observed shape of the interface (fig. 6.21). The situation shown in figure 6.21 represents conditions where the Biscayne aquifer is highly permeable. Farther northward, where the aquifer contains more sand and is less permeable, saltwater encroachment does not extend as far inland, and the surface of the interface is more steeply inclined (Klein and Hull 1978). As shown in figure 6.21, freshwater moves upward along the interface and is discharged to Biscayne Bay near the shoreline.

Although saltwater contamination is the most serious threat to the quality of water in the Biscayne aquifer, the aquifer also is highly susceptible to contamination inland. Because the aquifer is highly permeable and is at or near the land surface everywhere, any contaminants spilled or applied at the land surface can move quickly into the aquifer. Water from canals also recharges the aquifer, and contaminants in the canal water can easily infiltrate the aquifer. The natural quality of the water in the aquifer is

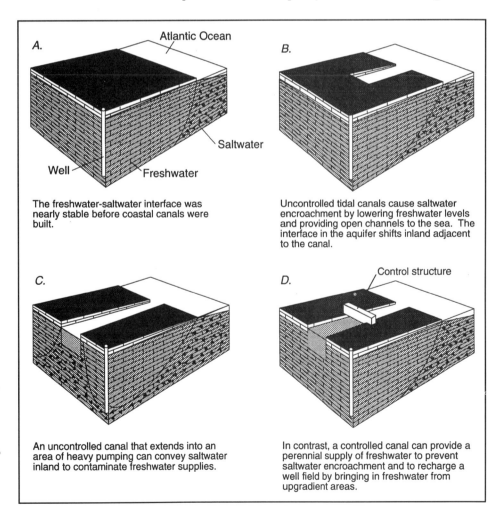

Figure 6.20. Effect of canals and control structures on the position of the freshwater/saltwater interface. A: Predevelopment. B: After canal construction, with normal rainfall. C: Uncontrolled canal extended into an area of heavy pumping. D: After addition of control structure (Klein et al. 1975).

suitable for most purposes, except locally where the water contains objectionable concentrations of dissolved iron.

During 1985, about 786 Mgal/d of water was withdrawn from the Biscayne aquifer. More than 72% of this water, or about 569 Mgal/d, was used for public supply (fig. 6.22). The Miami-Dade Water and Sewer Department withdrew more than 292 Mgal/d from the aquifer during 1985, making that department the largest single user of groundwater in Florida (U.S. Geological Survey 1990). Almost 23% of the total water withdrawn, or about 181 Mgal/d, was used for agriculture. Combined withdrawals for domestic and commercial use and for industrial and mining use were about 37 Mgal/d, divided about equally between the two use categories.

Intermediate Aquifer System

Aquifers that lie beneath the surficial aquifer system and above the Floridan aquifer system in southwestern peninsular Florida are grouped within the intermediate aquifer system (fig. 6.23). The intermediate aquifer system does not crop out, and it contains water under confined conditions. Although this aquifer system does not yield as much water as the Floridan, it is the main source of supply in Sarasota, Charlotte, and Lee counties where the underlying Floridan contains brackish water or saltwater. The intermediate aquifer system consists of beds of sand, sandy limestone, limestone, and dolomite of Miocene and Pliocene age (fig. 6.24). These strata are assigned to the upper part of the Tampa Member of the Arcadia Formation, the undifferentiated Arcadia Formation, and the Peace River Formation—all parts of the Hawthorn Group—and to the Tamiami Formation. Clay confining units that bound the permeable strata at top and bottom are included in the aquifer system. As the facies of the intermediate aquifer system change by a decrease in sand and limestone accompanied by an increase in clay, they become a part of the upper confining unit of the Floridan aquifer system. Local water-yielding beds in the Hawthorn Group in Brevard, Clay, and Indian River counties are not continuous with the intermediate aquifer system, and are not considered part of it. Likewise, local aquifers that are between the surficial aquifer system and the Floridan in places outside the area mapped in figure 6.23 are not considered part of the intermediate aquifer system in this chapter.

The intermediate aquifer system slopes gently and thickens to the south and southwest from the updip limit of the system (fig. 6.25). In many places, an unnamed confining unit divides the aquifer system into an upper aquifer, called the Tamiami–upper Hawthorn aquifer, and a lower aquifer, called the lower Hawthorn–upper Tampa aquifer (Duerr and Wolansky

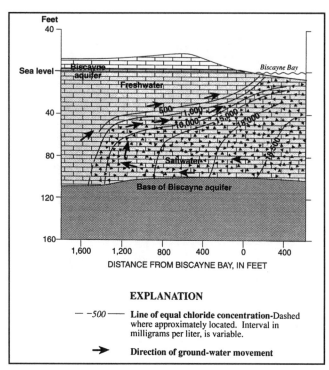

Figure 6.21. Approximate position of the freshwater/saltwater interface in the Biscayne aquifer under natural conditions (Cooper 1959).

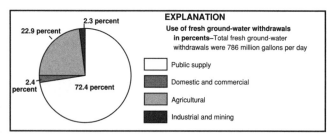

Figure 6.22. Withdrawals of freshwater from the Biscayne aquifer, by category of use, 1985 (U.S. Geological Survey 1990).

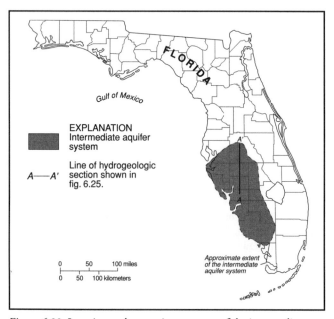

Figure 6.23. Location and approximate extent of the intermediate aquifer system.

1986). In Collier County, the lower Hawthorn–upper Tampa aquifer becomes predominantly a low-permeability clay, and accordingly the intermediate aquifer system is thinner there. In Charlotte and Glades counties and southward, the Tamiami–upper Hawthorn aquifer is the principal part of the system, whereas north of these counties the lower Hawthorn–upper Tampa aquifer is more productive.

Water in the intermediate aquifer system is under confined conditions, except locally near the northern limits of the aquifer system, where the upper confining unit is thin or absent and the system is in direct hydraulic contact with the surficial aquifer system. Most recharge to the intermediate aquifer system is by downward leakage through the upper confining unit. In updip areas, some of the water percolates downward through the lower confining unit of the intermediate aquifer system and recharges the underlying Floridan aquifer system. Regional movement of water in the intermediate aquifer system (fig. 6.26) is outward in all directions from high areas on the potentiometric surface in Polk County, and then generally coastward. The depression in the potentiometric surface in southern Sarasota County marks a pumping center. Most wells completed in the intermediate aquifer system yield 200 gal/min or less, but local yields of 2,000 gal/min are reported (Healy 1982).

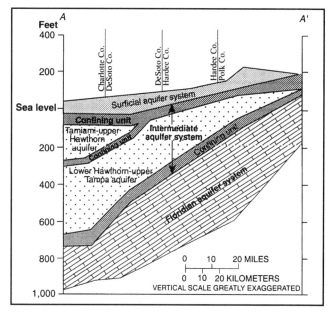

Figure 6.25. Hydrogeologic cross-section from central Polk County to central Charlotte County. See figure 6.23 for map location (Lewelling 1988).

Series	Stratigraphic unit			Hydrogeologic unit		Lithology
Holocene and Pleistocene	Surficial deposits			Surficial aquifer system		Undifferentiated sand with some limestone
Pliocene	Tamiami Formation			Confining unit	Intermediate aquifer system	Sand, limestone, and shell beds. Thick clay near top
Miocene	Hawthorn Group	Peace River Formation		Tamiami–upper Hawthorn aquifer		
				Confining unit		Mostly limestone, sandy limestone, and sand. Phosphatic in part. Dolomite beds common. Clayey in middle part
		Arcadia Formation		Lower Hawthorn–upper Tampa aquifer		
			Tampa Member	Confining unit		Limestone, sandy limestone, and sand. Clay beds in middle part
				Floridan aquifer system		

Figure 6.24. Stratigraphic units included in the intermediate aquifer system (Duerr and Wolansky 1986; Scott 1988b).

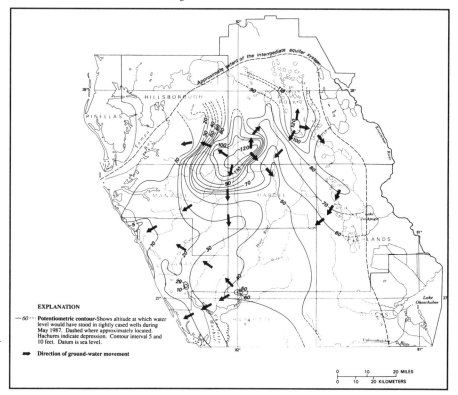

Figure 6.26. Potentiometric surface of the intermediate aquifer system, May 1987 (Lewelling 1988).

Transmissivity values reported for the aquifer system are 10,000 ft²/d or less.

The quality of water in the shallower parts of the intermediate aquifer system generally makes it suitable for most uses. In the deeper parts of the aquifer system, dissolved-solids and chloride concentrations may be excessive, especially near the coast. Chloride concentrations greater than 1,800 mg/L were measured locally in water from wells completed in the intermediate aquifer system in Manatee County (Healy 1982).

Of the 298 Mgal/d of water withdrawn from the intermediate aquifer system during 1985, almost 80%, or 233 Mgal/d, was used for agricultural purposes (fig. 6.27). About 10%, or 31 Mgal/d, was withdrawn for public supply. Domestic and commercial uses accounted for withdrawals of about 19 Mgal/d (some 6%), and about 15 Mgal/d (5%) was pumped for industrial and mining uses.

Floridan Aquifer System

The highly productive Floridan aquifer system underlies all of Florida, but contains water with high concentrations of dissolved solids in many areas (fig. 6.28). The aquifer system also extends over a large part of the coastal plain of Georgia, and smaller areas of coastal Alabama and South Carolina. Many cities in Florida—including Jacksonville, Gainesville, Tallahassee, Orlando, Clearwater, Tampa, and St. Petersburg—depend on the Floridan for water supplies. This aquifer system also is the source of water for many smaller communities and rural households. Total withdrawals from the Floridan aquifer system in Florida were more than 2.5 billion gallons per day during 1985 (U.S. Geological Survey 1990). Even though withdrawals are extremely large, hydraulic heads in the aquifer system have declined very little, except in the vicinity of pumping centers, or where the permeability of the aquifer system is uncommonly low.

Where the Floridan aquifer system contains freshwater, it is the principal source of water supply; however, where the Floridan contains saltwater, the aquifer system also has been used as a repository for treated sewage and industrial wastes emplaced by injection wells. Drainage wells have been used in several counties in central Florida to divert excess surface runoff into the Floridan aquifer system. More than 17 Mgal/d of saltwater was withdrawn from the aquifer system during 1985, mostly for desalinization, mixing with potable water, or cooling purposes (U.S. Geological Survey 1990).

In Florida, the Floridan aquifer system consists of a thick sequence of carbonate rocks of Tertiary age. The strata composing the aquifer system can be assigned to several formations or parts of formations (fig. 6.29). The most permeable formations in the aquifer system are the Ocala Limestone and the upper part of the Avon Park Formation. Where present, the Suwannee Limestone is a principal aquifer; however, it is thinner and less extensive than the Ocala Limestone and the Avon Park Formation (see chapter 4). Where it is highly permeable, the Tampa Member of the Arcadia Formation is included in the aquifer system. Erosion has completely removed both the Suwannee and Tampa limestones in places. Where they are sufficiently permeable, beds in the lower part of the Avon Park Formation, the Oldsmar Formation, and the upper part of the Cedar Keys Formation are included in the aquifer system. Some researchers consider limestone beds in the lower part of the Hawthorn Group to be part of the Floridan, but this author (Miller 1986) excluded those limestones from the aquifer system.

The Floridan aquifer system is defined on the basis of its permeability. The permeability of the carbonate rocks that compose the aquifer system is at least an order of magnitude higher than that of the confining units that

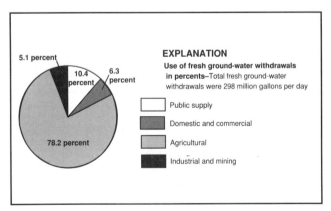

Figure 6.27. Withdrawals of freshwater from the intermediate aquifer system, by category of use, 1985 (U.S. Geological Survey 1990).

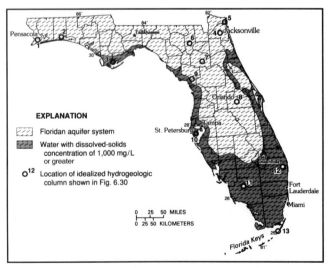

Figure 6.28. Extent of the Floridan aquifer system and saline water in the system (Sprinkle 1989; Miller 1990).

overlie and underlie the system. Accordingly, the top and base of the aquifer system do not coincide everywhere with the top or base of rocks of any single geologic formation, or strata of any single geologic age (fig. 6.30). The permeability contrasts that define the aquifer system also can occur within rocks of a certain age. For example, in places the top of the aquifer system is within the Oligocene sequence, rather than at the top or base of these beds. Likewise, the base of the aquifer system does not coincide with any single stratigraphic boundary everywhere, but rather is the base of high-permeability rocks. Regionally, the top of the aquifer system in Florida is at the top of the Late Eocene Ocala Limestone, and the base of the system is marked by massively bedded anhydrite in the Paleocene Cedar Keys Formation. In panhandle Florida, low-permeability siliciclastic Eocene rocks form the base of the aquifer system.

In most places, the Floridan aquifer system can be divided into the Upper Floridan and Lower Floridan aquifers (Miller 1986), separated by a less-permeable middle confining unit (fig. 6.30). The Upper Floridan aquifer is the best-known, most permeable, and most productive part of the aquifer system, and consists mostly of the Suwannee and Ocala limestones and the upper part of the Avon Park Formation; locally, the Tampa Member of the Arcadia Formation also is included. Unnamed middle confining units separate the Upper and Lower Floridan aquifers in most places. These confining units vary in stratigraphic position (fig. 6.30) and lithology. They may consist of clay, micritic limestone, or anhydrous dolomite (Miller 1986), and they range in age from Oligocene to Early Eocene. Regardless of their stratigraphic position or lithology, the rocks of the confining units retard the movement of water between the two aquifers. The Lower Floridan aquifer

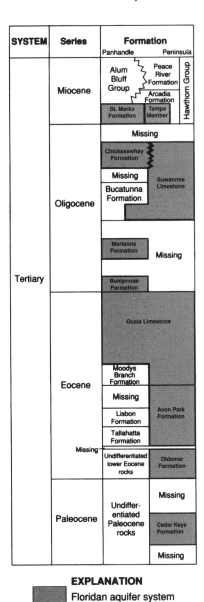

Figure 6.29. Stratigraphic units included in the Floridan aquifer system (Miller 1986).

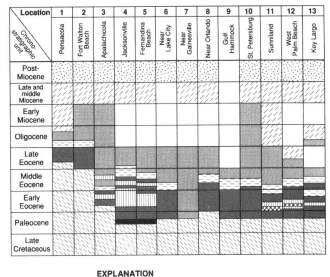

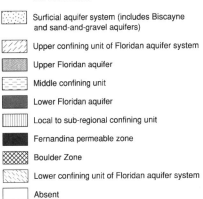

Figure 6.30. Relation of time-stratigraphic units to subdivisions of the Floridan aquifer system (Miller 1986). Location numbers refer to figure 6.28.

mostly consists of the lower part of the Avon Park Formation, the Oldsmar Formation, and the upper part of the Cedar Keys Formation.

In panhandle Florida, the Late Eocene Ocala Limestone is part of the Lower Floridan aquifer in places (fig. 6.30). Because it is deeply buried in most places and commonly contains saltwater, the Lower Floridan aquifer is not well known; however, an important, highly permeable zone occurs in the Lower Floridan in two areas. One of these is in the Fernandina Beach–Jacksonville area, and is called the Fernandina permeable zone (Krause and Randolph 1989). This zone is the source of a large volume of fresh to brackish water that leaks upward through the middle confining unit and provides part of the recharge to the Upper Floridan aquifer. The second area, southeast of a line from Cocoa Beach, in Brevard County, to Port Charlotte, in Charlotte County (Miller 1986), contains an extremely permeable and cavernous zone called the "Boulder Zone" (see chapter 4). The Boulder Zone is overlain by a confining unit and everywhere contains saltwater. The zone is used to receive treated sewage and other wastes emplaced by large-diameter injection wells, chiefly in the Miami–West Palm Beach area.

The Floridan aquifer system ranges from less than 200 feet thick in places along the Alabama/Florida border to more than 3,400 feet thick locally in central and southern peninsular Florida (fig. 6.31). The mapped thickness shown in figure 6.31 is the total for the upper and lower Floridan aquifers and all confining units within the aquifer system, including those areas where the aquifer system contains saltwater. Some of the thinning and thickening trends shown on figure 6.31 are related to geologic structures. For example, the aquifer system is thick in Gulf and Franklin counties along or near the axis of the Southwest Georgia Embayment, and in Nassau and Duval counties on the flank of the Southeast Georgia Embayment. The aquifer system thins along the axis of the Peninsular Arch of Florida (see figure 5.1). In western panhandle Florida, the aquifer system thins because of a facies change from highly permeable carbonate rocks to low-permeability clastic rocks. In Leon and Gadsden counties, the aquifer system is only about 600 feet thick (fig. 6.31), because gypsum has filled the original pore space in the rocks that compose the Lower Floridan aquifer. These low-permeability, gypsiferous rocks grade downward without a break into the siliciclastic rocks that form the lower confining unit of the Floridan aquifer system.

Some of the complexity of the Floridan aquifer system is shown on a cross section drawn from northwest to southeast approximately along the center of the Florida peninsula (fig. 6.32). The entire aquifer system thins over the Peninsular Arch, and thickens greatly to the southeast. The number of regional confining units increases to the southeast: for example, near the southeastern end of

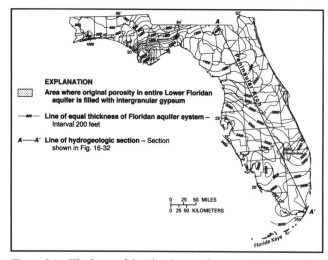

Figure 6.31. Thickness of the Floridan aquifer system (Miller 1986).

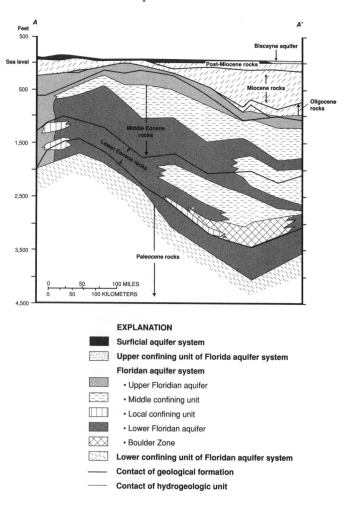

Figure 6.32. Hydrogeologic cross-section from Columbia County to Dade County (Miller 1990).

the section there are three regional confining units separating four permeable zones within the aquifer system. These confining units consist of low-permeability carbonate rocks, and the shallowest of the three separates the Upper Floridan aquifer from the Lower Floridan aquifer, whereas the two deeper confining units are within the Lower Floridan. The Boulder Zone and its overlying confining unit also are shown in the cross section. The regional confining units pinch out to the northwest and all are absent at the extreme northwest end of the cross section, where all the carbonate rocks are part of the Upper Floridan aquifer; however, local confining units may be present in the Upper Floridan.

The carbonate rocks of the Floridan aquifer system are generally soluble in slightly acidic groundwater. Dissolution begins as the water moves through original pore spaces and along joints and bedding planes in the rocks, and it may proceed until some of these openings are enlarged to caverns. These enlarged solution openings and other types of karst features form more readily where groundwater circulation is most vigorous. In the case of the Floridan aquifer system, groundwater enters, flows through, and discharges from the aquifer system more rapidly where the clayey upper confining unit of the system is absent or less than 100 feet thick (fig. 6.33). Sinkholes (see chapter 13) are more common where the aquifer system is unconfined or thinly confined. The distribution of large springs that issue from the Floridan aquifer system (fig. 6.34) is likewise directly related to the thickness of the upper confining unit, and in Florida all of the first-magnitude springs—those discharging 100 cubic feet per second or more—issue from the Floridan aquifer system (Rosenau et al. 1977). Comparison of figures 6.33 and 6.34 shows that these springs occur where the upper confining unit of the Floridan is thin or absent. In contrast, where the upper confining unit is thick, little dissolution of the aquifer system has occurred, except for deeply buried zones of paleokarst, such as the Boulder Zone.

The sinkholes, springs, and other karst features in central and northern Florida are the result of several intervals of dissolution. Some of the caverns from which first-magnitude springs issue are thought to be very old, and to have formed by downward-percolating water when the hydraulic head in the Floridan aquifer system was much lower than at present. It is possible that some of the major karst features are reactivated, and some may predate the clayey upper confining unit of the Floridan, which is composed mostly of sediments of the Hawthorn Group (see chapters 1, 4, 5, and 13).

The distribution of transmissivity values estimated for the Upper Floridan aquifer is shown in figure 6.35. All areas having transmissivity values greater than 1 million ft^2/d, and most of the areas having values greater than 250,000 ft^2/d, are where the aquifer system is unconfined or the upper confining unit is less than 100 feet thick (fig. 6.33). Where the aquifer system is thickly confined, transmissivity values are lower. However, transmissivity values also are directly related to the thickness of the aquifer system, and where it is thin in western panhandle Florida transmissivity values are less than 10,000 ft^2/d (fig. 6.35).

The major features of groundwater flow in the Floridan aquifer system are shown on a map of the potentiometric surface of the Upper Floridan aquifer (fig. 6.36). In May 1980 there were four prominent high areas on the potentiometric surface of the Upper Floridan, two in central peninsular Florida from which water moved outward in all directions, and two in panhandle Florida from which water moved toward the Gulf of Mexico. Water also moved into the aquifer as lateral flow from

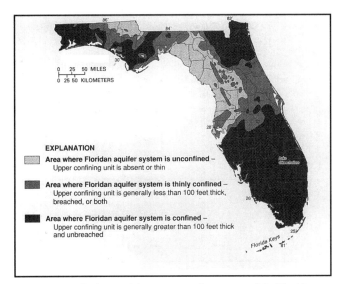

Figure 6.33. Thickness of the upper confining unit of the Floridan aquifer system (Miller 1986).

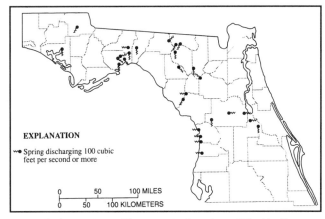

Figure 6.34. Locations of first-magnitude springs of Florida (Rosenau et al. 1977).

adjacent areas in Alabama and Georgia that are not shown in figure 6.36. Most of the water moved generally coastward or toward major streams and springs, which are low areas on the potentiometric surface; however, some of the water moved toward major pumping centers, including those at Fort Walton Beach, Fernandina Beach, Daytona Beach, New Smyrna Beach, and Clearwater.

Intense pumping in some areas has lowered the potentiometric surface to the point that the prepumping, coastward groundwater flow has been reversed, and water is moving from coastal areas toward pumping centers. This reversal creates the potential for saltwater encroachment. Pumping has lowered the potentiometric surface of the Upper Floridan aquifer more than 80 feet at Fort Walton Beach, where withdrawals are mainly for public supply, and at Fernandina Beach, where withdrawals are mostly for industrial (paper mill) uses (fig. 6.37). Withdrawals in west-central peninsular Florida for combined public-supply, agricultural, and mining uses have lowered the potentiometric surface by 10 to 50 feet or more over a broad area (fig. 6.37). Water-level data collected during 1985 (Bush et al. 1987) showed little change in the potentiometric surface from the 1980 maps presented here. Although large volumes of water were pumped from the Floridan aquifer system during 1980, springflow and discharge to streams remained the dominant means of discharge from the aquifer system.

Development of water from the Floridan aquifer system began in Florida during the 1880s, when wells were constructed for municipal supply in Jacksonville. By the early 1900s, several other Florida cities began withdrawing water from the Floridan. Many of the early wells flowed, because heads in the aquifer system were high, but the heads declined with increased withdrawals, requiring installation of pumps. During the 1930s, withdrawals for phosphate mining, citrus processing, and pulp and paper manufacture began, and these withdrawals soon became large. During 1985, withdrawals from the Floridan aquifer system in Florida were more than 2,500 Mgal/d (fig. 6.38). About 47% (1,181

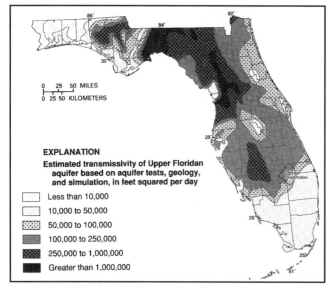

Figure 6.35. Estimated transmissivity of the Upper Floridan aquifer (Bush and Johnston 1988).

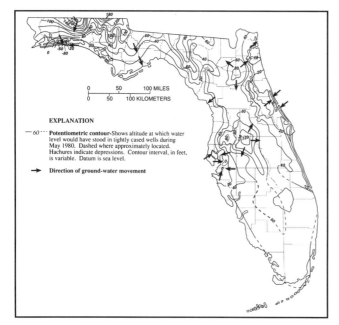

Figure 6.36. Potentiometric surface of the Upper Floridan aquifer in May 1980 (Bush and Johnston 1988).

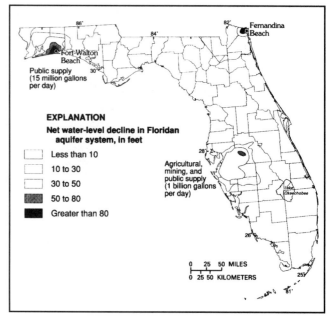

Figure 6.37. Net decline between estimated predevelopment potentiometric surface and observed potentiometric surface in May 1980 for the Upper Floridan aquifer (Bush and Johnston 1988).

Mgal/d) of the water was withdrawn for agricultural use, and about 28% (693 Mgal/d) was withdrawn for public-supply use. Industrial and mining uses (including about 15 Mgal/d withdrawn for thermoelectric-power use) accounted for withdrawals of about 23% (571 Mgal/d). The remainder of some 2% (60 Mgal/d) was pumped for domestic and commercial use. Pumpage from the Floridan aquifer system is not evenly distributed. A map of withdrawals from the aquifer system by county (fig. 6.39) shows that withdrawals during 1985 were much greater in central peninsular Florida than elsewhere. Polk County was the largest user of groundwater from the Floridan aquifer system, followed by Duval, Hillsborough, and Orange counties. Withdrawals were least in sparsely settled counties in the panhandle and Big Bend areas, and in southern Florida, where the Floridan aquifer system contains brackish water or saltwater.

Injection wells are used in several places in Florida to dispose of municipal and industrial wastes (fig. 6.40). The wastes are mostly injected into permeable zones in the Lower Floridan aquifer, such as the Boulder Zone, where the aquifer contains saltwater. About 208 Mgal/d was injected by these wells during 1988 (Miller 1990), about 97% of which was treated municipal sewage. Some wells, such as those shown in Polk County (fig. 6.40), inject wastes into permeable zones below the Floridan aquifer system (the entire aquifer system contains freshwater in Polk County). Drainage wells (not shown in fig. 6.40) have been used in parts of Florida since the early 1900s to dispose of excess storm runoff. Many of these wells, particularly in and near Orlando, are completed in the Upper Floridan aquifer. Public-supply wells in Orlando are drilled through the middle confining unit and produce water from the Lower Floridan aquifer.

Because of reactions of water with minerals in some places, and mixing of waters in other places, different ions dominate the dissolved-solids content of groundwaters in the Upper Floridan aquifer in different areas (fig. 6.41). The classification of waters shown in figure 6.41 is based on the dominant cation(s) and anion(s) in the water. For example, a sodium chloride water is one in which sodium ions account for more than 50% of the total cations in the water and chloride ions account for more than 50% of the total anions. A sodium chloride classification does not necessarily mean that the water is saltwater (although saltwater is included in this water type)—it merely indicates that these ions are dominant. In recharge areas, the water is a calcium bicarbonate type, generally having dissolved-solids concentrations of less than 500 mg/L (Sprinkle 1989).

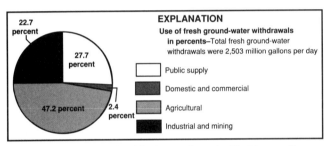

Figure 6.38. Withdrawals of freshwater from the Floridan aquifer system, by category of use, 1985 (U.S. Geological Survey 1990).

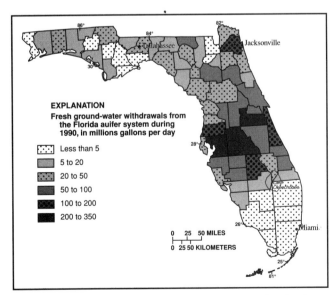

Figure 6.39. Estimated withdrawals of freshwater from the Floridan aquifer system, by county, 1990. No Floridan withdrawals in unpatterned counties (Marella 1992).

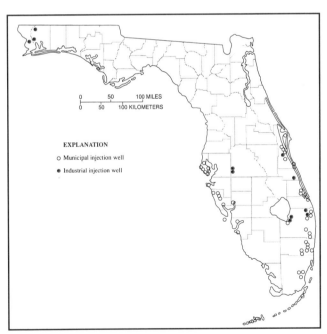

Figure 6.40. Locations of injection wells in Florida, 1988 (Miller 1990).

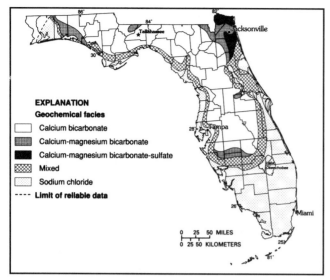

Figure 6.41. Hydrochemical-facies classification of waters in the Upper Floridan aquifer (Sprinkle 1989).

Locally, as in Gadsden County, southeast Lake County and adjacent areas, and across a wide area of northeast Florida, dissolution of dolomite adds sufficient magnesium to the water to produce a calcium-magnesium bicarbonate facies (fig. 6.41). In northern Nassau, eastern Duval, and northern St. Johns counties, gypsum dissolution adds sulfate, and results in a calcium-magnesium bicarbonate-sulfate facies. Sufficient gypsum is present in southern Hardee County and adjacent areas to produce a calcium-magnesium sulfate facies. In the area designated as "mixed" on figure 6.41, small amounts of saline water are mixed with the freshwater, resulting in a facies with approximately equal proportions of all major chemical constituents (Sprinkle 1989). As the amount of saline water in the aquifer increases seaward, a sodium chloride facies results. The general progression of hydrochemical facies in the water as it moves downgradient is from a calcium bicarbonate type to a mixed-water type, and finally to a sodium chloride type. Sprinkle (1989) postulated ion exchange to account for the hydrochemical facies mapped in panhandle Florida, where the Upper Floridan aquifer includes little dolomite. The area of sodium chloride water inland along the St. Johns River might be a result of upward migration of mineralized water along faults, or it might represent unflushed seawater.

Summary

Florida's abundant fresh groundwater resources are in five principal aquifers or aquifer systems, four of which are completely or partially exposed at the land surface and thus receive recharge directly from precipitation on aquifer outcrop areas. A fifth principal aquifer system, the intermediate, is completely buried and is recharged by downward leakage. The undifferentiated surficial aquifer system and one of its named parts, the sand and gravel aquifer, are composed largely of siliciclastic rocks which are relatively insoluble; thus, water from these aquifers is not highly mineralized. The Biscayne aquifer consists largely of carbonate rocks, and the Floridan aquifer system is composed entirely of carbonate rocks, except for minor evaporites. Water in the carbonate-rock aquifers is more highly mineralized than water in the aquifers composed of siliciclastic rocks. The intermediate aquifer system contains both siliciclastic and carbonate rocks, but the quality of the water in this aquifer system is more like that of the carbonate-rock aquifers. Low-permeability confining units separate the principal aquifers and aquifer systems in most places.

The permeability of the aquifers in siliciclastic rocks is determined chiefly by the texture and degree of sorting of the original sediment. In contrast, the permeability of the aquifers in carbonate rocks is greatly increased where dissolution has enlarged primary openings in the rock; however, filling of original pore space by dolomite or evaporite minerals that have been dissolved elsewhere, transported by groundwater, and reprecipitated can decrease the permeability of carbonate-rock aquifers in places.

More than 4 billion gallons of water were withdrawn from all aquifers in Florida during 1985 for all uses. The Floridan aquifer system produced more than half of the withdrawn water. The Biscayne aquifer, the principal source of water for the large urban areas along Florida's southeastern coast, is the second-most heavily pumped aquifer.

The natural quality of waters in the shallow parts of all the principal aquifers permits use of these waters for most purposes. Because all the aquifers contain saltwater in coastal areas, saltwater encroachment can occur where the aquifers are heavily pumped near the coasts. The aquifers exposed at the land surface, particularly the highly permeable carbonate-rock aquifers, are susceptible to contamination from human activities.

7

The Marine Invertebrate Fossil Record of Florida

Douglas S. Jones

Florida is recognized for its rich record of Cenozoic fossils. The shallow-water marine strata exposed across the state include some of the most densely fossiliferous deposits known. The abundance, diversity, and spectacular preservation of these fossils have made them the objects of study for almost two centuries. Florida's unique geographic position—with modern carbonate (reef) environments at its southern extremity, and subtropical to transitional climatic regimes—has resulted in tremendous faunal diversity throughout the Tertiary and Quaternary. An abundance of vertebrate fossils (see chapter 8) adds to the overall paleobiologic significance.

The general features of Florida's fossil record have been appreciated for many years. In fact, paleontologic assessments of fossil marine invertebrates were important components of the earliest geologic investigations undertaken in Florida (Whiting 1839; Allen 1846; Conrad 1846a, 1846b; and others). Paleontology even played a fundamental role in shaping the initial hypotheses concerning the geologic origin and formation of the Florida peninsula. Yet, despite the long history of study, many significant paleontologic questions remain unanswered.

Early Ideas: Fossils and the Origin of Florida

Before direct geologic observations, it was widely supposed in the early nineteenth century that Florida represented a southward extension of Eocene strata known in Alabama and Georgia (LeConte 1857). This view was reinforced by the observations of Conrad (1846a, 1846b), who made the first clear attempt to determine the age of "the Florida limestone" when he identified "Upper Eocene" limestones at Tampa Bay on the basis of their fossil content; he speculated that this rock extended throughout the peninsula. Based upon observations along all coasts, Conrad reported that the eastern and western shores were covered with a Pleistocene formation composed largely of recent shells, and that the elevated deposits encountered along much of Florida's coastal margin, including the Keys, were slightly older, of post-Pliocene age.

Conrad's initial observations were supported by those of other early investigators (Bailey 1850, 1851; Tuomey 1851; and others). However, the prevailing view would change by mid-century, following a visit by the distinguished natural historian and Harvard professor Louis Agassiz to the reefs and keys of southern Florida during the winter of 1851 with his protégé, Joseph LeConte. Noting that the coral species of the reef tract were identical to those forming the bedrock of the Keys and southeastern coast of the mainland, Agassiz theorized that the Florida peninsula was of comparatively recent origin, and that it had extended southward into the ocean by successive annexations of consecutive series of coral reefs, the most recent of which were still actively growing (Agassiz 1851). Examination of fossiliferous (coralline) rocks from Florida's eastern and southern coasts by Agassiz (1880) convinced him that "one and the same geological formation, identical in lithological character with the reef of Florida, and presenting only slight local modifications, extends all over the peninsula of Florida, at least as far north as St. Augustine." LeConte (1857) endorsed Agassiz's theory, but believed that the coral-growth agency alone was insufficient to account for the development of the peninsula without foundational sedimentary materials supplied by the Gulf Stream to produce a bank or platform upon which the corals could grow. The hypotheses of Agassiz and LeConte prevailed for several decades, obtaining such influence that the official geologic map of the United States accompanying the publication of the Ninth Census in 1872 represented the whole of Florida as alluvial (Smith 1881). Eventually these ideas succumbed to modification (LeConte 1883) as the coral-extension theory was limited to the southernmost (youngest) part of the peninsula. Evidence from investigations throughout the state reaffirmed the older opinions espoused by Conrad, Tuomey, and others concerning the geologic age and nature of the rocks that underlie the greater part of the peninsula (Smith 1881; Dall 1887; Heilprin 1887).

A Modern Perspective

Whereas early workers such as Conrad first drew attention to Florida fossils, it was not until the late nineteenth century that Angelo Heilprin (1887)—and especially William Healey Dall, with his *Contributions to the Tertiary Fauna of Florida* (1890–1903)—finally brought the extraordinary fossil deposits of the state (particularly the southern part) to the attention of the scientific community. These studies formed the foundation for a growing number of paleontologic investigations in the twentieth century that greatly expanded our understanding of Florida's fossil record.

The proliferation of studies involving Florida fossils in this century has more or less paralleled population growth and development because natural exposures of fossil-bearing strata are limited in Florida with its low topographic relief and heavy vegetation. In the absence of major tectonic activity, most sedimentary units are nearly flat-lying; hence, natural outcrops are for the most part restricted to riverbanks, sinkholes in the karst terrain, ancient beach ridges, and the like. For these reasons, exposures produced by human activities (e.g. quarries and mines, construction sites, canals, road cuts) play a critical role in Florida paleontology.

New fossil discoveries continue to be made throughout Florida by paleontologists from both professional and amateur ranks (Brown 1988; and others). These specimens augment large collections of Florida fossils housed in such institutions as the U.S. National Museum of Natural History in Washington, D.C., the Academy of Natural Sciences in Philadelphia, the Paleontological Research Institution in Ithaca, N.Y., and the Florida Museum of Natural History in Gainesville. An appreciation of the major patterns of ancient life in Florida through space and time is emerging. These patterns, continually refined as new data and reinterpretations appear, are the subject of the remainder of this chapter.

The Paleozoic

Paleontologists think of Florida primarily in terms of its abundant Cenozoic fossils, organisms not too unlike those living around its coasts today. This is not surprising, as the fossil-bearing rocks in the state are fairly young, of Tertiary or Quaternary age. Much less familiar are the older fossils known only from cores and cuttings from deep boreholes that penetrate Paleozoic and Mesozoic strata in the subsurface; yet, the faunas and floras obtained from these wells are diverse, and have proved to be diagnostic in age determinations and in biogeographic contexts.

A DIVERSE FAUNA FROM THE FLORIDA BASEMENT

The first speculation that Paleozoic rocks existed in the Florida subsurface appeared during the late 1920s and 1930s (Gunter 1928; Campbell 1939a); however, no fossils were recovered until improvement of drilling techniques during the 1940s provided cores with enough fossil material to formulate a tentative Paleozoic stratigraphic framework and to develop an outline of the regional setting (Applin 1951). Early attempts at correlating the Paleozoic rocks (Bridge and Berdan 1952) have been revised in light of subsequent refinements in fossil identifications and more recent data from microfossils. Of more than fifty wells penetrating lower Paleozoic sedimentary rocks in Florida, eighteen have yielded either megafossils or acid-resistant microfossils (Pojeta et al. 1976).

Florida's Paleozoic sedimentary rocks are basically confined to two regions, a small area in the panhandle near the junction of Florida, Georgia, and Alabama, and a larger area in the northern part of the peninsula, north of a line connecting Old Tampa Bay in the southwest to a point between Jacksonville and St. Augustine in the northeast (see chapter 2). These strata consist of

Figure 7.1. Photograph of linguloid brachiopod (UF 67807) in dark-gray, micaceous Paleozoic siltstone recovered in core from Alachua County at approximately 800 m depth. Scale has centimeter divisions. Specimens with UF numbers in this and subsequent figures are deposited in the Invertebrate Paleontology Collection at the Florida Museum of Natural History, University of Florida, Gainesville.

quartzitic sandstones, dark-gray, often micaceous shales, and red and gray siltstones, ranging in age from Early Ordovician to Middle Devonian. Fossils are scarce in the lower part of the sequence, and only a few wells have yielded moderately diverse faunas (Pojeta et al. 1976), which usually are in the part of the section dated as Silurian or Devonian.

The oldest and most widely distributed Paleozoic sedimentary unit consists of quartzitic sandstones and highly micaceous shales characterized by inarticulate brachiopods (fig. 7.1) and vertical *Skolithos* borings (Pojeta et al. 1976). Based on the occurrence of graptolites (*Didymograptus deflexus* and ?*D. protoindentus*) and a linguloid brachiopod identified by Howell and Richards (1949) as *Lingulepis floridaensis*, this unit appears to be Early Ordovician (Pojeta et al. 1976); it is interesting that neither of these graptolites is known from elsewhere in North America, but they are known from localities in Europe (Berry, in Pojeta et al. 1976).

Overlying the Lower Ordovician rocks are black shales with some interbedded sandstones that contain a diverse group of Middle to Late Ordovician fossils. The only trilobite from Florida, *Plaesiacomia exsul*, was recovered from these beds at a depth of 1,571 to 1,573 m in Madison County; it is considered to be of Llanvirnian-Llandeilian age, or early Middle Ordovician (Whittington 1953, 1992). Occurring with this trilobite are conularids and numerous small brachiopods, most likely obolids. Whittington and Hughes (1972) consider *P. exsul* to be a part of the *Selenopeltis* faunal province, which also includes northern Africa and eastern Europe. Conodonts (*Drepanodus* sp.), phosphatic brachiopods, and chitinozoans (Andress et al. 1969) from other wells also help to constrain the age of these rocks.

A thick sequence of black shales dated by fossils as Silurian to Devonian presumably overlies the Ordovician section in peninsular Florida; however, no wells are known with certainty to pass from the Silurian-Devonian shales into the Ordovician shales, so that the nature of this transition is unknown (Pojeta et al. 1976). In addition to an assemblage of acid-resistant microfossils identified by Schopf (1959), Goldstein et al. (1969) described a robust suite of chitinozoans from some wells that Cramer (1973) suggested were in mid- to latest Silurian rocks. The Florida palynomorph spectra are quite similar to age-equivalent material from Portuguese Guinea (West Africa) and from North Africa (Cramer 1973). Also known from this interval are arthropod megafossils, including eurypterids and phyllocarids (Kjellesvig-Waering 1950, 1955) and ostracodes (Swartz 1949). Pojeta et al. (1976) described an assemblage of Late Silurian to Middle Devonian shallow-water marine pelecypods from several wells, many of which show affinities to pelecypods of similar age in Bohemia and Poland (and, to a lesser extent, in North Africa and Turkey). These authors also mentioned a few additional macrofossils from these shales, including rhynchonellid brachiopods, orthoconic cephalopods, crinoids, hyolithids, *Tentaculites*, and a bellerophontacean gastropod.

THE GONDWANAN CONNECTION

Although the Paleozoic fauna and flora from the Florida subsurface are modest, those taxa that can be identified with some precision appear to fit an interesting paleobiogeographic pattern. Several groups of fossils (e.g. mollusks, arthropods, graptolites, palynomorphs) share affinities with fossil assemblages of similar age in South America, northwestern Africa, and Eurasia—not the southeastern U.S. (Whittington and Hughes 1972; Cramer 1973; Pojeta et al. 1976; Whittington 1992). This observation has led paleontologists and geologists to suggest that Florida was biotically isolated from North America during the early Paleozoic, perhaps at a relatively high paleolatitude.

Paleomagnetic and geochronologic evidence from Paleozoic core material recovered in north-central peninsular Florida supports this faunal interpretation and favors a Gondwanan origin for early Paleozoic Florida (Opdyke et al. 1987). Accretion of this exotic terrane, known as the "Suwannee terrane," occurred during the collision between Laurentia and Gondwana during the late Paleozoic, probably along a suture zone corresponding to the Brunswick-Altamaha magnetic anomaly in southern Alabama and Georgia (Dallmeyer 1987). Tectonic reconstructions indicate that, by the onset of early Mesozoic rifting associated with the breakup of Pangaea, the Paleozoic sedimentary sequence was already welded to North America as an integral part of the Florida basement complex (Ross and Scotese 1988; Scotese et al. 1988; Sheridan et al. 1988; see chapters 2 and 3).

The Mesozoic

Overlying the Paleozoic sequence in some places, and igneous and metamorphic rocks of the basement complex in others, is a thick, lithologically variable section of Mesozoic sedimentary rocks (see chapters 4 and 11). The oldest of these (Late Triassic to Early Jurassic) are largely nonfossiliferous, terrigenous, clastic rocks which presumably represent the filling of graben basins as Pangaea rifted apart (Walper et al. 1979; Salvador 1987). Jurassic strata in the Florida subsurface are largely confined to the panhandle and to the west-central part of the peninsula

(Miller 1982; Salvador 1987). Deposition began with evaporites during the late part of the Middle Jurassic, and, except for the Louann Salt and the Upper Jurassic Smackover Formation (transgressive marine limestone), the Jurassic sequence in Florida is predominantly fluvial and sparsely fossiliferous. Of some 135 wells that penetrated the sequence in western Florida, only two yielded fossils (spores) (Miller 1982).

CARBONATE ENVIRONMENTS AND FAUNAS

Deposition of the Smackover Formation during the Late Jurassic (Oxfordian) marked the first widespread marine transgression of the northern Gulf Coast (Walper et al. 1979). The carbonates of this unit contain skeletal limestones and late Oxfordian ammonites (Salvador 1987). In addition, small sponge, algal, and coral patch reefs have been recognized in the upper part of the unit (Baria et al. 1982).

Jurassic sedimentation in the panhandle of Florida came to a close in the Tithonian with the deposition of the Cotton Valley Group, a thick wedge of coarse-grained, terrigenous clastics that grades offshore into shallow-water marine clastic rocks. In southern Florida, however, several wells have penetrated a thick section of arkosic sandstones and shales overlain by evaporites, skeletal limestones, and dolomites. This unit directly overlies the basement rocks of the region and was named the Fort Pierce Formation by Applin and Applin (1965), who tentatively assigned a latest Jurassic age to the lowermost part; more recently, Applegate et al. (1981) referred to this unit as the Wood River Formation. A palynologic study (cited in Salvador 1987) indicates that only the lowermost 30 to 50 m may be of Tithonian age. Applin and Applin (1965) reported that foraminifer assemblages (miliolids, valvulinids, lituolids) are the most abundant fossils (fig. 7.2), with lesser amounts of dasyclad algae, ostracodes, and mollusk fragments. Parts of a Jurassic stromatoporoid and fragmental specimens of rudists and oysters with Cretaceous affinities also are known from this unit.

CRETACEOUS CARBONATE-PLATFORM FAUNAS

The Lower Cretaceous sedimentary sequence (Coahuilan Series and Comanchean Series of the Gulf Coast) thickens to the west and south across the Florida Platform (Applin and Applin 1944), where carbonate accumulation continued from the Late Jurassic into the Cretaceous, producing a thick (more than 2,500 m), uninterrupted sequence in southern Florida known as the Marquesas Supergroup (Walper et al. 1979; Braunstein et al. 1988). A composite series of carbonate shelf-margin reef buildups formed the Early Cretaceous shelf edge across the northern margin of the Gulf (e.g. the rudist-dominated Edwards, or Stuart City, reef trends of Texas), and this feature extends as a submarine scarp to form the western edge of the Florida Platform (Wu et al. 1990b). Drilling of the platform near the margin has encountered shallow-water marine foraminiferal (miliolid) limestones; however, the expected rudist-dominated shelf-margin facies were not found along the West Florida Escarpment (probably because of erosional retreat), although rudist-bearing limestones were obtained from talus below the escarpment (Winker and Buffler 1988). In contrast, rudistid reefs have been reported along the eastern margin of the Florida-Bahamas Platform (Dillon et al. 1988). Behind this shelf-margin buildup, shallow-water marine carbonate rocks, reflecting restricted conditions in the platform interior, grade landward (generally northward) into nonfossiliferous clastic rocks of alluvial and deltaic origin. The disappearance of framework builders (primarily rudists) that produced the rimmed platform margin during the early to mid-Cretaceous has been related to changes in nutrient availability that favored pelagic carbonate-fixers, thus leading to drowning of the margin and ramp development in western Florida during the Late Cretaceous (Winker and Buffler 1988; see chapter 11).

Paleontologic investigations show that the sequence of

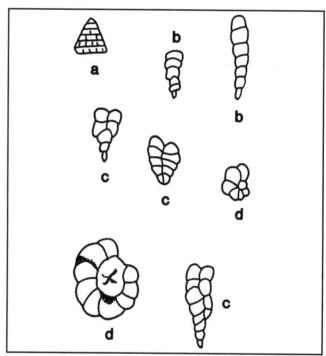

Figure 7.2. Line drawings of representative foraminifers from cores penetrating the Upper Jurassic to Lower Cretaceous Wood River (or Fort Pierce) Formation of peninsular Florida. A: *Coskinolina* sp. B: Uniserial forms. C: Textularian forms. D: *Pseudocyclammina* sp. All ×40–60.

microfossil assemblages in the Lower Cretaceous of Florida closely resembles the sequence in the standard outcropping and subsurface stratigraphic sections of age-equivalent rocks in Texas (Applin and Applin 1965). Diagnostic foraminifers that distinguish the Trinity, Fredericksburg, and Washita groups in Texas provide the basis for correlation of the approximately synchronous units of the Marquesas Supergroup in Florida. For example, the microfauna of the marine beds of the Trinity in Florida is characterized by several genera of larger benthic foraminifers—*Orbitolina, Dictyoconus, Choffatella,* and *Pseudocyclammina* (fig. 7.3)—that are also used to subdivide Trinity rocks in Texas and Arizona.

Whereas miliolids and several groups of smaller benthic foraminifers are most common in the Lower Cretaceous rocks of Florida, planktic foraminifers are rarely found in these shallow-water carbonates. Specimens of *Globigerina* spp. from limestones of Fredricksburg age (Albian) in wells near the southern end of the peninsula are a noted exception. Identifiable macrofossil remains are also rare. Among the few known examples are Early Cretaceous ammonites (*Dufrenoya texana*), a pectinid bivalve (*Neithea*), and brachiopods (*Kingena, Cyclothyris*) from a deep well on Key Largo; these forms are also known from surface exposures in Texas, and from the subsurface in Texas, Louisiana, and Arkansas (Applin and Applin 1965).

According to Haq et al. (1988), the late Early Cretaceous was characterized by a rapid fall of sea level, soon followed by two major rises in sea level (fig. 4.8). These eustatic changes resulted in extensive unconformities which mark the mid-Cretaceous worldwide where carbonate platforms, including the Florida Platform, were drowned (see chapters 4 and 11). The fauna of the Upper Cretaceous sedimentary sequence of Florida differs from that of Lower Cretaceous rocks, perhaps most notably by the absence of carbonate framework builders (rudists) and associated organisms from the platform-margin facies. While shallow-water carbonate production ceased on the west Florida margin, it continued over the interior of the platform.

The lowermost unit of the Gulfian Series (inside front cover), the Atkinson Formation (equivalent to the littoral-nonmarine Tuscaloosa Group in Alabama and Mississippi), contains a Woodbine (Cenomanian) microfauna in the lower, clastic part, and an Eagle Ford (Turonian) microfauna toward the top (clastics and carbonates). Miliolids and other benthic foraminifers (*Trocholina, Cuneolina*) and ostracodes are particularly common. Tests of *Globigerina* and some other planktic foraminifers are locally common; however, they probably do not represent deeper-water conditions, but rather episodic events involving storm-driven waters (Applin and Applin 1967). Macrofossil remains include abundant fragments of bryozoans, gastropods, bivalves (fig. 7.4), and echinoids (e.g. *Porpitella micra*).

Above the Atkinson Formation is a carbonate sequence (Pine Key Formation and Lawson Limestone) with clear Austin, Taylor, and Navarro affinities (Coniacian-Maastrichtian). Although microlithologic differences exist, the same diagnostic foraminifer assemblages used in Texas to distinguish Woodbine-Navarro units are used in the Florida subsurface (Applin and Applin 1967); however, the diagnostic foraminifers of the latest Cretaceous Lawson Limestone are an interesting exception, more closely resembling assemblages in the Maastrichtian of Cuba than Texas. Planktic foraminifers (e.g. *Globigerina, Gumbelina, Globotruncana*), the principal component of the microfauna of the Austin beds, are numerous throughout the sequence, and their diversity generally increases up-section. Prisms of *Inoceramus* and other mollusk fragments (*Ostrea* sp.) are also very common, as is the tiny pelagic crinoid

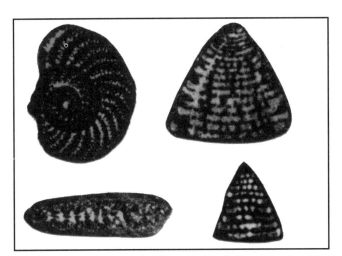

Figure 7.3. Photomicrographs of sectioned Early Cretaceous benthic foraminifers from Florida core material: *Choffatella decipiens,* a lituolid (left, top and bottom); *Dictyoconus floridanus* (top right) and *Coskinoloides texanus* (bottom right), orbitolinids. All ×30.

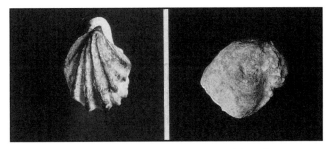

Figure 7.4. Late Cretaceous bivalves reconstructed from fragments recovered in drill cores from central Florida *Inoceramus* sp., 20 mm (left); *Exogyra* sp., 25 mm (right).

Saccocoma from the lower Lawson Limestone (Applin and Applin 1967).

The upper member of the Lawson Limestone (inside front cover) is chiefly an algal-rudist biostrome, composed of calcareous algae (dasycladaceans) and rudistid fragments. This early- to mid-Maastrichtian unit caps the Mesozoic section in Florida and is disconformably overlain by the Paleocene Cedar Keys Formation, the lowermost unit of the Cenozoic. Available evidence indicates that latest Cretaceous and earliest Tertiary faunas known from more complete Cretaceous/Tertiary boundary sections elsewhere in the world are not represented in Florida.

The Cenozoic

Tertiary faunas are readily distinguished from those of the Cretaceous, largely because of the major biotic crisis and mass-extinction episode that marked the Cretaceous/Tertiary boundary. Although the cause, or causes, of this extinction event are still debated (exogenic vs. endogenic; Officer et al. 1992), the effect upon faunas and floras is widely acknowledged. In Florida the transition from Mesozoic to Cenozoic faunas is obscured (1) by a substantial hiatus in the rock record corresponding to the late Maastrichtian to the mid-Paleocene, and (2) by incomplete knowledge of the fossil biota, which is known only from cores and cuttings.

By the Late Paleocene, carbonate-platform sedimentation resumed across the peninsula with the deposition of the Cedar Keys Formation. This shallow, broad, carbonate bank has been likened to the present-day Bahamas and Campeche banks (Chen 1965). During the Paleocene and early Eocene, central evaporite lagoons extended over much of the south-central region of the peninsula, while calcareous algae and marine invertebrates produced biogenic carbonates. As limestone and evaporites accumulated over most of the platform, a thinner sequence of mixed carbonates and clastics, represented by the Midway and Wilcox groups, was deposited in the panhandle. This depositional phase persisted into the earliest Eocene. After a brief hiatus, platform carbonate accumulation continued with the Oldsmar and Avon Park formations, while in the panhandle clastic rocks of the Claiborne Group were deposited. The former units are recognized chiefly on the basis of their diagnostic benthic-foraminifer faunas (Applin and Applin 1944). Evaporite sedimentation stopped by the middle Eocene, but shallow-water carbonate-bank deposition continued across the Florida Platform through the Late Eocene, interrupted only by sea-level oscillations. The Georgia Channel System (Huddlestun 1993), including the deep Suwannee Channel across northern Florida and southern Georgia (Cheetham 1963; Chen 1965; Carter and McKinney 1992; see chapter 4), effectively isolated the carbonate platform from siliciclastic sources to the north.

EOCENE FAUNAS AND FLORAS

The dolomitized peritidal sediments of the Middle Eocene Avon Park Formation are the oldest rocks exposed in Florida. Consequently, the fauna and flora of the Avon Park and younger units are better known than any of those discussed previously, which are known only from the subsurface. The Avon Park Formation is exposed by quarry operations in the area of Citrus and Levy counties, near the core of the Ocala Arch. There, diagenesis and dolomitization have selectively left elements of what was once a fairly diverse biota. The Avon Park is characterized by abundant foraminifers such as *Coskinolina, Lituonella, Dictyoconus,* and others (Chen 1965); however, at one locality Randazzo et al. (1990) reported several species of vertebrate, invertebrate, trace, and plant fossils from beds representing shoreline to open-marine environments. Consistent with this paleoenvironmental interpretation for at least the upper part of the Avon Park are trace fossils (*Ophiomorpha* sp.) that suggest water depths of 3 to 4 m in a shoreline transition zone, and the small irregular echinoid *Neolaganum dalli*, a key taxon for this unit (fig. 7.5).

Perhaps the most spectacular fossil assemblage recovered from the Avon Park Formation involves an extensive seagrass community (Ivany et al. 1990). This occurrence is particularly significant because seagrass fossils are so rare in the fossil record. The extensive seagrass bed, dominated by the wide-bladed genera *Thalassodendron* and *Cymodocea,* contains several additional seagrass taxa, as well as a diverse array of epibionts, mollusks, and echinoderms, many of which are represented by juveniles (fig. 7.6). The presence of so many juveniles among the seagrass blades suggests that the role of seagrass communities as "nurseries" had already been established by the Eocene. Co-occurring fossils of dugongs ("sea cows") and sea turtles, both heavy grazers on seagrass, complement the paleoecologic picture. Ivany et al. (1990) noted that the Avon Park seagrass assemblage conforms to the general view that modern seagrasses and all fossil occurrences show distinct Tethyan affinities in their distribution (den Hartog 1970). This distribution is echoed by the coralline alga *Archaeolithothamnium parisiense*—also known from the Eocene of the Paris Basin and from the late Middle Eocene of Florida (Johnson and Ferris 1948)—and by many of the fossil invertebrates discussed below.

The overlying Ocala Limestone occurs throughout Florida (Puri and Vernon 1964). It is exposed in a small part of the panhandle near the Florida-Georgia-Alabama junction, and much more widely across the northwestern part of the peninsula, where it borders the Gulf of Mexico (fig. 7.7).

The rich fauna of the Ocala Limestone has resulted in numerous published descriptions of the fossils (Fischer 1951; Harris 1951; Richards and Palmer 1953; Puri 1957; Toulmin 1977; and others), despite the fact that primary aragonitic skeletal material generally is not preserved. Many fossils, including most mollusks, are

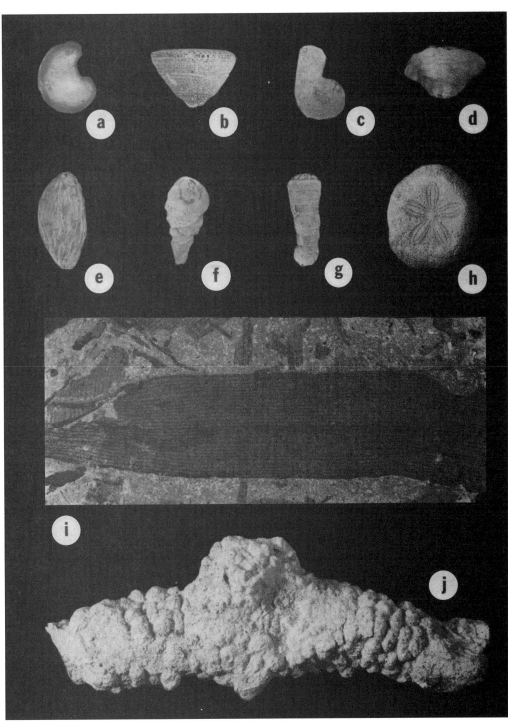

Figure 7.5. Micro- and macrofossils from the Eocene Avon Park Formation. A-G are benthic foraminifera. A: *Gunteria floridana*, ×7. B: *Dictyoconus americanus*, ×7. C: *Spirolina coyensis*, ×8. D: *Fabiania (Pseudorbitolina) cubensis*, ×7. E: *Fabularia vaughani*, ×19. F: *Cribobulimina (Valvulina) cushmani*, ×18. G: *Lituonella floridana*, ×9. H: irregular echinoid (*Neolaganum dalli*, UF 67808, 18 mm). I: seagrass blade (*Thalassodendron auricula-leporis*, 80 mm). J: trace fossil (*Ophiomorpha* sp., UF 17631, 210 mm).

Figure 7.6. Fossil seagrass blades and associated epibionts from the Eocene Avon Park Formation at the Dolime Quarry. The upper photograph shows a juvenile ophiuroid (UF 28103, 15 mm). The lower photograph shows both juvenile asteroids (UF 28135-28137, 10 mm) and ophiuroids (UF 28129-28130) on blades of *Thalassodendron* (Ivany et al. 1990).

1972; Hunter 1976; Williams et al. 1977; Zachos and Shaak 1978; Nicol and Jones 1982; McKinney and Jones 1983; Croft and Shaak 1985; McKinney and Zachos 1986; Carter and Hammack 1989; Nicol et al. 1989; and others). The Ocala Limestone exhibits minimal lithologic variability, so that attempts to subdivide this thick carbonate unit often have focused upon its fossil content.

Macrofaunal studies of the Ocala Limestone involving biogeography, paleoecology, and evolutionary relationships have lagged dramatically behind biostratigraphic investigations (Jones 1982); however, the last decade or so has witnessed an increase in such studies. Studies of the Ocala Limestone echinoid fauna (Croft and Shaak 1985), some of its unusual bivalve mollusks (Nicol and Jones 1984; Jones and Nicol 1989; Krumm and Jones 1993; and others), and even its trace fossils (Randazzo et al. 1990; Diblin et al. 1991; and others) provide a clearer picture of paleoecologic conditions and biogeographic patterns (discussed below).

Although the generic composition of Cenozoic Caribbean reefs was firmly established during the Eocene (Budd et al. 1992), no large reef trends or reef masses have been encountered in the Paleocene and Eocene carbonate section in Florida (Chen 1965; see chapter 4). This is verified by extensive fossil collecting throughout the Ocala Limestone, where fossil corals are uncommon, and has sparked debate among geologists. Some argue that reef corals were abundant during this interval, but that their aragonitic skeletons simply were not preserved; others believe that reef corals were never a major component of Florida's early Tertiary fauna, and, indeed, Paleocene and early Eocene reef corals are rare throughout the Gulf Coast and Caribbean regions (Vaughan 1900; Wells 1934). Middle to Late Eocene Caribbean coral assemblages (e.g. the Gatuncillo Formation of Panama; Budd et al. 1992) represent the first true reef-building faunas following the major extinction of reef-builders at the end

known only as molds and casts. The most common fossil groups include the benthic (larger) foraminifers, ostracodes, bryozoans, mollusks, and irregular echinoids (fig. 7.8), all of which have been used in various biostratigraphic schemes to zone the Ocala Limestone (Puri 1957; Cheetham 1963; McCullough 1969; Hoganson

Figure 7.7. Map of the outcrop area of the Ocala Limestone, and a photograph showing dissolution pipes in the Upper Ocala at a quarry near Newberry.

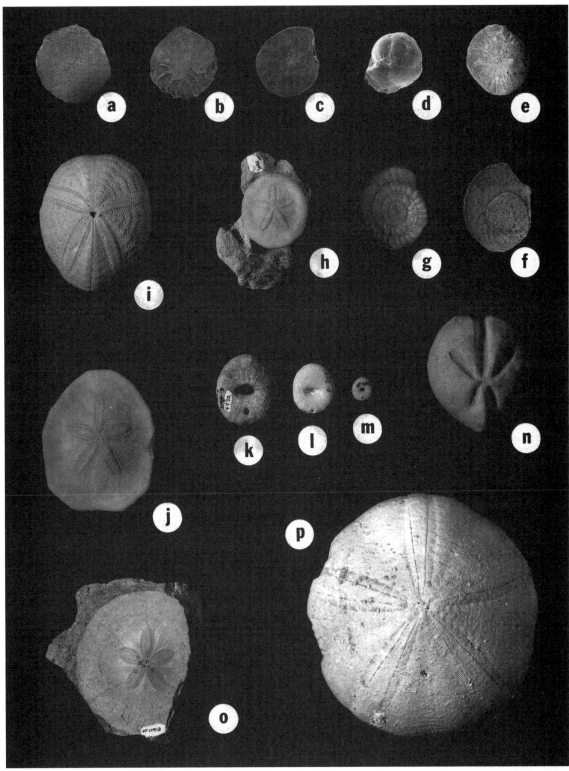

Figure 7.8a. Micro- and macrofossils from the Eocene Ocala Limestone. A-G are large benthic foraminifers. A: *Lepidocyclina ocalana*, ×5. B: *Amphistegina pinarensis cosdeni*, ×25. C: *Nummulites (Operculinoides) ocalanus*, ×11. D: *Eponides jacksonensis*, ×25. E: *Nummulites (Camerina) vanderstoki*, ×7. F: *Heterostegina ocalina*, ×7. G: *Nummulites (Camerina) vanderstoki*, ×7. H-P are irregular echinoids. H: *Neolaganum durhami* (UF 13101, 40 mm). I: *Eupatagus antillarum* (UF 3913, 65 mm). J: *Wythella eldridgei* (UF 3304, 42 mm). K: *Oligopygus wetherbyi* (UF 4930, 40 mm). L: *Oligopygus haldemani* (UF 67809, 28 mm). M: *Oligopygus phelani* (UF 1645, 13 mm). N: *Schizaster armiger* (UF 3302, 55 mm). O: *Periarchus lyelli floridanus* (UF 17913, 100 mm). P: *Amblypygus americanus* (UF 2544, 115 mm).

Figure 7.8b. Micro- and macrofossils from the Eocene Ocala Limestone. Q-T are mollusks. Q: *Amusium ocalanum* (UF 67810, 45 mm). R: *Aturia alabamensis* (UF 13061, 165 mm). S: *Xenophora* sp. (UF 18131, 45 mm). T: *Hyotissa podagrina* (UF 5826, 75 mm). U: Crab *Ocalina floridana* (UF 5263, 55 mm). V: Annelid *Rotularia vernoni* (UF 3300, 14 mm). W-ZZ are trace fossils (all *Lithoplaision ocalae*). W: 115 mm; X: 65 mm; Y: Longitudinal section 125 mm; Z: 77 mm; ZZ: 53 mm. Additional mollusks from the Ocala Limestone are illustrated in figure 7.11.

of the Cretaceous. In Florida, the distribution of coral-boring mollusks within the Ocala Limestone (fig. 7.9) suggests that the broad, shallow, carbonate-bank depositional environment envisioned by Cheetham (1963) and Chen (1965) contained a complex network of hardgrounds, as well as firm muds, and included small coral patches (Krumm and Jones 1993). Massive reefs, if they existed at that time in Florida, were confined to the eastern and southern edges of the platform margin.

Probably the most notable biogeographic pattern displayed by the Eocene faunas of Florida is the strong Tethyan influence. During the Paleogene, one of the main routes of interoceanic circulation was at low latitudes (fig. 7.10), extending from Indonesia westward across southern Asia, the Middle East, through the Mediterranean, and across a smaller Atlantic Ocean into the Caribbean, Central America, and on into the Pacific. This Tethyan Seaway linked the shallow-water marine environments of seemingly disjunct regions, as revealed by the numerous faunal elements shared between Florida and the rest of the Tethyan Province. Richards and Palmer (1953) first called attention to Tethyan mollusks in the Florida Eocene. Their work was soon followed by others, and the list of Tethyan taxa grew. Givens (1989) recently listed twenty-one Tethyan mollusk genera from the Eocene of Florida, and Nicol (1991) noted nineteen (fig. 7.11). Although mollusks have been studied most extensively with respect to their Tethyan affinities, seagrasses and calcareous algae show similar patterns (Ivany et al. 1990). Other marine invertebrate groups, such as the echinoids, might show similar patterns once their biogeographies are better constrained (Nicol 1991).

As the Eocene drew to a close, many of the prominent Tethyan elements disappeared. Species with Oligocene (Vicksburgian) affinities appeared, and the biota assumed a more provincial, tropical American aspect.

THE OLIGOCENE: DEMISE OF THE CARBONATE-PLATFORM FAUNAS

With only minor exceptions, the entire Florida Platform was submerged during the Paleocene and Eocene (see chapter 4). A great section of carbonates, more than 2,000 m thick in the central part of the peninsula (and greater to the south), had accumulated since the Late Cretaceous; however, this episode of carbonate production across the platform was to change by the

Figure 7.9. Cluster of thirty lined boreholes of the coral-boring bivalve *Lithophaga palmerae* from the Eocene Ocala Limestone. Most boreholes are still occupied and bear the septal pattern of the host coral, *Astrocoenia incrustans*. Overall height of this specimen (UF 43651) is about 20 cm.

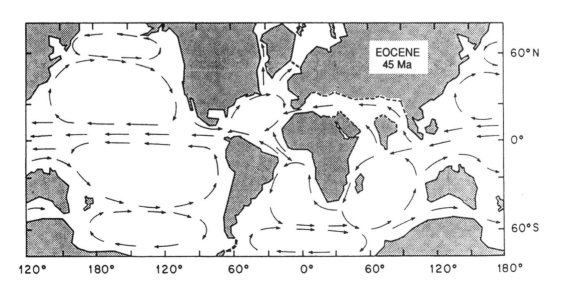

Figure 7.10. Distribution of continents, oceans, and major oceanic currents during the Middle Eocene, indicating the low-latitude, circumglobal circulation associated with the Tethyan Seaway (modified slightly from Kennett 1982).

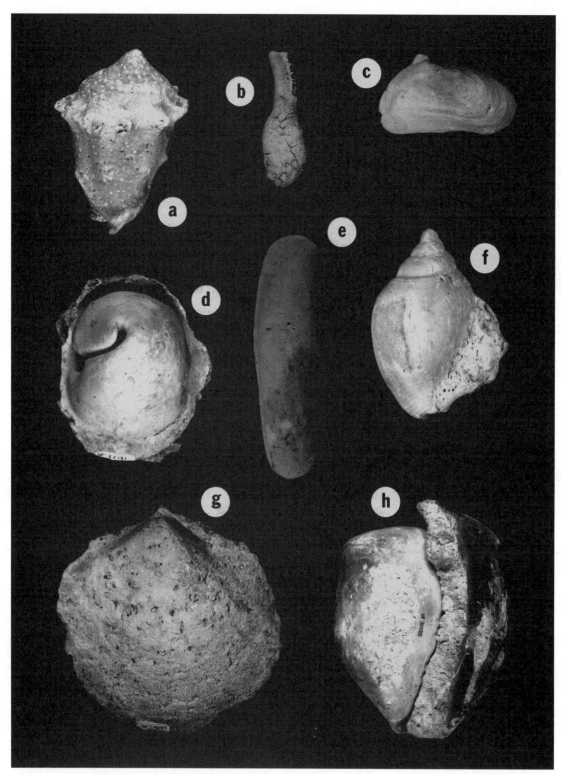

Figure 7.11. Some familiar mollusks from the Florida Eocene with Tethyan affinities (adapted from Nicol 1991). A: *Eovasum vernoni* (UF 19123, 68 mm). B: *Clavagella* sp. (UF 20125, 31 mm). C: *Nayadina* (*Exputens*) *ocalensis* (UF 15876, 38 mm). D: *Velates floridanus* (UF 22131, 69 mm). E: *Lithophaga palmerae* (UF 47250, 63 mm). F: *Caricella obsoleta* (UF 19129, 42 mm). G: *Pseudomiltha megameris* (UF 24603, 140 mm). H: *Gisortia harrisi* (UF 30585, 135 mm).

Oligocene with the disappearance of the northeast-trending Suwannee Channel across northern Florida and southern Georgia (Chen 1965; McKinney 1984; Carter and McKinney 1992). This trough, which may have been 150 to 300 m deep (Cheetham 1963; Carter and McKinney 1992) and current-swept (Pinet and Popenoe 1985b; Huddlestun et al. 1988), had effectively isolated the Florida Platform from the rest of the Gulf Coastal Plain and the siliciclastic sediments shed from the southern Appalachians (see chapters 4 and 11). Although it seems that the Suwannee Channel functioned as a physical barrier during much of the Eocene, some of the distinctiveness of the Florida fauna (when compared with the rest of the eastern Gulf Coast) can be attributed to facies and stratigraphic mismatching by paleontologists studying these regions (Carter and McKinney 1992).

Other major changes that were to profoundly affect Florida's marine faunas also occurred near the Eocene/Oligocene boundary: (1) Global circulation patterns that characterized the Cretaceous to Eocene were transformed to the modern mode; for example, the eastern Tethys almost completely closed, severely restricting equatorial circulation and westward flow of the Tethys Current. (2) Pronounced climatic cooling occurred, as indicated by oxygen-isotope evidence (fig. 4.6). (3) The abyssal ocean filled with cold water, establishing the psychrosphere. (4) Profound lowering of sea-level caused extensive erosion around the Gulf Coast. (5) Major extinctions of marine taxa occurred (Kennett 1982; Miller et al. 1987). Sea-level lowering allowed for development of Florida's terrestrial vertebrate fauna, the oldest site in Florida dating from the Oligocene (see chapter 8).

Climatic changes in the Caribbean Basin had dramatic effects on the biota. For instance, early Oligocene reef-coral faunas are very rare throughout the Caribbean (as well as Florida), indicating substantial faunal turnover since the Late Eocene (Budd et al. 1992). In contrast, late Oligocene reef corals are known to be large, abundant, and diverse, and to have formed prominent barrier reefs throughout the Caribbean (Vaughan 1919; Frost and Langenheim 1974). Vestiges of this latter fauna are seen in the composition of the latest Oligocene to earliest Miocene biota of the Tampa Member of the Arcadia Formation (discussed in the following section).

Placement of the Eocene/Oligocene boundary in Florida's shallow-water marine carbonate sequence remains controversial (Nicol et al. 1976; Hunter 1976, 1981; Bryan 1991, 1993; and others). The macrofossils typically used in biozonation schemes for this part of the record (e.g. the gastropod *Turritella martinensis,* the annelid *Rotularia vernoni,* and the large foraminifer *Asterocyclina*) are facies-controlled and lack the temporal resolution of planktic foraminifers or calcareous nannofossils, which are missing from these rocks. Regional data point to continuous deposition and transitional facies across this time interval (Bryan 1993).

The culmination of major limestone accumulation over the Florida peninsula is represented by the Oligocene Marianna and Suwannee limestones. The Marianna crops out in the panhandle (Jackson County); the Suwannee extends from the panhandle ("Duncan Church facies," or Bridgeboro Limestone of Bryan and Huddlestun 1991) down the west-central part of the peninsula. In northernmost Florida the coralgal Bridgeboro Limestone contains massive scleractinian corals and coralline algae indicative of reefs (Bryan 1991; Bryan and Huddlestun 1991). These coralgal reefs probably flanked the northeast-trending Gulf trough (see chapter 4) of the Georgia Channel System (Huddlestun 1993), separating the shelf province (foramol/bryomol Marianna Limestone) of northwestern panhandle Florida from the miliolid, peloidal, chlorozoan limestone with local patch reefs and coral thickets (Suwannee Limestone) of the Florida Platform province, which included the Florida peninsula and southeastern Georgia (Bryan 1991, 1993).

All of these Oligocene carbonate units have extensive benthic-foraminifer faunas (Cole and Ponton 1930; Applin and Jordan 1945) (fig. 7.12). Most also contain macrofossils (fig. 7.12) that have been studied since the late 1800s (Dall and Harris 1892); for example, Mansfield (1937) described a diverse molluscan fauna from the Suwannee Limestone, although most specimens are preserved as molds and casts, or occasionally silicified; among the more diagnostic taxa are *Glycymeris suwannensis* and *Orthaulax pugnax hernandoensis.* Puri and Vernon (1964) also listed numerous bryozoans and ostracodes from the Suwannee Limestone. Perhaps the most familiar macrofossils of the Suwannee are the irregular echinoids *Rhyncholampas gouldii* and *Clypeaster rogersi* (fig. 7.12), two of several echinoids which typically occur throughout these Oligocene carbonates.

THE NEOGENE

Whereas carbonate paleoenvironments prevailed in Florida during the Mesozoic and early Cenozoic (Paleogene), clastic sedimentation regimes prevailed during the Neogene (see chapters 4 and 5). Limestones are volumetrically small and of local significance in the Miocene and Pliocene. Florida's Neogene sediments, having a maximum thickness of barely 300 m, suggest accumulation rates slower than in previous times. Because of these paleoenvironmental differences, and

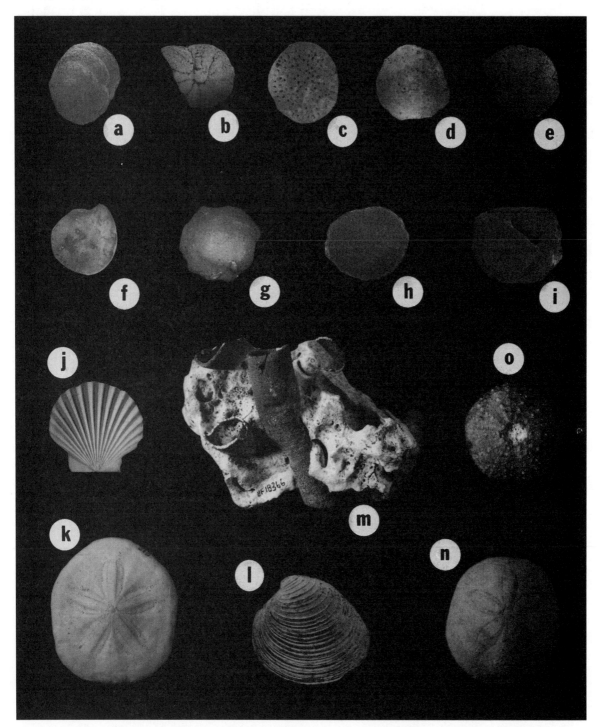

Figure 7.12. Micro- and macrofossils from the Oligocene Suwannee (A–E, L–O) and Marianna (F–K) limestones. A–I are large benthic foraminifers. A: *Discorinopsis gunteri*, ×20. B: *Pararotalia* (*Rotalia*) *mexicana*, ×20. C: *Dictyoconus cookei*, ×14. D: *Miogypsina intermedia*, ×10. E: *Pararotalia* (*Rotalia*) *byramensis*, ×24. F: *Nummulites* (*Operculinoides*) *dius*, ×9. G: *Lepidocyclina mantelli*, ×10. H: *Lepidocyclina parvula*, ×11. I: *Eponides mariannensis*, ×24. J: bivalve *Pecten perplanus* (UF 67811, 32 mm). K: echinoid *Clypeaster rogersi* (UF 3314, 55 mm). L: bivalve *Chione craspedonia* (UF 67812, 22 mm). M: mollusk *Kuphus incrassatus* (UF 18366, 130 mm). N: echinoid *Rhyncholampas gouldii* (UF 67813, 37 mm). O: echinoid *Gagaria mossomi* (UF 28245, 40 mm).

major, global changes associated with the Paleogene/Neogene transition (e.g. establishment of steep, permanent, latitudinal temperature gradients; continued restriction and final closure of the Tethys Seaway; growth of Antarctic ice sheets and increased glaciation), the marine faunas underwent dramatic changes throughout the 25 million years of the Neogene.

The shallow-water marine fossil assemblages of the Florida Miocene and Pliocene are chiefly molluscan. They span the entire spectrum from depauperate faunas and poorly preserved faunas of molds and casts to highly diverse, rich shell beds with exquisite preservation (including original color patterns). While the fossil content of most lithologic units is modest, the remarkable preservation and diversity of molluscan faunas in the Miocene Chipola Formation and Pliocene Pinecrest beds rank them among the most species-rich deposits in the world. Although neither fauna has been completely monographed, estimated molluscan diversity for each unit exceeds 1,000 species (Vokes 1989; Olsson 1968).

By the beginning of the Miocene, the major ocean basins of the world had more or less assumed their modern proportions (Kennett 1982). Low-latitude circulation via the Tethys Seaway was restricted, although open communication still existed between the Caribbean and the eastern Pacific across submerged parts of Central America (Jones and Hasson 1985). The Miocene and Pliocene molluscan faunas of Florida belonged to the Caloosahatchian Province (Petuch 1982a, 1988), one of four major provinces recognized for the tropical western Atlantic during the Neogene. The Miocene Caloosahatchian Province extended across the southeastern United States from North Carolina, including all of Florida and the Gulf Coast (including the Mississippi Embayment) to Texas and the east coast of Mexico, southward to the Yucatan (fig. 7.13). It was bounded on the south by the Gatunian Province, which included Cuba, the islands of the Caribbean, Central America, and most of northern South America.

The dominant lithologic unit of the Florida Neogene is the Hawthorn Group (Scott 1988b; chapter 5). Unfortunately, the Miocene Hawthorn sediments have yielded comparatively few dateable fossil assemblages, and diagenetic overprinting has obliterated the vast majority of fossils, leaving mainly molds and casts (Scott 1988b). The most familiar fossil assemblages from the Miocene Caloosahatchian province of Florida are those of the Chipola, Oak Grove, and Shoal River formations of the Alum Bluff Group in northwestern Florida; these units cluster in the late Early to Middle Miocene. The other famous fossil fauna of the Florida Miocene is known from the earliest Miocene (or latest Oligocene?) Tampa Member of the Arcadia Formation (Scott 1988b).

In 1842, T. A. Conrad visited the Tampa region and collected invertebrate fossils, mostly from Ballast Point, that he subsequently described and illustrated (Conrad 1846b). Conrad's discussion of Tampa fossils—which followed Allen's (1846) brief mention of their existence by a few months—was among the earliest accounts of Florida paleontology, and the first to assign a Late Eocene age to the Tampa fauna. The excellent fossils (fig. 7.14) and their unusual preservation inspired the term "silex beds" to designate the silicified parts of the unit.

Since then, the fossil beds around Tampa Bay, particularly in the classical type areas at Ballast Point and Six Mile Creek (Dall and Harris 1892), have been the subject of numerous studies (Heilprin 1887; Dall 1890–1903, 1915; Mansfield 1937; Weisbord 1973; and others). In 1892 Dall recognized subdivisions of the Tampa Limestone that he designated the "*Orthaulax* bed" (later, in 1915, calling it the "*Orthaulax pugnax* zone") and the "Tampa limestone." The former subdivision, named for gastropods of the genus *Orthaulax,* was introduced to replace the local name "Tampa silex bed," which nevertheless remained in use well into the twentieth century. Matson and Clapp (1909) united Dall's subdivisions into a single Tampa Formation, a move also advocated by Mansfield (1937) and later authors because of lithologic and faunal variability, as well as the limited surface exposure.

The fauna of the Tampa Member is chiefly marine, although some land snails (fig. 7.14) are known from the Ballast Point site. Mollusks are most common, but corals, barnacles, and foraminifers also are present. Both Dall (1915) and Mansfield (1937) reported more than

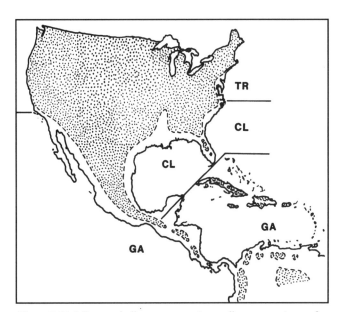

Figure 7.13. Miocene shallow-water marine molluscan provinces of the tropical western Atlantic. TR: Transmarian. CL: Caloosahatchian. GA: Gatunian (modified from Petuch 1988).

Figure 7.14. Macrofossils from the Tampa Member of the Arcadia Formation. A and E-F are corals. A: *Montastrea canalis* (UF 8986, 155 mm). E: *Montastrea canalis* (UF 8954, 90 mm). F: *Desmophyllum willcoxi* (UF 8290, 30 mm). B-D, I, and K are marine gastropods. B: *Turritella tampae* (UF 9203, 41 mm). C: *Cerithium praecursor* (UF 67814, 12 mm). D: *Urosalpinx? inornata* (UF 67815, 34 mm). I: *Turbo (Senectua) crenorugatus* (UF 3592, 33 mm). K: *Conus designatus* (UF 67816, 25 mm). G and L are land snails. G: *Hyperaulax americana* (UF 67817, 15 mm). L: *Cepolis crusta* (UF 67818 [top view], 13 mm. H and J are marine bivalves. H: *Ventricolaria glyptoconcha* (UF 3591, 20 mm). J: *Venericardia serricosta* (UF 67819, 26 mm).

300 species of mollusks from the Tampa. Based upon the percentage of molluscan species surviving into the Holocene (the "Lyellian percentage"), Dall placed the age of the Tampa fauna near the transition between the Oligocene and Miocene. Mansfield (1937) believed that the mollusks suggest an Early Miocene age. The twenty-eight species in the coral fauna discussed by Weisbord (1973) also indicated a latest Oligocene to earliest Miocene ("Aquitanian") age. In the absence of age-diagnostic species, and based upon faunal correlations with other units in Georgia and Mississippi, the age designation has not changed in recent years (Scott 1988b).

Except for the Tampa Member of the Arcadia Formation, the remaining Hawthorn Group of Scott (1988b) is not noted for its fossil content. Preservation is generally poor. Aragonitic fossils are rarely preserved except as molds or casts, and calcitic forms are only slightly more common. Diatoms, ostracodes, and occasional foraminifers provide limited stratigraphic control. Nevertheless, certain sites have produced fossil faunas of moderate quality and diversity. One example is the recently described fossil fauna from the Parachucla Formation at White Springs, which includes more than sixty-five taxa of Early Miocene marine invertebrates, mostly mollusks (Portell 1989). Another fauna from the Lower to Middle Miocene Marks Head and Coosawhatchie formations at Brooks Sink yielded approximately forty recognizable taxa (Jones and Portell 1988).

Probably the best-known and most fossiliferous unit of the Florida Miocene is the Chipola Formation of the Alum Bluff Group in Calhoun and Liberty counties, in the panhandle. The Chipola is a blue-gray to yellowish-brown, highly fossiliferous marl, with biofacies ranging from shoreline beach, to lagoonal with oyster reefs, to coral patch reefs in an offshore back-reef setting probably nowhere deeper than 30 m (Vokes 1989). The Oak Grove and Shoal River units of the Alum Bluff Group, which are also fossiliferous, are sandier and probably represent near-shore conditions. The Chipola Formation has been considered late Early or Middle Miocene, and younger than the Torreya Formation of the Hawthorn Group (which it overlies at the famous Alum Bluff exposure); however, recent chronostratigraphic evidence (Bryant et al. 1992) favors a slightly older Chipola Formation (18.3 million years) and indicates an overlap in age (15.3–19 million years) and possible interdigitation with the Torreya .

The fauna of the Chipola (fig. 7.15) is one of the most species-rich, ecologically diverse, and well preserved in the western Atlantic (Vokes 1989). Gardner (1926–50) monographed more than 400 species in her massive effort. Since then, Drs. Emily and Harold Vokes of Tulane University have described many new species, and discovered countless more. In a recent overview of the Chipola Formation, Vokes (1989) estimated that there are probably more than 1100 mollusk species in the fauna. Many of these represent classic Tethyan groups that give the malacofauna a modern South Pacific appearance (Petuch 1988). A high proportion of endemism also characterizes the fauna.

The great number of mollusk species approaches that which might be expected for a tropical fauna. Besides mollusks, similar species abundances probably characterize other Chipola invertebrate groups that have not received comparable attention from paleontologists. Vokes (1989) suggested that the Chipola reflects the last truly tropical climatic conditions of the Miocene. The succeeding Shoal River Formation represents a cooler, more temperate climate that corresponds to Middle Miocene growth of the Antarctic ice sheet and worldwide steepening of latitudinal temperature gradients. Continued global cooling led to a distinctly cooler Late Miocene world with a climatic regime that was definitely in the glacial mode (Kennett 1982). Extinctions, probably related to this cooling, characterized the Late Miocene Caloosahatchian Province (Stanley and Campbell 1981).

By the beginning of the Pliocene, climatic cooling had reduced Petuch's (1988) four molluscan faunal provinces of the tropical western Atlantic to two large provinces (superprovinces, *sensu* Stanley 1986), the Gatunian to the south (the Caribbean Province of Woodring 1959) and the Caloosahatchian to the north; Florida was part of the latter. Florida's Pliocene Caloosahatchian faunas are almost entirely distinct from those of the adjacent Gatunian Province (Caribbean, Bahamas, Central America, and northern South America; fig. 7.16). This faunal contrast has been attributed to a predominance of carbonate over siliciclastic sedimentation in the Gatunian Province (Petuch 1982a), and/or to oceanographic barriers related to cold, upwelling zones, with local strengthening of Gulf Stream circulation (Stanley 1986).

It is important to note that the age designations of many of Florida's Neogene and Pleistocene faunas have changed within the last decade or so. Since the days of Heilprin and Dall, the ages of Tertiary faunas were based on the proportion of fossil species that survived to the present (Lyellian percentages). Dall (1887) himself lamented that this method was "impracticable, illogical, and misleading." He thought it highly improbable that faunas from southern Europe, designated as Pliocene by this method, were synchronous with those in America. Ages based on planktic foraminifers correlated to the geomagnetic-polarity time scale are younger than those suggested by the molluscan faunas from many units in

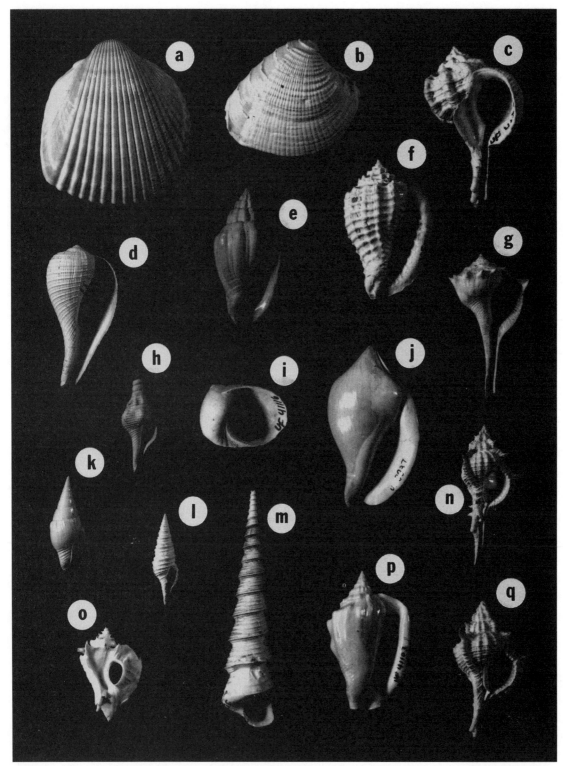

Figure 7.15. Fossil mollusks from the Miocene Chipola Formation. A-B are bivalves. A: *Cardium chipolanum* (UF 67820, 40 mm). B: *Chione burnsii* (UF 67821, 26 mm). C-Q are gastropods. C: *Dermomurex vaughani* (UF 41582, 35 mm). D: *Ficus eopapyratia* (UF 3124, 39 mm). E: *Falsilyria pycnopleura* (UF 46089, 40 mm). F: *Morum chipolanum* (UF 8571, 28 mm). G: *Busycon sicyoides* (UF 42260, 40 mm). H: *Turbinella chipolanum* (UF 35791, 27 mm). I: *Sinum chipolanum* (UF 41116, 24 mm). J: *Orthaulax gabbi* (UF 8837, 68 mm). K: *Strombina aldrichi* (UF 38128, 14 mm). L: *Polystira albidoides* (UF 67822, 30 mm). M: *Turritella subgrundifera* (UF 42750, 75 mm). N: *Haustellum gilli* (UF 40232, 45 mm). O: *Typhis obesus* (UF 42282, 15 mm). P: *Strombus aldrichi* (UF 41103, 50 mm). Q: *Chicoreus chipolanus* (UF 67823, 38 mm).

the Caribbean region. For example, microfossils from the Gatun Formation of Panama yielded a Pliocene age, although Woodring (1982) prefered to retain the Middle Miocene age suggested by the percentage of extant mollusks. Stanley and Campbell (1981) and Stanley (1986) argued that the age discrepancy is a result of Plio-Pleistocene mass extinctions of mollusks concentrated in the tropical western Atlantic, giving the faunas the appearance of greater age. They argued that this regional mass extinction extended northward from the tropics to Virginia, that cooling probably was the cause, and that the impoverishment of the molluscan faunas of the region relative to those of the tropical eastern Pacific and other areas is a result. Although both the patterns of this extinction and mechanisms surrounding it recently have been challenged (Allmon et al. 1993; Jackson et al. 1993), the conclusion that faunas from the Gatun (Panama), Bowden (Jamaica), Esmeraldas (Ecuador), and Tamiami-Pinecrest (Florida) formations belong in the Pliocene, and not in the Miocene, appears sound.

The Pliocene Caloosahatchian Province in Florida is characterized by two major faunas, both of which are renowned for their diverse, abundant, and well preserved molluscan fossils: (1) the fauna of the Jackson Bluff Formation in the central panhandle (fig. 7.17), and (2) the fauna of the Pinecrest beds of the Tamiami Formation in the southern part of the peninsula. The Jackson Bluff mollusks were monographed by Mansfield (1930, 1932), who described nearly 200 species; many more have been added since, and estimates of species richness now are considerably higher (e.g. 171 gastropods alone reported by Campbell et al. 1975). However, mollusks are not the only fossil group represented in the Jackson Bluff: diverse assemblages of corals (Weisbord 1971), ostracodes, and foraminifers (Puri 1953; Cushman and Ponton 1932) have been reported, but other invertebrate groups remain largely unstudied.

SHELLY FAUNAS OF SOUTHERN FLORIDA

The Pinecrest fauna (fig. 7.18) of southern Florida is considered by many to be a temporal equivalent of the Jackson Bluff fauna. Petuch (1988) believed that the faunas are biogeographically distinct enough to warrant subprovincial status, and he has referred them respectively to the Buckinghamian and Jacksonbluffian subprovinces of the Caloosahatchian Province (fig. 7.16). Stanley (1986), like many others, correlated these two faunas of the Florida Pliocene with the molluscan faunas of the Duplin and Yorktown formations of the Carolinas and Virginia, for which Petuch (1988) had reserved the Yorktownian subprovince. In general, the Pinecrest contains more tropical mollusks (e.g. Spondylidae, Lucinidae, Carditidae, Conidae, Turbinellidae, Strombidae, Terebridae), as well as hermatypic corals; however, the evidence indicates that the Pinecrest fauna represents a mixture of tropical and temperate elements, varying according to collection location (Olsson 1964).

Before discussing the Pinecrest fauna and its relation to fossils from other Pliocene and Pleistocene units, it is important to note that major controversies and differences of opinion surround the late Cenozoic stratigraphy of southern Florida (see Lyons 1991; Scott 1992; Scott and Allmon 1992). Pliocene and Pleistocene deposits in the region consist of siliciclastic and carbonate lithologies whose lateral and temporal relationships are obscured by (1) thinness and discontinuous distribution of units, (2) limited exposures, (3) rapid facies change, and (4) repeated advance and retreat of the sea over this low-elevation region in response to the the many sea-level oscillations of the Plio-Pleistocene. Compounding the problem has been the practice of stratigraphers to recognize formations (lithostratigraphic units) on the basis of their fossil content (Scott 1992). To a certain extent this arises from the frustration of observing significant faunal changes in sequences of sand or carbonate without distinctive lithologies, where the fossils themselves often constitute the principal sedimentary components (i.e. "shell beds" or condensed faunas). Further complicating the situation has been the tendency of investigators to erect units based upon collections of fossils mixed from several horizons at one locality (e.g. specimens picked off spoil piles dredged from strata below water level), or

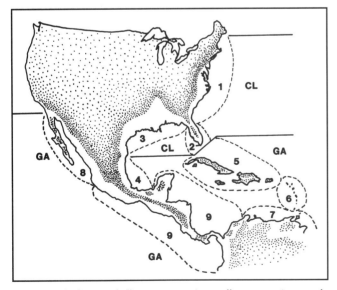

Figure 7.16. Pliocene shallow-water marine molluscan provinces and subprovinces of the tropical Americas. CL: Caloosahatchian. GA: Gatunian. 1: Yorktownian. 2: Buckinghamian. 3: Jacksonbluffian. 4: Agueguexitean. 5: Guraban. 6: Carriacouan. 7: Puntagavilanian. 8: Imperialian. 9: Limonian (after Petuch 1988).

Figure 7.17. Fossils (mostly mollusks) from the Pliocene Jackson Bluff Formation of the Florida Panhandle. A, D, F, and J are bivalves. A: *Semele alumensis leonensis* (UF 67824, 25 mm). D: *Macrocallista reposta* (UF 67825, 47 mm). F: *Anadara propearesta* (UF 67826, 44 mm). J: *Chione ulocyma* (UF 67827, 36 mm). B, E, G-H, and I are gastropods. B: *Turritella etiwaensis* (UF 7683, 130 mm). E: *Diodora cattiliformis alumensis* (UF 7018, 13 mm). G: *Calliostoma philantropus pontoni* (UF 67828, 16 mm). H: *Calophos wilsoni* (UF 7634, 30 mm). I: *Ecphora quadricostata umbilicata* (UF 2621, 52 mm). C is a coral. C: *Phyllangia blakei* (UF 67829, 40 mm).

Figure 7.18a. Selected macrofossils from the Pliocene Tamiami Formation of southern Florida, emphasizing specimens from the Pinecrest beds near Sarasota. A-N are gastropods. A: *Cancellaria propevenusta* (UF 30800, 42 mm). B: *Calophos wilsoni* (UF 35158, 33 mm). C: *Subpterynotus textilis* (UF 35500, 55 mm). D: *Trigonostoma hoerlei* (UF 60245, 10 mm). E: *Conus druidi* (UF 18059, 124 mm). F: *Siphocypraea carolinensis* (UF 58961, 98 mm). G: *Calophos wilsoni* (UF 35158, 33 mm). H: *Dicathais handgenae* (UF 51682, 31 mm). I: *Hexaplex hertweckorum* (UF 13877, 155 mm). J: *Vasum locklini* (UF 30713, 110 mm). K: *Cassis floridensis* (UF 32183, 122 mm). L: *Ecphora bradleyae* (UF 47732, 83 mm). M: *Strombus hertweckorum* (UF 34502, 160 mm). N: *Melongena consors* (UF 30386, 172 mm).

Figure 7.18b. Selected macrofossils from the Pliocene Tamiami Formation of southern Florida, emphasizing specimens from the Pinecrest beds near Sarasota. O, R, and U are echinoids. O: *Encope tamiamiensis* (UF 8619, 73 mm). R: *Mellita aclinensis* (UF 28207, 65 mm). U: *Rhyncholampas evergladensis* (UF 24524, 67 mm). P, S, and V are corals. P: *Septastrea crassa* (UF 67830, 165 mm). S: *Solenastrea bournoni* (UF 55329, with coral-inhabiting barnacle, *Eoceratoconcha weisbordi*, 92 mm). V: *Septastrea marylandica* (UF 67831, encrusting gastropod shell 50 mm). Q, T, W-Z, YY, and ZZ are bivalves. Q: *Plicatula hunterae* (UF 34869, 54 mm). T: *Hyotissa haitensis* (UF 7844, 138 mm). W: *Placunanomia plicata* (UF 33989, 75 mm). X: *Anadara notoflorida* (UF 31230, 70 mm). Y: *Chesapecten septenarius* (UF 13838, 93 mm). Z: *Perna conradiana* (UF 34589, 80 mm). YY: *Carolinapecten eboreus* (UF 52139, 164 mm). ZZ: *Mercenaria corrugata* (UF 3606, 138 mm).

mixed from several localities without stratigraphic-provenance information. Such practices have contributed to an almost intractable set of stratigraphic-nomenclature problems for southern Florida, and they have impeded, more than they have enhanced, the recognition of broad paleontologic patterns for the region.

A convenient example is provided by the "Pinecrest beds," a name proposed by Olsson (1964)

> for certain strata composed largely of sand, barren or highly fossiliferous, encountered directly below a surface limestone in the general region of the 40 mile bend on the Tamiami Trail (Route 41) west of Miami in the western part of Dade County and extending across its boundary into Collier County, Florida. The name is taken from an old settlement on the Everglades road (which branches off from the present highway at 40-mile bend) about one mile west of the Dade-Collier County line. A small collection of fossils taken from the ditches or pits in this area was described by Mansfield in 1931, who considered their age as late Miocene. The Acline fauna, first described by Tucker and Wilson (1932, 1933) from a few pits in the Punta Gorda area, belongs to the same stratigraphic unit and is considered as a facies development.

Olsson goes on to describe fossils, from many localities throughout southern Florida, that belong to the Pinecrest fauna; however, in not mentioning a specific (measured) stratigraphic section when proposing the unit, he left the door open for subsequent interpretation. Realizing the need for a type section, Hunter (1968) designated the lowest bed in Mansfield's (1931) brief section (where most material had been dredged from below water level) as the type section of the "Pinecrest sand member" of the Tamiami Formation. This expanded the concept of the Pinecrest beds beyond that originally expressed by Olsson (Waldrop and Wilson 1990), who nevertheless later treated the Pinecrest beds as a separate formation, which he considered older than the Caloosahatchee Formation, but younger than the Tamiami Formation (Olsson 1968). Although many agree with this assessment, controversy abounds as to the relationship between the Pinecrest beds and the Buckingham and Tamiami formations described by Mansfield (1939), as well as to the age of the Pinecrest fauna (Early or Late Pliocene) and its temporal equivalents along the Atlantic Coastal Plain. Field excursions and discussions to address some of these questions (see papers in Scott and Allmon 1992) have not resolved major stratigraphic issues, but have exposed the diversity of opinion that exists among stratigraphers. Lyons (1991) reviewed the history and status of the more important stratigraphic controversies of southern Florida, and readers desiring more background are referred to him (see also chapter 5).

Today the Pinecrest fauna is best exposed in large quarry operations in the vicinity of Sarasota County (Petuch 1982b; Stanley 1986; Jones et al. 1991; Allmon 1992, 1993). There it occurs in strata variously referred

DESCRIPTION OF UNITS AT APAC SHELL PIT (from Petuch, 1982)			
15m	0	Yellow Quartz Sand	CALOOSAHATCHEE FORMATION ?
	1	Shell Fragments	PINECREST BEDS
	2	*Hyotissa*	
	3	Mytilids	
	4	"BLACK LAYER"	
	5	*Vermicularia* bed	
10m	6	Mixed *Hyotissa* and shells	
5m	7	Mixed shells	
	8	*Vermicularia* bed	
	9	*Hyotissa* layer	
	10	*Mercenaria* layer	
	11	*Ecphora* and *Balanus* fauna	

Figure 7.19. Dense shell accumulations in the APAC shell pit in Sarasota. Stratigraphic section modified slightly from Petuch (1982b).

to as the "Pinecrest Formation" (Weisbord 1972), "Pinecrest beds" (Petuch 1982b), "Pinecrest Sand Member of the Tamiami Formation" (Weisbord 1981), "Buckingham Formation" (Petuch 1986, 1988), or "Fruitville Formation" (Waldrop and Wilson 1990). However, this fauna also occurs in different facies across southern Florida, southward from Sarasota and eastward to the Miami region. The faunal composition is also variable, with higher proportions of corals and associated tropical mollusks to the south and east, where the sandy lithology gives way to predominantly limestone facies. It is quarry exposures in the Sarasota region, however, that provide our best look at the dense shell accumulations of the Pinecrest (fig. 7.19). These deposits, which formed in supratidal, intertidal, lagoonal, and deeper, open-bay paleoenvironments during one or more transgressive episodes, rank among the most highly fossiliferous beds in Florida (Petuch 1982b).

The profusion of shells and their taxonomic diversity have intrigued paleontologists and shell-collectors for generations, yet the fauna has never been monographed. Stanley (1986) listed 212 bivalve species from the Pinecrest. The gastropods have not been formally tallied; however, Abbott (1974) and Keen (1971) both reported gastropod/bivalve ratios of about 3:1 for modern American faunas. Therefore, one may reasonably expect about 800 to 850 molluscan species in the Pinecrest; other speculations place the number at about 1,000 (Campbell et al. 1975) to around 1,200 (Olsson 1968). The Pinecrest thus shares notoriety with the Chipola as the most diverse molluscan assemblage in Florida.

THE "CALOOSAHATCHEE," THE PLEISTOCENE, AND BEYOND

The earliest discussions of the richly fossiliferous beds discovered along the Caloosahatchee River (Heilprin 1887; Dall 1887, 1890–1903; Dall and Harris 1892), which were termed the "Caloosahatchee beds" or "Caloosahatchee marl," noted their late Tertiary position and referred them to the Pliocene. By the end of the nineteenth century, these beds were widely recognized as *the* fossiliferous unit of southern Florida. The Caloosahatchee Formation (Matson and Clapp 1909) was eventually restricted to the lower strata along the river when Sellards (1919) described two overlying shell-bearing units, the Fort Thompson Formation and the Coffee Mill Hammock Formation, which were both considered to be Pleistocene. However, it was not until the 1930s that older fossils (considered to be "upper Miocene"), different from the Caloosahatchee invertebrates, were recognized by Mansfield (1931, 1932, 1939) and Tucker and Wilson (1932, 1933). These older fossils (see foregoing discussion of the Pinecrest fauna), as well as the younger, Pleistocene faunas, clearly showed that the paleontologic record of southern Florida is more complex than originally anticipated in the last century.

Wilson (in Kier 1963) and Olsson (1964) presented the first arrangement of the beds and faunas in southern Florida as they are generally recognized today. In their schemes, the Caloosahatchee Formation overlies the Pliocene Tamiami Formation (relationships to the Pinecrest, Buckingham, and Fruitville of Waldrop and Wilson, 1990, remaining unclear), and has historically been considered Pliocene. Although DuBar (1958a, 1958b, 1974), Hazel (1983), Petuch (1988), and others favor a Pleistocene age based on faunas, Lyons (1991) concluded that most faunal and geochemical evidence supports placement in the latest Pliocene. Allmon et al. (1993) held that a consensus of investigators consider the Caloosahatchee to span the Pliocene/Pleistocene boundary. Olsson (1964) considered the Caloosahatchee to be the most geographically restricted of the southern Florida Tertiary units, its principal area of development extending in a troughlike belt southeastward from North St. Petersburg to the Caloosahatchee River and beyond. On the basis of its faunal content (primarily mollusks), the Caloosahatchee has been reported to be contemporaneous with the Chowan River and James City formations of Virginia and North Carolina, as well as the Bear Bluff and Waccamaw formations of North and South Carolina, although these correlations are disputed (see Lyons 1991).

Olsson (1964) observed that the Caloosahatchee fauna (fig. 7.20) is decidedly tropical, whereas the Pinecrest fauna includes a mixture of both temperate and tropical elements. The preservation of the fossils is excellent. The size of the fauna, although still imprecisely known, was historically considered to be less than that of the Pinecrest. Olsson and Harbison (1953) recorded 505 mollusk species or subspecies from the Caloosahatchee in their monograph of the fauna at St. Petersburg; however, they went on to speculate that once the entire molluscan fauna was known it might approach 1000 species. More recently Campbell et al. (1975) reported nearly 500 species of gastropods alone. Assuming a 3:1 gastropod/bivalve ratio, the Caloosahatchee molluscan fauna might contain nearly 700 species. In a re-analysis of molluscan diversity changes throughout the Pliocene-Holocene of the western Atlantic, Allmon et al. (1993) concluded that the diversities of the Caloosahatchee and Pinecrest molluscan faunas are very similar; however, at any one site (e.g. the APAC quarry in Sarasota), alpha diversity shows a sharp decline from the Pinecrest to the overlying Caloosahatchee. Abundance, on the other

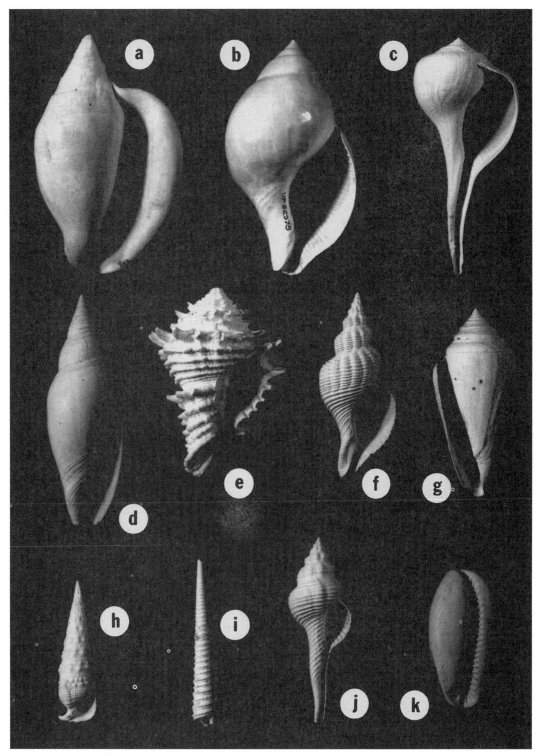

Figure 7.20a. Selected macrofossils from the Caloosahatchee Formation of southern Florida. A-K are gastropods. A: *Strombus leidyi* (UF 50548, 180 mm). B: *Liochlamys bulbosa* (UF 50078, 90 mm). C: *Busycon rapum* (UF 36073, 137 mm). D: *Mitra lineolata* (UF 58535, 106 mm). E: *Vasum horridum* (UF 47018, 90 mm). F: *Fasciolaria scalarina* (UF 58642, 138 mm). G: *Conus adversarius* (UF 14995, 125 mm). H: *Cerithium caloosaense* (UF 58646, 75 mm). I: *Turritella perattenuata* (UF 67832, 83 mm). J: *Fusinus caloosaensis* (UF 58111, 91 mm). K: *Siphocypraea problematica* (UF 15627, 63 mm).

Figure 7.20b. Selected macrofossils from the Caloosahatchee Formation of southern Florida. L-R are bivalves. L: *Miltha caloosaensis* (UF 35422, 79 mm). M: *Panopea floridana* (UF 36159, 157 mm). N: *Cardium dalli* (UF 46522, 110 mm). O: *Anadara rustica* (UF 3734, 60 mm). P: *Arca wagneriana* (UF 36767, 128 mm). Q: *Stralopecten caloosaensis* (UF 14117, 70 mm). R: *Mulinia sapotilla* (UF 35146, 25 mm). S is a coral. S: *Dichocoenia caloosahatcheensis* (UF 42110, 103 mm). T-U are regular echinoids. T: *Lytechinus variegatus plurituberculatus* (UF 12895, 57 mm). U: *Echinometra lucunter* (UF 12937, 65 mm).

hand, remains high from Pliocene shell beds into the Pleistocene.

Probably the greatest disruption to the biogeographic patterns of the western Atlantic occurred with the development of the Central American Isthmus during the Pliocene (Keigwin 1978; Jones and Hasson 1985; Coates et al. 1992). Gradually restricting and eventually eliminating marine circulation between the Atlantic and Pacific by the Late Pliocene, the land bridge connecting North and South America bisected Petuch's Gatunian province into Atlantic and Pacific components. Superimposition of the climatic changes of the Pleistocene upon this event led directly to the modern configuration of faunal provinces within the region (fig. 7.21). Circulation patterns also were altered: currents that previously flowed westward were deflected, and the configuration of the Gulf Stream was changed as a prelude to Northern Hemisphere glaciation (Kennett 1982). The severed marine connection has been implicated in the differential extinction between molluscan faunas of the tropical eastern Pacific and the Caribbean (Dall 1892, part 3), the Caribbean (both Gatunian and Caloosahatchian provinces) suffering disproportionately and supporting relatively depauperate faunas throughout the late Neogene to Holocene (Olsson 1961, 1972; Woodring 1966; Vermeij 1978; Vermeij and Petuch 1986; Stanley 1986; Petuch 1988; and others). This pattern, which has been attributed to differences in (1) areal restriction of shallow seas, (2) temperature, and/or (3) nutrients or food supply, has recently been challenged on several fronts by Allmon et al. (1993) and Jackson et al. (1993), who contended that high origination rates accompanied high extinction rates during the Plio-Pleistocene, and that species diversity does not differ appreciably between the Pacific and Atlantic regions.

Also coming under fire by Allmon et al. (1993) is the conventional wisdom dictating that faunal diversity, as indicated by the rich molluscan fossil assemblages in southern Florida, shrank throughout the Pleistocene in response to climatic deterioration (Stanley and Campbell 1981; Stanley 1986; Petuch 1988). Petuch (1988) indicated that the Pliocene was marked by the evolution and proliferation of new genera, whereas the Pleistocene was a time of destruction and elimination of entire faunas, linked to continental glaciation. In contrast, Allmon et al. (1993) marshaled evidence from several sources suggesting that Pleistocene climatic changes were not particularly significant along the coast of the southeastern U.S., and had relatively little impact on the fauna. Krantz (1990), for example, concluded from isotopic studies of fossil bivalves that there was no significant cooling along the Virginia–North Carolina coastal plain during the Pliocene (from 4 to 2 million years ago, approximately). Cronin (1990) found no evidence of colder climates along the coast south of Cape Hatteras, even during the Pleistocene (Hazel 1968; Cronin 1981, 1988). Therefore, the molluscan faunas of successively younger Pleistocene units in Florida may not represent stepwise reductions in diversity related to cooling. Allmon et al. (1993) questioned the reality of diversity declines altogether.

Authors have argued about faunal-diversity changes through the Pleistocene section in southern Florida since the days of the earliest investigations. Certainly the stratigraphic confusion in the region (Scott 1992) has exacerbated the problem for paleontologists. In the standard arrangement of Pleistocene units of southern Florida, the Bermont, Fort Thompson, and Coffee Mill Hammock formations overlie the Caloosahatchee (Lyons 1991). Various authors (Cooke and Mossom 1929; DuBar 1958a, 1958b, 1974; Ward and Blackwelder 1987; Petuch 1991; and others) have favored combining the Fort Thompson and Coffee Mill Hammock formations, which were established by Sellards (1919) from exposures along the Caloosahatchee River, into a single formation, often with the Coffee Mill Hammock as a member of the Fort Thompson. Lyons (1991) maintained that the Coffee Mill Hammock Formation is clearly younger, representing the approximate time span 0.13 to 0.11 million years before the present. The faunas of these two formations, however, show few appreciable differences, and are almost indistinguishable from modern faunas of the region. The Coffee Mill Hammock

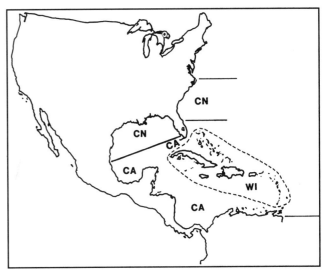

Figure 7.21. Modern shallow-water marine faunal provinces of the tropical western Atlantic. CN: Carolinian. CA: Caribbean. WI: West Indian (after Briggs 1974). Provincial boundaries for coastal Florida are drawn at Cape Canaveral on the Atlantic coast and at Cape Romano, just south of Naples, on the west coast.

Formation and its fauna are considered to be contemporaneous with the Miami and Key Largo formations of the Florida Keys, as well as the Anastasia Formation (coquina) of the east coast of Florida. All probably formed during the sea-level high stand about 125,000 years ago (isotope stage 5e).

A distinct biostratigraphic unit between the Caloosahatchee and Fort Thompson formations was named the Bermont Formation by DuBar (1974). Earlier authors (Mansfield 1939; and others) had hinted at its existence, and by the 1960s and 1970s it was well known under such names as the "Glades formation" (Vokes 1963), "unit A" (Olsson 1964), or the "Glades unit" (McGinty 1970; Hoerle 1970). The fauna of the Bermont (fig. 7.22) was initially considered to be Early Pleistocene (Vokes 1963; Hoerle 1970), or perhaps even Late Pliocene to Early Pleistocene (Waller 1969). However, DuBar (1974) placed the Bermont in the medial Pleistocene, possibly influenced by his belief in the Early Pleistocene age of the upper beds of the Caloosahatchee (Lyons 1991). Petuch (1989, 1990, 1991) added to the confusion by "informally" naming additional units (to the Bermont and other formations), recognized principally by key taxa and without adequate stratigraphic control, further clouding the distinction between biozonation and lithostratigraphy. Although Petuch referred to the Bermont as "middle" Pleistocene, a variety of biochronologic evidence (Hulbert and Morgan 1989) and isotopic evidence (Webb et al. 1989) favors an Early Pleistocene age (see discussion in Lyons 1991), in keeping with the historical designation.

Faunal differences between the Bermont and overlying units are small. Most taxa range from the Bermont through younger formations and are recognized as part of the modern fauna. In practice, a few diagnostic taxa are used in the field to distinguish one shelly Pleistocene unit from another, and all are dominated by modern species (fig. 7.22). Olsson (1964) and subsequent authors contended that the Pleistocene faunas are not as diverse as the Pinecrest and Caloosahatchee faunas. Again, Allmon et al. (1993) argued that, although major extinctions occurred, origination rates were also high, and as a consequence diversity did not fall appreciably in a series of steps through the Pinecrest, Caloosahatchee, Bermont, Fort Thompson, and into the Holocene.

The nomenclature, the attributed ages, and even the very concepts of Plio-Pleistocene stratigraphic units of southern Florida are in a state of flux (see Scott 1992; papers in Scott and Allmon 1992). These problems are far from being solved, and little consensus has been achieved. Until this situation stabilizes, until units are correlated to the geomagnetic-polarity time scale by means of microfossil zonations or chemostratigraphy, and until fossils are routinely collected within the framework of a proper lithostratigraphic context, it will be difficult to evaluate the biotic changes that occurred in southern Florida through the Neogene and Quaternary.

Summary

Florida's rich fossil record of Cenozoic marine life has been studied for almost two centuries and is well known to paleontologists worldwide. The oldest rocks exposed in Florida date from the Eocene, so that the vast majority of Florida's fossils reflect the changing conditions of the Tertiary in the tropical to subtropical western Atlantic. Less familiar to paleontologists are the Paleozoic and Mesozoic faunas, which are known entirely from the subsurface, but which nevertheless yield important clues about the tectonic evolution of the region and the development of the Florida-Bahamas Platform. With its low topographic relief, the broad Florida Platform lies near sea level, where it continues to be influenced by eustatic fluctuations, much as it did during the Pleistocene and earlier epochs. Hence, an appreciation of the present-day biota of the littoral and neritic environments along each coast is almost a prerequisite for interpreting Florida's marine fossil record.

Major paleontologic challenges face investigators in the years ahead, ranging from basic collection and field documentation of new fossil faunas to more sophisticated, synthetic studies. Among the more urgent of these challenges are (1) correlation of marine and nonmarine faunas (especially in light of Florida's rich vertebrate record), linking events in the terrestrial realm with those in the sea; (2) establishment of a sound chronologic framework for Florida's Cenozoic marine faunas by correlation to the geomagnetic-polarity time scale, using microfossils or other appropriate stratigraphic tools; (3) unravelling the biostratigraphic conundrum of the Plio-Pleistocene in southern Florida; (4) basic systematic treatment (monographing) of most fossil faunas, even those familiar ones that have escaped proper attention (e.g. Pinecrest beds); and (5) comparison of Florida's faunas with contemporaneous faunas around the Caribbean and along the Gulf and Atlantic coastal plains. All of these refinements are necessary before evolutionary and paleoecologic studies of Florida fossils can explain biodiversity in the region through time.

Figure 7.22. Selected macrofossils from the Pleistocene Bermont Formation of southern Florida. A: *Strombus mayacensis* (UF 49420, 165 mm). B: *Haustellum anniae* (UF 55452, 42 mm). C: *Fasciolaria okeechobensis* (UF 19505, 135 mm). D: *Vasum floridanum* (UF 41682, 115 mm). E: *Turbinella hoerlei* (UF 50440, 185 mm). F: *Latirus maxwelli* (UF 14773, 63 mm). G: *Fusinus watermani* (UF 53267, 68 mm). H: *Anadara aequalitas* (UF 67833, 63 mm). I: *Planorbella* sp. (UF 51368, 20 mm). J: *Phalium inflatum* (UF 14105, 69 mm). K: *Cypraea cervus* (UF 14648, 135 mm). L: *Clypeaster rosaceous* (UF 42000, 115 mm). M: *Miltha carmenae* (UF 6556, 107 mm). N: *Hippoporidra* sp. (UF 58036, 80 mm). O: *Chione cancellata* (UF 67834, 25 mm). P: *Manicina areolata* (UF 55106, 145 mm). A-G and I-K are gastropods; H, M, and O are bivalves; L is an irregular echinoid; N is a bryozoan; and P is a coral.

8

Fossil Mammals of Florida

Bruce J. MacFadden

OUTCROPS OF CENOZOIC SEDIMENTS throughout Florida contain the richest record of fossil vertebrates east of the Mississippi River, rivaling the classic localities in western North America. The extensive interbedding of marine and nonmarine units and a unique paleobiogeographic position make Florida important to the understanding of ancient mammals in North America.

In contrast to the classic Cenozoic sediments and faunas of western North America, the corresponding sequence in Florida until recently has had few direct radiometric calibrations. Age determinations principally have been by faunal correlations with fossiliferous sections elsewhere that contain interbedded, dated volcanic units. The strength of the Florida record of fossil vertebrates has been the proximity of marine and terrestrial deposits and faunas, which provides a virtually unique opportunity for correlation between these two major sedimentary realms as they are exposed in eastern North America. With the recent application of the $^{87}Sr/^{86}Sr$ dating method (Webb et al. 1989; Jones et al. 1991; Bryant et al. 1992; see also chapter 3), the Florida sequence provides an important opportunity to correlate between major sedimentary environments.

This chapter presents the history of Cenozoic mammals in Florida, including the temporal context and significance of this fossil record. Given the great richness of this sequence and the intended scope of this chapter, all facets of this subject cannot be discussed. The better-known and better-studied part of the fossil record of mammals—in particular the rich sequence of land mammals—is addressed. Interested readers can also consult popular books on this subject (Thomas 1968; Brown 1988), as well as the references cited here and in those books.

History of Investigations

Simpson (1942) showed that the antiquity of collecting fossil mammals in western North America extends back almost two millennia, and unpublished evidence also indicates that the same is true in Florida. Florida's fossil vertebrates (including mammals) were collected during prehistoric times, as is evidenced at several archaeological sites across the state where fossil horse (*Equus*) teeth are found. The reasons for these fossil collections are still enigmatic, but one plausible explanation is that the teeth might have been used as weights for fish seines (L. Kozuch, pers. comm. 1992).

It is not until the late nineteenth century that we have the next definitive evidence of fossil-mammal collecting in Florida. During that time, fossil land mammals, mostly of late Cenozoic age, were brought to the attention of Dr. Joseph Leidy, a Philadelphia physician turned paleontologist (and regarded by many people as the founder of the science of vertebrate paleontology in North America). In a series of articles (see Hay 1901), Leidy described, for example, important early finds of fossil mammals along the Peace River in central Florida, fossil *Equus* from near Ocala, and a large Miocene land-mammal fauna from Mixson's Bone Bed near the town of Williston (Leidy and Lucas 1896). In contrast to the ubiquitous Pleistocene land mammals in southeastern North America known from prior discoveries, Leidy's Florida studies represented a virtually unique view of Tertiary land mammals in this region.

The next important phase of vertebrate paleontology came during the first three decades of the twentieth century. Sellards (1916) was the first to present a detailed faunal description of fossil mammals from the late Tertiary Bone Valley phosphate deposits in central Florida. Also during that time, additional Miocene land mammals were discovered in the panhandle and north-central parts of Florida, including the now well known Thomas Farm locality in Gilchrist County. One of the premier twentieth-century paleontologists, George Gaylord Simpson, published the original descriptions of these faunas (Simpson 1930, 1932). During the 1930s, field crews from Harvard University, mostly under the direction of Bryan Patterson, worked the Thomas Farm site.

It was not until after World War II that the first "resident" vertebrate paleontologists were hired at the University of Florida, Thomas Farm was deeded to the university, and research collections were begun at the Florida Museum of Natural History. Since that time, the Museum has maintained a growing and active research program in Florida vertebrate paleontology, with emphasis on Cenozoic mammals.

In addition to university-based research, Florida vertebrate paleontology has benefited greatly from the contributions, enthusiasm, and support of amateur fossil-hunters throughout the state. Relevant scientific collections have undergone dramatic growth during the last few decades, and many important specimens have been donated by the network of fossil enthusiasts in Florida. Popular books that discuss this subject include those of Thomas (1968) and Brown (1988).

The Geologic Context of Florida's Fossil Mammal Deposits

The oldest fossil mammals in Florida are from the Upper Eocene Ocala Limestone (fig. 8.1a), which crops out extensively across the northern part of the state; these deposits represent shallow-water marine paleoenvironments. The Oligocene fossil record of mammals is relatively poor, only a few smaller localities having yielded fragmentary faunas from isolated outcrops.

The classic Middle Miocene Thomas Farm locality in north-central Florida seems to be a rare fragment of a paleokarst accumulation in a Tertiary sinkhole. Many Middle Miocene fossils, including both marine and land vertebrates, come from clastic deposits of the Hawthorn Group, which are exposed over much of the state (Scott 1988b; Bryant et al. 1992). Miocene and Early Pliocene mammalian faunas are well represented, particularly from the marginal-marine and fluvial phosphatic deposits of the Bone Valley Member in central Florida (fig. 8.1b). These deposits have drawn much attention to the rich and varied land-vertebrate fauna. Other Late Miocene localities of the "Alachua clays" represent fluvial deposits—for example, the Love Bone Bed near the town of Archer (Webb et al. 1981), McGehee Farm (Hirschfeld and Webb 1968), and Moss Acres near Williston (Lambert 1990).

Middle to Late Pliocene deposits, younger than about 3 million years, are widespread in Florida, particularly in the southern part of the state. Also notable are the fossil-mammal occurrences in shell mines (e.g. the APAC pit) developed in the Pinecrest and Caloosahatchee formations east of Sarasota (Jones et al. 1991).

Pleistocene deposits are ubiquitous throughout the state, including hundreds of documented fossil localities representing diverse sedimentary environments. One of the most spectacular of these localities is the early Pleistocene Leisey Shell Pit in Hillsborough County, where an extraordinarily rich terrestrial mammal fauna comes from strata interbedded with the dominantly marine Bermont Formation (Webb et al. 1989). In addition to the many coastal localities, later Pleistocene fossil verte-

Figure 8.1. A: Field crew collecting fossil mammals from the Upper Eocene Dell Limerock Pit near Mayo. Archaeocete whales have been found at this locality. B: Miocene and Pliocene Bone Valley phosphate deposits representing a coastal paleoenvironment. These sediments are highly fossiliferous and preserve a rich record of marine and terrestrial mammals from about 4.5–5.0 million years ago.

brates are known from the extensive systems of filled solution-enlarged fissures—features of the karst topography that formed particularly in the Ocala Limestone.

Sequence of Mammal Faunas

Florida's fossil mammals are found throughout the state in sediments ranging in age from 45 million years to about 10,000 years ago. Although there are some major hiatuses in the Florida stratigraphic sequence, every Cenozoic epoch after the Paleocene is represented. The highlights of this sequence follow.

EOCENE

Eocene mammalian faunas in Florida are exclusively marine and include whales and sirenians. Numerous shark teeth are found with the mammals. The whales are of the most primitive group, the archaeocetes, and are relatively well represented in coastal-plain deposits of the southeastern U.S. Fragmentary remains of Eocene archaeocetes—mostly teeth, vertebrae, and ribs of *Zygorhiza* and *Basilosaurus*—have been collected from the upper Ocala Limestone at several localities in northern Florida (fig. 8.1). Domning et al. (1982) described some of the earliest known sea cows, including important specimens of *Protosiren*, also from the Ocala Limestone.

OLIGOCENE

In contrast to the widespread Upper Eocene Ocala Limestone, Oligocene deposits are relatively limited in Florida, and so too is the fossil record of mammals from that time. The oldest occurrence of land mammals, and first clear paleontological indication of the emergence of the Florida peninsula, is from the I-75 Local Fauna in northern Florida. This is a serendipitous occurrence, the deposit having been discovered in southwestern Gainesville during excavations for the interstate highway in 1965. In addition to other marine and terrestrial vertebrates, this small assemblage of land mammals includes insectivores, bats, rodents, carnivores, horses, and artiodactyls. These taxa are diagnostic of the mid-Oligocene (probably Whitneyan; Patton 1969), about 30 million years old according to the time scale of Woodburne (1987) (fig. 8.3). This locality and fauna correlate to the upper part of the classic White River Group of western North America (e.g. the Brule Formation of the section exposed at the Big Badlands of South Dakota; Emry et al. 1987).

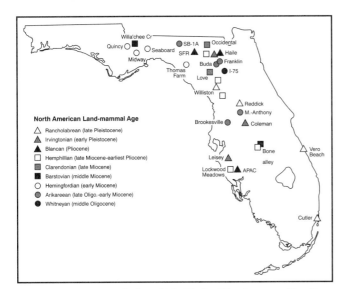

Figure 8.2. Locations of selected Florida fossil land-mammal localities and North American land-mammal ages, Oligocene to Pleistocene (see also fig. 8.3).

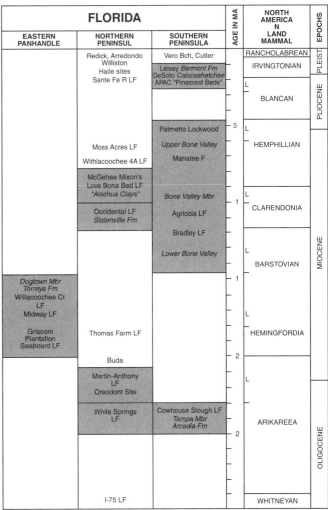

Figure 8.3. Correlation chart showing major fossil-mammal-bearing sedimentary units (shaded) and localities from the Late Oligocene to Pleistocene. Correlations to the Cenozoic time-scale and North American land-mammal age boundaries follow Woodburne (1987). Fm: formation; LF: local fauna; Mbr: member (see also fig. 8.2).

MIOCENE

The Miocene epoch is abundantly represented in Florida. Fossil marine mammals, although abundant and widely distributed, have been studied less than contemporaneous land mammals. Interesting occurrences of all extant classes of vertebrates, principally including bony fish, sharks, sirenians, and cetaceans (including whales, porpoises, and dolphins), offer much potential for future research. In contrast, the very rich fossil record of Miocene land mammals is better known and is discussed more fully here.

Arikareean (age 29 to 20 million years) Simpson (1932) described Arikareean mammals—including carnivores, perissodactyls, and artiodactyls—from Franklin phosphate pit no. 2 near Newberry. Patton (1969) described another small Arikareean local fauna from near Brooksville. Other fossil assemblages in the older literature have also been referred to the Arikareean, but revised chronologic interpretations indicate that only the two aforementioned faunas, until relatively recently, represented the extent of our knowledge of this important land-mammal age.

During the last few decades, however, several new localities in northern Florida containing Arikareean mammals have been collected. The Buda Local Fauna of Alachua County was discovered in 1965 and subsequently worked by crews from the Florida Museum of Natural History. This assemblage consists of some 13 taxa of land mammals, including an insectivore, carnivores, perissodactyls, and artiodactyls (Rich and Patton 1975; Frailey 1979). Another diagnostic fauna of late Arikareean age, the SB-1A Local Fauna from north of Live Oak, includes a bear-dog, a true dog, a weasel-like mammal, squirrel, horse, and camel (Frailey 1978). MacFadden (1980) described a primitive artiodactyl, the oreodont *Phenacocoelus luskensis,* from the Martin-Anthony road cut north of Ocala. A small faunule containing a carnivore, rhinoceros, and camelid also occurs stratigraphically above the oreodont-bearing unit. The stage of evolution represented in these occurrences correlates to the late Arikareean as it is known from classic localities in northwestern Nebraska and southeastern Wyoming (Harrison Formation, about 21 million years old) (MacFadden 1980). Several partial skeletons of Arikareean oreodonts are also known from the White Springs Local Fauna.

Hemingfordian (age 20 to 16 million years) Simpson (1930, 1932) described some of the original Hemingfordian land mammals from sites in the Florida panhandle, including Griscom Plantation and the Midway and Quincy fullers-earth mines. Olsen (1964a, 1964b, 1968) reported Hemingfordian land mammals in marginal-marine to nonmarine deposits in the Florida panhandle around Tallahassee (e.g. the Seaboard Local Fauna), which were correlated to planktic-foraminiferal stage N6 (see also Tedford and Hunter 1984). A $^{87}Sr/^{86}Sr$ age determination that approximates the age of the Seaboard Local Fauna at about 18.4 million years is consistent with previous correlations to the marine zones (Bryant et al. 1992).

Discovered in 1931, the Thomas Farm locality in Gilchrist County is unquestionably the richest Hemingfordian site in eastern North America. Field crews from the Florida Museum of Natural History have done much to advance our understanding of the paleoecology of this site, which is predominantly a sinkhole fill (Pratt 1990), and to expand the faunal list (particularly the small mammals), which now includes more than 80 vertebrate taxa, including about 30 mammals (Webb 1981b; Pratt and Morgan 1989). Individual taxa within this fauna have many interesting aspects, which are discussed in detail below.

Barstovian (age 16 to 11.5 million years) Relative to the other parts of the Miocene, the Barstovian is poorly represented in Florida. MacFadden (1982) described specimens of the three-toed browsing horse *Hypohippus* from the Bone Valley phosphate deposits that presumably were collected from lower parts of the section exposed in the open-pit mines. Similarly, Webb and Crissinger (1983) and Hulbert and MacFadden (1991) described fossil mammals collected from the rarer, older (i.e. pre-Hemphillian) units within the Bone Valley Formation.

An interesting and important assemblage, the Willacoochee Creek fauna, consisting of sixty-eight vertebrate taxa, including twenty-nine mammals, has been recovered from fullers-earth mines in the Florida panhandle north of Quincy. The stage of evolution of this fauna indicates an earliest Barstovian age of about 15 to 16 million years (Bryant 1991b; Bryant et al. 1992).

Clarendonian (age 11.5 to 8 million years) Clarendonian deposits were not well known in Florida until the discovery in 1974 of the Love Bone Bed in Archer. Tens of thousands of vertebrate specimens have been recovered from this site, and the extraordinarily rich fauna consists of more than eighty identified vertebrate taxa, of which more than half are mammals (Webb et al. 1981). Other taxa undoubtedly will be added as further studies are done, particularly lower vertebrates and microfauna. Although there is a small marine-vertebrate component (e.g. sharks and shore birds), the other taxa indicate a fluvial and riparian paleoenvironment, with both browsing and grazing herbivores.

The land mammals are of particular importance because they represent a high species diversity, similar to contemporaneous faunas found elsewhere in the U.S., and comparable to the present-day African savannas. This continent-wide, extinct land-mammal assemblage has been termed the "Clarendonian chronofauna," and represents a maximum diversity for the late Tertiary of North America (Webb 1969). Mammals of the Love site represent a diverse assemblage including carnivores, gomphothere proboscideans, artiodactyls, rhinoceroses, tapirs, and horses—the latter representing the time of maximum diversity of the group (MacFadden 1992).

Hemphillian (age 8 to 4.5 million years) With the discovery of Mixson's Bone Bed in Williston (Leidy and Lucas 1896), fossil land mammals of Hemphillian age were among the first studied in great detail in Florida. Mixson's also has been of great importance because before discovery of the sites mentioned below, it represented the extent of our knowledge of Late Miocene fossil mammals from eastern North America. The Mixson's Bone Bed Local Fauna includes important representatives of horses, rhinoceroses, carnivores, sloths (see below), camelids (including lamines), and proboscideans.

Several important early Hemphillian sites have been worked in northern Florida within the last several decades. These include the Withlacoochee 4A site southwest of Ocala, the McGehee Farm site near Newberry, and the Moss Acres site near Williston. These contain important and diagnostic land mammals, including a diverse assemblage of taxa endemic to North America. McGehee is of great interest because, like Mixson's, it contains megalonychid sloths, representing an early wave of immigration from South America (Marshall et al. 1979).

PLIOCENE

Pliocene land mammals represent the Blancan land mammal age (4.5 to 1.8 million years). As with the Miocene, Blancan land mammals are found at numerous localities throughout Florida. Particularly rich deposits include some of the Haile (Robertson 1976; and others) and Santa Fe River sites in northern Florida, and the APAC pit east of Sarasota, the latter having yielded estimated ages of about 2.5 to 2.0 million years, by a variety of techniques (Jones et al. 1991). A major faunal change is apparent in Blancan sites: numerous endemic mammals (including rhinoceroses and several taxa of horses) are extinct, and another wave of South American immigrants appeared, including sloths, glyptodonts, armadillos, and porcupines. The Late Pliocene (late Blancan to early Irvingtonian) represents the height of the "Great American Interchange" about 2 million years ago, which resulted from the newly formed land connection between the Americas in the Isthmus of Panama (individual immigrant taxa are discussed below).

PLEISTOCENE

There are literally hundreds of known Pleistocene fossil localities in Florida, only a few of which are depicted in figure 8.2. These localities represent diverse sedimentary environments, and they have yielded faunas varying from a few fragmentary bones to tens of thousands of specimens. A few of these localities—those that are well represented in the collections, have been actively studied, or are of historical interest—are the Inglis, Leisey (fig. 8.4), Haile, Coleman, and DeSoto sites of Irvingtonian (Early to mid-Pleistocene) age, and the Reddick, Arrendondo, Williston, Vero Beach, and Cutler sites of Rancholabrean (Late Pleistocene) age (figs. 8.2, 8.3, and 8.4). These vertebrate faunas are of importance from an evolutionary and biogeographic perspective because they continue to show the effects of the Great American Interchange, and they record the great megafaunal extinction that occurred prior to 10,000 years ago, when various edentates, rodents, carnivores, proboscideans, perissodactyls, and artiodactyls became extinct in this region and elsewhere in North America.

Figure 8.4. Fossil bones exposed during excavations at the middle Pleistocene (Irvingtonian) Leisey Shell Pit, Hillsborough County. Tens of thousands of important fossil mammal specimens have been collected from Leisey.

Highlights of Fossil Mammals from Florida

The exceedingly rich fossil record of Cenozoic mammals can only be highlighted here. Coverage of vertebrate groups, reflecting the state of knowledge, is not even: micromammals, with a poorly known fossil record, and marine mammals, which are not well studied, are given less attention. The classification and temporal distribution follows the excellent recent synthesis by Hulbert (1992).

MARSUPIALS (MAGNORDER MARSUPIALIA)

Marsupials are primarily a Southern Hemisphere group with a relict Gondwanan distribution, but one family, the Didelphidae (opossums), also is widespread in Cenozoic deposits of North America. In Florida the first record of didelphids is an undescribed new taxon from the I-75 Local Fauna. The common middle Tertiary genus *Peratherium* occurs at several Miocene localities in the northern part of the state, including one of the youngest North American occurrences for this genus at Thomas Farm. There is a hiatus from the Middle Miocene until the mid-Pleistocene in which marsupials are absent from Florida, and elsewhere in North America. The extant opossum, *Didelphis virginiana*, which is widely distributed today, occurs in the middle and Late Pleistocene record from Florida and represents a northward dispersal from South America across the Panamanian land bridge during the later phases of the Great American Interchange about 1 million years ago.

EDENTATES (SUPERORDER EDENTATA)

Edentates have an extensive fossil record in the New World, including the two major groups, the cingulates, or shelled forms, and pilosans, or hairy forms. Edentates originated in South America and underwent a major adaptive radiation there during most of the Tertiary. The presence of the various subgroups (i.e. sloths, anteaters, glyptodonts, and armadillos) in North America indicates northward dispersal during the late Cenozoic before and during the height of the Great American Interchange.

Glyptodonts and Armadillos (Order Cingulata)
Glyptodonts, with bony carapaces composed of ornamented plates, are found at many Florida Pliocene and Pleistocene sites. This group is principally represented by two species of the common North American genus *Glyptotherium*.

The common nine-banded armadillo, *Dasypus novemcinctus*, a familiar extant mammal, was introduced into southern Florida during the first half of this century and has since expanded its range northward up the peninsula, into Georgia, and beyond (Humphrey 1974). The fossil record of armadillos in Florida is extensive since the Pliocene; like many other fossil mammals found in the southeast, the armadillos represent a natural dispersal into this region during the Great American Interchange. Two main groups are represented during that time. The dasypodids, represented by the extinct *Dasypus bellus*, is significantly larger than the extant *D. novemcinctus*. The other group of armadillos in Florida, wholly extinct today, are the pampatheres, or giant armadillos, including *Holmesina* (fig. 8.5). The evolutionary sequence of giant armadillos in Florida is interesting because there is a size increase that seems to have been gradual during some intervals and rapid at other times (Edmund 1987; Hulbert and Morgan 1993). Both *Dasypus bellus* and *Holmesina* became extinct at the end of the Pleistocene, about 10,000 years ago.

Sloths (Order Phyllophaga) The earliest record of sloths in Florida is during the early Hemphillian. One group, the megalonychids, represented by *Pliometanastes protistus*, has been recovered from the McGehee site in Alachua County (Hirschfeld and Webb 1968). The other occurrence involves the mylodontids, represented by *Thinobadistes* from Mixson's Bone Bed in Levy County (Webb 1989). About 8 to 9 million years old, these fossils represent some of the earliest occurrences of South American immigrants in North America, predating by 5 to 6 million years the formation of the Panamanian land bridge and the Great American Interchange during the late Cenozoic. Without any known land bridges during the Late Miocene, the mode of dispersal of these sloths has been somewhat enigmatic, but almost certainly must have involved dispersal across water, possibly island-hopping through what is now Central America, or across the Caribbean.

During the latest Cenozoic, about 2.5 million years ago, sloths were very abundant in Florida, where they are represented by a more advanced suite of mylodontids

Figure 8.5. Reconstruction of the extinct armadillo *Holmesina septentrionalis* based on composite material from the Pleistocene of Florida (Edmund 1985). This giant armadillo stood roughly 1 meter tall and was about 2 meters long.

(*Glossotherium* and *Paramylodon*) and megalonychids (*Megalonyx*). The third principal group, the megatheres, is represented in Florida by *Eremotherium* and *Nothrotheriops*. The presence of these taxa in North America indicates northward dispersal across Central America. Sloths became extinct in Florida, and elsewhere in North America, by about 10,000 years ago.

The fossil record of edentates in Florida is one of the richest in North America and preserves the remains of the great northern migration from South America. Like many mammals, sloths, glyptodonts, and armadillos all disappeared from North America by the end of the Pleistocene, some 10,000 years ago.

INSECTIVORES (GRANDORDER INSECTIVORA)

Fossil insectivores are characteristically rare in Florida, as they also are at other fossil localities. This rarity is primarily a result of the fact that small fossil mammals (including insectivores, bats, rodents, and lagomorphs) require special recovery techniques (screenwashing) not frequently employed at the majority of fossil sites. In Florida, the first known insectivore is a shrew from the I-75 Local Fauna. From younger deposits, the hedgehog *Amphechinus* from the Early Miocene Buda Local Fauna (Rich and Patton 1975) is a noteworthy and rare occurrence in North America from a group restricted today to the Old World (Walker 1975). Shrews and moles are known from the Early to Middle Miocene of Florida, particularly as a result of screen-washing efforts at Thomas Farm. The extant shrew genera *Blarina* and *Sorex* and the mole *Scalopus* are known from the Florida Pliocene and Pleistocene (Hulbert 1992). Our knowledge of Florida's fossil insectivores (along with the rodents and bats) is poor, but it undoubtedly will improve when more screen-washing is accomplished.

BATS (ORDER CHIROPTERA)

The preferred natural habitats of many kinds of bats are caves and sinkholes, and because of the paucity of cave and sinkhole deposits in the geologic record and the small size and fragility of bats, they are rarely preserved as fossils; however, there are some notable exceptions.

The earliest record of fossil bats in Florida comes from the mid-Oligocene I-75 Local Fauna; the fossils have not been described, but are known to include five species representing four extant families. In younger deposits, the first abundant record of chiropterans in the Florida sequence has been brought to light by intensive microfaunal work at the Thomas Farm site. This largely undescribed fauna in the collections of the Florida Museum of Natural History includes more than 2,000 specimens, probably representing eight to ten taxa, making it the largest and most diverse bat fauna from any single locality in the world (Lawrence 1943; Morgan 1989).

In the subsequent fossil record, bats are poorly represented until the Plio-Pleistocene, where intensive microfaunal work again has resulted in recovery of some important samples, representing particularly the later part of this time interval. For example, Gut (1959) described the vampire bat *Desmodus* from the Late Pleistocene Reddick Local Fauna of northern Marion County. Morgan et al. (1988) described another species of this genus from the Inglis and Haile sites, extending the range of the genus back into the Late Pliocene, and expanding the known geographic range northward from its present Neotropical distribution. According to these authors, *Desmodus* preyed upon large mammals like sloths, and followed the prey northward during the Great American Interchange.

CARNIVORES (ORDER CARNIVORA)

Carnivores are characteristically rare in Florida, as they also are elsewhere in the fossil record. This stems from the fact that carnivores are usually at the top of the food chain, and the number of individuals expectably is about an order of magnitude less than the number of individual primary consumers, the herbivorous mammals. There are certain exceptions, however, and these special accumulations result from some special concentration as a result of the process of fossilization.

Dogs, Wolves, and Foxes (Family Canidae) The canids comprise three subfamilies, two with a relatively good fossil record in Florida (the third, the Hesperocyoninae, is known only from the very early Miocene). The first of these better-represented subfamilies, the Borophaginae, are bone-crushing dogs from the Miocene and Pliocene of North America, first represented in Florida by *Tomarctus* from the Thomas Farm site. Late Miocene borophagine canids are well represented by a large sample of *Epicyon* from the Love site. Dogs of this subfamily referred to as *Osteoborus* have been found in the Bone Valley Member. All of these canids were wide-ranging in North America and are good guide fossils to the land-mammal ages they represent.

The other canid group, the Caninae, includes the canines (or "true" dogs), wolves, and foxes. Although they are known from the Miocene of Florida, they are not abundant (relative to the normal rarity of carnivores in general) in deposits older than the Plio-Pleistocene; in these younger sediments, the modern genera *Urocyon* (gray fox) and *Vulpes* (red fox) are represented, as well as seven extinct species of the extant genus *Canis,* including the common large dire wolf of the Late Pleistocene, *Canis dirus.* These genera were very widely distributed in North America during that time.

"Bear-dogs" (Family Amphicyonidae)   Amphicyonids were generally medium-sized to large bear-dogs, an extinct carnivore group found throughout Holarctica during the Miocene. The oldest amphicyonids in northern Florida come from the Early Miocene (Arikareean sites) and are taxonomically similar to those found in western North America (Frailey 1979). Two species of bear-dogs are found at Thomas Farm and are referable to the genera *Cynelos* and *Amphicyon.* Olsen (1958, 1960) presented a detailed morphologic analysis of *A. longiramus* from Thomas Farm (fig. 8.6). He found that this bear-dog had lost the characteristic carnassial shear of many other carnivores, and it probably was rather omnivorous; the postcranial skeleton was very robust. These features are more characteristic of bears than of dogs, which led Olsen to conclude that amphicyonids and ursids are very closely related. Later Miocene bear-dogs also are known from Florida, including specimens of *Ischyrocyon* from the Love site.

Bears (Family Ursidae)   Bears first appear in Florida with the occurrence of *Hemicyon johnhenryi* from Thomas Farm. Other Miocene bears occur at some of the larger localities, including *Indarctos* from the Withlacoochee 4A site, and *Agriotherium* from the Bone Valley Formation. Kurten (1966, 1967) described abundant remains of bears from several Florida Pleistocene localities. He speculated that the extinct spectacled bear *Tremarctos floridanus* was adaptively similar to the Pleistocene cave bear, *Ursus speleaus,* whose remains are abundant in Europe, and they probably occupied similar niches. The Florida record of *Tremarctos* is significant because it represents a considerable range extension for this genus, which is known today only in mountainous, forested regions of South America (Walker 1975). *Ursus,* the genus that includes extant North American bears, is represented by fossils at several Late Pleistocene localities in Florida.

Walruses (Family Odobenidae)   Walruses are represented by a single living genus and species, *Odobenus rosmarus,* which is restricted mostly to shallow, coastal, circum-Arctic waters (Walker 1975). It is therefore surprising from a biogeographic point of view to find fossil walruses in Florida. Nevertheless, the extinct genus *Trichecodon* is known from the upper Bone Valley Formation, and there also are fragmentary remains from the early Pleistocene.

Raccoons (Family Procyonidae)   Fossil racoons and their close relatives, all grouped within the Procyonidae, have a complex biogeographic history in the Americas. Some of the best-preserved fossil procyonids come from Florida (Baskin 1982). The earliest records from Florida are from the Late Miocene, including a relatively good sample from the Love site (fig. 8.7). This occurrence slightly predates or is roughly contemporaneous with, the dispersal of procyonids into South America about 9 million years ago (Marshall et al. 1983), where the group underwent a significant adaptive radiation, ultimately resulting in the present-day diversity on that continent. *Procyon,* the extant raccoon, first occurs in Late Pliocene deposits of Florida and apparently represents a north-

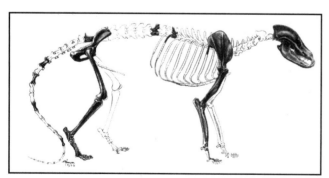

Figure 8.6. Composite skeletal reconstruction of the extinct bear-dog *Amphicyon longiramus* from the middle Miocene Thomas Farm site, Gilchrist County (from Olsen 1960). This animal was about 2 meters long. Known elements are indicated by shading (reprinted by permission of the author and the Museum of Comparative Zoology, Harvard University).

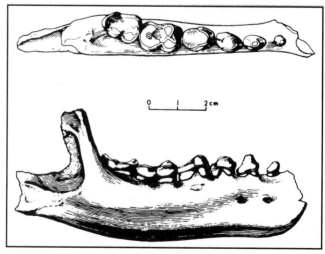

Figure 8.7. Dorsal (top) and lateral (bottom) views of right mandible with cheek teeth (p1-m2) of the procyonid *Arctonasua floridana* from the late Miocene Love site, Alachua County (Baskin 1982; reprinted by permission of the Society of Vertebrate Paleontology).

ward "back migration" during the Great American Interchange.

Weasels, Skunks, Otters, and Their Relatives (Family Mustelidae) This very diverse group of carnivores is well represented in the fossil record of Florida. The oldest mustelids are from Arikareean localities in northern Florida (Frailey 1979). Several taxa, including the fairly widespread *Leptarctus* and one new taxon, are known from the Early to Middle Miocene deposits at Thomas Farm. Late Miocene and Early Pliocene mustelids are even more diverse, and include taxa referable to the weasels (*Trigonictus*), wolverines (*Plesiogulo*), otters (*Enhydritherium*), and skunks. Extant mustelid genera are relatively common for carnivores in the Pleistocene, and are represented by *Mustela* (minks and weasel), *Lutra* (otter), and three skunks, *Mephitis, Conepatus,* and *Spilogale.*

False Sabercats, or Paleofelids (Family Nimravidae) Traditionally, saber-toothed cats were joined together into a single group, the machairodonts, because they shared the distinctively enlarged "saber-tooth" canine. Paleontologists have since analyzed cranial morphology, particularly in the ear region, and have concluded that the evolution of enlarged, saber-tooth canines occurred several times in different carnivores. The extinct Nimravidae, for the lack of a better term, is a family of "false sabercats," with a morphologically primitive ear region, that is less closely related to the common extinct sabercats (e.g. *Smilodon*) than the latter is to non-saber-toothed cats (e.g. lions, panthers, bobcats, and domestic cats). Nimravids have their origins in the Old World, and their presence in North America during the Miocene represents an immigration event, apparently over the Bering land bridge. The earliest record of nimravids in Florida (and an early New World record for this group) is based on a single specimen from the Arikareean Buda Local Fauna (Frailey 1979). By far the best sample of a nimravid in the Florida fossil record (and one of the best in North America) is the excellent cranial, dental, and postcranial material of *Barbourofelis* (fig. 8.8) from the Late Miocene Love site (Baskin 1981).

"True" Cats (Family Felidae) Within the true cats, both the saber-toothed and other cats are relatively well represented in the Florida fossil record. The earliest record of the felid sabercats is *Nimravides* from the late Miocene Love site. In earliest Pliocene deposits of the upper Bone Valley Member, felid sabercats are represented by *Machairodus* and *Megantereon.* The genus *Smilodon* is found at Florida Pliocene and Pleistocene localities and includes the cosmopolitan latest Pleistocene species *S. populator* or *S. fatalis,* especially well known from the La Brea Tar Pits in California. In her descriptions of Florida *Smilodon,* Berta (1987) concurred with other research suggesting that the large canines were probably used to puncture critical blood vessels (e.g. in the neck), rather than the classic interpretation that these weapons were used to pierce the thick flesh of larger-bodied prey species. Based on postcranial remains, Berta also concluded that *Smilodon* was probably more powerful and flexible than other cats, but not as highly cursorial. Another, less well known (in the popular literature, at least) felid sabercat of the Pliocene and Pleistocene in Florida is *Homotherium,* represented by a large undescribed sample from the Irvingtonian "Hog Heaven" site (Haile 21A) in northern Florida.

The oldest conical-toothed or non-saber-toothed felids in Florida are from the Early Pliocene sequence of the upper Bone Valley Member. MacFadden and Galiano (1981) described several specimens (fig. 8.9) from the phosphate deposits that were later referred to the extant bobcat genus *Lynx* (Werdelin 1985). During the Pliocene and Pleistocene there was considerable diversity among these true cats, which included a large, lion-sized jaguar, *Panthera atrox,* a cheetah (*Miracinonyx,* closely related to

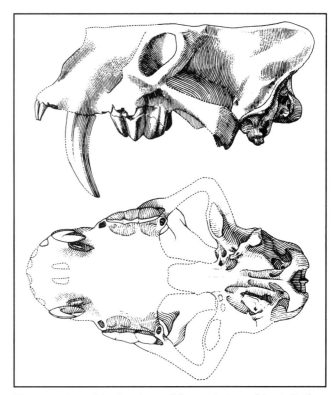

Figure 8.8. Lateral (top) and ventral (bottom) views of the skull of a false sabercat *Barbourofelis* from the Late Miocene Love site, Alachua County (from Baskin 1981 and reproduced with permission of the America Society of Mammalogists).

Old World forms), and other cats no longer native to this region (e.g. the ocelot). The extant puma, *Puma concolor,* was a relatively late arrival in Florida and is known only from the Late Pleistocene (Kurten 1965; Hulbert 1992).

Hyenas (Family Hyaenidae)   Hyenas are an almost exclusively Old World group of predominantly scavenging carnivores, with a notable exception of relevance to Florida: Berta (1981) described the hyena *Chasmaporthetes* from two Late Pliocene to Early Pleistocene faunas of Florida (and elsewhere in North America). The initial the dispersal of this form from Eurasia began about 3.5 million years ago. In contrast to our modern concept and knowledge of hyenas of the African savannas, *Chasmaporthetes* was a powerful, highly cursorial predator with well developed, bladelike carnassials (Berta 1981).

RODENTS (ORDER RODENTIA)

Because of their generally small size, rodents are not frequently encountered in surface prospecting for fossils. Thus, these and other elements of the "microfauna" are now routinely collected by screen-washing (McKenna 1962). Although a few fossil rodent taxa were known from Florida prior to the widespread use of screen-washing, the technique has greatly advanced our understanding of this most diverse order of mammals. Because of the incredible diversity of the rodents, only the broadest highlights of selected taxa can be discussed here.

Extinct Burrowing Rodents (Family Mylagaulidae)   This extinct family of burrowing rodents has its closest relatives in the extant aplodontids (mountain beavers from the Pacific northwest). Mylagaulids have a distinctive dentition that is very high-crowned, and the number and shape of the enamel folds (lakes) are taxonomically diagnostic. Mylagaulids were fairly widespread (although usually not very abundant) during the Early and Middle Miocene of North America, and because of their rapid evolution they are good guide fossils when found. The Florida record of mylagaulids is scanty, and there are only a few occurrences. Late Miocene specimens of *Mylagaulus* in a temporal sequence from Florida show examples of evolutionary reversals in both size and morphology (Baskin 1980) (fig. 8.10).

Squirrels (Family Sciuridae)   Other than Late Pleistocene fossils referable to the modern genera *Tamias, Sciurus,* and *Glaucomys,* the most notable occurrence of fossil squirrels in Florida has come from intensive screen-washing at Thomas Farm. Within this microfaunal sample, much of which is still undescribed,

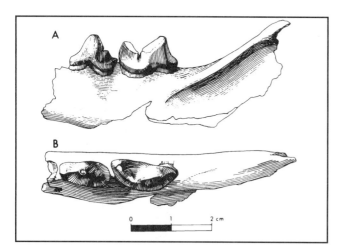

Figure 8.9. Lateral (A) and dorsal (B) views of jaw with 4th premolar and 1st molar of *Lynx rexroadensis* from the Early Pliocene Upper Bone Valley deposits, Polk County (MacFadden and Galiano 1981). In lateral view note the slicing carnassial notch of the 1st molar, which is characteristic of felids.

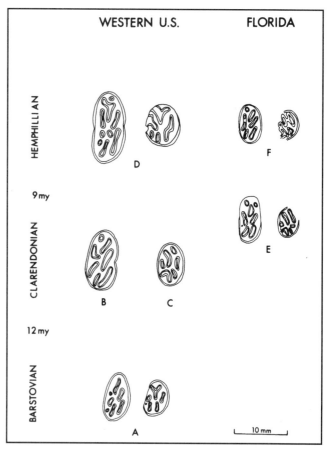

Figure 8.10. Evolutionary reversal in the extinct rodent *Mylagaulus* from Florida. In the western United States this genus is characterized by an increase in size and complexity (in enamel folds), whereas in Florida there is a slight size decrease and negligible increase in complexity of enamel folds (from Baskin 1980 and reproduced with permission of the *American Midland Naturalist*).

three taxa of squirrel-like rodents have been reported, including a flying squirrel and a chipmunk (Pratt and Morgan 1989).

Beavers (Family Castoridae) Although beavers are known from the Miocene of Florida, they are not abundant. The most interesting occurrence in Florida is the giant *Castoroides* from many Irvingtonian and Rancholabrean localities (Martin 1969b). With the size of a black bear, *Castoroides* was the largest rodent in North America during the Pleistocene (Kurten and Anderson 1980).

Pocket Gophers (Family Geomyidae) The fossil record of pocket gophers in Florida extends back to a single occurrence from the Early Miocene (Hulbert 1992), but by far the best-known occurrences are from the Pliocene and Pleistocene. Two genera are represented in deposits of this age, *Geomys,* which has a fairly continuous fossil record spanning the last 2 million years (Wilkins 1984), and *Thomomys,* whose presence in Florida represents an extraordinary range extension from its present-day distribution in western North America (Wilkins 1985) (fig. 8.11).

Pocket Mice and Kangaroo Rats (Family Heteromyidae) Although the Family Heteromyidae is first represented by an undescribed genus from the late Oligocene I-75 Local Fauna (Patton 1969), its only other known representative in Florida is the relatively common extinct genus *Proheteromys*. At Thomas Farm, the best-known occurrence, Bryant (1991a) studied the population dynamics of a large sample of more than sixty-three individuals of *P. floridanus*. Based on wear classes in the teeth, he found that average longevity of this fossil rodent was about 6.7 months, only 6% survived past one year, and fecundity was 8.5 offspring per female. These population parameters are similar to those for modern heteromyid rodents, and suggest a fairly conservative behavioral ecology for this group during the last 18 million years.

Mice and Voles (Family Muridae) This principally Old World group, represented by some 100 known genera, is the most successful in terms of overall mammalian diversity. In Florida, the first murid is represented by an undescribed form in the I-75 Local Fauna, and the group is represented by *Copemys* from several later Miocene localities. *Sigmodon,* the cotton rat, was fairly common in North America during the Plio-Pleistocene, and the several described species present a chronoclinally evolving sequence in the Florida localities (e.g. Haile, APAC, and Leisey); thus, the stage of evolution of *Sigmodon* is a good temporal index (Martin 1969a; Morgan and Ridgway 1987; Hulbert and Morgan 1989).

Voles are generally an advanced, rapidly evolving group with a more northern distribution. The presence in Florida of forms such as *Microtus* during the Late Pleistocene is taken to represent cooler glacial intervals (Martin 1968; Webb 1974a).

Porcupines (Family Erethizontidae) and Capybaras (Family Hydrochoeridae) Both of these caviomorph groups originated in South America, dispersed northward during the Plio-Pleistocene, and are relatively well represented in the Florida sequence. The modern distribution of the porcupine *Erethizon* in North America excludes the Southeast, but Frazier (1981) has described fossils referable to this genus from the Plio-Pleistocene of Florida. Frazier believed that rather than forming a local ancestral-descendant sequence, these species probably are found in the Florida fossil record as a result of multiple dispersals from South America.

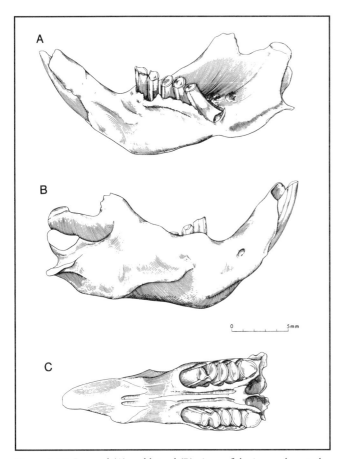

Figure 8.11. Internal (A) and lateral (B) views of the jaw and ventral (C) view of the palate of the pocket gopher *Thomomys orientalis* from the Pleistocene of Florida (Wilkins 1985).

Frazier also found that these extinct *Erethizon* already possessed dentitions capable of stripping bark off conifers, which is a distinctive adaptation of the modern forms (Walker 1975). *Erethizon* became locally extinct by the end of the Pleistocene.

Capybaras are a group of large rodents confined to Panama and the lowlands of northern South America, represented by the extant *Hydrochaeris hydrochaeris,* which can attain a length of almost 1.5 m and a weight of 50 kg, making it the largest living rodent (Walker 1975). Capybaras extended their range northward during the Late Pliocene and Pleistocene, spreading throughout southern North America, including Florida. Two genera, *Neochoerus* and *Hydrochaeris*, are known from this region; they are differentiated by their highly distinctive dentitions of adjacent plates (laminae) and cementum (Ahearn and Lance 1980; Ahearn 1981). Perhaps the most noteworthy feature of these extinct Pleistocene forms is the diversity of their body sizes, the two species of *Neochoerus* being larger than the modern capybara, and the one extinct species of *Hydrochaeris* being smaller (Ahearn 1981). Capybaras became extinct throughout North America by the end of the Pleistocene.

Most of the middle and late Cenozoic rodent families that commonly occur in North America are also found in Florida. These families represent the evolutionary history of indigenous groups, such as the sciurids and heteromyids, as well as many immigrant taxa, including the murids and caviomorphs. As screen-washing continues at diverse sites in Florida, important new samples will be recovered that will allow a better understanding of this highly successful group of mammals.

RABBITS AND HARES (ORDER LAGOMORPHA)

Lagomorphs have a relatively good fossil record in North America in middle and late Cenozoic deposits. In Florida there is a possible early record from the Late Oligocene I-75 Local Fauna, and from some Early Miocene sites. *Hypolagus,* the common Miocene and Pliocene lagomorph found elsewhere in North America, also occurs at several Florida sites (White 1987). Extinct species of the modern genera *Sylvilagus* and *Lepus* are known from numerous Pleistocene sites in Florida (Hulbert 1992).

PERISSODACTYLS (ORDER PERISSODACTYLA)

Perissodactyls, the odd-toed ungulates, represent one of the dominant groups of hooved herbivorous mammals that lived in terrestrial environments during the Cenozoic. Although today they only are represented by equids (horses, zebras, and their relatives), tapirs, and rhinoceroses, this modern diversity is just a fraction of what is known from the rich fossil record of the group.

Rhinoceroses (Family Rhinocerotidae) Rhinoceroses first appeared in Florida during the Miocene and are represented by two groups from that epoch: *Menoceras,* similar to Arikareean forms from the western United States, and *Floridaceras whitei,* an endemic species from Thomas Farm. Late Miocene rhinoceroses are well represented in samples from the Love site, the McGehee site, and the upper Bone Valley Member, including the two forms *Aphelops* (fig. 8.12) and *Teleoceras,* which show niche specializations in these coexisting taxa. *Aphelops* was a more highly cursorial form, whereas *Teleoceras,* with its short and stout limbs, may have been semi-aquatic (ecologically similar, but unrelated to the present-day hippopotamus). Teeth of both were relatively high-crowned, and these animals probably were grazers. In terms of total biomass, rhinoceroses (like the horses) were a dominant component of the community of larger-bodied terrestrial mammals preserved in Florida (MacFadden 1992), as they were also elsewhere in North America during the Miocene (Voorhies 1981; and others). Although some of the principal chapters in the evolutionary history of rhinoceroses occurred in North America, the family became extinct there during the earliest Pliocene; its youngest recorded occurrence in Florida is from the upper Bone Valley Member.

Horses (Family Equidae) The fossil record of horses is abundantly represented in Florida during the late Cenozoic. The oldest occurrence of the Equidae, which provides the first evidence for emergence of the Florida peninsula, is *Mesohippus* from the Whitneyan I-75 local fauna, a genus that was very common elsewhere in North America during the Oligocene. In the Arikareean of Florida, fossil horses are relatively rare, the only

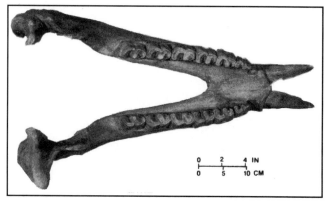

Figure 8.12. Dorsal view of jaw of the rhinoceros *Aphelops* from the Late Miocene Love site, Alachua County (from MacFadden 1979).

reported occurrence representing a few taxonomically undiagnostic specimens of a *Parahippus*-like form from the SB-1A and Buda local faunas (Frailey 1978, 1979). This rarity may represent sampling bias, because there are only a few faunas, and they are neither abundantly represented nor diverse.

The first abundant fossil equids in Florida come from Hemingfordian sites in the panhandle (Olsen 1964a, 1964b, 1968), and from Thomas Farm (Bader 1956; and others). The former sites contain the common Hemingfordian three-toed horse *Parahippus leonensis*, whereas the latter site also contains the rarer dwarf horse *Archaeohippus* and the browsing *Anchitherium*. The Thomas Farm horses, which account for about 80% of all macrofossils collected there, have many interesting aspects. The Middle Miocene was an important time in their evolution. For example, the dominant three-toed horse *Parahippus leonensis* is the outgroup for the explosive adaptive radiation of hypsodont (high-crowned) horses that occurred in North America during the Middle Miocene (MacFadden and Hulbert 1988; Hulbert and MacFadden 1991) (fig. 8.13). Correlations with panhandle localities containing *P. leonensis* and new $^{87}Sr/^{86}Sr$ age estimates indicate that this explosive radiation occurred after about 17.5 million years ago (MacFadden et al. 1991). The other common horse at Thomas Farm, *Archaeohippus blackbergi*, is very small (body mass estimated at about 50 kg; MacFadden, 1986a); this species represents one of the first examples of phyletic dwarfing in fossil horses.

Hulbert (1984) studied the population dynamics of *Parahippus leonensis* from Thomas Farm by analyzing the wear and age classes of a sample of jaws representing eighty-nine individuals. The dental wear was not discrete (relative to other fossil-horse populations studied), and suggested that breeding was not seasonal, but probably occurred throughout the year. There was some sexual dimorphism: the average body size of males within the population was estimated at 116 kg, and that of females at 91 kg, which, as an example, is comparable to the white-tailed deer, *Odocoileus virginianus*. Individuals had an estimated average life span of three to four years, and a maximum potential life span of nine years, significantly less than that of modern natural populations of *Equus*.

The early Barstovian, from about 16 to 14 million years ago, was a very important time in equid evolution, because two major groups diversified: the three-toed hipparions, and the mostly one-toed equines. One of the first-known instances of high-crowned (hypsodont) equid diversification in North America is recorded by at least two species of *Merychippus* from the Willacoochee Local Fauna of the Florida panhandle, which has been dated at about 16.6 to 14.7 million years (Bryant et al. 1992). In younger deposits, Barstovian horses from Florida referable to *Merychippus* have been described from the lower Bone Valley Formation (Webb and Crissinger 1983; and others). In addition, a primitive and very rare three-toed browsing horse, *Hypohippus*, has also been described from what are probably contemporaneous units in the lower Bone Valley Member (MacFadden 1982).

In North America, the Clarendonian (age about 11.5 to 8.0 million years) was a time of maximum diversity of horses. By far the most extensive representation of this equid fauna in Florida comes from the latest Clarendonian Love site. At least six genera are represented at that site, including diverse body sizes and feeding types. Hulbert (1982) studied the population dynamics of one of these horses, *Neohipparion trampasense*, based on a study of 229 individuals. In contrast to the Thomas Farm *Parahippus* (which had continuous tooth wear and

Figure 8.13. Reconstruction of *Parahippus leonensis*, a three-toed horse from the Middle Miocene Thomas Farm site in Gilchrist County. The social organization of this fossil horse species is depicted as a small harem with a male, three females, and two colts. This species is estimated to have weighed between 90 and 115 kg, with a size comparable to the modern white-tailed deer (MacFadden 1992; reprinted by permission of Cambridge University Press).

a potential longevity of about 9 years; see above), this species shows discrete tooth wear, suggesting seasonal breeding, and potential longevity had increased to about 13.5 years.

The early Hemphillian Moss Acres site also has a significant equid fauna, representing a time just before the collapse of the highly diverse Clarendonian chronofauna (Webb 1969). The late Hemphillian equid fauna from the upper Bone Valley Member contains a diverse assemblage of three-toed hipparionine and one-toed equine forms with many interesting aspects (fig. 8.14). Webb and Hulbert (1986) described the species *Pseudhipparion simpsoni,* a tiny, gazelle-sized form, and the only known horse with partially evergrowing (hypselodont) cheek teeth. Another Bone Valley horse, *Dinohippus mexicanus,* is the closest Pliocene relative of the modern genus *Equus* (MacFadden 1986b).

A major extinction occurred in the equid faunas throughout North America during the Early Pliocene, after about 4.5 million years ago, and only three genera —*Equus, Nannippus,* and *Cormohipparion*—are known from younger deposits in Florida. By the Pleistocene, the latter two genera had become extinct, and two or three species of *Equus* were abundant (e.g. a large sample from the Leisey Shell Pit), until about 10,000 years ago, when equids become extinct throughout the New World.

Hermanson and MacFadden (1992) studied the evolution of posture in fossil horses, and some critical specimens came from Florida localities (see also MacFadden 1992). Modern horses spend most of the day standing; they can expend much energy in this posture, and are thus prone to fatigue. Dissections of modern *Equus* reveal that the upper-arm (biceps) muscle has a tendon that "locks" onto a uniquely evolved bony ridge on the humerus (the intertubercular crest). When this locking mechanism is engaged, it conserves muscular energy in the forelimb; an analogous structure is found in the hind limb. Because this apparatus contains a bony structure, the intertubercular crest, Hermanson and MacFadden (1992) were able to trace its origin and evolution in fossil horses. They found that the intertubercular crest is a relatively recent innovation; it is absent in Late Miocene specimens of *Dinohippus* from Moss Acres, and it is first well developed in Plio-Pleistocene *Equus,* including the excellent sample from the Leisey Shell Pit (MacFadden 1992).

Tapirs and Chalicotheres (Families Tapiridae and Chalichotheridae) Chalicotheres are extinct, clawed, browsing herbivores related to tapirs, horses, and rhinos (i.e. other perissodactyls); they lived from the Eocene to the Miocene in the Northern Hemisphere, and to the Pleistocene in Africa. With a few exceptions, they are usually rare, and this probably reflects their forest habitats, where preservation of fossils was less likely than in the more common open-country, flood-plain deposits. Although chalicotheres are otherwise absent from eastern North America, Frailey (1979) described specimens of *Moropus* from the Arikareean Buda Local Fauna in Alachua County.

Like chalicotheres, tapirs are characteristically rare in most late Cenozoic localities in North America. The first record of tapirs in Florida comes from the Early Miocene, and is mostly represented by fragmentary remains. One of the best samples of a late Tertiary tapir from anywhere in North America is that of *Tapirus simpsoni* from the 9 million year old Love site (Yarnell 1980). Fossil *Tapirus* is also found in the Bone Valley Pliocene and numerous Pleistocene sites, including *T. veroensis,* first described from the latest Pleistocene Vero Beach site (Ray and Saunders 1984).

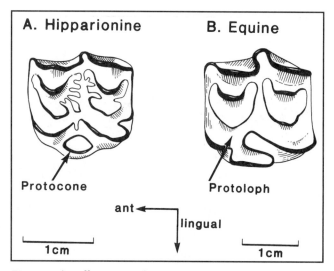

Figure 8.14. Different enamel patterns in two major groups of horses from the Bone Valley Member. A: Hipparionine, represented by *Nannippus minor,* is dominantly a three-toed grazing group in which the protocone generally remains an isolated enamel column.
B: Equine, represented by *Astrohippus stocki,* is dominantly a one-toed group (particularly in later forms) in which the protocone connects to the protoloph. Ant: anterior; ling: lingual (MacFadden 1992; reprinted by permission of Cambridge University Press).

ARTIODACTYLS (ORDER ARTIODACTYLA)

Artiodactyls, the even-toed ungulates, are the most diverse group of extant large mammals. The fossil record of artiodactyls is particularly rich in Florida, where late Cenozoic sediments preserve primitive artiodactyls, the Suoidea (including pigs, peccaries, and extinct families), as well as the advanced Selenodontia (including camels, tragulids, antilocaprids, deer, cattle, and extinct families).

Highlights of those artiodactyls with a good fossil record in Florida are presented below.

Extinct "Hogs" (Family Entelodontidae) Although entelodonts are known in Florida only from a very limited sample, and only from the Early Miocene (Arikareean), their gigantic size makes them noteworthy. The described specimens from Florida are all referred to the genus *Dinohyus* (literally, "terrible pig"). These specimens are closely comparable to *Dinohyus* from the classic Arikareean sequence of western Nebraska. Joeckel (1990) studied the adaptations of entelodonts, including *Dinohyus*. He estimated that this genus probably attained a body mass of about 750 kg, was highly cursorial, and was both an omnivore and a scavenger.

Peccaries (Family Tayassuidae) Peccaries are fairly common in modern Neotropical forests and lower-latitude, drier biomes, but they are not found in Florida today (the nearest part of their current range is western Texas). The first record of a fossil peccary from Florida is in the undescribed sample from the Late Oligocene I-75 Local Fauna. Other samples are known from the Miocene (e.g. Thomas Farm, Love site, Bone Valley Member). Two Plio-Pleistocene genera, *Mylohyus* and *Platygonus*, are well represented in Florida. At certain sites there are large, undescribed samples of these peccaries, including the Irvingtonian "Hog Heaven" (Haile 21A) and the Rancholabrean Cutler site in Miami. Peccaries became extinct in Florida by about 10,000 years ago.

Oreodonts (Family Merycoidodontidae) Oreodonts are among the most common fossil mammals encountered in Oligocene deposits in the western United States, (e.g. in the classic Big Badlands of South Dakota). In younger deposits, oreodonts are generally rare, and they became extinct during the Late Miocene. For some reason, probably related to ancient climate and habitats, oreodonts were never very common in Florida. The oldest oreodont fossils from this region are a few tooth fragments from the late Oligocene I-75 Local Fauna (Patton 1969). An oreodont has been reported from an Early Miocene near-shore marine limestone (MacFadden 1980) (fig. 8.15), and Hemingfordian oreodonts have been reported from Thomas Farm (Maglio 1966) and the Florida panhandle (Bryant 1991b). Several partial skeletons of an oreodont (currently undescribed) have been collected from the Arikareean White Springs Local Fauna. Despite their presence in other Barstovian and Clarendonian localities in western North America, oreodonts are unknown in Florida after the Middle Miocene.

Extinct Horned Artiodactyls (Family Protoceratidae) Protoceratids were moderately common and wide-ranging artiodactyls during the Miocene, particularly along the Gulf Coast and in the midcontinent. The males in advanced members of this group evolved prominent nasal and frontal horns. As with other artiodactyls, the horns were probably used for visual display and sexual selection (i.e. in competition among males for reproductive females during mating season). Perhaps the most spectacular protoceratid discovered in Florida is *Kyptoceras* from the earliest Pliocene of the upper Bone Valley Formation (Webb 1981a) (fig. 8.16).

Camels and Llamas (Family Camelidae) The first group of camelids to appear in the Florida record are the so-called giraffe-camels (Subfamily Aepycamelinae) from the Early Miocene. Advanced members of this group evolved very elongated (stilt-legged) limbs and elongated necks (by expanding the cervical vertebrae), and thus are convergent with modern Giraffidae, which are wholly Old World in fossil and present distribution. Some of the giraffe-camels attained the size of modern giraffes; for example, *Aepycamelus major*, from the Love site, had a mass of about 1,000 kg (MacFadden and Hulbert 1990) and stood about 5 m tall. In addition to their height, their low-crowned dentition also suggests a feeding habit of browsing from parts of the forest canopy.

Llamas are indigenous to South America today and comprise the llama, guanaco, and alpaca—all conservatively placed in the genus *Lama*—and the vicuna, re-

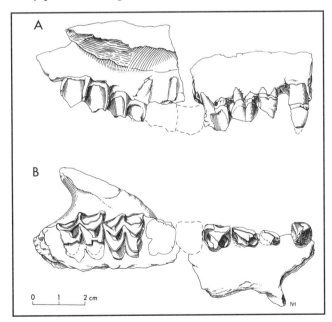

Figure 8.15. Lateral (A) and ventral (B) views of the right upper dentition of the oreodont *Phenacocoelus luskensis* from the Early Miocene of northern Florida (from MacFadden 1980 and reproduced with permission of the Paleontological Society).

ferred to *Vicugna* (Walker 1975). The dentitions of llamas are very distinctive, with a so-called llamine buttress, and the origins of this group can be traced back into the late Cenozoic. In fact, *Hemiauchenia* from the Bone Valley Member is one of the earliest representatives of the group, and this genus is known also from the Pliocene and Pleistocene (Webb 1974b). Pleistocene llamas in Florida include a second genus, *Palaeolama*, which is very abundant at certain sites (e.g. the Leisey Shell Pit). Llamine camels became extinct in Florida (as they did elsewhere in North America) at the end of the Pleistocene.

Pronghorns (Family Antilocapridae)   North American antilocaprids (sometimes incorrectly called "antelopes," which are an exclusively Old World group) are known by the single living monotypic genus *Antilocapra*, common to drier climes in the West. In Florida, the oldest antilocaprids are from the upper Bone Valley Member and are referred to *Hexobelomeryx* (previously *Hexameryx*), the six-horned antilocaprid (Webb 1973; Hulbert 1992) (fig. 8.17). Pliocene antilocaprids are known from the genus *Capromeryx,* an excellent sample of which has been collected from the Santa Fe River sites.

Deer, Elk, and Moose (Family Cervidae)   Deer and their relatives are advanced ruminant artiodactyls in which the males have antlers (i.e. deciduous cranial appendages that are shed annually). One of the earliest records of cervids in North America comes from an undescribed new genus and species from the upper Bone Valley Member, which is known mostly from fragmentary antlers and jaws. The extant white-tailed deer, *Odocoileus virginianus,* has a fossil record in Florida extending back to the late Pliocene.

Cattle and Bison (Family Bovidae)   The bovids represent the most advanced and diverse group of artiodactyls and are principally an Old World group that has undergone an extensive adaptive radiation during and since the Miocene. In Florida, bovids are relatively widespread and are represented by the genus *Bison*, comprising two extinct species known from the Pleistocene. These two species, if they indeed form an ancestral-descendant series, represent an example of phyletic dwarfing (Robertson 1974). Their presence in Florida is usually taken to indicate middle and Late Pleistocene (Rancholabrean and possibly back into the Irvingtonian) time; they arrived from the Old World after one or more dispersals across the Bering land bridge. The geographic range of *Bison* decreased dramatically during the late Quaternary, with extinction in Florida and the rest of eastern North America. In addition to the extremely large size (by

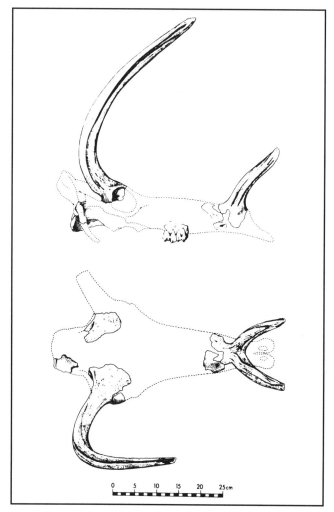

Figure 8.16. Right lateral (top) and dorsal (bottom) views of a cranial reconstruction of the ruminant artiodactyl *Kyptoceras amatorum* from the Early Pliocene Upper Bone Valley Member of Polk County (Webb 1981a; reprinted by permission of the Society of Vertebrate Paleontology).

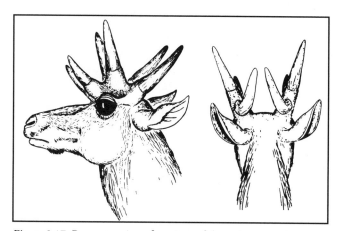

Figure 8.17. Reconstruction of cranium of the extinct pronghorn *Hexobelomeryx simpsoni* from the Early Pliocene Upper Bone Valley Formation of central Florida (from Webb 1973 and reproduced with permission of the American Society of Mammalogists).

modern standards), *Bison* from Florida is interesting because in some specimens there is an interesting paleopathology: a cranial bone degeneration resulting from advanced stages of syphilis.

PROBOSCIDEANS

Proboscideans, represented today by the Indian elephant *Elephas maximus* and the African elephant *Loxodonta africana*, originated in the Old World. The dispersal of primitive members of this group into the New World during the Barstovian about 14 to 15 million years ago represents a major immigration datum in the Miocene (Tedford et al. 1987).

"Shovel-Tuskers" (Family Amebelodontidae) Excellent Miocene proboscidean samples come from the Love and Moss Acres sites and are referred to *Amebelodon*, a shovel-tusker with distinctively flattened lower tusks (Webb et al. 1981; Lambert 1990). The possible function of flattened shovel-tusks has been the topic of much discussion in the literature. Originally it was believed that the tusks were used to scoop up aquatic vegetation, and many classic reconstructions of shovel-tuskers show them in this pose. However, other work, which included many Florida specimens, has shown that this functional interpretation is oversimplified. Lambert (1992) hypothesized that the tusks in extinct proboscideans served a variety of functions related to food procurement and/or to sexual selection. Lambert (1992) concluded that *Amebelodon* from Florida was a more generalized feeder than was previously believed (fig. 8.18); he stated that

> They might be thought of as walking Swiss army knives having the tools to deal flexibly with a variety of feeding situations. Thus, at different times

the tusks could have functioned as classically believed to scoop up aquatic vegetation, but they also could have been used for other kinds of food procurement, such as stripping bark off trees.

Mastodonts and Mammoths (Families Mammutidae and Elephantidae) In the Late Pliocene (Blancan), another Old World immigrant appears in the fossil record of Florida, the American mastodont (*Mammut americanum*). It was followed during the Early Pleistocene (Irvingtonian) by the mammoth (*Mammuthus*). Thereafter, mammoths and mastodonts are relatively common in the Florida record. They are easily distinguished, because in mastodonts the large molar teeth are cuspate, whereas in the mammoth the grinding teeth are composed of thick plates covered with cement (fig. 8.19). It is fairly well accepted that these two coexisting groups of proboscideans partitioned their ecosystem to minimize competition. Thus, the mastodonts probably were browsers, feeding on shrubs and trees in a variety of forest communities, whereas the mammoths were grazers, feeding in savanna and grassland communities (Dudley 1987). There is considerable discussion about the cause of extinction of large-bodied mammals at the end of the Pleistocene, in Florida and elsewhere in the world. Although there is some evidence of coexistence of Late Pleistocene mammals and native Americans, and even predation or scavenging by humans (Bullen et al. 1970), changes in climate and vegetation probably were at least equally important factors in the megafaunal extinction in this region than was human impact, which is often termed the "Pleistocene overkill hypothesis."

SIRENIANS (ORDER SIRENIA)

Modern sea cows and their extinct relatives are a group with Tethyan origins during the early Tertiary. Domning et al. (1982) described fragmentary remains of primitive North American sirenians, including several discoveries of *Protosiren* from the Eocene Ocala Limestone of central

Figure 8.18. Reconstruction of the shovel-tusked proboscidean *Amebelodon* from the Late Miocene of northern Florida. Species of this genus had a considerable size range, but all were somewhat to considerably larger than the modern African elephant (from Lambert 1992 and reproduced with permission of the Paleontological Society).

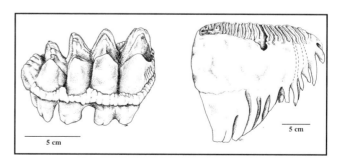

Figure 8.19. Comparison of teeth of Pleistocene proboscideans from Florida. Left: molar of mastodon (*Mammut*); right: molar of mammoth (*Mammuthus*) (modified from Brown 1988).

and northern Florida that are important because they represent early occurrences for this group.

Middle Miocene sirenians are very abundant in the marine facies of the Hawthorn Group and contemporaneous deposits, usually represented by the dense rib bones, which are relatively amenable to fossilization (fig. 8.20). Along with shark teeth, these rib bones are among the most common fossils encountered in the creek beds of northern Florida around Gainesville, for example. These sirenians are mostly referable to the dugong genera *Metaxytherium* and *Dioplotherium*.

Late Miocene sirenians are very abundant in marginal-marine sediments throughout much of Florida. Complete skeletons of the common form, *Metaxytherium*, have been collected from the Middle Miocene of the lower Bone Valley Member. In deposits younger than about 5 million years, sirenians are absent from the Florida fossil record, until the appearance in the late Pleistocene of the extant *Trichechus*, which represents a relatively recent dispersal into this region.

CETACEANS (MIRORDER CETE)

Because of the abundance of marine deposits, the Cenozoic fossil record of cetaceans (whales, porpoises and dolphins) is very extensive throughout Florida; it is, in fact, one of the better fossil records anywhere in the world, but, with a few exceptions, it has been poorly studied.

Archaic Toothed Whales (Suborder Archaeoceti) The phylogenetic position of archaeocetes is at the base of the adaptive radiation of the order, and they present an evolutionary mosaic of primitive and advanced characters. It is fairly well accepted that the closest primitive relatives of cetaceans are the mesonychids, an extinct group of terrestrial carnivorous mammals. Retaining the ancestral feeding habit, archaeocetes were carnivorous fish-eaters that inhabited early Tertiary shallow seas. Two genera of these primitive whales are represented in Florida: the very large *Basilosaurus* (as long as 20 m), and the smaller *Zygorhiza* (as long as 6 m) (fig. 8.21).

Dolphins, Porpoises, and Toothed Whales (Suborder Odontoceti) The first record of dolphins and porpoises from Florida is from the middle Miocene, which includes two extinct long-nosed genera, *Pomatodelphis* and *Schizodelphis*. River dolphins, no longer indigenous to the southeast, occur in Florida in the upper Bone Valley Formation. The earliest record of an extant dolphin genus in Florida is *Stenella* from the early Pleistocene. Toothed whales, which include sperm whales and beaked whales, first occur in Florida in the Miocene. The extant sperm whale *Physeter* is found in Pleistocene deposits of Florida.

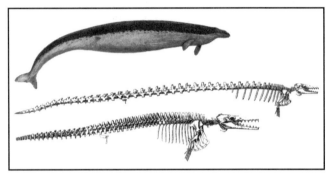

Figure 8.21. Reconstruction of a Late Eocene archaeocete whale (top), and skeletons of the archaeocete whales *Basilosaurus* (length 20 meters) and *Zygorhiza* (6 meters) from the Late Eocene of the southeastern United States (modified from Savage and Long 1986).

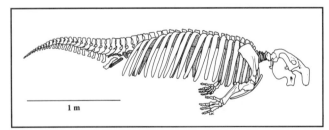

Figure 8.20. Reconstructed skeleton of the extinct sirenian *Metaxytherium floridanum* from the Middle Miocene deposits of the Bone Valley Member of central Florida (modified from Domning 1988).

Figure 8.22. Pleistocene megafaunal extinctions: possible cause. The wave of extinctions of large-bodied mammals that culminated about 10,000 years ago usually is attributed to climate change on the one hand and human "overkill" on the other, the latter of which is depicted in this painting by Chris Williams (property of the Florida Museum of Natural History).

Baleen Whales (Order Mysticeti) Including the right, rorqual, and humpback whales, this group comprises the largest known mammals, some species approaching 30 m in length. All mysticetes are characterized by their unique baleen structure, by which they sieve plankton and tiny fish (Walker 1975). The earliest fossil record of baleens in Florida comes from Miocene deposits and includes smaller forms. The gigantic forms referable to the extant genus *Balaenoptera* have lived in this region since the latest Pliocene. Fossil bones of some of these larger whales show extensive evidence of tooth grooves, indicating predation or scavenging by sharks.

The fossil record of cetaceans in Florida is extensive, and, like micromammals in the terrestrial realm, it represents one of the most understudied parts of our record. When these groups are better known, the richness of Florida's marine deposits will yield many important systematic insights as well as interesting paleobiologic interpretations.

Summary

The fossil record of mammals in Florida from both the marine and terrestrial realms represents one of the most continuous and rich Cenozoic sequences in the world. Although it traditionally has been calibrated by stage of evolution and marine/nonmarine comparisons, the potential for isotopic calibrations (e.g. $^{87}Sr/^{86}Sr$) has been demonstrated more recently. From this important fossil record we have gained much insight into the diversification of major mammalian groups through time, as well as a better understanding of such interesting paleobiologic aspects as population structure, evolutionary patterns, changing biogeography, and extinction (fig. 8.22). As vertebrate fossils continue to be discovered and existing museum collections are studied more fully, additional insight will be gained into this phase in the history of ancient life in southeastern North America.

9

The Economic and Industrial Minerals of Florida

Guerry H. McClellan and James L. Eades

In 1991, the value of Florida's estimated nonfuel mineral production was $1.4 billion, a decline of $172 million from the estimate for 1990 (White and Schmidt 1993). The decrease resulted from decline in several sectors of the state's economy, especially in construction. Two sectors particularly affected were the cement (portland and masonry) and crushed stone industries, providers of major construction commodities. In 1989, increases in these same sectors resulted in an increase in produced-mineral values. These cyclical economic trends are typical of Florida's mineral industries, which rise and fall in response to national and international economic factors (fig. 9.1). The value of the products of Florida's mineral industries has generally been increasing over the past two decades, and has been over the $1 billion mark since 1974. Florida ranks fourth among the states in total value of nonfuel minerals, contributing about 5% of the nation's total. The state ranks second nationally, behind California, in industrial mineral sales. Florida leads the world in the production of phosphate rock; leads the nation in heavy-mineral production; is ranked among the top three states in crushed stone, masonry cement, and peat; and is ranked twentieth in crude-oil and gas production. Direct employment by the mining industries is about 8,000 to 9,000, and several times that number benefit from the economic multiplier effect.

Unlike many states where one industry, such as construction materials, dominates the minerals sector, Florida has rather diversified mineral industries. The state has several large phosphate producers that contribute about 50% of the state's mineral value, and a number of cement, stone, sand, and gravel producers add another 33% to the sector from construction materials. Important contributions also come from the several clay and heavy-mineral operations in the state. This diversity of production decreases the impact of regional, national, and international economic downturns on the state's mining industry, but some cyclical variation is observed.

Issues confronting the state's mining industry include rapid depletion of readily available, high quality raw materials for the construction, phosphate, and fullers earth industries; increasing foreign competition in many of these markets; competition for land and water resources; and increasing regulatory and environmental factors. The mining industry as a whole has become more attentive to some of these issues in the recent past. Minerals producers are very aware of the present and future importance of these influences on their well-being. Industry, environmental groups, and Florida's universities are involved in discussions to identify problems and to attempt to find mutually acceptable solutions.

In the preparation of this chapter, four references were repeatedly consulted, and readers are referred to these for more detailed information on geology, stratigraphy, petrology, and mineralogy; these references are Campbell (1986), Spencer (1989), White and Schmidt (1991, 1993), and Boyle and Schmidt (1992).

For the purposes of this chapter, reserves of either fuel or minerals are those deposits that can be economically produced using existing technology. Deposits that do not meet this definition of reserves are considered resources.

Petroleum and Natural Gas

Florida's oil and gas are produced from two widely separated fields (fig. 9.2). Florida's first oil discovery was in September 1943 at the Sunniland field in Collier County. The fourteen southern Florida fields in Collier, Hendry, and Lee counties produce from structural and stratigraphic traps in the Sunniland Formation (Lower Cretaceous). Production is from porous, algal plate, foraminiferal, pelletal limestone reef pods in an algal bank facies (Tyler and Erwin 1975). These northwest-southeast trending reefs were cut into pods by channels that later were filled by low permeability carbonate muds. The local reservoir rocks for the Sunniland-Felda fields are the upper 30 m of the reef pods of the Sunniland Formation (Lloyd 1991).

Production in northwestern Florida began in June 1970, with the discovery of the Jay field in Santa Rosa County. Eight oil and gas fields in northwestern Florida are located in Escambia and Santa Rosa counties and produce from a combination of structural and stratigraphic traps in the Upper Jurassic Smackover Formation carbonates and from the Norphlet sands (Sigsby 1976; Lloyd 1991). The productive interval of the Smackover is a porous dolomite associated with transgressive muds and algal mats and a regressive unit of hardened grainstones (Ottman et al. 1973).

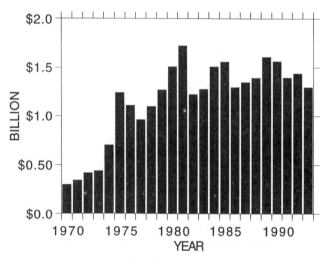

Figure 9.1. Variation in value of Florida's mineral industry with time.

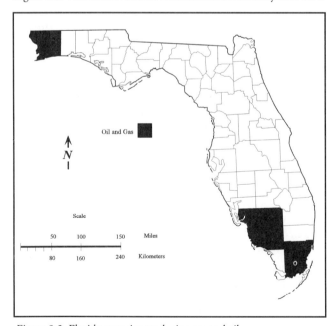

Figure 9.2. Florida counties producing gas and oil.

In 1990, Florida had dropped to twentieth among the states in gas and oil production (White and Schmidt 1993). The Jay field of northwestern Florida was producing 62% of the state's oil and 86% of its natural gas. Seismic and gravity exploration surveys are continuing in both fields.

Offshore drilling in state waters through 1983 showed only one significant oil show near the Marquesas Keys off Monroe County. Effective July 1990, all drilling was prohibited in Florida's state waters. Exploratory drilling in federal waters, offshore Florida, has shown some encouraging results for the Smackover Formation and Norphlet Sandstone in wells located in the Destin Dome off the northwest coast (Gould 1989). Chevron reported indications of gas from the Norphlet sands in their well no. 6404 (Oil and Gas Journal 1989), extending the seaward and eastward trend of this offshore gas from the Mobile area into offshore Florida.

Analysis of Florida's estimated oil and gas reserves shows that less than 20% of the remaining resources can be produced using current technology and under current economic conditions (Tootle 1991). Annual production data for oil and gas (fig. 9.3) show the mature condition of the industry in the state.

Gas and oil from the northwest Florida fields contain as much as 10% hydrogen sulfide (Ottman et al. 1973). When these products are burned, the sulfur compounds contribute significant energy values; however, it may be necessary to clean gas and oil containing this much sulfur before it is used as an ordinary fuel. Refinery processing to desulfurize 12,000 barrels per day of this oil (more than 90% of 1991 production) would produce as much as 80 metric tons per day of byproduct elemental sulfur that could be used in a variety of industrial processes. The price of sulfur and the market for sulfur-

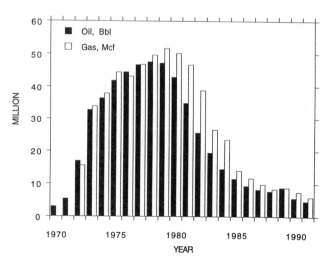

Figure 9.3. Temporal variation in Florida gas and oil production.

9
The Economic and Industrial Minerals of Florida

Guerry H. McClellan and James L. Eades

In 1991, the value of Florida's estimated nonfuel mineral production was $1.4 billion, a decline of $172 million from the estimate for 1990 (White and Schmidt 1993). The decrease resulted from decline in several sectors of the state's economy, especially in construction. Two sectors particularly affected were the cement (portland and masonry) and crushed stone industries, providers of major construction commodities. In 1989, increases in these same sectors resulted in an increase in produced-mineral values. These cyclical economic trends are typical of Florida's mineral industries, which rise and fall in response to national and international economic factors (fig. 9.1). The value of the products of Florida's mineral industries has generally been increasing over the past two decades, and has been over the $1 billion mark since 1974. Florida ranks fourth among the states in total value of nonfuel minerals, contributing about 5% of the nation's total. The state ranks second nationally, behind California, in industrial mineral sales. Florida leads the world in the production of phosphate rock; leads the nation in heavy-mineral production; is ranked among the top three states in crushed stone, masonry cement, and peat; and is ranked twentieth in crude-oil and gas production. Direct employment by the mining industries is about 8,000 to 9,000, and several times that number benefit from the economic multiplier effect.

Unlike many states where one industry, such as construction materials, dominates the minerals sector, Florida has rather diversified mineral industries. The state has several large phosphate producers that contribute about 50% of the state's mineral value, and a number of cement, stone, sand, and gravel producers add another 33% to the sector from construction materials. Important contributions also come from the several clay and heavy-mineral operations in the state. This diversity of production decreases the impact of regional, national, and international economic downturns on the state's mining industry, but some cyclical variation is observed.

Issues confronting the state's mining industry include rapid depletion of readily available, high quality raw materials for the construction, phosphate, and fullers earth industries; increasing foreign competition in many of these markets; competition for land and water resources; and increasing regulatory and environmental factors. The mining industry as a whole has become more attentive to some of these issues in the recent past. Minerals producers are very aware of the present and future importance of these influences on their well-being. Industry, environmental groups, and Florida's universities are involved in discussions to identify problems and to attempt to find mutually acceptable solutions.

In the preparation of this chapter, four references were repeatedly consulted, and readers are referred to these for more detailed information on geology, stratigraphy, petrology, and mineralogy; these references are Campbell (1986), Spencer (1989), White and Schmidt (1991, 1993), and Boyle and Schmidt (1992).

For the purposes of this chapter, reserves of either fuel or minerals are those deposits that can be economically produced using existing technology. Deposits that do not meet this definition of reserves are considered resources.

Petroleum and Natural Gas

Florida's oil and gas are produced from two widely separated fields (fig. 9.2). Florida's first oil discovery was in September 1943 at the Sunniland field in Collier County. The fourteen southern Florida fields in Collier, Hendry, and Lee counties produce from structural and stratigraphic traps in the Sunniland Formation (Lower Cretaceous). Production is from porous, algal plate, foraminiferal, pelletal limestone reef pods in an algal bank facies (Tyler and Erwin 1975). These northwest-southeast trending reefs were cut into pods by channels that later were filled by low permeability carbonate muds. The local reservoir rocks for the Sunniland-Felda fields are the upper 30 m of the reef pods of the Sunniland Formation (Lloyd 1991).

Production in northwestern Florida began in June 1970, with the discovery of the Jay field in Santa Rosa County. Eight oil and gas fields in northwestern Florida are located in Escambia and Santa Rosa counties and produce from a combination of structural and stratigraphic traps in the Upper Jurassic Smackover Formation carbonates and from the Norphlet sands (Sigsby 1976; Lloyd 1991). The productive interval of the Smackover is a porous dolomite associated with transgressive muds and algal mats and a regressive unit of hardened grainstones (Ottman et al. 1973).

In 1990, Florida had dropped to twentieth among the states in gas and oil production (White and Schmidt 1993). The Jay field of northwestern Florida was producing 62% of the state's oil and 86% of its natural gas. Seismic and gravity exploration surveys are continuing in both fields.

Offshore drilling in state waters through 1983 showed only one significant oil show near the Marquesas Keys off Monroe County. Effective July 1990, all drilling was prohibited in Florida's state waters. Exploratory drilling in federal waters, offshore Florida, has shown some encouraging results for the Smackover Formation and Norphlet Sandstone in wells located in the Destin Dome off the northwest coast (Gould 1989). Chevron reported indications of gas from the Norphlet sands in their well no. 6404 (Oil and Gas Journal 1989), extending the seaward and eastward trend of this offshore gas from the Mobile area into offshore Florida.

Analysis of Florida's estimated oil and gas reserves shows that less than 20% of the remaining resources can be produced using current technology and under current economic conditions (Tootle 1991). Annual production data for oil and gas (fig. 9.3) show the mature condition of the industry in the state.

Gas and oil from the northwest Florida fields contain as much as 10% hydrogen sulfide (Ottman et al. 1973). When these products are burned, the sulfur compounds contribute significant energy values; however, it may be necessary to clean gas and oil containing this much sulfur before it is used as an ordinary fuel. Refinery processing to desulfurize 12,000 barrels per day of this oil (more than 90% of 1991 production) would produce as much as 80 metric tons per day of byproduct elemental sulfur that could be used in a variety of industrial processes. The price of sulfur and the market for sulfur-

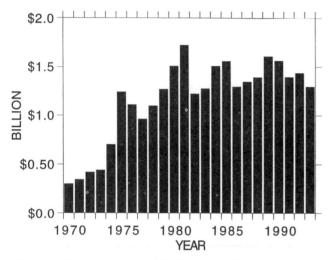

Figure 9.1. Variation in value of Florida's mineral industry with time.

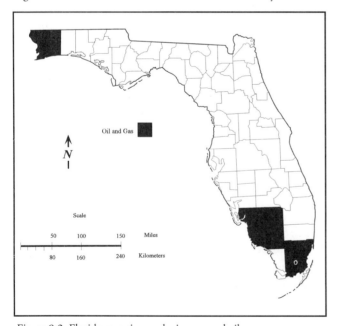

Figure 9.2. Florida counties producing gas and oil.

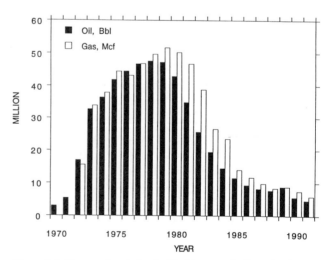

Figure 9.3. Temporal variation in Florida gas and oil production.

containing fuels will determine the economic potential for recovering this byproduct.

A sulfur removal facility in Santa Rosa County recovers 35,000 to 50,000 tons per year (tpy) from natural gas produced from the Jay and Blackjack fields in the Panhandle (White and Schmidt 1993).

Potential Energy Resources

Florida is the nation's second leading producer of peat. Although most of this material is currently used for agriculture and horticulture, use of peat as an alternative energy source for generation of electricity is a significant, developing trend. Present plans call for more than 700 megawatts of electricity to be produced from 2.4 million tpy of dry Florida peat by the mid- to late 1990s, which would increase the state's peat production tenfold, while quadrupling current national production. Peat-fueled power plants are reported to be economically and environmentally competitive with conventional coal-fired plants, and Florida has large resources and reserves of peat. The fuel value of peat can be affected by the geology: "dry" peats may have undergone partial oxidation above the water table, resulting in decreased fuel values.

The phosphate ores of central Florida contain an average of about 100 ppm of uranium, which means that each metric ton of phosphate concentrate will contain about one pound of U_3O_8 for each ton of P_2O_5. This industry produces 35 to 40 million tpy of phosphate concentrate. In 1979, Florida phosphate reserves were estimated to contain 225,000 tons of U_3O_8 (Sweeney 1979). Thus, these ores represent a significant potential energy resource for the nation's power industry. The technology for recovering much of the uranium from the chemically processed ores has been available for several years (Huwyler 1983). The depressed state of construction in the nuclear power industry during the 1980s and early 1990s has decreased the commercial demand for uranium and resulted in depressed prices (Pool 1991). By 1991 this economic downturn had caused Florida's byproduct uranium industry to shrink to a single uranium producer and two plants producing a total of 2.1 million pounds of U_3O_8 per year (Pool 1992). In mid-1992, the announcement was made that these plants were shut down because their long-term contracts had expired (Power 1992).

Phosphate Rock

Phosphate rock is the largest mineral industry in the state, accounting for about 50% of Florida's produced-minerals value. Florida ranks first in the nation in the production of phosphate rock, and produces about 30% of the world total. Stowasser (1992) has forecasted a production increase of as much as 20% by the year 2000, which is of interest because the U.S. Bureau of Mines has predicted a steady decline in the state's phosphate-rock production during the twenty-first century, and termination of mining around the year 2010. Reorganization and consolidation seem to have brought a new vitality to the industry.

The origin and detailed geology of the Florida phosphate deposits are treated in chapter 12, and the environmental considerations are addressed in detail in chapter 13.

Phosphate mining began in Florida in 1879 in Miocene sediments near Hawthorne (Cooke 1945). Commercial exploitation began in the early 1880s when companies began to mine phosphatic pebbles from the Peace River near Fort Meade, in Polk County. As time passed, technology and economics allowed the miners to move from the river-pebble to the land-pebble and hard-rock (replaced limestone) phosphates, and then to mining the finer-grained "phosphate matrix" (the admixture of clay, quartz sand, dolomite, and phosphate that occurs over a wide area of west-central Florida including southeastern Hillsborough County and southwestern Polk County (fig. 9.4). The hard-rock district was located in parts of Alachua, Citrus, Dixie, Gilchrist, Hernando, Lafayette, Levy, Marion, Sumter, and Taylor counties. In the 1960s, hard-rock mining ceased for a variety of technical and economic reasons. At the same time, mining began in the northern phosphate district, located mainly in Hamilton and Columbia counties.

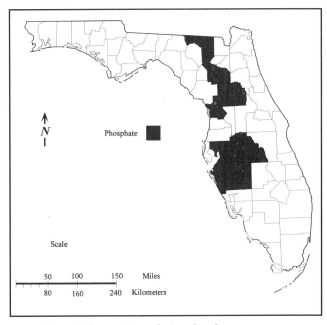

Figure 9.4. Florida counties producing phosphate.

Starting in the late 1970s, the central Florida phosphate companies began moving their mining operations into the "southern extension," located in parts of DeSoto, Hardee, and Manatee counties (Opyrchal and Wang 1981; Zellars and Williams 1978).

The classic central Florida phosphate district consists of phosphate deposits that are highly reworked and weathered marine and estuarine sediments along the southern and eastern flanks of the Ocala Arch. The main ore zone belongs to the Peace River Formation and Bone Valley Member of the Hawthorn Group, and is believed to have been deposited in a warm, shallow sea in a nearshore environment (Riggs 1984). The deposit consists of nearly equal parts of quartz, phosphate, and clay, with some associated dolomite (Altschuler et al. 1956). Postdepositional lateritic alteration of the carbonate-fluorapatite in the Bone Valley sediments has been severe, resulting in formation of iron-aluminum phosphates (Van Kauwenbergh et al. 1990). These are the so-called "leached zone ores" that contain high contents of uranium associated with the resistant phosphate minerals, wavellite and crandallite.

The phosphate matrix is covered by Pleistocene quartz sands whose origin has been the subject of some debate. Altschuler and Young (1960) postulated that some of these sands represent a weathered residuum of the Bone Valley sediments. Cathcart (1962), Pirkle et al. (1965), and Pirkle et al. (1967) believed that the sands are primary clastics deposited during a Pleistocene transgression.

The southern extension, or south Florida phosphate district, differs from the central Florida district in several respects: it is a more marine facies with lower P_2O_5; high MgO in francolite, dolomite, and clay; lower pebble content; and less fines. The overburden-to-ore ratios also are higher in this area. The sediments occur in the Peace River Formation and Bone Valley Member of the Hawthorn Group, and samples show less evidence of reworking and alteration than the materials in the central Florida district.

The north Florida district is controlled by the Ocala Arch. There, as in central Florida, the Miocene sediments lap onto the structure to the west. The uppermost clastic unit (a mixture of sand, clay, dolomite, and phosphate) contains the ore. These deposits most likely belong to the Statenville Formation of the Hawthorn Group in this area (Scott 1988b).

Phosphate mining in Florida is relatively simple. After land preparation, large draglines are used to remove the overburden. The same draglines then mine the ore and drop it in a pit, where the ore is slurried by hydraulic monitors and pumped to the mineral-processing plant.

Before 1930, the ores were coarse enough to be beneficiated by simple washing and sizing. This method was not very efficient, and could not be applied to all of the various types of ores.

Later, anionic flotation was added to the washing and sizing to separate the finer particles of phosphate from the quartz sand left after the ore was washed free of clay (deslimed). As the demand for higher-grade concentrates increased, cationic flotation was introduced to remove the last vestiges of quartz and other impurities from the samples. Two-stage flotation was more costly, and was used only to prepare the highest-grade concentrates for export and specific chemical processes. Since about 1970, manufacturers of diammonium phosphate, the main ingredient in most solid fertilizers, have been concerned about the presence of magnesium in phosphate concentrates. This magnesium is derived from isomorphic substitutions in carbonate-fluorapatite, from Mg-containing clays in the ore (smectites and palygorskite), and from dolomite (McClellan and Lehr 1982). Significant effort has gone into developing processes to remove dolomite by flotation and/or density separation from ores of the south Florida district. Although several technically feasible processes have been developed, for economic reasons none has yet been brought into general commercial use in Florida. Flotation can reject as much as 99% of the liberated quartz sand while recovering about 80% of the phosphate grains from the flotation feed (Zellars and Williams 1978). The final flotation concentrate may represent only 10% to 25% of the weight of the mined ore.

Of the mined rock, 90% is used to make fertilizer. Another 5% is used in livestock-feed supplements (primarily for chickens and cattle). The balance is used in food products, chemicals, and ceramics. Both products and rock are moved by rail and trucks to ports in Tampa, Manatee, and Jacksonville for marketing in domestic and international sectors. Florida's exports may be sold directly or bartered for raw materials such as nitrogen, sulfur, and potash, which are used in fertilizer production. This low-cost, readily available, domestic source of fertilizer is a key factor in the ability of the United States to produce surplus food for internal and international use at reasonable prices.

The phosphate industry has more than $10 billion in capital invested in Florida, and new construction, expansion, and replacements of plants and equipment run into hundreds of millions of dollars per year. About 8,000 people are directly employed by the phosphate companies, with an annual payroll of more than $300 million. It is estimated that an additional 50,000 jobs are created in supporting industries.

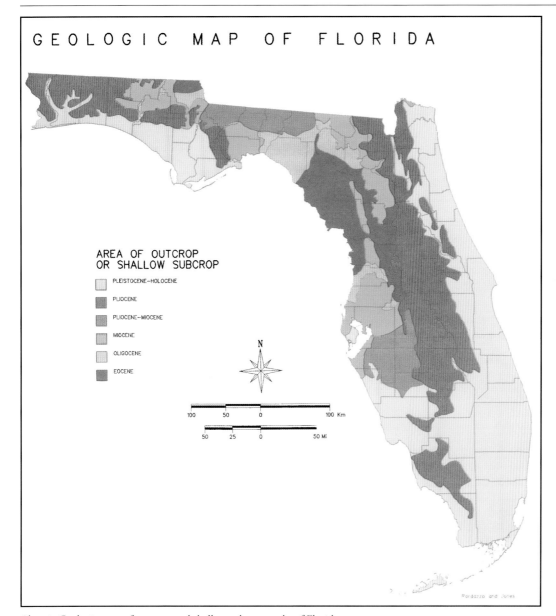

Plate 1. Geologic map of outcrop and shallow subcrop rocks of Florida.

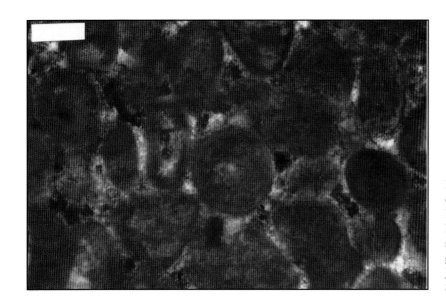

Plate 2. Photomicrograph of an oolitic packstone from the Avon Park Formation (Middle Eocene). Pore-filling calcite cement surrounds concentric oolitic grains. Sample provided by J.-G.-Duncan, Florida Geological Survey. Bar scale = 100 μm.

Plate 3. The Upper Eocene Ocala Limestone exposed by quarry operations. The extensive chimney-shaped dissolution cavities can coalesce to form larger and more bulbous cavities. Settlement of the younger, overlying sediments often results in sinkhole development (supplied by W. Wisner, Florida Department of Transportation).

Plate 4. The Anastasia Formation, a Pleistocene beach deposit that is widespread on the East Coast. It is also found on the West Coast near Sarasota.

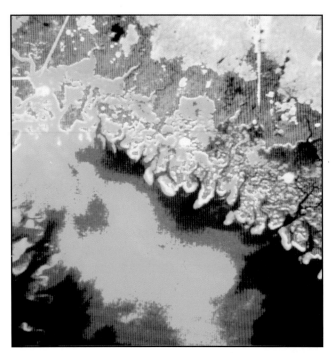

Plate 5. Satellite image of the southwest Florida coast, showing the Ten Thousand Islands and adjacent areas. The linear trends represent old Quaternary shorelines.

Plate 6. Redfish Pass, separating North Captiva and Captiva islands in Lee County. The pass was formed by a hurricane in 1921.

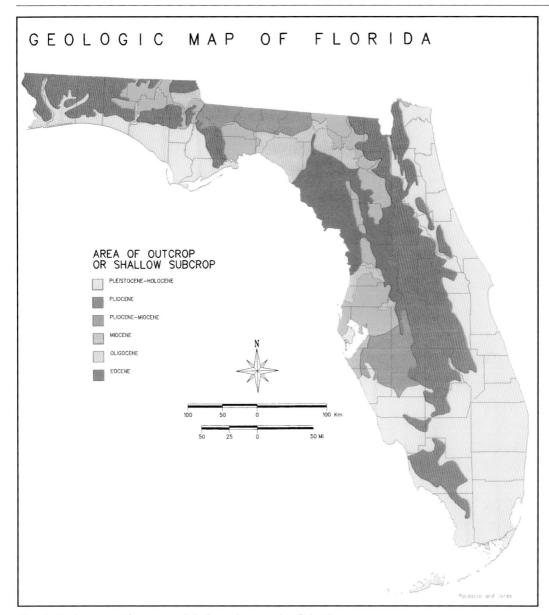

Plate 1. Geologic map of outcrop and shallow subcrop rocks of Florida.

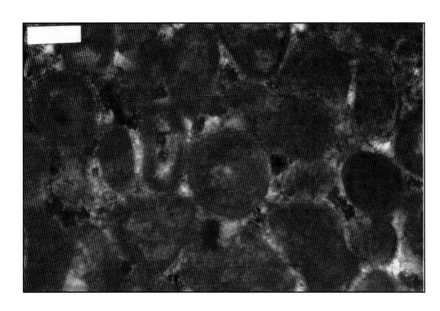

Plate 2. Photomicrograph of an oolitic packstone from the Avon Park Formation (Middle Eocene). Pore-filling calcite cement surrounds concentric oolitic grains. Sample provided by J.-G.-Duncan, Florida Geological Survey. Bar scale = 100 μm.

Plate 3. The Upper Eocene Ocala Limestone exposed by quarry operations. The extensive chimney-shaped dissolution cavities can coalesce to form larger and more bulbous cavities. Settlement of the younger, overlying sediments often results in sinkhole development (supplied by W. Wisner, Florida Department of Transportation).

Plate 4. The Anastasia Formation, a Pleistocene beach deposit that is widespread on the East Coast. It is also found on the West Coast near Sarasota.

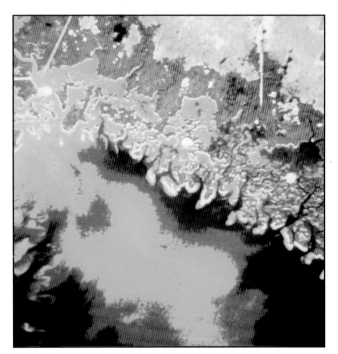

Plate 5. Satellite image of the southwest Florida coast, showing the Ten Thousand Islands and adjacent areas. The linear trends represent old Quaternary shorelines.

Plate 6. Redfish Pass, separating North Captiva and Captiva islands in Lee County. The pass was formed by a hurricane in 1921.

Plate 7. Oblique aerial photo of Bunces Pass, showing two recently formed barrier islands in southern Pinellas County: North Bunces Key (left), which initially emerged in 1962, and South Bunces Key (right), which initially emerged in 1977.

Plate 9. Nutrient-rich waters in Alaska Sink, Hillsborough County, which has an urbanized drainage basin.

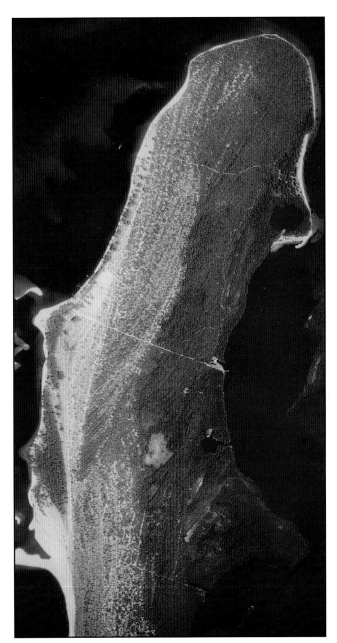

Plate 8. Northern part of Cayo Costa Island showing patterns of numerous beach ridges.

Plate 10. Spur-and-groove development at Sand Key, Florida (photo courtesy of E. A. Shinn).

Plate 11. Underwater photograph of spur-and-groove development at Looe Key, Florida (photo courtesy of D. F. McNeill).

Plate 12. A living reef community, including *Acropora palmata* (photo courtesy of D. J. Cook).

Florida phosphate rock production peaked at 43 million tpy in 1980 (fig. 9.5) and has had some highs and lows since that time. The twenty-year trend has shown an increase in production. Stowasser (1992) stated that increased demand for phosphates will require new investment in mines or plants, or a tightening of supplies will occur by the mid-1990s. Mine production plans indicate a growth of Florida phosphate rock to 40 million tpy or more by the year 2000.

Florida's phosphate resources have been estimated at 4.1 billion tons (Zellars and Williams 1978). This resource base should last hundreds of years at the current rate of mining. About 21% of these resources occur in the central Florida district, 45% are in the south Florida district, and 34% occur in the north Florida district.

Florida's phosphate reserves, based on current economics and technology, are estimated to be at least 520 million tons (Stowasser 1985). This comparatively small reserve base has been the basis for predicting the closing of the industry around the year 2010 at mining rates of 40 million tons per year.

In addition to uranium, fluorine is an economic byproduct of phosphoric-acid production. The fluorine from the rock reacts with silica to form SiF_4 gas. During acid production this gas is recovered as fluorosilicic acid (H_2SiF_6) in wet scrubbers that are part of the environmental-protection equipment. Fluorosilicic acid is widely used in the preparation of chemical compounds and in the treatment of public drinking water. Fluorine can also be used in the manufacture of aluminum metal, where it is converted to synthetic cryolite for use both as an electrolyte and as a solvent for the reduction of alumina (the Bayer process). Fluorosilicic acid also can be used in phosphate rock acidulation to replace as much as 25% of the sulfuric acid. Hydrofluoric acid, a valuable alternative product, can be prepared by the decomposition of fluorosilicic acid to HF and SiO_2. Demand for fluorine is expected to continue growing at 3% to 4% per year.

Environmental problems include water usage, power consumption, radiation, water quality, air quality, slimes disposal, and wetlands protection. Since the 1970s, the industry has concentrated efforts on water conservation and recirculation. The result has been that, while rock production has increased by 35%, water use has declined by 53% (from 3,500 gallons per ton of concentrate to 1,200 gallons per ton). Much of this progress can be related to improved settling of the clay waste using sand-clay mixing and/or chemical flocculation to speed the release of water. Power consumption in phosphoric acid production has been greatly reduced by using wet rock in processes and grinding, thus avoiding costly drying. Also, many plants have added cogeneration capability to their sulfuric acid plants and are producing significant portions of their power. In some cases, cogeneration allows plants to sell electric power to local grid systems.

Radon, a daughter product of the uranium-series decay, remains a major environmental problem. Uranium occurs naturally in the phosphate ores, and after mining some remains in the overburden spoil piles, waste sands, and clay tailings. In addition, some uranium remains in the phosphogypsum waste from phosphoric acid production. These occurrences all affect the postreclamation use potential for much of the land. Use of this land for recreation (lakes, parks, and golf courses) is generally acceptable, but some precautions may be needed during building construction to assure attenuation of radon gas concentrations to acceptable levels (see chapter 13).

In addition to water quantity, water quality problems have persisted, principally with respect to process waters from ore beneficiation and chemical processing. Beneficiation may result in suspended solids or residual reagents in the process water. Chemical processing may result in dissolved solids or thermal pollution in the process water. Phosphate mines and plants must meet strict water quality requirements specified in their water discharge permits. The industry has accepted its responsibilities in protecting the state's environment.

Air quality problems are associated with dust from mining operations and chemical plants, and gases from the chemical processes. Timely reclamation has greatly reduced dust problems from mining operations. The decrease in the use of dryers for rock and products also has reduced the number and types of dust sources. Gases such as ammonia, sulfur oxides, and fluorine species have caused problems in the past. New technology and

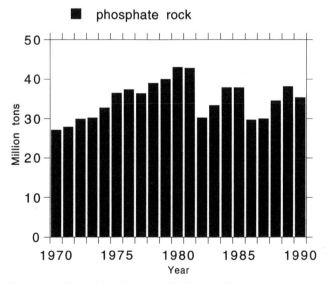

Figure 9.5. Temporal variation in Florida phosphate production.

improved environmental equipment have greatly improved the air quality around phosphate operations.

Clays (mostly smectite and palygorskite) average about 25% by weight of typical phosphate ore in Florida. During disaggregation of the ore to separate the phosphate particles from the gangue minerals, the minus 75 μm clay-containing fraction (called slimes) is highly dispersed. The hydration envelopes associated with these dispersed clay particles cause a significant increase in volume and account for nearly 90% of the water lost from the process (Zellars and Williams 1978). Clay waste (slimes) disposal has been a problem for more than forty years. The main slimes problems are retention of water, large areas occupied for long periods of time during dewatering, and periodic failures of containment structures. The latter problem has largely been resolved by state-mandated dam construction standards, and no major failures have been reported since 1971. Techniques for dewatering slimes have been developed, and are being evaluated in large-scale tests by the industry. No official assessment of the success or failure of these methods has been made.

Wetlands remain a divisive issue for phosphate companies, regulatory agencies, and environmental groups. These areas serve as wildlife habitats, retain surface water, remove sediment and excess nutrients from surface waters, and can aid in aquifer recharge. Questions concerning mining and restoration of wetlands have not been resolved. The economic value of the phosphate and the environmental value of the wetlands are still being debated at the state and national levels.

Crushed Stone

Florida ranks second in the nation in crushed-stone production. The value of the produced stone ranks second behind phosphate rock in the state's mineral sector, providing some 20% of the total. Over the past twenty years, crushed-stone production has steadily increased from about 40 million tons to about 70 million tons (fig. 9.6). Stone is produced in twenty-nine counties; the five leading producers—Broward, Dade, Hernando, Lee, and Sumter counties—account for about 70% of the output (fig. 9.7). Limestone accounts for nearly 90% of the production, shells (5%), and dolomite (2%); marl and unspecified stone account for the rest. Some of the largest limestone quarries in the U.S. are located in southern Florida. Limestone production, in terms of tonnage, is concentrated in Broward, Collier, Dade, and Lee counties in southern Florida, and Hernando County in central Florida. Shell production is concentrated in Hillsborough, Manatee, and Sarasota counties along the west-central Florida coast. Dolomite is mined in Jackson, Suwannee, and Taylor counties in north-central Florida, and in Citrus County on the Gulf Coast. Marl is produced in Collier and Lee counties in southwestern Florida, and in Indian River County on the Atlantic Coast.

These mined limestones vary in geologic age from Middle Eocene to Pleistocene. Mineability of limestones depends on lithology, structure, and geomorphology. Lithology is the most important factor, because it determines the physical characteristics that control the final use and value of the product. Hardness, shape, porosity, permeability, texture, mineral composition, and gangue minerals are all factors related to lithology. Structure and geomorphology control the areal extent, thickness, karst development, groundwater conditions, and overburden thickness that may impact the economic geology of a deposit.

Purely economic factors may be even more important than these technical considerations. Transportation is the principal factor that determines the price and market for low-value bulk materials such as crushed stone (and sand and gravel). The high cost of truck transportation (approximately $0.50/ton for the first mile, and $0.10/ton for each additional mile) normally limits haul distances to a radius of about fifty miles from the quarry. Longer distances require access to less-expensive transport such as rail, barge, or ship. Barge shipments may only be $0.01 to $0.02/ton per mile. Rail costs are only slightly greater (Langer 1988). Therefore, ready access to local markets and/or low-cost transportation are essential for a viable crushed stone industry. Uncertainty in fuel prices always makes the transportation cost of such materials difficult to forecast. The concept of regional markets for

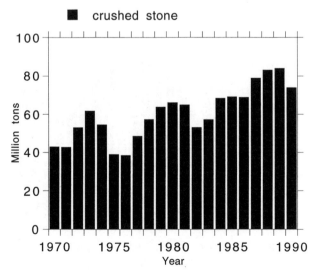

Figure 9.6. Temporal variation in Florida crushed-stone production.

crushed stone within the state is clearly shown in the U.S. Bureau of Mines data collection by districts for Florida producers (White and Schmidt 1991; Boyle and Schmidt 1992).

In northwestern Florida, limestone and dolomite crop out in Holmes, Jackson, Walton, and Washington counties. Crushed stone production from a principal producer is reported in Jackson County (White and Schmidt 1991). The stratigraphic units that make up these outcrops are the Upper Eocene Ocala Limestone, the Oligocene Marianna and Suwannee limestones, and the Upper Oligocene and Miocene Chattahoochee Formation (Campbell 1986). The case-hardened parts of the Marianna Limestone and the indurated Suwannee Limestone are the most desirable stones in this area for use as aggregate.

In northwestern and central peninsular Florida, stone is mined in an area extending from Wakulla and Jefferson counties southward to Manatee County. The stratigraphic units in this area include the Middle Eocene Avon Park Formation, the Upper Eocene Ocala Limestone, the Oligocene Suwannee Limestone, the Miocene St. Marks Limestone, and the Miocene Hawthorn Group. The Avon Park Formation, where it is being mined, is a thin-bedded, tan to brown dolomite. The Ocala Limestone in this area is soft and largely used as construction fill. In the past, some of these high-Ca limestones have been calcined to produce lime (CaO). The Oligocene Suwannee Limestone is the main stone of commercial interest and is mined in Hernando, Pasco, Sumter, and Taylor counties for use as limestone aggregate.

Limestones and coquina are mined discontinuously along the Atlantic Coast from St. Johns County to the Florida Keys in Monroe County. The Pleistocene Anastasia Formation, the Miami Limestone, and the underlying Fort Thompson Formation are the main stones of commercial interest in this area. In the upper Keys, the Key Largo Limestone is of local interest.

In Lee and Collier counties of southeastern Florida, stone is mined from the Pleistocene Fort Thompson Formation and from the Pliocene Tamiami Formation. The limestones in this region are variable in composition and often are interbedded with sands.

All stones are mined in open pits. The mining method depends on the nature of the stone and the local elevation of the water table. Land preparation and overburden removal are usually done by bulldozers or draglines. Some rocks are soft enough to be ripped by bulldozers. Flooded pits are mined by draglines or shovels, depending on the hardness of the rock. Blasting is needed in some of the more indurated stones. After removal, the stones are usually transported by truck to a crushing and sizing plant where the product grades are prepared. Stones with high contents of clays or fines also may require washing to meet specifications.

The major uses of crushed stone in Florida are for road-base materials and concrete aggregates. Other uses include chemical, agricultural, and metallurgical applications. Most crushed stone is moved by truck, smaller amounts moving by rail and water.

Florida's limestone reserves and resources are very large. Most of the peninsula is underlain by limestone, and reserve estimates for Dade, Broward, and Palm Beach counties alone are more than 34 billion tons, with a local reserve base of more than 100 billion tons (Edgerton 1974). Much of the stone is unavailable because of urban development or statutory constraints.

The major environmental problems of limestone mining are air, water, and noise pollution. Dust from mining and processing can be controlled by a variety of techniques. Water problems include contamination of penetrated aquifer systems and drainage problems around the mine and plant. Noise can be controlled by natural or artificial screens and by minimizing blasting.

Cement

Portland cement is produced from limestone, silica, and iron and aluminum oxides and hydroxides. The latter generally are derived from clays, shales, or fly ash. This mixture is calcined to form cement clinkers containing

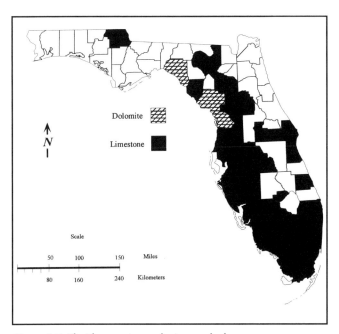

Figure 9.7. Florida counties producing crushed stone.

about 65% CaO; these clinkers are finely ground (to minus 44 μm) to produce cement. The proportions of ingredients are carefully controlled to yield a satisfactory product. The main reagent in cement is CaO, produced from the calcination of limestone. The CaO is hydrated when the cement is mixed with water prior to use in construction, and the resultant $Ca(OH)_2$ then reacts with the aluminum, silica, and iron derived from the shale, clay, or fly ash constituents to form cementitious compounds. The raw materials for cement production can be found within the state, although some companies may import one or more of the components. Limestone is available from mines in several counties, as are quartz sand and clay. Low-Mg clays suitable for cement production are either in short supply or are located too far from the production plants to be economical. Fly ash from coal burning power plants may be used as a substitute.

Masonry cement is made by mixing portland cement containing specified quantities of $Ca(OH)_2$ with either baghouse dust, fly ash, mill scale, or (sometimes) smectite clay minerals. The latter materials are used as sources of iron and aluminum, but, more important, they also improve the moisture-holding capacity and wet workability of the mixture. Staurolite from the heavy-mineral mines in the Trail Ridge area of northern Florida has been used as an alumina source in some cements. The resources and reserves of the mineral components of cements are discussed in the various commodity sections.

Calcined clinkers are imported and ground into cement in Bradenton and Tampa. An operation at Fort Pierce serves as a marketing and distribution center for imported cement.

Portland and masonry cements are the third-largest component of the state's mineral industry, based on the value of products sold. Florida was first in 1990 and third in 1991 among states producing masonry cement, and went from sixth to seventh in the production of portland cement. The six plants located in Dade, Hernando, Hillsborough, and Manatee counties (fig. 9.8) all produce portland cement; three of the plants also produce masonry cement. All of these plants are located close to rapidly growing areas of central and southern Florida.

Cement production is closely tied to construction activities. Demand rises and falls with the economic fortunes of the state, region, and nation. The rapid growth of the state has sustained a twenty-year increase in production of portland and masonry cements (figs. 9.9 and 9.10).

Air and water pollution are two concerns of the cement industry. Air pollution includes fugitive dust from clinker kilns, coolers, and stacks. This dust can be controlled by electrostatic precipitators and/or wet scrubbers. Dust disposal is strictly regulated by the Environmental Protection Agency. Recycling opportunities for the dust are limited by state agency regulations on alkali content of cement used in concrete for public roads and structures. Conversion of plants to coal-burning also can involve air-quality problems with gaseous and particulate emissions. Water pollution concerns have focused on temperature, suspended solids, and pH of effluents.

The costs of environmental responsibilities and regulation and high fuel costs have created an opportunity market for imported cement in Florida. Cement clinkers

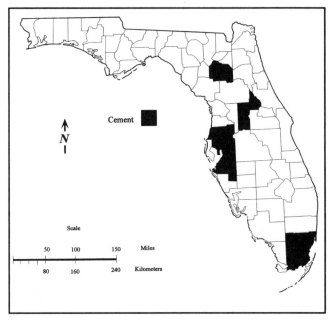

Figure 9.8. Florida counties producing cement.

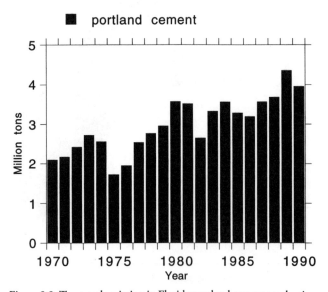

Figure 9.9. Temporal variation in Florida portland cement production.

are imported from France, Spain, and Mexico. Florida has historically imported cement from Europe and Latin America.

Sand and Gravel

Quartz sand is one of Florida's most abundant natural resources. Florida ranks sixteenth in the nation in production of sand and gravel, which accounts for about 4% of the value of the state's mineral production. The materials were produced by some ten companies operating eighteen mines in more than ten counties scattered around the state (fig. 9.11). Most of this material is sold for use in concrete (as coarse and fine aggregate). These data do not include manufactured sand (limestone fines) used as fine aggregate in areas where quartz sand is not available or is in short supply.

The commercial sand deposits of Florida are associated with marine terraces and ancient shorelines. According to Cooke (1945), the older, topographically high terraces contain the coarsest material, whereas the younger, low terraces contain finer sands and more impurities (clays and carbonates). Scott et al. (1980) divided the sand and gravel deposits of Florida into four types based on origin: modern beach deposits, river alluvium, marine-terrace deposits, and sand and gravel from a particular geologic formation.

Most of the sand and gravel mined in northwestern Florida is from Pleistocene marine-terrace deposits in Leon and Wakulla counties, and from the Pliocene to earliest Pleistocene Citronelle Formation in Escambia County. The Citronelle and sediments derived from it are the only significant source of gravel in the state. In northern peninsular Florida, several sand units occur in Hawthorn Group and younger sediments, but they are too fine-grained or impure for uses other than road fill. Lake Wales Ridge (Pliocene) extends from western Clay County southward to Highlands County and is composed of loose surface sands that overlie red, yellow, and white clayey sands. Locally, quartz gravel and discoid quartzite pebbles—similar to those reported in the Citronelle of northwestern Florida—are present (Pirkle et al. 1964).

In central Florida, the numerous sand ridges of the Central Highlands (White 1970) contain the poorly sorted (fine sand to pebbles) Mio-Pliocene sand deposits of greatest economic importance. These sands often contain a kaolinite matrix. The sands of the Gulf Coastal Lowland (White 1970) are found on Pleistocene terraces. Although these deposits are too fine-grained for construction and aggregate sands, they have been mined for glass sand in the Plant City area. In southern Florida, the sand deposits are of local importance only and are used for construction and fill materials. The Pleistocene terrace sands, Anastasia Formation, Fort Thompson Formation, Plio-Pleistocene Caloosahatchee Formation, and Pliocene Tamiami Formation all contain sand deposits of local importance (Scott et al. 1980).

Florida sand mines are all open-pit surface mines. Typically, the sand is mined by hydraulic methods, using a suction dredge to slurry the sand for transport to the classification plant. At the plant, vibrating screens are used to separate the coarse and fine sand fractions. The

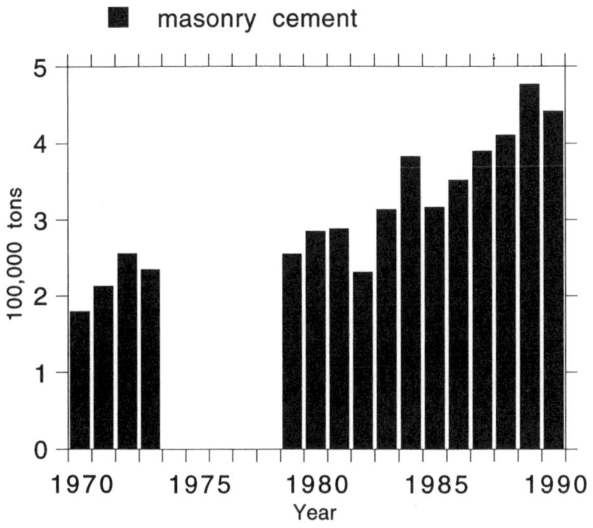

Figure 9.10. Temporal variation in Florida masonry cement production.

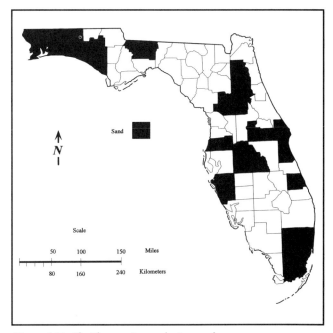

Figure 9.11. Florida counties producing sand.

coarse material is stockpiled for loading, and the fines are sent to a settling area.

Construction sand and gravel is the main use for these products in Florida, most of that going into concrete and related products. Nearly 80% of Florida's sand and gravel is produced in Hendry, Lake, Marion, Polk, and Putnam counties. Industrial sand is used for glass, foundry, and abrasive raw materials. This is a relatively small volume business (about 500,000–600,000 tpy), with mines located in Escambia, Glades, Lake, Marion, Polk, and Putnam counties. Some byproduct industrial sand is produced from kaolin operations.

The overall trend in sand and gravel production for the past twenty years has been up (fig. 9.12). Like many construction related industries, it can be affected by external economic factors (inflation, recession, interest rates), and cycles can cause short-term highs and lows. The predicted growth in population for the state indicates continued growth in this area.

Florida's reserves and resources of sand are large. Gravel resources are much more limited, and they occur in only a few geologic environments in the state. Low unit values of sand and gravel restrict their marketing area, and they should be located as close as possible to the final consumer. This often brings them into land use conflicts in the areas with rapidly growing populations that have the greatest demand for construction raw materials.

The environmental concerns of sand and gravel companies are competition for land, water, and air resources. Water quantity and quality often are locally important issues where hydraulic mining is done. Although a large part of the water is recycled during mining operations, the permits requested by sand companies for high consumptive water use can create conflict with local property owners. Process water also may contain suspended and dissolved solids that decrease water quality. Discharged water must meet permitted specifications. Land reclamation is an important issue in many areas where the old mine site may be left as a lake or other topographic depression.

Clays

Florida ranked second in the nation among the forty-four states reporting clay output during 1991 (Boyle and Schmidt 1992). Clay sales were the sixth-largest component of the state's minerals industry. Clay deposits occur in many parts of Florida, but only a few deposits have the proper mineralogy, purity, and quantity for commercial exploitation. Economic factors such as transportation, power, labor, and marketing also must be considered in evaluating a deposit. Production in the clay industry of Florida peaked in 1973 at more than 1 million tons (fig. 9.13). Between 1975 and 1990, production averaged more than 600,000 tpy. The twenty-year trend showed a slight general decline.

The types of clays produced in Florida are common clay, fullers earth, and kaolin. Common clay in this sense is a clay or claylike material that is plastic enough for easy molding. In Florida, such clays often are mixtures of kaolinite with mica and/or smectite. Local deposits of common clay, used for fill, occur in geologic formations exposed at the surface in nearly every county in the state.

A commercial deposit of a bloating clay occurs in Clay County and consists of a mixture that is dominantly smectite, with some kaolinite and enough associated organic matter to produce bloating when calcined. When

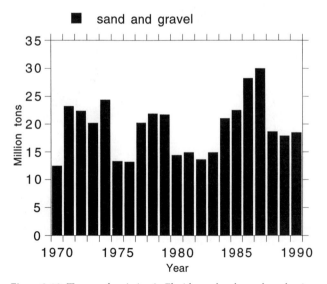

Figure 9.12. Temporal variation in Florida sand and gravel production.

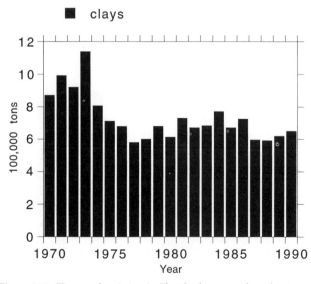

Figure 9.13. Temporal variation in Florida clay mineral production.

fired in a rotary kiln at about 1100°C, the clays sinter and become plastic over a narrow temperature range. During this plastic stage, gases evolved from water in the clay and from the decomposition of the associated organic matter cause the clay to bloat, producing low-density, glassy masses. These glassy materials are air-quenched during discharge from the kiln to produce lightweight aggregates. This clay occurs in the Coosawhatchie Formation (Charlton Clay Member?) of the Hawthorn Group. During exploitation, a sandy overburden and an upper unit of weathered, sandy clay are removed to expose the lower, smectite-rich clay unit. This lower unit is preferentially mined and mixed with varying amounts of the upper sandy-clay unit to obtain a calciner feed that yields a good lightweight aggregate. The aggregate is marketed for construction, drainage, and ornamental uses.

Common clays mined in Lake County are derived from Plio-Pleistocene sediments associated with Lake Wales Ridge. These clays occur as lenses associated with quartz sand containing minor amounts of mica and heavy minerals. Principal sales are to construction companies and cement and ceramic producers.

Fullers-earth clays derive their name from the use of clays and other fine-grained materials to remove oil and lanolin from animal skins. No specific mineralogic composition is implied in this definition. Instead, desirable physical properties related to absorption, decolorizing, and purifying are the criteria for selection. Today, the term fullers earth generally describes clays containing calcium montmorillonites and/or palygorskite (the latter often referred to in commercial literature as attapulgite). Florida ranks second in the nation in the production of fullers earth, which is mined in Gadsden and Marion counties (fig. 9.14). The Gadsden County deposits are predominantly palygorskite mixed with some smectite, sepiolite, and fine-grained quartz. The fullers-earth deposits are found in two distinct stratigraphic intervals (Merkl 1989): the Dogtown Clay Member of the Torreya Formation, and the slightly younger Meigs Member of the Coosawhatchie Formation, both in the Hawthorn Group. The Coosawhatchie Formation unconformably overlies the Torreya Formation. The clay beds are less than one to three meters thick and are separated by as much as three meters of hard sands. In some areas, a thin (up to 25 cm), dark, organic zone, described as a paleosol by Weaver and Beck (1977), occurs in association with the clay beds. The Dogtown Clay Member consists of dolomite, long-fiber palygorskite, smectite, minor quartz, and pyrite. The Meigs Member is reported to contain smectite, short-fiber palygorskite, sepiolite, and abundant opal-CT (Merkl 1989). In some areas, the overlying Pleistocene sands and Hawthorn material combine to produce overburden thicknesses ranging from less than one meter to nearly 30 meters. Production from northwest Florida counties would be in the range of 300,000 to 400,000 tpy from three companies operating several mines.

The fullers-earth deposits in Marion County occur in the lower Hawthorn Group (Penney Farms Formation), and as outliers at the southern limit of the Hawthorn outcrop belt. These clay units are the only Hawthorn materials present, and are underlain by the Eocene Ocala Limestone. The overburden consists of undifferentiated Pleistocene sands. The clays consist of two units: an upper, highly weathered, mottled, sandy clay to clayey sand containing smectite and kaolinite in variable ratios, and a lower, massive, green clay consisting mostly of Ca-saturated smectite. This massive clay averages about 8 m thick and is further subdivided in the mining operation into an upper, high-density unit and a lower, low-density unit for processing purposes. These density differences may result from desiccation microcracks in clay layers that occur and persist above the water table. The overburden is highly variable, depending on the amount of erosion; a stripping ratio of 1:1 or less is typical.

The largest market for fullers earth is as pet-waste absorbents and industrial oil and grease absorbents. Pet litters now use more than 60% of fullers earth sold in the U.S., with a value of about $100 per ton. Some experts are predicting a domestic (and possibly an international) shortage of material to meet the rapidly growing demand for this product (H. B. Murray, personal communication

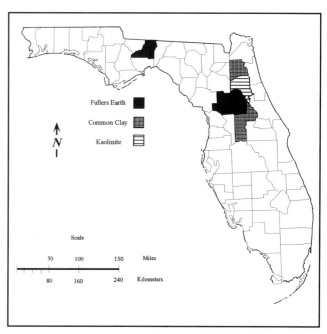

Figure 9.14. Florida counties producing clay.

1992). The other industrial uses, such as oil clarification, pesticide carrier, and salt-resistant drilling muds, have shown little growth over the past decade. Florida's two fullers earth producing areas have shown little growth in total tonnage over the past two decades (fig. 9.15).

The only active kaolin mine in Florida is in western Putnam County near the town of Edgar. The deposit occurs in Lake Wales Ridge and is thought to be Plio-Pleistocene. Kane (1984, 10) informally named these "fine to coarse, argillaceous sands and gravels" the "Grandin sands." They unconformably overlie the Hawthorn Group and are unconformably overlain by a thin veneer of Pleistocene sands in western Putnam County. These coarse clastics had been referred to as the Citronelle and Fort Preston formations in earlier literature (Cooke 1945; Pirkle 1960; Puri and Vernon 1964). The ore is predominantly quartz sand, with about 20% kaolinite and minor amounts of mica and heavy minerals. With respect to geologic age, the ore is probably the youngest deposit of kaolinite to be commercially exploited. The origin of the kaolinite is not well understood; it may have formed by weathering in place, or by deposition under deltaic conditions.

The extremely fine particle size of the clay (often minus 0.2 μm) is thought to give it its unusual plastic properties. It is used mainly in ceramic and tile manufacturing. The crude ore is hydraulically mined, and the slurry is pumped to a plant where sand is removed in two stages. Thickeners and filter pressing are used to concentrate the clay before final drying, grinding, and shipment. The coarse sand is sold as a byproduct, and heavy minerals may be removed from the quartz sands on spirals to improve their quality. This operation produces 40,000 to 50,000 tpy of kaolinite product.

All of Florida's clays are mined in open pits. The initial steps involve land clearing and overburden removal by combinations of bulldozers, draglines, and earthmovers. A dragline or shovel is used to remove the clay for transport by truck (sometimes by rail) to the processing plant. The kaolin-bearing sands are mined hydraulically by a suction dredge.

In general, clays are processed by physical means that include various combinations of drying, grinding, and sizing. The combinations of treatments will vary with the nature of the desired product. In the case of fullers earth, grinding may be followed by either single stage or dual stage drying, size classification to specifications, and packaging. Common clays often are used on an as-is basis for construction purposes. Those used in the cement industry may receive some processing to control the composition of the product.

Little has been published on the reserves and resources of clays in Florida. Common clays are considered abundant in northern Florida. Southern Florida has a deficiency of clays, especially those suited for manufacture of cement (high-alumina, low-magnesia). Transportation is a major economic factor for commodities with low unit value, including clays. Thus, location of materials with suitable properties close to the consumer is an essential condition for economic development. Identified resources of bloating clays are very limited. This may result from a lack of published data, exploration, and testing, rather than an absence of suitable materials. Ampian (1985) estimated Florida's fullers-earth resources at 300 million tons. There is inadequate geologic information to know what part of these resources could be converted to reserves under present or foreseeable economic and technologic conditions. Continued growth in demand for animal-waste absorbents could result in development of new mines. Growing interest in water conservation has created a potential market for these clays on golf courses and other recreational areas to improve moisture retention in very sandy soils. Technologic innovations to improve the product yield from mined ore also will receive attention in the future from several segments of this industry, in order to stretch existing resources and to improve cost-effectiveness.

Environmental concerns of the clay industries are primarily focused on air and water pollution. Control of dust during mining, transport, processing, and storage is a concern for most clay producers. Possible air pollution from calciner fuels also has attracted some attention over the past few years. Groundwater consumption and pro-

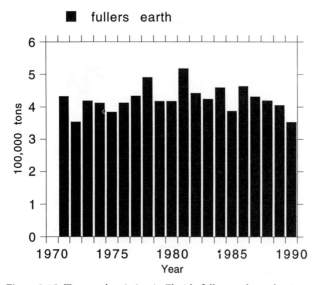

Figure 9.15. Temporal variation in Florida fullers earth production.

tection are concerns for industries where local dewatering is required to facilitate mining. Land reclamation is an agency-regulated activity that is a concern to all mining industries.

Heavy Minerals

The heavy-mineral sands of Florida are considered to have originated from the erosion of igneous and metamorphic rocks of the southern Appalachians (Gillson 1959). Weathering and erosion of these crystalline rocks provided the heavy-mineral grains that were transported southward by fluvial and marine currents. During this process, abraded detrital heavy-mineral grains were concentrated by winnowing and left in various types of coastal deposits in Florida and adjacent states. Heavy-mineral sands are common in the quartz-sand and clayey deposits that occur throughout northern Florida; however, economic deposits of heavy-mineral sands in Florida occur in rather limited environments of Trail Ridge in Bradford and Clay counties, and ancient beach ridges in southeastern Clay and northeastern Putnam counties. The sands of the Trail Ridge ore body contain an average of 4% heavy minerals (specific gravity greater than 2.9), of which 45% are the titanium minerals ilmenite, leucoxene, and rutile (Carpenter et al. 1964, 790). Other heavy minerals include staurolite, zircon, kyanite, sillimanite, tourmaline, spinel, topaz, and corundum. Detailed descriptions and relative percentages of heavy minerals are given by Spencer (1948) and Pirkle and Yoho (1970). Pirkle et al. (1977) stated that Trail Ridge is the topographically highest and oldest shoreline along which commercial concentrations of heavy-mineral sands are found in Florida.

The Green Cove Springs heavy-mineral deposits in Clay County occur at lower elevations and are believed to have formed in northwest-trending beach ridges developed along the shoreline of a generally regressing sea (Pirkle et al. 1974). Other depositional models suggest that the economic deposits developed during a stillstand of the regressing sea (E. A. Mallard, personal communication 1993). The sediments of the Green Cove Springs deposits are finer-grained and richer in zircon and monazite than the Trail Ridge deposit. The geologic age of all the heavy-mineral sands in northeastern Florida is Pleistocene, based on radiocarbon dating and their association with the sea-level stands and associated dated shorelines at various elevations (Pirkle et al. 1991; Elsner 1992).

After clearing the land of timber and other vegetation, heavy-mineral sands are mined by dredges, which may be bucket-wheel types or, more commonly, suction dredges with cutter heads. Rougher concentrates are separated from the quartz sand by wet-gravity methods, using combinations of spiral separators. This wet mill is barge-mounted and floats in the pond with the dredge. The preconcentrates from the wet mill, containing as much as 85% heavy minerals, are pumped to piles on shore, where they dewater and are trucked to the dry mill for final separation. Waste sand and rejects from the wet mill are back-filled into the pond behind the dredge. A typical mining operation will produce about 30 to 50 tons per hour of preconcentrate from a sand feed containing 4% heavy minerals. Thus, a large volume of material is moved to recover a small quantity of heavy mineral–sand preconcentrate.

The preconcentrate from the wet mill is brought to the processing plant, where it is causticized to remove organic coatings from the mineral grains. Following causticizing, the sands are dried in a rotary kiln and then sent for electrostatic separation. This separates the electrically conducting titanium minerals (ilmenite, rutile, and leucoxene) from the nonconducting silicate gangue minerals (such as staurolite, zircon, kyanite, sillimanite, tourmaline, and topaz). Ilmenite can then be removed from the titanium-rich concentrate by magnetic separation to produce the final ilmenite concentrate, which averages about 98% ilmenite and contains about 64% TiO2 (Kramer 1990). The nonmagnetic residue may be rich in rutile and leucoxene, and can be recovered as a very valuable coproduct in small tonnages.

The nonconductors from the first electrostatic separation also are separated magnetically to produce staurolite (magnetic) and zircon (nonmagnetic) concentrates. The staurolite is sent to the stockpile for sale as an aluminum-rich cement additive, nonferrous foundry sand, polishing compound for marble, or silica-free sandblasting agent. The zircon concentrate is passed over another set of wet spiral or table classifiers to prepare a special enriched grade that is then calcined at about 1000°C and electrostatically and magnetically separated to prepare a high-grade, very white concentrate and some coproducts. One coproduct from the table classifiers is a monazite concentrate that is high in thorium and rare earths.

The main use for titanium-containing heavy minerals is to produce titanium dioxide pigment for use in paint, paper, plastics, and related products. In 1990, these markets used more than 90% of the titanium consumed in the United States (Kramer 1990). Welding rods, fluxes, chemicals, glass fibers, and titanium metal were the other principal uses. Staurolite is used mainly in the cement industry, and as an abrasive.

Zircon is used primarily in refractories, abrasives, and foundry sands. Zirconium metal is used to clad nuclear fuel rods. The associated element hafnium is used in control rods of nuclear reactors, and as a constituent in superalloys (Templeton 1992).

Rare earth–containing monazite concentrates are produced by one of the companies mining heavy-mineral sands in Florida. These concentrates are exported from the state for processing. Monazite is a phosphate mineral with high contents of thorium, as well as lanthanum, cerium, praseodymium, neodymium, samarium, europium, gadolinium, and yttrium. The main uses of rare earths are in chemical catalysts, metallurgical additives, ceramics and polishing agents, and electronics.

Florida leads the nation in the production of titanium and zirconium concentrates, and ranks third among the states in production of rare earths. The existing mines in Clay County (fig. 9.16) can produce 200,000 to 300,000 tpy of ilmenite-leucoxene concentrate. A third mine—near Maxville, in southwestern Duval County, northwestern Clay, northeastern Bradford, and southeastern Baker counties—will add about 150,000 tpy of titanium-concentrate capacity. Florida reserves of titanium minerals have been estimated at 8.1 millon tons (Kramer 1990). Current information indicates that mining will continue at least until the year 2010 (Templeton 1992). Florida's zircon production varies from 100,000 to 120,000 tpy. Although no specific estimates have been made of Florida's reserves and/or resources of zircon and monazite, their coproduction will likely continue as long as there is demand for these minerals and the production of titanium minerals remains economic. Quantitatively, the reserves and resources of zircon and monazite in Florida have to be considered very limited.

The main environmental problems of the heavy-minerals mining companies in Florida are water quantity and quality. Although mining and processing operations consume a few million gallons of water per day, disruption and restoration of the surficial aquifer system during mining and reclamation may be a larger, long-term problem. Land reclamation, recontouring, and revegetation are required by state and local regulatory agencies to minimize the impacts of heavy-minerals mining.

Peat

Florida consistently ranks among the nation's leaders in peat production. Florida ranked first among the states before 1989, but since then has dropped to second place behind Minnesota. There has been an irregular upward trend in peat production over the past twenty years (fig. 9.17). As mentioned in the section on potential energy resources, plans to use peat in electric-power production could result in a tenfold increase in peat production over the next five to ten years. At present, ten companies produce both reed-sedge and humus peat in seven counties. A comprehensive review of the peat resources of Florida has been prepared by the Florida Geological Survey and should be consulted for detailed descriptions (Bond et al. 1984b).

Peats are generally considered to be of Holocene to Pleistocene age. Davis (1946) classified peat deposits of Florida into eight types, based on location: (1) coastal occurrences, including marshes, swamps, lagoons, estuaries, and dune depressions; (2) large, nearly flat, poorly

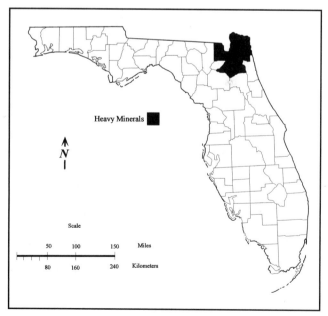

Figure 9.16. Florida counties producing heavy mineral sands.

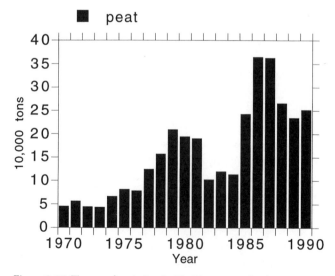

Figure 9.17. Temporal variation in Florida peat production.

drained areas like the Everglades; (3) river-valley marshes; (4) flatland swamps; (5) lake and pond marshes; (6) seasonally flooded depressions; (7) lake-bottom deposits; and (8) peat layers buried below other strata. Peat deposits may be wet or dry, and moisture content may affect their properties and uses. Peats may be naturally dry, or may have been drained by man. Dry peats may undergo oxidation that changes their physical and chemical characteristics. Peat can accumulate at a rate of nearly 10 cm per 100 years; the accumulation rate varies with climate, water-table position, and nutrient supply (Moore and Bellamy 1974).

Following land-clearing operations, trenching is required to dewater most peats before mining. This dewatering allows the peat to support the shovels, draglines, and earthmoving equipment used in the mining operations. After excavation, the peat is usually transferred to storage for drying, followed by shredding. Large-scale mining for use as a fuel may require the use of more sophisticated equipment of the type seen in Europe (e.g. bucket-wheel excavators with traveling belt systems).

The principal nonfuel use of Florida peat as a soil conditioner in gardens, greenhouses, and nurseries, as well as for lawns and golf courses. Peat improves the physical properties of soils (lightens textures), and increases their moisture-holding capacity (eight to twenty times the weight of the peat). Peat often is blended with fertilizers, sand, sawdust, or perlite before delivery to the job site; blends of this type have greatly diversified the demand for peat in landscaping.

Peat production in Florida has averaged more than 250,000 tpy since 1985 (fig. 9.17). The determination of the peat resource/reserve base in Florida has been described by the U.S. Bureau of Mines as "a credible job" compared to the efforts of many other states (Cantrell 1992). Resources have been estimated at 6.8 billion tons (U.S. Department of Energy 1979), and reserves have been estimated at slightly more than 600 million tons of fuel-grade peat (Griffin et al. 1982). Both figures show a significant reserve base for these materials. Peat occurs in a number of Florida counties (fig. 9.18).

The environmental concerns of peat mining involve a variety of issues related to surface waters and biota. Lowering the surficial water table can affect the flora and fauna associated with the habitat, increase discharge, and cause problems of water quality and quantity. Water quality may be affected by suspended and dissolved inorganic and organic materials. Surface subsidence due to increased drainage has been documented in the Florida Everglades (Stephens 1974). Any significant expansion of peat mining will require an assessment of the impact on the local environment. The burning of peat as a fuel produces emissions similar to those from the combustion of coal, including carbon dioxide, sulfur oxides, nitrogen oxides, particulates, trace elements, and heavy metals.

Trends

The mineral industries of Florida are evolving in several respects, including regulatory activity by various agencies, technical developments, increased competition, and reputation. Potential future changes might include authorization of the Resource Conservation and Recovery Act to include mine and mill waste, more air quality and particulate regulations, acid-rain abatement, toxic substance notification, control of storm water runoff, recycling, recovery, and landfill regulations.

Differences in federal and state regulation of energy resources will continue to be discussed. Differences in state law and regulatory policies will need to be clarified as interest in offshore energy and mineral resources grows. The proposed tenfold expansion of peat mining for fuel in Florida may have impacts that are not fully understood at present.

Environmental programs often create opportunities for mineral industries. For example, acid-rain abatement can increase the demand for reactive, high-Ca limestones for use in scrubbers to reduce sulfur-dioxide emissions. Fly ash from power station electrostatic precipitators can be used as an additive in or substitute for cement in concrete. The disposal or use of sludge scrubber wastes from air pollution control units may present economic opportunities that cannot be seen at present.

The use of clay liners alone or as part of a composite

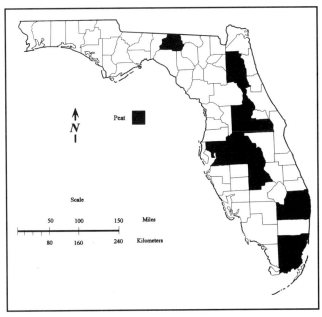

Figure 9.18. Florida counties producing peat.

system for landfills is a topic of much technical, regulatory, and legal discussion. The design and construction of closure structures also may be an opportunity. The use of water-retaining clays in the management of golf courses and parks is under evaluation. The development of sorbent clay systems for treatment of storm-water runoff is a potentially large market where technology is beginning to develop.

Mining operations that use dredges to mine sand, clay, and heavy-mineral sands may encounter new regulations for water use and water quality, as well as preservation and protection of resources and wetlands. The dewatering of quarries and mine pits will be more carefully monitored and regulated. Recycling water from waste treatment plants can decrease the industrial needs of some mining operations. As the population of the state grows, competition for its resources also will intensify.

Technical developments by the industry will continue to make it more efficient. These developments will include the use of chemical and urban waste as fuels for cement, clay, and lightweight aggregate plants (Michaud 1988). Urban garbage, waste oils, and combustible organic liquids will receive the most attention. Alternative fuels may be used alone or in combination with conventional fuels. It will be necessary to comply with air- and water-quality regulations in the use of these materials.

Recycling will become important as materials from old interstate highways, streets, roads, buildings, tires, and roofs are processed to cut costs and alleviate depletion of natural resources (Rukavina 1989). The Florida Department of Transportation has built a test strip of asphalt pavement containing used tires, is evaluating recycled concrete in pavements, and has a development program for recycling plastic for highway hardware.

Florida's laws governing underground storage tanks have created opportunities for some mineral processors. Soil from fueling sites and hot-mix plants may become contaminated with a variety of petroleum-derived materials. The organic matter must be removed before the soil can be replaced or sent for disposal. Plants with high temperature dryers and associated air pollution control equipment may be able to treat these soils to remove the contaminants, which could become a profitable business for operators with seasonal equipment use. Disposal of the treated soil in mined-out quarries also might be possible.

Foreign competition has begun for some of Florida's mineral industries, particularly aggregate and cement. Competitors have established operations at strategic locations in the Bahamas and Mexico to compete in regional markets. The state's mineral industries must concentrate on increased efficiency and high-quality products to sustain themselves and their market share.

The mineral industries of Florida have shown sustained growth for the past twenty-five years. Given the projected population growth of the state and the world, the industry's prognosis seems good. All interested parties should be prepared to work together to assure its continued growth and success.

10

Geology of the Florida Coast

Richard A. Davis Jr.

THE FLORIDA COAST REPRESENTS A WIDE variety of morphologies that have developed in response to a spectrum of processes and geologic settings. Among the contiguous forty-eight states, Florida has the longest and most diverse coast. The adjacent narrow (Atlantic) and wide (Gulf) continental shelves, the range in tidal and wave conditions, the variety of coastal orientations, the diverse sediment types, and the range in availability of sediment give rise to this broad spectrum of coastal geologic environments. An important additive is the relatively frequent occurrence of hurricanes, which typically leave their imprint on the coast.

Florida's coastal zones can be subdivided into five geomorphic elements: (1) the east-coast barrier system, (2) the mangrove coast of southwestern Florida, (3) the central Gulf barrier-island system, (4) the marsh coast of the Big Bend area, and (5) the panhandle, including the Apalachicola Delta (fig. 10.1); the Florida Keys are treated separately in chapter 14. Each of these elements is well defined and internally similar, except for the east-coast barrier system, where the sector north of Cape Canaveral is the southern part of the Georgia Bight; it is adjacent to a broad shelf, in contrast to the southern part.

The settlement of Florida and much of its continuing growth have concentrated on the coast. More than half of its 14 million residents live in the coastal zone, which is also the main attraction for the majority of visitors. Human activities have had a great impact on the coast, and there are indications that the trend will continue.

This chapter will address the process-response systems that give rise to the present coast and its diverse components. Because the coast is geologically very young, only the late Holocene will be considered in detail. Related information on the Florida Platform and the influences of sea-level changes can be found in chapter 11.

Coastal Processes

The geography of Florida contributes greatly to the character of its coast, and to the variety of conditions to which the coast is subjected both seasonally and over the long term. The peninsular nature of the state contributes to an extensive maritime climate. The north-south attitude of the Florida peninsula and the east-west orientation of the panhandle coast provide diverse conditions as weather systems pass, and the state is situated at a latitude where passage of frontal systems dominates the winter conditions and trade winds prevail during the summer. Tropical storms irregularly punctuate the summer conditions.

WEATHER

Weather is a fundamental factor influencing the coastal environments of Florida. The interrelation varies with the coastal reach in question, and with the season. For example, during summer the prevailing winds are southeasterly and southwesterly, at low to moderate velocities (fig. 10.2). This produces small waves along most of the peninsular coast, but generally larger ones along the panhandle coast, due primarily to the relatively long fetch. The result is that summer conditions in general produce south-to-north longshore drift along the peninsula, and east-to-west drift along the panhandle.

Winter conditions are very different. Cold fronts are generated in Canada and move southward across the plains and offshore from Texas. Because of the warming effect of the water, they lose some of their strength as they move over the Gulf. As a frontal system approaches Florida, the wind is out of the southwest, increasing in strength as the front moves closer. Once the front passes over the Gulf Coast, the wind reverses direction rapidly and is at its strongest from the north or northeast. The succession of events is the same as the system moves across the Atlantic Coast. The result is that these frontal systems have little impact on the panhandle coast, but are the dominant weather factors on the west and east coasts of the peninsula.

The combination of these weather conditions gives rise to overall north-to-south longshore transport of sediment along the peninsula, and east-to-west transport along the panhandle. Volumes of transported sediment range from about 500,000 cubic meters per year (m^3/yr) on the northeast coast (Dean and O'Brien 1987) to less than 50,000 m^3/yr on the Gulf Coast (Walton 1976). Longshore transport in the panhandle increases toward the west, where it reaches more than 200,000 m^3/yr along Santa Rosa Island (Walton 1976).

WAVES

The east coast of Florida is bounded by a shelf that is moderately broad and gently sloping in the north, and both narrower and steeper to the south toward Cape Canaveral. South of this cuspate foreland and continuing through the Florida Keys, the adjacent shelf is narrow and steep, permitting relatively high wave energy to impact the coastal zone, with mean annual wave heights of as much as 70 cm (Nummedal et al. 1977). The west coast shows a marked contrast, in that the shelf is about 200 km wide, with shoreface gradients that range from 1:1,300 in the central area to less than 1:3,000 in both the north and south. These conditions, coupled with the limited fetch of the Gulf of Mexico, cause wave energy reaching the coast to be very low, and mean annual wave heights of 10 to 25 cm are typical (Tanner 1960a). The panhandle coast is bounded by a fairly narrow and steep shelf, except along the Apalachicola Delta, where it is very shallow for many kilometers offshore.

TROPICAL STORMS

Florida is in the path of Atlantic hurricanes (see chapter 13) and has been subjected to many of these devastating storms. A single storm can result in billions of dollars in property damage and loss of life, and can cause striking changes to the coast. In a few days, a single storm can produce modifications to the shoreline and related enviroments that are equivalent to a century or more of "normal" conditions.

The most susceptible areas of Florida are the southeast coast and the panhandle. The northeast coast is less likely to be affected, and the peninsular Gulf coast is least affected by tropical storms. The reason for this variance of hurricane influence is that the storms approach Florida and the northern Gulf of Mexico from the southeast and move toward the northwest. The peninsula protrudes across this path. If a storm enters the Gulf, it commonly will continue toward the northwest—thus the high incidence of landfalls along the Louisiana and Texas coasts. Some storms may move toward the north and impact Alabama or the western panhandle. It is rare that the storms will turn strongly toward the northeast and thereby have severe impact on the west coast of the peninsula. The most recent major hurricane to do so occurred in 1921.

TIDES

Tidal range is microtidal (less than 2 m) throughout the Florida peninsula, except on the northeast section, where spring tides are higher than 2 m (Nummedal et al. 1977). In general, the Atlantic Coast experiences semidiurnal tides, whereas tides are mixed on the Gulf Coast. There is, however, considerable variation in the effect of

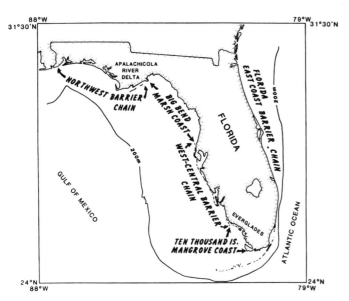

Figure 10.1. Florida's five major coastal sections.

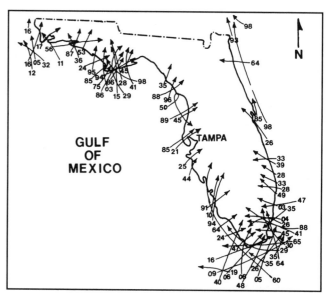

Figure 10.2. Hurricane landfalls in Florida, 1880–1980.

tidal processes on coastal development. Regardless of absolute tidal ranges, it is the relative influence of tides and waves that molds coastal morphology (Davis and Hayes 1984). The Florida peninsula provides several examples of this important relationship between wave and tidal energies. The tide-dominated sections are the mangrove coast of southwestern Florida and the marshy coast of the Big Bend area (fig. 10.1), where spring tides are no higher than 1.3 m. A broad, gently sloping shoreface produces very low wave energies in these areas, and tidal effects dominate. Most of the east coast and the Florida Keys are wave-dominated. Mixed-energy coastal morphology prevails in the northeastern part of the peninsula, where tidal range is greatest, and in the west-central barrier system, where tides are just under 1 meter. Tides on the panhandle are also about 1 m, and wave energy is higher than on the peninsular Gulf Coast.

Another important aspect of the tidal climate is the tidal prism (i.e. the tidal budget during each cycle). The tidal prism is a volume of water that is the product of the tidal range and the area. Large estuaries like Tampa Bay and Charlotte Harbor have tremendous tidal prisms because of their area, even though the tidal range is less than a meter (fig. 10.3). Small bays have small tidal prisms. The prism has great influence on the size and shape of tidal inlets. There is great diversity in the nature of tidal inlets and the coastal bays that they serve, both estuaries and lagoons.

Late Holocene Sea Level

Holocene sea-level data are scarce for much of the Florida peninsula; however, the best data are for the southwest coast and the Florida Keys. The period of interest begins about 8,000 years before present (ybp) (fig. 10.4). The rate of rise has been modest over that time, with a rather distinct slowing 3,000 to 3,500 ybp (Scholl et al. 1969; Enos and Perkins 1979; Robbin 1984). It is estimated that rates of sea-level rise were 25 cm per 100 years before about 3,000 ybp, and have been only 4 cm per 100 years since that time (Wanless 1982). Both of these rates are slow relative to the conditions of the early Holocene, when sea-level rise was about 1 cm per year.

Because of the general lack of relief and the karstified carbonate strata that dominate the Florida peninsula, there is a poorly developed drainage system that carries

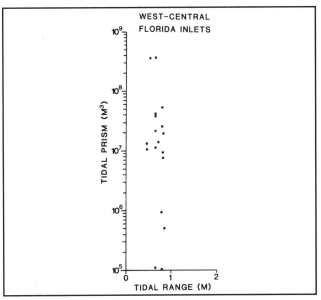

Figure 10.3. Plot of tidal range versus tidal prism for several west-central Florida inlets, showing the great range in prism. This range is due primarily to range in area of the coastal bays and is a major factor in determining the stability of the inlets and the size and shape of the ebb-tidal delta (Davis 1989; reprinted by permission of Kluwer Academic Publishers).

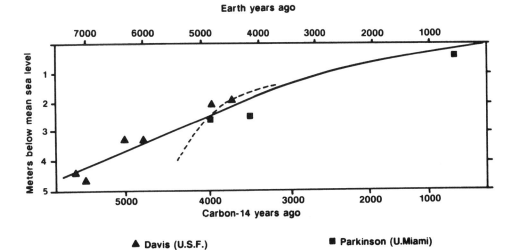

Figure 10.4. Generalized late Holocene sea-level curve (after Stuiver and Scholl 1967) with additional dates from southwest Florida.

only a modest sediment load to the coast. Virtually all of the sand-sized component of that load is trapped in the numerous estuaries. Because of the slow rate of sea-level rise and the general dearth of terrigenous sediment supplied to the Florida coast, most of the terrigenous sediments of the Holocene coastal sequence are reworked older Quaternary sediments, except near the Apalachicola Delta, where sediment has been supplied during the Holocene.

The rate of sea-level rise is very important to the development of the coast of Florida. The great change in the rate of rise during the past 12,000 years has produced variety in the nature of the coast. In general, the more rapid the rate of sea-level rise, the less stable are the coastal conditions. During the early Holocene, when sea level was rising by as much as a centimeter per year, equilibrium of the coast could not be maintained, because of rapid migration of the shoreline. Consider, for example, the west coast near Sarasota, where the offshore shelf gradient averages about 1:1000. If sea level were rising at a rate of 1 cm per year, after a century the total rise would be 1 m, and the shoreline would have moved inland 1,000 m, or 1 km, which would be catastrophic. During the early Holocene, sea-level rise was so rapid that no stable coastal morphology could be developed.

When sea-level rise slowed to about 3 mm per year about 7,000 ybp, the shoreline moved about 300 m in 100 years—also a rapid migration that did not permit formation of large and relatively stable barrier islands. About 3,000 ybp sea level was not much lower than at present; in fact, some investigators believe that it was fluctuating about its present position (Fairbridge 1961; and others). This condition provided a significant amount of time for waves, tides, and longshore transport to mold the coast into a reasonably stable network of environments including barrier islands, inlets, estuaries, and marshes. Today it appears that sea level is rising by nearly 3 mm per year (see chapter 14). Coasts are becoming relatively unstable and are migrating landward in many areas of the state. The sea-level trend may be toward the condition that prevailed about 5,000 to 6,000 ybp, when the rate of rise was at a moderate rate (fig. 10.4).

East-Coast Barrier System

The barrier-island and tidal-inlet system of Florida's east coast (fig. 10.1) is the longest (550 km) in the U.S. There are twenty-two inlets, and all but one (Matanzas) have been significantly modified by engineering activities (Marino and Mehta 1986). Located approximately in the center of this coast is Cape Canaveral, a very large cuspate foreland that projects 20 km seaward of the main coastal trend. Seaward of this cape is a well developed cape-retreat massif (Swift et al. 1972; Field and Duane 1974). Cape Canaveral strongly influences the shoreline orientation and sedimentation patterns along at least 125 km of Florida's east coast.

The north end of this coastal reach is the southern part of the Georgia Bight, or the Georgia Embayment, as it is sometimes called. Here barrier morphology is influenced by a mixed regime of wave and tidal energies. This major embayment has its center in South Carolina near St. Helena Sound, where tidal ranges approach 3 m. There is a general decrease in tidal range toward both the Outer Banks of North Carolina and Cape Canaveral in Florida, the distal limits of the Georgia Bight (Nummedal et al. 1977; Hayes 1979). From Cape Canaveral to the south end at Key Biscayne just north of the Florida Keys, the barrier system is distinctly wave-dominated.

GEOMORPHOLOGY

The present east coast of Florida is characterized by Holocene quartz-sand barrier islands; however, there are restricted, but significant areas (e.g. Anastasia Island) where carbonate-rich remnants of Pleistocene barriers face the open Atlantic Ocean. North of Cape Canaveral about 20% of the coast is Pleistocene (Hayes 1994). Foremost among these Pleistocene units is the Anastasia Formation, a cemented, quartzitic, molluscan grainstone that formed in beach and shallow-water nearshore environments (Stauble and McNeill 1985). Exposures of this formation (plate 4), along with modern beach rock and worm reefs (Kirtley and Tanner 1968), create the only rocky coasts in eastern Florida outside the Florida Keys. Early settlers used this material for construction, including the old fort at St. Augustine. The Florida east-coast barriers are typically anchored by the underlying Pleistocene Anastasia Formation, and locally they develop high foredunes that prevent overwashing and landward migration. Some barriers are welded to the mainland and represent an important type of microtidal, wave-dominated barrier-island system that contrasts markedly with the wave- and overwash-dominated barrier-island model presented by Hayes (1979).

Related coastal environments, such as tidal inlets, tidal flats, marshes, and back-barrier bays, are all present and show a distribution that reflects the tidal range and the tide/wave energy balance. In northeastern Florida, tidal inlets are wide and deep, and tidal flats and marshes are relatively extensive, which is typical of a mixed-energy coast (Davis and Hayes 1984).

An example of this mixed-energy situation is seen in

the area of Anastasia Island and Matanzas Inlet south of St. Augustine (Hayes 1994). There, barriers are relatively short, dunes are well developed, and marshes and tidal flats are extensive (fig. 10.5). Matanzas Inlet is the only tidal inlet along the entire east coast without jetties or other stabilizing structures. It has experienced considerable change, mostly by southward migration (Mehta and Jones 1977). Its relatively natural state and the tidal range of about 1.7 m produce a morphology typical of a mesotidal, mixed-energy inlet. The ebb-tidal delta is distinctly wave-influenced, as evidenced by the terminal lobe (fig. 10.5). Modest lateral flood channels are present, with some evidence of seaward sediment transport (Gallivan and Davis 1981). Various channels within the inlet and tidal complex are either flood- or ebb-dominated. The flood delta contained all the typical elements of a mesotidal delta (Hayes 1975), including an ebb shield, spillover lobes, and ebb spits in the late 1970s (Davis and Fox 1981). Now the flood delta has become merged with the south end of the island.

Barrier islands and related inlets south of Matanzas Inlet are distinctly wave-dominated, with long and continuous barriers and rather high dune ridges. Beaches range from narrow and steep to the wide, gently sloping profiles of the famous beaches at Daytona. A prominent longshore bar and trough system is present in most places (Allen 1991). The height of the dune ridges is typically 8 to 12 m, which prevents overwash and is not typical of wave-dominated barriers. The unstable inlets and the long barriers are both quite characteristic of wave-dominated barriers.

The widely spaced inlets and the absence of washover produce a back-barrier area that is narrow and contains little in the way of marsh and tidal-flat environments; this is due to the restricted back barrier flat environment typically produced by washover fans, and the greatly reduced tidal range behind the barriers. The attenuation of tidal flow through the small inlets and the long distances between inlets produce a distinct gradient in tidal range from about a meter inside the inlet to less than half that midway between adjacent inlets. Consequently, water in the "lagoons" behind the barriers, away from the inlets, is essentially fresh, the Indian River being an excellent example.

SEDIMENTARY PROCESSES

The Florida east coast faces the open Atlantic and receives long-period swell with mean values of about 8 to 9 seconds. The relatively broad shelf off the northeast Florida coast and the presence of the Bahamas Bank off the remainder of the east Florida coast provide protection from Atlantic waves. The mean significant wave height during winter months is about 1.2 m (Stauble and DaCosta 1987).

North of Cape Canaveral, the general direction of longshore sediment transport in Florida is to the south (Davis et al. 1992); volumes of transported sediment decrease from north to south (fig. 10.6). Along several reaches, littoral drift is hypothesized as being reversed, forming littoral cells (Stapor and May 1983; Stauble and DaCosta 1987) (fig. 10.7), some of which are large. The convergences are areas of abundant sand supply resulting

Figure 10.5. Southern end of Anastasia Island, showing Matanzas Inlet and fairly extensive back-barrier marsh and tidal-flat environments.

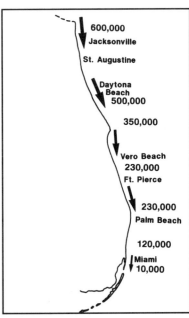

Figure 10.6. East coast of Florida, showing general trends in longshore sediment transport (after Dean and O'Brien 1987). Values are cubic meters yr^{-1}.

in beach-ridge growth, whereas the divergences are areas of erosion indicated by calcarenite pebbles, presumably from the Pleistocene Anastasia Formation. Cape Canaveral itself may have been formed by converging littoral transport (Hayes 1994). Local erosion may be a serious problem where diverging transport occurs (Stauble and DaCosta 1987). Wave energy focused at the beach by topographic variations on the shelf could account for these littoral cells (Mehta and Brooks 1973). These reversals and convergences are partly responsible for the tremendous diminution in rate of transported sediment along the entire east coast, from 500,000 cubic meters yr^{-1} on the north to only 10,000 cubic meters yr^{-1} near Miami (fig. 10.6).

Southwest Florida Mangrove Coast

The coast of peninsular Florida between Cape Sable, on the north side of Florida Bay, and Cape Romano, at the southern end of the west-central barrier coast (fig. 10.1), is characterized by mangrove swamps, tidal channels, small open bays, and abundant oyster and sabellarid-worm reefs. This coastal segment is tide-dominated due to the broad, gently sloping continental shelf and the very low mean annual wave height (10 cm), coupled with the broadly embayed coastal configuration. The mangrove coast is contained between two wave-dominated cuspate forelands, Cape Romano and Cape Sable, and it can be divided conveniently into two sections: a northern part, commonly called the Ten Thousand Islands, and a southern part that is adjacent to the Everglades and lacks the numerous, discrete mangrove islands (fig. 10.8).

GENERAL GEOLOGY AND GEOMORPHOLOGY

This coast is adjacent to a uniformly sloping inner shelf that has an average gradient of 1:3,000 out to a depth of 10 m (fig. 10.8). Cape Romano on the north is a quartz-sand cuspate foreland that is closely associated with an extensive bank of linear, tide-dominated sand ridges (Davis and Klay 1989; Davis et al. 1989a). Cape Sable, on the south, is a carbonate-sand foreland area containing

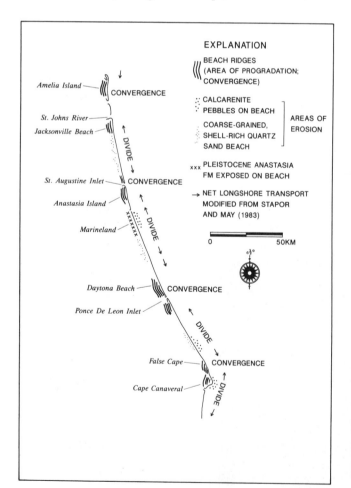

Figure 10.7. Variation in beach-sand texture and mineralogy and hypothesized littoral-drift cells for the coast of Florida north of Cape Canaveral (Stapor and May 1983).

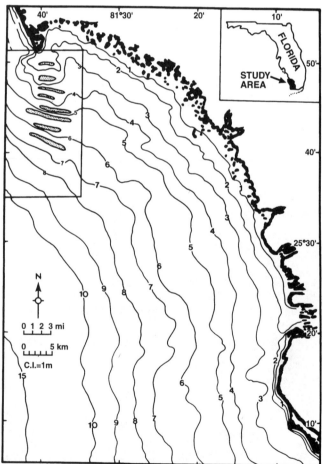

Figure 10.8. Southwest Florida coast. The elongate stippled areas just south of Cape Romano represent tidal sand ridges. Bathymetry in meters.

three cuspate accretionary-ridge complexes (Roberts et al. 1977) (fig. 10.9). The Ten Thousand Islands coastal area is characterized by mangrove islands separated by numerous tidal channels (plate 5). The few beaches are small, discontinuous, and composed of shell debris. Modest upland runoff enters the coast through rivers that empty behind the mangrove islands—in contrast to the southern part of this coastal area, where several small rivers empty directly into the open Gulf.

This area is underlain by Pliocene and Pleistocene carbonates of the Tamiami and Miami formations (Enos and Perkins 1977). Above the carbonates is a thin, well sorted Pleistocene quartz sand that is locally clayey and contains root casts (Parkinson 1989). This unit is interpreted as a terrestrial sand sheet that is at least partly eolian (Davis and Klay 1989). It may be time-equivalent to the Anastasia Formation beach-rock unit that is widespread on the east coast.

PRESENT MORPHOLOGY AND PROCESSES

The modern coastal morphology here is a result of about 3,000 years of very slowly rising sea level, a limited sediment supply, and extremely low physical-energy conditions along the coast. On the north, around Gullivan Bay just east of Cape Romano, spring tides are about 1.4 m, decreasing to 1.1 m at Cape Sable. Although this range of tides is small, tidal influences are very different from north to south. Tidal currents in the linear sand ridges near Cape Romano and to the east (fig. 10.8) flow at rates as high as 70 centimeters per second (cm/s), but they diminish greatly to the southeast and flow at rates of only 17 cm/s a few kilometers offshore of the central part of the area (Davis et al. 1989a). Wave energies are low throughout, but increase toward the south because of exposure to strong northerly winds associated with passage of winter weather fronts. The northern part of the area is in the lee of the coast, and so it is protected from these winds. Cape Sable owes its wave-dominated configuration (fig. 10.9) to a combination of these northerly winds and the prevailing southwesterly winds.

The Ten Thousand Islands area displays an unusual arrangement of mangrove islands. Adjacent to the irregular mainland coast is a relatively open area called the "chain of bays" (Parkinson 1989). This relatively open-water coastal environment is wide on the north and narrow in the southern part of the area (plate 5). On the north, the Ten Thousand Islands are discrete mangrove islands separated by numerous tidal channels (fig. 10.10), whereas to the south mangrove stands are continuous, and the coast is interrupted only by a few streams from the mainland. This marked difference in geomorphology is due to a combination of tidal influence and availability of fine-grained sediment. The northern area has relatively strong tidal currents and a dearth of land-derived mud, whereas the southern area has sluggish tidal currents and receives more fine-grained terrigenous sediment from the several streams that drain the Big Cypress Swamp and the Everglades. This difference in geomorphology is responsible for the distribution of fine sediments along the present outer coast. Mud is relatively abundant south of the Ten Thousand Islands, where it is carried directly to the open coast; it is only

Figure 10.9. One of the cuspate-foreland areas that make up the Cape Sable complex. These wave-dominated features are composed mostly of biogenic sand and gravel.

Figure 10.10. Typical mangrove coast of southwest Florida. Here a tidal channel and its relatively strong currents have developed a tidal delta composed of shell gravel.

about half as abundant in a corresponding area to the north, where the supply is generally lower and most mud—trapped in the bays and mangrove swamps—does not reach the coast.

LATE HOLOCENE HISTORY

The present geomorphology of the mangrove coast is in part a result of seaward progradation of the coast coincident with the slowing of sea-level rise about 3,000 ybp. The Cape Sable area apparently has prograded by as much as 8 km (Enos and Perkins 1979), achieving its present seward limit about 1,200 to 1,500 ybp (Roberts et al. 1977). Vermetid-gastropod reefs began to develop about 3,000 ybp along much of this coast, and their growth kept pace with slowly rising sea level (Shier 1969). These vermetid reefs provided the topographic highs for mangrove colonization, which eventually led to the domination of the coast by mangrove islands (Shier 1969; Parkinson 1989). These islands and the sediment accumulation that they foster produced a 2 m thick progradational sequence (fig. 10.11) as sea level rose very slowly during the late Holocene.

West-Central Barrier System

The barrier-inlet system along the west-central part of the Florida peninsula has the most diverse morphology of any barrier system in the world (Davis 1994). This system extends for about 300 km and includes twenty-nine barrier islands and thirty tidal inlets (fig. 10.12). Included are long, narrow, wave-dominated barriers (fig. 10.13a) of both spit and upward-shoaling

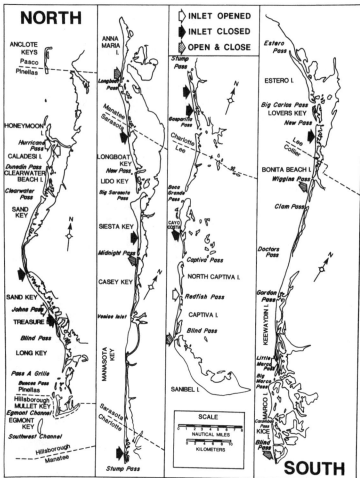

Figure 10.12. Islands and inlets of west-central Florida's barrier system. Several inlets have been opened and closed during the past century (Davis 1989; reprinted by permission of Kluwer Academic Publishers).

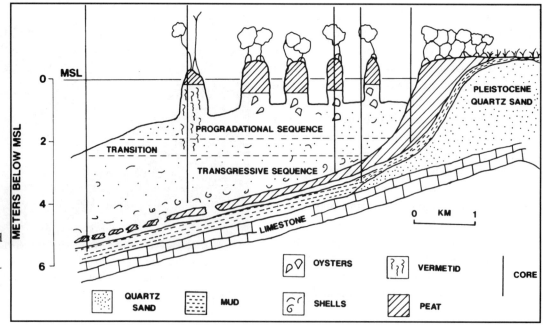

Figure 10.11. Idealized cross-section of the Ten Thousand Islands area, showing stratigraphy and late Holocene progradation after an initial transgression (after Parkinson 1989). MSL: mean sea level.

origins, and mixed-energy drumstick barriers (fig. 10.13b) in a range of sizes and shapes. Spring tidal range is just under 1 m, and mean annual wave height is about 25 cm at the shore. Barriers extend from Anclote Key on the north to Kice Island at Cape Romano (fig. 10.12).

This coastal reach is fairly straight and oriented west of north throughout most of its extent; there is one major dislocation about two-thirds of the way to the south at Sanibel Island. Subsurface data from coring and seismic surveys indicate that this change in the coast is bedrock-influenced (Evans et al. 1989). Two subtle coastal headlands are present, one comprised of Miocene exposures (Tampa Formation) in the northern part of the barrier system in central Sand Key, and one comprised of Miocene strata (Hawthorn Group) on the central part of the coast near Venice Inlet (fig. 10.12). The coastal bays protected by the barriers and served by the inlets vary greatly in size and shape, and these bays influence tidal prisms, inlet size, morphology, and stability. There is considerable control of barrier islands by the antecedent geology of Neogene and Pleistocene units (Davis and Kuhn 1985; Evans et al. 1985; Davis et al. 1989a; Evans et al. 1989). At least in the northern part of the barrier system, the underlying carbonate strata are karstified.

MODERN PROCESSES AND GEOMORPHOLOGY

This coastal barrier system is characterized by mixed influences of tidal and wave energy. Some individual barriers (fig. 10.13) are indicative of more wave than tidal influence (e.g. Anclote Key, Casey Key, and Captiva Island); others show substantial tidal influence (e.g. Caladesi Island, Siesta Key, Lovers Key, and Marco Island). The morphology of the barriers results from interplay between these coastal processes. Slight modifications to either tidal or wave parameters can cause a significant shift in the morphology of the barrier-inlet system along this low-energy coast (Davis 1994). Wave refraction across ebb-tidal deltas is a major cause of such changes (Gibeaut and Davis 1993) and gives rise to drumstick barriers.

Hurricanes are fairly common along this barrier coast and have caused important changes during historical time. Similar changes likely took place throughout the Holocene. Relatively stable tidal inlets such as Hurricane Pass (formed in 1921), Johns Pass (formed in 1848), and Redfish Pass (formed in 1921; plate 6) resulted from the breaching of barrier islands by hurricanes; they are stabilized by pre-Holocene strata. Data from the past century for this reach of barriers show four inlets being opened, nine being closed, and five both opened and closed (fig. 10.12). Closure is generally related to decrease in tidal prism, as at Midnight Pass and Dunedin Pass (fig. 10.14).

Figure 10.13a. Anclote Key, an example of a wave-dominated barrier and the northernmost island in this barrier system.

Figure 10.13b. Caladesi Island, an example of a mixed-energy (drumstick) barrier and a state park in Pinellas County.

Figure 10.14. Dunedin Pass, an inlet that was 500 m wide in the early twentieth century. It was essentially closed by 1987 as the result of a steady decrease in tidal prism.

Another important feature of this barrier system is the development of three new barrier islands during a twenty-five-year period beginning in the early 1960s—Three-Rooker Bar, north of Honeymoon Island, and North Bunces Key and South Bunces Key, adjacent to Bunces Pass (fig. 10.12). Each of these islands was formed by upward shoaling, without benefit of an event such as a storm. All occur in an area of sediment abundance, such as inlet-related shoals or barrier spits, and all have a similar morphology. As soon as supratidal conditions persisted, colonization by opportunistic vegetation took place, followed shortly by coppice mounds, and then dunes, which have reached elevations of more than 1 m above the back beach. These wave-dominated barriers are 1.5 to 3 km long and have recurved spits at each end (plate 7). Further extension is presently prevented by tidal channels at each end. It is likely that several of the older barriers in this system had similar origins.

HOLOCENE HISTORY

The barriers and related coastal environments along the west-central part of the Florida peninsula were developed during the late Holocene. The oldest supratidal accumulations on any of the barriers are about 3,000 years old (Stapor et al. 1988), an age that is persistent in many barriers along this coast. Nearshore subtidal Holocene sediments beneath the present barrier islands are 4,200 to 4,500 years old. Nearby sub-barrier deposits have been dated at 4,250±70 ybp (Davis and Kuhn 1985).

The stratigraphic data base for this coast is spotty—there is good control in the northern third of the area, and little to the south. Multiple beach-ridge systems have been analyzed and dated extensively in the southern area (Stapor et al. 1988). One can only compare by analogy, which seems to be appropriate, in consideration of the apparent similarities throughout this coastal reach.

In general, this is a sediment-starved coast, and virtually all Holocene terrigenous sediments have been supplied by reworking during the transgression of the past 7,000 years, which continues today. The Holocene sediment prism is thickest under the barriers, and thins offshore, so that at depths of 5 to 6 m there is little or no quartz sand overlying the pre-Holocene strata. There is also a shoreward thinning, and the marine Holocene sediments pinch out at the mainland coast. The total Holocene sequence at any given location along this coast is typically less than 7 or 8 m thick, and may be as thin as 3 m (fig. 10.15). In most areas the Holocene sediments rest unconformably on pre-Pleistocene strata; however, in places a thin, clayey Pleistocene sand is present (Davis and Kuhn 1985; Davis et al. 1985).

Evans et al. (1985) have suggested that these barriers formed at some distance offshore and migrated to the break in slope of the bedrock, where stability was achieved; however, these authors do not provide data on the origin of the barriers, and neither vibracoring data nor analysis of time-series of aerial photos (Hine et al. 1987a) substantiate the idea of significant landward migration. Undoubtedly there was modest migration due to overwashing during the early stages of barrier development, which has been observed in barriers of this area that developed during historical times (Davis and Hine 1989).

Areas along and adjacent to the present-day islands where sediment was relatively abundant became emergent and developed prograding beach-ridge systems relatively early. These 3,000-year-old ridges have been identified at Siesta Key, Gasparilla Island, Cayo Costa, Sanibel Island, and Marco Island (Stapor et al. 1988). Younger beach-ridge sets also have been identified on these islands (plate 8; fig. 10.16). The younger islands are generally narrow, with a single beach-dune ridge, and are about 1,500 to 2,000 years old (Davis and Kuhn 1985; Stapor and Mathews 1980; Stapor et al. 1988).

Once the islands emerged, they were frequently overwashed until beach-foredune ridges became elevated enough to prevent it. Prograding ridge systems were built on islands where sediment was abundant, whereas islands in areas of little sediment were extended alongshore and remained narrow (plate 8; figs. 10.12 and 10.16). Associated tidal inlets also were a means for capturing considerable sediment brought to the ebb-tidal deltas by littoral drift. Although some interpretations invoke a transgressive history (Evans et al. 1985), the barriers in this system probably formed close to their present positions. If they formed some distance offshore and mi-

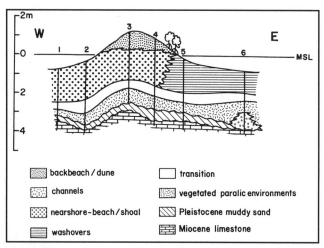

Figure 10.15. Cross-section of Anclote Key, showing the typical stratigraphy of late Holocene barrier islands of the west-central barrier chain (modified from Davis and Kuhn 1985; see fig. 10.12 for location).

grated landward, the record of this migration has been destroyed.

Marshy Coast of the Big Bend Area

A coastline characterized by a marine to brackish-marsh plant community extends for 350 km from the Apalachicola River Delta to Anclote Key (fig. 10.1). This coastline, commonly referred to as the Big Bend Coast, is morphologically complex because of variations in the surface of the limestone bedrock, and because of actively discharging freshwater springs, large oyster bioherms, a modern river delta (the Suwannee), and possible paleoshorelines. Two key factors have led to development of this heavily vegetated coast: (1) a lack of siliciclastic sand and mud, and (2) the low energy of waves that reach the coast. The shoreline constitutes a sizeable part of the entire Florida coastline, and along with the inner shelf system is probably the best modern U.S. example of an epicontinental marine system. It was still more extensive during the high sea-level stands of the geologic past.

GENERAL SETTING

Within the southern Big Bend area during the Pleistocene, quartz sand was concentrated considerably inland from the present coast, along the Brooksville Ridge and the Ocala Upland. This siliciclastic accumulation thins toward the Gulf, and is essentially absent on the present coast (Hine et al. 1988).

Holocene sea-level rise has flooded a low-gradient, karstified, bare limestone surface where Eocene-Oligocene skeletal grainstones are exposed (fig. 10.17). The quartz sand is situated on the uplands to the east, and the lack of ancient large streams flowing across peninsular Florida in this area helps to explain the lack of quartz sand on the coast. The Suwannee River is the only significant river that discharges along this coast today; other rivers, such as the Crystal, the Homosassa, and the Withlacoochee, emanate from springs fed by the Floridan aquifer system and travel only short distances before reaching the Gulf of Mexico. The Suwannee River is a geologically young stream, formed when the Okefenokee Swamp developed in southern Georgia during the late Neogene to Quaternary (Carver et al. 1986); its delta is small, but it extends into the Gulf, and includes a few large distributary channels.

Oyster-reef complexes are abundant and extensive along this coast. They are essentially parallel to the coast-

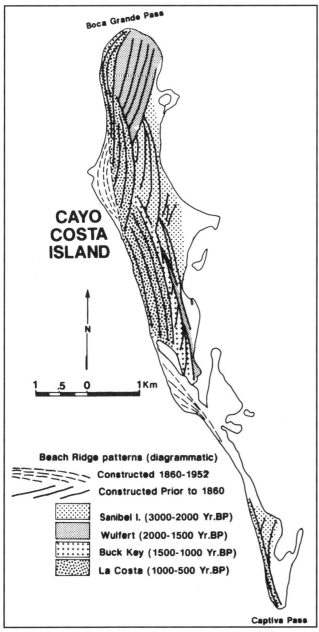

Figure 10.16. Cayo Costa Island, showing numerous patterns of beach ridges (modified from Stapor et al. 1988; see fig. 10.12 for location).

Figure 10.17. Small scale Karstic surface of Eocene limestones exposed near Ozello in Citrus County.

line, and are breached by numerous small tidal channels that have small tidal deltas at each end.

COASTAL PROCESSES

The Big Bend area has historically been considered a low-wave-energy coast. Although it is not a zero-energy coast as described by some (Tanner 1960a), data from wave gages and other observations clearly show that it has a wave climate that is low in comparison to virtually any other part of the Gulf of Mexico. The mean annual wave height is about 10 cm (Tanner 1960a). Frequent extra-tropical storms during the winter and infrequent tropical storms during the summer and fall provide a local storm surge and a sea state that is capable of flooding and eroding the marsh (Hine et al. 1987b).

The general low-wave-energy environment is due to the extreme width of the shelf (more than 150 km) and the very low seaward-dipping gradient (1:5,000) inherited from the ancient underlying carbonate platform. Also, the dominant extra-tropical storm winds are from the north-northwest, blowing alongshore or offshore.

HYDROGEOLOGIC PROCESSES

Eocene and Oligocene limestones are at or near the surface along the southern part of the marsh coast (plate 1), and the karst topography has strongly influenced local coastal morphology and sedimentation. Two basic karst processes have occurred in this area: (1) surface dissolution due to downward percolation of acidic waters from overlying marsh sediments; and (2) regional dissolution and subsequent collapse due to mixing-zone undersaturation (Back and Hanshaw 1970; Plummer 1975; Back et al. 1986) and older, lower sea-level stands (see chapter 4). The combination of these factors has produced karst topography on three easily recognizable horizontal scales. Features at the smallest scale (centimeters to a few meters; fig. 10.17) are less important than medium-scale (tens to hundreds of meters) or large-scale (kilometers) karst-induced coastal features. Medium-scale features include rectilinear tidal creeks whose courses are controlled by joint patterns, and rock-cored hammocks that form marsh islands (fig. 10.18). Large-scale features include broad, shallow bedrock depressions that form shelf embayments; elevated rocky areas between embayments, forming marsh-island archipelagos; and linear channel structures etched in bedrock by laterally migrating spring-discharge sites (Hine et al. 1987b). The numerous islands in the marsh archipelago have developed on a flooded, elevated, topographically irregular bedrock surface. The marsh islands, or hammocks, are underlain by localized bedrock highs, or nubs. Marsh sediments are thin and discontinuous, and limestone bedrock exposures are abundant.

Panhandle Coast

The panhandle coast of Florida is the only part of the state's coastal regime that has received high rates of sediment influx during the Holocene. This coastal reach is conveniently subdivided into two parts: the Apalachicola Delta area, which is the primary sediment source, and the remaining, extensive, sandy coast, which is a sediment sink. Both areas are wave-dominated, with barriers, well developed beaches and foredunes, and widely spaced tidal inlets (fig. 10.19).

COASTAL GEOMORPHOLOGY

The Apalachicola area is dominated by a sizable fluvial delta where sediment has accumulated from a drainage basin that includes much of Georgia and Alabama. Sediment sources include the southern Appalachian Mountains, the Piedmont, and the Coastal Plain. Most of the delta accumulated during the Quaternary, including the Holocene. As sea level has risen over the past several thousand years, sediment input has decreased, and reworking by waves and currents has become the dominant process. The open Gulf Coast in this area is characterized by a nearly continuous barrier-island system that extends from Dog Island on the east to the tip of St. Joseph Island on the west (fig. 10.20). The adjacent Gulf shoreface is extremely shallow, with a gradient of about 1:1,800. It is this relict delta lobe that is producing the barriers as a result of reworking by waves and currents.

Within this local barrier system there is a substantial range of morphologies, reflecting variation in processes and sediment availability. The eastern part of the barrier system is typically wave-dominated, with elongate, narrow islands, some washovers, and narrow foredune complexes. Some progradation is present locally on Dog

Figure 10.18. Vertical aerial photo of the estuary of the Crystal River, showing extensive oyster reefs with tidal channels (after Hine et al. 1988).

Island and at Cape St. George. By contrast, St. Vincent Island is very wide as a result of long-term progradation of beach ridges, coupled with its mainland origin (Stapor 1973; Donoghue 1991) (fig. 10.20). The oldest part of the barrier is about 4,400 years old (Schnable 1966), making it one of the oldest barriers in Florida.

The area of Cape San Blas and St. Joseph Island is also very interesting and dynamic. At present, the young, prograding beach-ridge complex at Cape San Blas (fig. 10.21) is being truncated by erosion. Much of the southern end of St. Joseph Spit also is being eroded, with extensive littoral transport toward the north; the north end of the spit is rapidly prograding, and large foredunes are prominent.

The western part of the panhandle coast is a combination of elongate, wave-dominated barrier islands and spits and mainland beaches. Large foredune complexes are present throughout, including the largest dunes in the state. This area also has the best beaches in the state with respect to pure quartz composition, lack of organic debris, and water clarity. This coast is wave-dominated because of the relatively steep gradient—about 1:60 to a depth of 20 m, decreasing to 1:350 out to 40 m. The only large barrier is Santa Rosa Island, which extends 70 km from the mouth of Choctawhatchee Bay to the mouth of Pensacola Bay (fig. 10.19). This Holocene barrier has developed over a Pleistocene core (Otvos 1985).

Tidal inlets along this coast are few, because there are few barrier islands; those off the Apalachicola Delta, such as West Pass and East Pass (fig. 10.20), carry large tidal prisms and are tide-dominated. This condition results in deep and relatively stable inlet channels. Inlets associated with the western part of the panhandle are modified by navigational structures. Except for East Pass, at Destin, they tend to lack a developed flood-tidal delta. Prominent ebb deltas are present at each of the three major

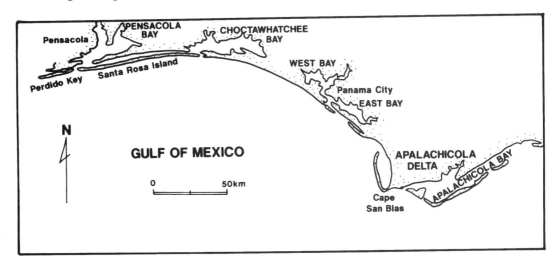

Figure 10.19. The western Panhandle coast (Hine et al. 1988).

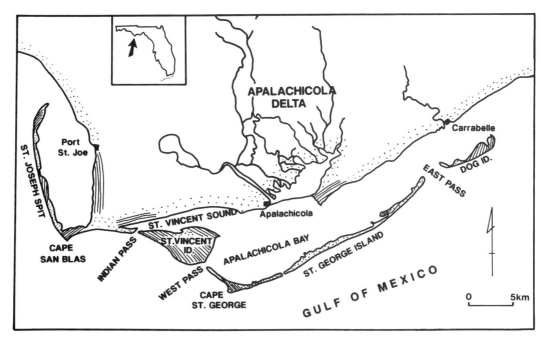

Figure 10.20. The Apalachicola Delta area, showing major geomorphic elements.

inlets (fig. 10.19), resulting from the large tidal prisms that are a function of the large estuaries served by these inlets.

COASTAL PROCESSES

This wave-dominated coast is characterized by a distinct east-to-west littoral drift from Cape San Blas to the Alabama line. Littoral drift in the Apalachicola Delta area is in local cells that result from a combination of shoreline orientation and wave refraction.

Mean annual wave heights, although only about 50 to 60 cm, are the highest of the entire U.S. Gulf Coast (Tanner 1960a; Hine et al. 1988). In general, the panhandle coast is subjected to southeasterly winds, producing waves that increase in size westward and produce westerly littoral drift. Walton (1973) estimated the net drift at about 200,000 cubic meters per year to the west.

Summary

The coast of Florida, notwithstanding its microtidal range, comprises diverse coastal regimes. There is wide variation in morphology of adjacent shelves, sediment availability, sediment composition, and dominant coastal processes. The result is a fascinating coastal system that has developed during only the past few thousand years. Its large geographic extent and its variety make the Florida coast the most diverse in the world.

Included in the spectrum of coastal segments are wave-dominated barrier systems (east coast and panhandle); tide-dominated, vegetated paralic systems (southwest and Big Bend areas); and a mixed-energy barrier system (Gulf central). This diverse suite of morphologies and related stratigraphic sequences has developed largely during the past 3,000 years under conditions of limited sediment supply (except on the east coast). The tidal range is rather uniform throughout. However, the wave climate and shoreface bathymetry varies, providing a combination of coastal processes that produce a wide range of coastal morphologies, in spite of overall low energy.

The balance among sea-level change, sediment availability, and coastal processes is a delicate one. A modest shift in any one of these factors would likely produce considerable coastal response. The prognosis of further rise in sea level, along with ever-increasing human influence on the coast, gives cause for concern about considerable change in the future for all of these coastal environments.

Figure 10.21. Oblique aerial photo of Cape San Blas, looking toward the north area and displaying truncated beach ridges.

11

Structural and Paleoceanographic Evolution of the Margins of the Florida Platform

Albert C. Hine

This chapter presents a broad view of the geologic history of the margins of the Florida Platform to illustrate how the early underpinnings provided a template on which depositional processes built the Mesozoic and Cenozoic stratigraphic record, and describes those depositional processes in a paleoceanographic context. The Florida Platform is placed in a larger geologic setting to demonstrate how global geologic events have been recorded locally. Process is emphasized—not a static recantation of product. A key theme is the understanding of integrated oceanographic systems in order to comprehend Earth history.

This chapter defines the geologic boundaries of the Florida Platform and portrays the major geologic events that have built and shaped the edges of the carbonate edifice, with emphasis on the western part of the Florida Platform. The west coast of the State of Florida falls midway across the platform (fig. 11.1), and the western half of the Florida Platform presently lies below sea level—hence the emphasis on that area. The West Florida Shelf and the West Florida Slope are among the broadest and most extensively submerged continental-margin systems in the world. Also outlined in this chapter is the Quaternary development of the shelf and the slope depositional systems, both being subdivided into western, southern, and eastern geographic sectors.

Early Platform Development

During the Late Jurassic to Early Cretaceous, the Florida-Bahamas Platform complex was part of one of the most extensive carbonate systems in the Earth's geologic history (fig. 11.2). Stretching nearly 7,000 km from the north-central Gulf of Mexico along the eastern North American continental margin to Canada, this

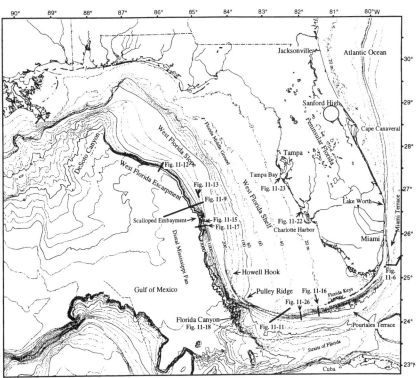

Figure 11.1. Bathymetric map of the modern Florida Platform, bounded on the east and south by the Straits of Florida and on the west by the Florida Escarpment. The emerged State of Florida (peninsular Florida) lies on the eastern part of the platform, and the expansive West Florida Shelf and Slope occupy the western part. The map shows the locations of many figures and geographic names mentioned in the text.

"great Mesozoic carbonate bank," or "gigaplatform," developed on basement structures associated with early Mesozoic rifting and sea-floor spreading between North America and Africa, involving the early formation of the North Atlantic Ocean (Schlee et al. 1979; Schlee et al. 1988; Grow et al. 1979; Jansa 1981; Sheridan et al. 1988; Klitgord et al. 1988; F. O. Meyer 1989; Poag 1991), as well as the Late Triassic to Early or Middle Jurassic rifting and sea-floor spreading that formed the Gulf of Mexico (Sawyer et al. 1991; Salvador 1991b).

It has not been precisely determined how most of this huge system was terminated, but a number of processes were capable of shutting down widespread carbonate production on the Florida Platform. Such processes include (1) eutrophication of seawater by nutrients from terrestrial runoff and from coastal upwelling (Hallock and Schlager 1986; Poag 1991); (2) plate migration into colder climates (Jansa 1981); (3) burial by river deltas where siliciclastic sediments shed from the eroding Appalachian Mountains, Piedmont, and continental interior were accumulating (Schlager 1981, 1989; Tyrell and Scott 1989); and (4) multiple sea-level fluctuations (eustasy coupled with local subsidence; Schlager 1981; Hine and Steinmetz 1984; Schlager 1989). Most likely, a combination of these factors terminated the gigaplatform.

The relatively large size of the late Mesozoic to Cenozoic carbonate-producing system of the Florida-Bahamas megabank—compared to the rest of the eastern North American gigaplatform and its Gulf of Mexico counterpart (fig. 11.2)—is ultimately related to the initial large size of the Jurassic basement terrane exposed to shallow marine waters. This basement included a northwest-trending structural prong called the Peninsular Arch (fig. 11.3), separating the Atlantic and Gulf basins. This structural element consisted of Paleozoic continental crust and Middle to Late Jurassic transitional rift crust (fig. 11.4) (Sheridan et al. 1988; Sawyer et al. 1991; DeBalko and Buffler 1992).

Today, active areas of carbonate sedimentation are restricted to the south-southwestern parts of the Florida Platform and the Bahamas Bank. These areas persist as broad, extensive carbonate platforms as a result of (1) long-term residence in a tropical to subtropical climate, (2) separation from a siliciclastic-sediment source (the southeastern U.S. continental mainland) by an open seaway (Bahamas) or by distance (southern Florida), and

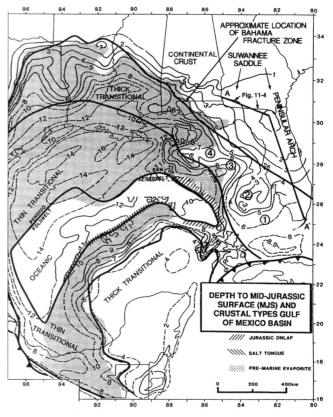

Figure 11.3. Structure contour map of the mid-Jurassic surface, illustrating crustal types and distribution of salt (shaded area) in the Gulf of Mexico basin. Depth is in kilometers. Crustal types are continental, thick transitional, thin transitional, and oceanic. Much of the west Florida margin rests upon the boundary between thick transitional and thin transitional crust. Note the presence and distribution of major basement structural highs and lows. 1: South Florida Basin. 2: Sarasota Arch. 3: Tampa Embayment. 4: Middle Ground Arch–Southern Platform. 5: West Florida Basin. Also shown are the Peninsular Arch and the Suwannee Saddle (DeBalko and Buffler 1992).

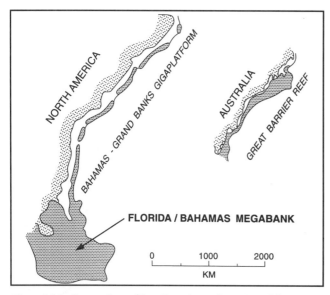

Figure 11.2. Comparison of Late Jurassic configuration of the Florida-Bahamas-Grand Banks gigaplatform with the modern Great Barrier Reef of Australia. The gigabank, when the Early Cretaceous counterparts in the Gulf of Mexico are included, is more than four times longer than the modern world's largest reef tract (from Poag 1991). Note the relatively large size of the Florida-Bahamas megabank part of the gigaplatform.

(3) the absence of persistent environmental stress (i.e. nutrient overload from upwelling, schizohaline waters).

NORTHERN BOUNDARY OF THE FLORIDA PLATFORM

The northern boundary of the basement structure supporting the Florida Platform was a linear structural basin located between the Peninsular Arch and the Paleozoic rocks of the southeastern U.S. This structural low originally was related to a suture zone and accreted continental terrane associated with the final closing of the Iapetus Ocean (proto-Atlantic Ocean) in the late Paleozoic. The northeast-southwest orientation, linear basin shape, location, and size were caused by Mesozoic extensional faults associated with regional Late Triassic to Early Jurassic rifting (McBride et al. 1989; Sawyer et al. 1991).

During the Late Jurassic, the initial carbonate stratigraphic sequences onlapped from the south onto the Peninsular Arch basement rocks (fig. 11.4). However, as the Peninsular Arch became covered with shallow-water carbonates during the Early Cretaceous (probably due to subsidence and extended sea-level rise), the structural low forming the northern boundary of the Peninsular Arch also formed the northern boundary of the Florida Platform. This structural low has been called the Suwannee Saddle (fig. 11.3), and the seaway that flooded it (fig. 11.5) has been referred to variously as the Suwannee Strait, Channel, or Seaway; the Gulf Trough; or the Georgia Channel System (Chen 1965; McKinney 1984; Klitgord et al. 1984; Pinet and Popenoe 1985a; Popenoe

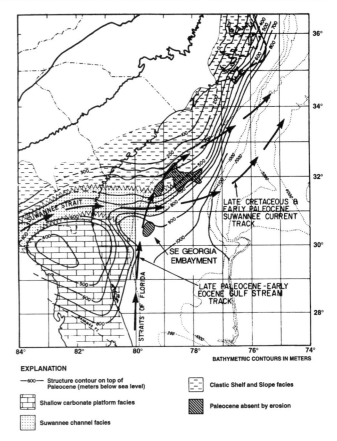

Figure 11.5. Structure-contour map of the top of the Paleocene, showing sedimentary facies, the track of the early Paleocene Suwannee Current through the Suwannee Strait, and the early track of the Florida Current through the Straits of Florida. Note the dominance of carbonates on the Florida Platform and the relatively rapid transition to siliciclastics across the Suwannee Strait. The Suwannee Current allowed carbonate sedimentation to flourish by providing this dynamic barrier (from Dillon and Popenoe 1988).

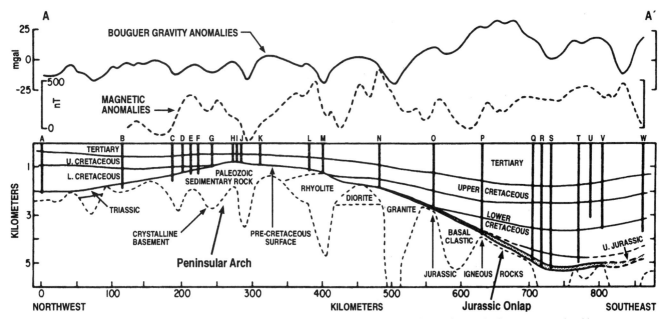

Figure 11.4. Geologic cross-section oriented roughly north-south through peninsular Florida (see fig. 11.3), showing generalized basement structures and overlying sedimentary units (mostly limestones). Also included are magnetic and gravity profiles and borehole locations (A–W). Variations in thickness of sedimentary cover is largely due to differential subsidence (Klitgord et al. 1984).

et al. 1987; Dillon and Popenoe 1988; Popenoe 1990; Huddlestun 1993). From the mid-Cretaceous to the late Paleogene, the basin and its seaway were of paramount importance in maintaining the carbonate sediment-producing environment to the south (see chapter 4).

During the Early Cretaceous, the western arm of the North Atlantic gyre flowed to the north seaward of the eastern margin of the Florida-Bahamas megabank (Berggren and Hollister 1974; Pinet and Popenoe 1985a). By the mid-Cretaceous—and certainly by Campanian time—a strong northeasterly flow was passing through the Suwannee Strait from the northern Gulf of Mexico to the Atlantic Ocean (fig. 11.5) (Chen 1965; Pinet and Popenoe 1985a), depositing sediments on the inner Blake Plateau. Perhaps there was little to no flow earlier because of the relatively shallow floor of the seaway, and because the eustatic sea-level rise associated with the second-order mid-Cretaceous sea-level high stand (Haq et al. 1987a) had not reached an elevation sufficient to flood the seaway.

For most of the Late Cretaceous and Paleogene, the Suwannee Strait produced a sharp facies boundary between the siliciclastics shed off the southern Appalachian Mountains to the north and the huge carbonate megabank to the south. Small reefs developed along the southern margin of the seaway (Antoine and Harding 1965). The flow through this seaway prevented siliciclastic sediments from inundating and smothering the evolving carbonate bank to the south. In addition, this dynamic boundary prevented nutrients carried by streams from reaching the Florida Platform.

Eventually, the seaway filled during Oligocene sea-level low stands, allowing a flood of siliciclastics that covered the eastern part of the Florida Platform; this part of the platform remained elevated relative to the western part, which had drowned by that time. Most of the siliciclastic influx came from north-to-south longshore transport of sediment from Piedmont and coastal-plain rivers reaching the Atlantic Ocean and the northeastern Gulf of Mexico (Meade 1969; Meisburger and Field 1975). Most likely, there were no south- or east-flowing sediment-bearing streams on the Florida Platform south of the present St. Johns River.

Siliciclastic sediments (as well as authigenic sediments; Riggs 1979a, 1979b, 1984) were also distributed east-to-west by numerous marine transgressions and regressions (Field 1974; and others), and the consequence of this process can be seen in the numerous Neogene to Quaternary paleo-shorelines, scarps, bluffs, and beach ridges that characterize present-day, exposed peninsular Florida (White 1970). Eventually, siliciclastic sediments extended all the way to the Florida Keys (Enos and Perkins 1977).

EASTERN BOUNDARY OF THE FLORIDA PLATFORM AND SEPARATION FROM THE BAHAMAS BANK

There are two fundamentally different views concerning the topographic complexity of the early basement structure underlying the Florida-Bahamas region. Mullins and Lynts (1977) postulated that the Bahamas Bank formed during the Jurassic on top of rift-generated horst-and-graben topography (the so-called graben hypothesis); this hypothesis requires that the seaways now separating the banks originally formed as structural lows, and that the Florida-Bahamas megabank was situated on the structural highs. During long-term subsidence associated with the regional passive-margin setting, carbonate derived sedimentation on the megabank kept pace, forming great thicknesses (as much as 14 km) of shallow-water limestones. The basins—although they also accumulated great thicknesses of both shallow and deep-water carbonate sediment—lagged somewhat behind the banks, thus amplifying the original depositional relief.

Sheridan et al. (1988) and Leg 101 Scientific Party (1988) postulated that a great, contiguous carbonate megabank, extending from the West Florida Escarpment to the Blake-Bahamas Escarpment, had formed by the Late Jurassic on a basement terrane not segmented into the large horsts and grabens of Mullins and Lynts (1977) (the so-called megabank hypothesis). Although there may have been deep-water reentrants (Ball et al. 1985; and others), most of the area from western Florida to the Blake-Bahamas Escarpment, including the entire Florida Platform, was covered by shallow-water carbonate depositional environments that persisted until the mid-Cretaceous. Seismic evidence and Ocean Drilling Program borehole data from the southern Blake Plateau (Leg 101 Scientific Party 1988) suggest that basins (including the Straits of Florida) were underlain by shallow-water limestones before the Late Cretaceous—a finding that contradicts the graben hypothesis (fig. 11.6).

Starting in the mid-Late Cretaceous, the Cuban and Antillean orogenies developed left-lateral shearing between the Caribbean and North American plates. This motion produced faults and fold axes preferentially aligned with the margin of the banks, including the eastern margin of the Florida Platform (Mullins 1983a). It was then (post-middle Cenomanian to Late Cretaceous) that the megabank broke up to form a number of banks and basins in the Florida-Bahamas region (Eberli and Ginsburg 1987, 1989; Denny et al. 1994). Presumably, during this event the eastern margin of the Florida Platform became detached from the Bahamas Platform, forming the Straits of Florida.

Laterally restricted hyperbolic seismic reflections beneath the modern Straits indicate that the Florida

Figure 11.6. Cross-section from Key Largo borehole in the Florida Keys across the Straits of Florida to Great Isaac borehole on northwest Great Bahama Bank. This interpretation indicates that shallow-water limestones (Albian-Aptian) lie beneath the Straits of Florida. The hyperbolic reflections on top of the Albian-Aptian and lower Eocene to Paleocene indicate Florida Current activity. Numbers within stratigraphic units are seismic velocities in kilometers per second (Sheridan et al. 1988).

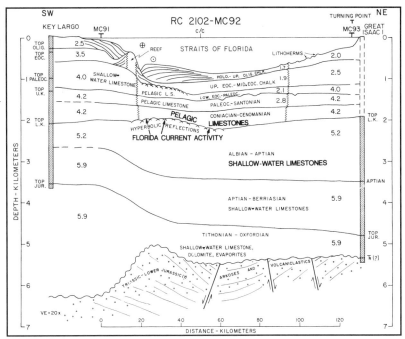

Current (the local component of the Gulf Stream) first appeared during the Cenomanian to Coniacian (fig. 11.6) (Ladd and Sheridan 1987; Sheridan et al. 1988). The eastern boundary of the Florida Platform developed into a windward, reef-dominated carbonate-bank margin (Popenoe et al. 1984). Seismic data and interpretations shown in figures 11.7 and 11.8 indicate a shallow-water carbonate bank with a marginal-reef facies passing eastward into slope and basin facies of the Straits of Florida.

Three major subsurface features probably affected Cenozoic sedimentation along the eastern margin of the Florida Platform. On the north is the Southeast Georgia Embayment (associated with the basement structure of the Blake Plateau basin; fig. 11.5). The modern, mesotidal coastal embayment (Hayes 1979) of northern Florida, Georgia, and southern South Carolina resulted from this underlying feature.

The second feature is the Sanford High, located in the subsurface just northwest of Cape Canaveral (fig. 11.1). How this feature formed is not clear, but Puri and Vernon (1964) suggested that fault movement during the early Miocene produced local uplift. This high area caused local upwelling of the Florida Current during middle Miocene high stands, producing local phosphorite deposits (Riggs 1979b; Compton et al. 1990). In addition, there has been speculation that this antecedent topographic high may have played a role in producing the cuspate-foreland complex that includes modern Cape Canaveral as the youngest component. A prominent cuspate foreland existed along this part of the Florida east coast during the Pleistocene—and perhaps earlier, considering the presence of ancient Cape Orlando (represented by beach ridges, scarps, and terraces of the eastern

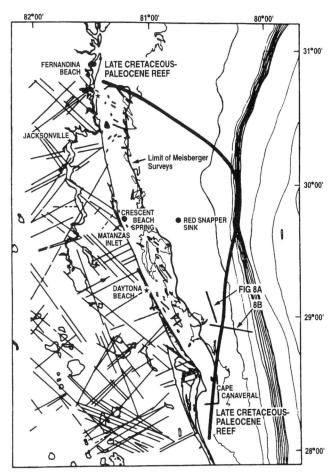

Figure 11.7. Map showing the seaward limit of subsurface Paleocene reef trend, downward flexures or folds beneath the inner shelf (small black areas just offshore), and fracture traces mapped on land by Vernon (1951). Features described by Vernon are similar to deformation features seen offshore in seismic data (Popenoe et al. 1984; reprinted by permission of A. A. Balkema, P.O. Box 1675, Rotterdam, The Netherlands).

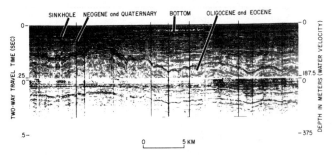

Figure 11.8. A: Part of a seismic line across the East Florida Shelf, illustrating folding of subsurface Eocene and Oligocene limestones due to dissolution of underlying Paleocene carbonates. Dissolution and folding were probably active during the late Oligocene sea-level low stand.

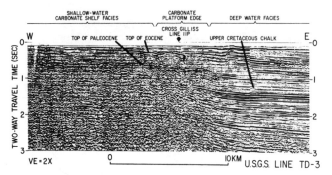

Figure 11.8. B: Part of a seismic-reflection profile just north of Cape Canaveral, showing major facies changes across the Paleogene carbonate-platform margin. Profile locations shown in figure 11.7 (Popenoe et al. 1984; reprinted by permission of A. A. Balkema, P.O. Box 1675, Rotterdam, The Netherlands).

Florida coastal plain; White 1958, 1970; Osmond et al. 1970).

Third among these subsurface features is the karst developed in Paleocene, Eocene, and Oligocene limestones, which produced both subsurface and exposed sinkholes and local stratigraphic deformation in the form of folds and sags (Meisburger and Field 1975; Popenoe et al. 1984). The folds have approximately 80 m of subsurface relief (fig. 11.8a). Red Snapper Sink off Matanzas Inlet is exposed at the sea floor and has more than 125 m of relief. These structures resulted from deep-seated dissolution in older rocks, and their distribution appears to be related to regional joint patterns and to the distribution of carbonate facies (i.e. these deformation structures and sinkholes lie on top of the buried Late Cretaceous to Paleocene reef; fig. 11.7). Popenoe et al. (1984) indicated that dissolution occurred during the late Oligocene to early Miocene sea-level low stands. The extent to which these subsurface structures control modern coastal morphology and shelf topography (e.g. Cape Canaveral, St. Lucie River estuary) can only be conjectured.

WESTERN BOUNDARY OF THE FLORIDA PLATFORM

The western boundary of the modern Florida Platform is defined by the West Florida Escarpment, a high-relief (as much as 2 km), north-south trending, largely erosional slope that is morphologically complex and vertical to overhanging in some places (Corso 1987; Winker and Buffler 1988; Corso et al. 1989) (fig. 11.1). Basement rocks beneath the west-central parts of the platform are Paleozoic to Triassic igneous and sedimentary rocks incorporated in grabens, block-faulted basins, and synclines (Dobson and Buffler 1991). The Bahamas Fracture Zone probably represents a major crustal boundary between continental and thick transitional rocks (fig. 11.3). An extension episode during the Middle Jurassic produced a series of arches and basins (Apalachicola basin, Tampa embayment, Middle Ground Arch, Sarasota Arch, south Florida basin, and west Florida basin). These basement rocks controlled the distribution of the Middle Jurassic Louann Salt and Upper Jurassic sequences (Dobson and Buffler 1991).

The western margin of the Florida-Bahamas Lower Cretaceous carbonate platform (specifically, the Florida Escarpment) coincides with, and most likely was controlled by, a prominent, westward-dipping flexure—a tectonic hinge zone separating thick transitional crust from thin (Buffler and Sawyer 1985; DeBalko and

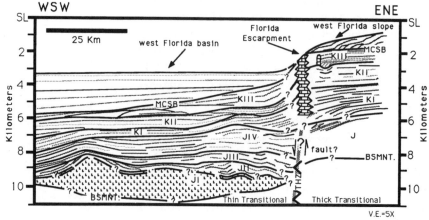

Figure 11.9. Cross-section of the West Florida Platform, showing ages and thicknesses of major stratigraphic units. The Upper Jurassic to Lower Cretaceous platform margin (JIV-KI) is situated over an antecedent topographic high marking the tectonic hingezone (THZ) that forms the boundary between thick and thin transitional crust. J1 is the Jurassic salt (landward extent of salt is unknown). KI-III are Lower Cretaceous sequences. MCSB is the mid-Cretaceous sequence boundary (from DeBalko and Buffler 1992). Post-MCSB rocks on the West Florida Platform are deepwater carbonates indicative of a drowned carbonate ramp.

Buffler 1992) (fig. 11.9). This flexure developed on the aforementioned basement structures and was related to trans-tensional tectonics associated with Late Jurassic sea-floor spreading in the Gulf of Mexico. Differential movement along this flexure focused the development of a paleo-shelf edge by concentrating skeletal and non-skeletal buildups (Corso et al. 1989; DeBalko and Buffler 1992), which marked the early transition from a Late Jurassic to Early Cretaceous ramp to an aggradational carbonate platform. This new, rimmed platform was constructed on the equivalent of the Knowles Limestone, a Late Jurassic to Valanginian carbonate-ramp system in east Texas (Salvador 1987; Winker and Buffler 1988). The base of the Lower Cretaceous marks the ramp-to-rim transition (Corso et al. 1989). The western margin of the Florida Platform reverted to a ramp system during the Late Cretaceous.

The western margin of the platform was also influenced by siliciclastic sedimentation. During sea-level low stands, siliciclastics bypassed the platform margin and were deposited as deep-sea fans in the eastern Gulf of Mexico basin. The siliciclastic lobes were reshaped into at least one contourite mound by intensified currents during a Cenomanian sea-level low stand, forming part of a widespread unconformity seen throughout the Gulf of Mexico; this is the mid-Cretaceous unconformity (Buffler and Sawyer 1985), now called the mid-Cretaceous Sequence Boundary (MCSB), because it is not everywhere an unconformity (Buffler 1991) (fig. 11.9). Schlager et al. (1984b) have indicated that this boundary represents a fundamental environmental change that produced widespread drowning of mid-Cretaceous carbonate platforms.

SOUTHERN BOUNDARY OF THE FLORIDA PLATFORM

The southern and southwestern margins of the Early Cretaceous Florida Platform extended across the present Straits of Florida to northern Cuba and eastward to the modern Bahamas, thus also defining part of the southern margin of the Florida-Bahamas megabank (Denny et al. 1994) (fig. 11.10). The location of this carbonate-bank margin may have been controlled by basement structures associated with fracture zones and transform faults linking sea floor-spreading processes in the Gulf of Mexico and the Atlantic Ocean (Sheridan et al. 1988), and associated with the distribution of oceanic and transitional rift crust (Dobson and Buffler 1991; Buffler 1991).

Along with the entire western part of the Florida Platform, this southern part of the megabank was segmented and drowned during the late Albian (?) to middle Cenomanian (Denny et al. 1994). This drowning is marked by the MCSB. The margin of this shallow-water megabank stepped back to the north during initial drowning, but then began to prograde back southward over the drowned part from the Late Cretaceous through the Cenozoic (Denny et al. 1994) (fig. 11.11). There also were numerous episodes of erosion in the southern Straits of Florida during this time, occasionally interrupting or altering progradation.

From the Late Cretaceous (late Cenomanian; Angstadt et al. 1985) to the Middle Eocene, downbuckling and flexing caused by the collision between Cuba and the foundering megabank produced a foredeep, the southern Straits of Florida, providing an avenue for the Florida Current (Denny et al. 1994). As mentioned above, the eastern margin of the Florida Platform was forming at the same time. This foredeep accumulated as much as 2 km of gravity-flow deposits, mostly shed from the evolving Cuban orogen to the south. The original southern margin of the shallow-water Lower Cretaceous megabank is now contained within the Cuba-Bahamas orogenic boundary. There is evidence that exposed carbonates along a major thrust fault in northern Cuba may represent original bank-edge deposits (Pardo 1975; E. Rosencrantz, pers. comm. 1992).

The margins of the Early Cretaceous Florida-Bahamas megabank appear to have been structurally controlled by rift and drift processes associated with the opening of the Gulf of Mexico and Atlantic Ocean. During the Late Cretaceous, the Florida and Bahamas platforms became

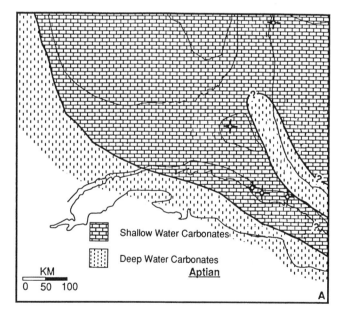

Figure 11.10. Paleogeographic map of the southern Florida Platform during the Aptian, before widespread drowning of the Platform. Shallow-water carbonates extended across the Straits of Florida to northern Cuba (Denny et al. 1994).

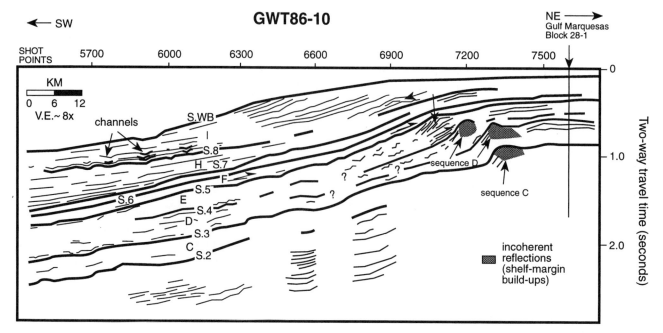

Figure 11.11. Interpretation of part of a seismic line across the modern southwest margin of the Florida Platform. Note the overall south-southwest progradation of the Late Cretaceous, post-drowning sequence (C-I). This prograding margin was dominated probably by skeletal (?) buildups (shelf margin) during the Late Cretaceous to Eocene (sequences C and D) (Denny et al. 1994).

distinct features separated by the Florida Straits. Long-term subsidence resulting from lithospheric cooling and sediment loading produced the great thicknesses seen in this carbonate province. Mesozoic and Cenozoic carbonates underlying the Florida Platform are less than 1 km thick over the Peninsular Arch, but more than 6 km thick in the South Florida Basin, indicating substantial differential subsidence.

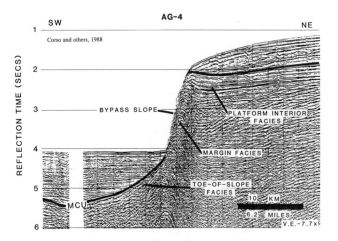

Figure 11.12. Multichannel seismic line across the Early Cretaceous margin, west Florida, illustrating key seismic facies and the West Florida Escarpment. At this location, the margin has not been severely eroded back, as the margin facies appear to be intact. Onlapping sediments at the base of the escarpment are Quaternary distal Mississippi River fan deposits. MCU is the mid-Cretaceous unconformity, which extends up onto the platform (Corso et al. 1988).

Drowning of the West Florida Margin

During the earliest Cretaceous, most of the west Florida margin developed into a high-relief carbonate bank characterized by sediment bypassing—probably much like the modern Bahamas Bank (Mullins et al. 1988a, 1988b) (fig. 11.12). Corso et al. (1989) pointed out that north of DeSoto Canyon the carbonate platform was narrower and closer to a siliciclastic source that inhibited shelf-edge buildups. Both siliciclastic and carbonate sediments were more efficiently shed from the northwestern part of the Florida Platform, allowing it to prograde along a lower-gradient slope. Ultimately—perhaps by the end of the Cretaceous—the Early Cretaceous platform in this northern sector became completely buried by siliciclastics (fig. 11.13).

South of DeSoto Canyon, the western part of the Florida Platform accumulated some 4.5 km of shallow-water sediments that terminated seaward in a constructional marginal escarpment that was steep, high (about 2 km), and characterized by sediment bypassing. Near the West Florida Escarpment, possible reef structures may represent extensions of well-known Lower Cretaceous reef trends through Texas and Louisiana (Mitchum 1978; Kirkland et al. 1987). The rapid accumulation of thick shallow-water carbonates and the high relief of the platform margin resulted from rapid subsidence of the underlying, relatively new continent-ocean crustal transition, and from the ability of the prolific shallow-water environments to produce and retain carbonate sediments.

LITHOLOGIES INDICATING DROWNING

By Cenomanian time, the Gulf of Mexico carbonate banks, including the west Florida margin, were drowned and covered by a neritic marl containing some siliciclastic debris (Mitchum 1978; Mullins et al. 1988b; Gardulski et al. 1991). The marls are pelagic foraminifer-coccolith carbonate muds that have undergone lithification, chertification, and dolomitization. The vertical juxtaposition of underlying mollusk-rich, shallow-water limestones with overlying pelagic facies indicates that drowning was rapid and terminal.

During the Maastrichtian, the transition to a pelagic, aggradational ramp was completed (Gardulski et al. 1991). The west Florida margin had subsided to bathyal depths, forming a deep-water plateau as the increasingly deeper-water sedimentary facies were unable to keep pace with subsidence (Bryant et al. 1969; Mullins et al. 1988b). By the end of the Late Cretaceous, water depths exceeded 900 m over the marginal parts of the former shallow-water carbonate bank (Mitchum 1978). Since the Maastrichtian—and up to the present—the western part of the Florida Platform has been a ramp system that has undergone phases of aggradation and progradation (Gardulski et al. 1991). During the Late Cretaceous, the central Florida Platform continued to accumulate shallow-water carbonates, and at least 600 m of differential subsidence occurred between the central platform and the western platform margin (Mitchum 1978).

DROWNING PROCESSES

The termination of carbonate platforms by drowning—not by tectonic uplift or transport to higher latitudes poleward of their "Darwin point" (Grigg 1982), and not by burial under siliciclastics—has been viewed as enigmatic and paradoxical, because of the ability of healthy carbonate-producing environments to keep pace with the fastest rates of geo- and glacioeustatic sea-level rise (Schlager 1981). However, if a once-favorable environment changes to something more inimical to carbonate-sediment production (Lighty et al. 1978; Hallock and Schlager 1986), or if a carbonate platform is unable to retain the sediments it produces due to physical-transport processes, as on the Blake Plateau (Dillon et al. 1985) and Pedro Bank in the Caribbean Sea (Glaser and Droxler 1991), then sediment accumulation on the platform will fall behind sea-level rise. Given long-term subsidence (as on a passive margin) and long-term sea-level rise (on the rising limb of a first- or second-order sea-level cycle), what was once a shallow-water bank becomes a deep-water margin plateau or ramp (Ahr 1973; Hine and Steinmetz 1984; Read 1985; Mullins et al. 1988b).

During the mid-Cretaceous (Albian-Coniacian), a number of major carbonate platforms around the world were drowned (Hallock and Schlager 1986; Vogt 1989). Also, Mullins et al. (1991, 1992) have shown that carbonate platforms of the southeastern Bahamas underwent extensive retreat during the mid-Cretaceous, a time of widespread anoxia, nutrients, high and rising sea levels, and mid-plate volcanism in the Pacific Ocean basin (Vogt 1972, 1979, 1989; Arthur and Schlanger 1979; Schlanger et al. 1981; Haq et al. 1987a; Larson 1991). Vogt (1989) and Jenkyns (1991) indicated that combinations of these factors are linked to platform demise.

These Early to mid-Cretaceous carbonate banks supported "Urgonian" faunal assemblages, which included

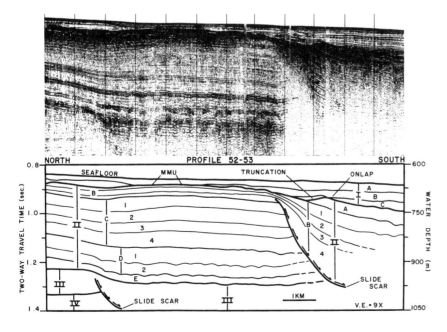

Figure 11.13. Along-slope (along-strike) single-channel air-gun seismic-reflection profile, showing a large submarine slide. Note the truncation of Miocene units IIC-IIE and the abrupt thickening of unit IIB. Also note earlier failure in lower left, which truncated units III and IV. The top of unit II is the Middle Miocene unconformity (MMU), formed by acceleration of the Loop Current (Mullins et al. 1988c).

rudists as the dominant reef-building organism (Jones and Nicol 1986; Vogt 1989). These assemblages also included hermatypic hexacorals, bivalves, foraminifers, echinoderms, coralline algae, stromatoporoids, and bryozoans. Features such as coral thickets, small banks, biostromal lenses, bioherms, and micritic stromatolites were part of the overall reef framework (Masse and Phillip 1981). It is speculated that similar faunal assemblages dominated the west Florida margin during the mid-Cretaceous as well. Albian to Aptian rudist reefs have been reported in the Lower Cretaceous along the eastern part of the Florida-Bahamas platform (Dillon et al. 1988).

The marginal-reef facies along the western Florida Platform has been eroded back as much as 6 to 26 km (Corso et al. 1988, 1989; Paull et al. 1990a, 1990b), leaving little direct evidence for the presence of a rudist community. Instead, rock dredgings along the exposed escarpment have recovered low-energy, lagoonal mudstones, wackestones, and packstones that were deposited behind the open margin (Freeman-Lynde 1983). Only a few skeletal packstones and grainstones have been recovered, indicating a more open and exposed depositional setting (Freeman-Lynde 1983). The large rudistid-bioherm complex exposed along the margins of an Early Cretaceous platform in the El Abra range in Mexico may be a suitable analog for what was deposited along the west Florida carbonate margin (Griffith et al. 1969; Wilson 1975; Paull et al. 1990a).

Hallock and Schlager (1986) pointed out that the primary carbonate producers on shallow-water carbonate platforms are highly adapted to oligotrophic, nutrient-deficient environments. With the input of nutrients such as nitrates and phosphates due to changes in ocean-circulation patterns (upwelling or convective overturn), the carbonate-producing community (autotrophic carbonate benthos) is stressed, outcompeted for space, and replaced by non-carbonate-producing, heterotrophic organisms, including bio-eroders that can actively destroy reef framework (Hallock and Schlager 1986). In general, carbonate banks that have been capped by a drowning unconformity (Schlager 1989)—the effects perhaps amplified by bio-erosion—are then covered by a condensed section (Loutit et al. 1988) consisting of organic-rich shales (no evidence of this on the west Florida margin) and/or phosphorite (present on the west Florida margin), indicating the availability of excess nutrients.

During the mid-Cretaceous, the oceans were substantially more stratified than they are today, because higher sea-surface temperatures created a stable, less-dense surface layer. As a result, the waters beneath the surface layer tended to be more anoxic, toxic, and nutrient-laden than today's relatively well ventilated oceans. At times during the mid-Cretaceous, anoxia was widespread. Long-term (or repeated short-term) exposure to these waters would have been deleterious to carbonate-producing communities. Such exposure could have occurred by elevation of the boundary between oxygenated and anoxic, nutrient-rich waters during short-period, rapid sea-level rises; by repeated bathing of reefs during decadal El Niño-type upwelling events; or by catastrophic buoyancy upwelling resulting from mid-plate volcanism that frequently generated large hydrothermal plumes (Vogt 1989).

Such mid-plate volcanism, as mentioned above, was a hallmark of the mid-Cretaceous in the Pacific. Plate reconstructions indicate that the mid-Cretaceous Gulf of Mexico and proto-Caribbean Sea had a direct connection to the eastern Pacific and to the global ocean, so that the west Florida margin could have been influenced by volcanogenic upwelling of anoxic, toxic, and nutrient-rich waters (Pindell 1985; Klitgord and Schouten 1986; Ross and Scotese 1988; Pindell et al. 1988). Regardless of the mechanism of drowning, shallow-water carbonate production on the west Florida margin ceased during the mid-Cretaceous—never to return—and ultimately an enormous, distally inclined carbonate ramp was created.

Development of the West Florida Ramp System

Seismic and borehole data show that development of the west Florida ramp since the mid-Cenomanian has been characterized by periods of aggradation and progradation caused by significant paleoceanographic events (Mullins et al. 1987, 1988a; Gardulski et al. 1991). The sedimentary lithologies constituting these progradational-aggradational depositional sequences have changed through time as well (Gardulski et al. 1986, 1990; Mullins et al. 1988b, 1988c). The earliest ramp was formed not long after the MCSB was formed (probably by the Campanian); it was progradational, and developed on top of the drowned, shallow-water platform. The clinoforms consisted of calcitic planktic and benthic foraminifers, occasionally punctuated by sandy turbidites. By the Maastrichtian, this early ramp had evolved into an aggradational system.

From the Maastrichtian to the Late Oligocene, the west Florida ramp aggraded by the persistent rain of planktic foraminifers and radiolarians, and abundant volcanic ash was introduced from the Caribbean, both by ocean currents and winds. Benthic foraminifers made a lesser contribution. There was little winnowing, and there is no evidence of erosional surfaces within this sequence. As a result, the north-flowing Suwannee Current (proto-Loop Current) off the west Florida margin (ultimately passing through the Suwannee Strait) probably did not impinge upon the sea floor, but was capable of transporting sand-sized volcanic sediments as far as 750 km from their source (Gardulski et al. 1991).

Within 5 million years or less (Late Oligocene to Early Miocene), this aggradational ramp began to prograde westward, as shown by west-dipping clinoforms evident in seismic profiles. Ramp progradation is recognized in cores by the presence of reworked sediments, small turbidites, mixed biostratigraphic zonal assemblages, and some shallow-water foraminifers (Gardulski et al. 1991). In addition, these clinoforms contain elevated concentrations of dolomite, organic matter, and clays (Mullins et al. 1987).

The reasons for this relatively abrupt change in depositional style during the late Paleogene to early Neogene are not known; however, several important regional and global events of those times may have been important. First, there was a transition in global climate from a "greenhouse" to an "icehouse" world (Miller et al. 1991). Glacioeustasy was becoming more pronounced. There was compression of low latitudinal climate belts, and probably changes in intensity of atmospheric circulation (Kennett 1982). Extensive east-to-west progradation filled in seaways within the Bahamas Bank and caused the leeward margin of Great Bahama Bank to prograde some 25 km out into the Straits of Florida (Eberli and Ginsburg 1989). Gardulski et al. (1991) noted that this Bahamian progradation occurred at the same time that the west Florida ramp began to prograde to the west, suggesting a strengthening of easterly trade winds that caused regional westward sediment transport across the banks and basins at these latitudes.

Second, the Georgia Channel System finally closed during that time (Popenoe 1990), deflecting the Suwannee Current to the south, and forming the precursor to the modern Loop Current (Pinet and Popenoe 1985a; Popenoe et al. 1987). This early Loop Current was not as strong as the modern system, because it had not yet been accelerated as a result of lower Caribbean tectonic and paleoceanographic events.

MASS WASTING

This Late Oligocene to Middle Miocene ramp contains an enormous slide scar, indicating a huge mass-wasting event that occurred during progradation (Mullins et al. 1986, 1988b) (fig. 11.13). Actually, en-echelon offsets were formed by three generations of slides—pre-Middle Miocene, Middle Miocene, and post-Middle Miocene—all in roughly the same area of the west Florida margin. The cumulative result of these catastrophic erosional events was a large, scalloped embayment complex within the trend of the West Florida Escarpment (fig. 11.1). Mullins and Hine (1989) cited the importance and relative frequency of scalloped embayments in the development (and demise) of carbonate platforms. On the West Florida Escarpment, the Middle Miocene event was the largest: the slide probably was at least 120 km long, 30 km wide, and 350 m thick. Undoubtedly, the remains of this slide and others lie buried in the eastern Gulf of Mexico basin (Brooks et al. 1986; and others). The cause of these massive slides is not known. Mullins et al. (1986) speculated that sediment overloading, earthquakes, cliff sapping by corrosive seeps, and sea-level low stands all could be responsible. These slides also could be related to the "Abaco event" (Leg 101 Scientific Party 1988), a postulated mid-Miocene tectonic event that caused pronounced mass-wasting in the Bahamas.

LOOP CURRENT INTENSIFICATION

During the Middle to Late Miocene, a new aggradational ramp developed as a result of intensification of the Loop Current (fig. 11.14). Significant local erosion, winnowing, and hardground development also accompanied current intensification. This aggradational system has

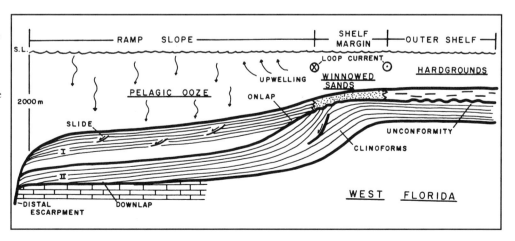

Figure 11.14. Schematic summary of the carbonate-ramp slope of west-central Florida. Major facies include pelagic ooze, shelf-margin winnowed sands, and outer shelf hardgrounds. Note the Loop Current-induced upwelling and winnowing. Note also the contrast between the modern sequence I, with its seaward-thickening wedge, and the prograding clinoforms of sequence II. The top of sequence II is the Middle Miocene unconformity (Mullins et al. 1988b).

persisted to the present, forming a sequence of parallel/laminated seismic facies as much as 250 m thick (Mullins et al. 1987). Aggradation occurred primarily seaward of where the Loop Current interacted with the sea floor.

Mullins et al. (1987) have pointed to formation of the Isthmus of Panama and the closing of the oceanic gateway leading from the Caribbean Sea to the Pacific Ocean as the primary reason for Loop Current intensification. This closing redirected the westerly flow of the Caribbean Current to the north across the Nicaraguan Rise, up through the Yucatan Straits, and into the Gulf of Mexico. However, Droxler et al. (1992) have indicated that the Middle Miocene foundering of a large, shallow-water carbonate platform covering the Nicaraguan Rise may have been responsible for this intensification. Probably, closing of the Caribbean/Pacific gateway further accelerated the Caribbean Current after initial intensification by foundering of the Nicaraguan Rise megabank. The relationship of this platform foundering to the formation of the Isthmus of Panama—if any—is unknown.

The best evidence for current intensification is seen in the extensive Middle Miocene marine unconformities that surround the Florida Platform. Beneath the modern, upper slope of the west Florida margin is a pronounced unconformity with high seismic amplitude that erosionally truncates the dipping clinoforms of the older, prograding ramp (Mullins et al. 1987) (figs. 11.13 and 11.14). This surface is onlapped/downlapped by the sequence forming the aggradational ramp.

Data from Exxon borehole CH 33-48 (fig. 11.15) also indicate significant lithologic and textural differences across this boundary (Mullins et al. 1987). Below the unconformity, within the earlier prograding sequence, calcite and authigenic dolomite dominate the mineralogy, and the sediments are 95% mud-sized. In addition, insoluble residues (quartz and palygorskite) are high (20%), as is organic matter. Above the unconformity, dolomite is rare, insoluble residues decrease, aragonite and high-Mg calcite are more common, and the

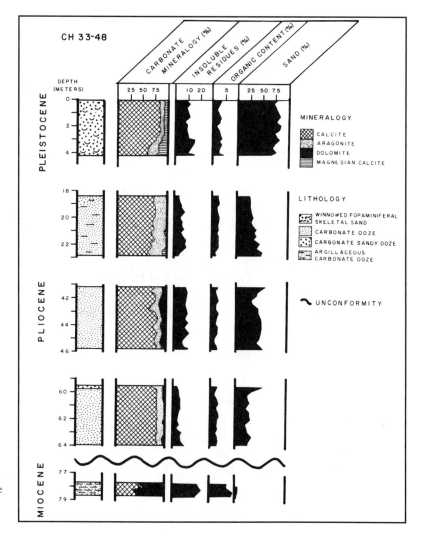

Figure 11.15. Sedimentologic data from Exxon borehole 48 on the West Florida Slope. Note the abrupt changes in lithology, mineralogy, and grain size above and below the Miocene/Pliocene unconformity. These data indicate an intensification of the Loop Current (Mullins et al. 1987).

sediments are 30% to 80% sand-sized, indicating an increase in current strength at this site. The sands consist of foraminifers, quartz grains, and reworked glauconite and phosphorite grains (Mullins et al. 1987).

Seaward of the Exxon borehole, slope sediments are more pelagic. This indicates that the strengthened Loop Current has had its greatest influence on the sea floor between water depths of 400 to 600 m, where a swath of winnowed foraminiferal sands and hardgrounds is now exposed (Mullins 1988a) (fig. 11.14). Seaward of this belt of coarser sediments, thick pelagic oozes have accumulated. The thickness of these oozes is due to stimulated biologic productivity caused by upwelling of nutrient-rich waters associated with Loop Current eddies (Vukovich and Maul 1985). All of these data led Mullins et al. (1987) to conclude that a distinct, major oceanographic event occurred during the Middle Miocene, manifest in the change from a prograding ramp (erosion terminated progradation) to an aggrading ramp.

Other Middle Miocene unconformities surrounding the Florida Platform occur at the Pourtales Terrace (Gomberg 1974, 1976; Locker et al. 1991) (fig. 11.16), the Miami Terrace (Mullins and Neumann 1979), and the Blake Plateau (Pinet and Popenoe 1985a). All of these unconformities formed in response to regional current intensification. All are topographically rugged, and produce prominent, high-amplitude seismic reflections resulting from extensive diagenesis, bio-erosion, phosphatization, possible dissolution by migration of internal platform fluids, and reworking-relithification.

Finally, topographically induced upwelling on the center of the Florida Platform was associated with current intensification during Middle Miocene sea-level high stands. This important process produced enormous phosphorite deposits, making Florida one of the largest producers of phosphate products (Riggs 1979a, 1979b, 1984; see chapters 9 and 12).

West Florida Escarpment

The exposed West Florida Escarpment is an enormous, steep-gradient, high-relief, submarine surface that extends approximately 750 km from DeSoto Canyon on the north to the western Straits of Florida on the south (figs. 11.1 and 11.12). Its relief ranges from about 1 km in the north to nearly 2 km in the south. The top of the escarpment lies under about 1.6 km of water and its base is 2.5 to 3.4 km below sea level. This feature defines the western boundary of the Florida Platform and represents most of the depositional relief of the Early Cretaceous shallow-water, rimmed platform. Original depositional relief was probably greater, because approximately 1 km of Pliocene and post-Pliocene hemipelagic sediments associated with the distal Mississippi River fan now onlap the base of the escarpment (Bouma et al. 1986; Buffler 1991) (figs. 11.1, 11.9, and 11.12).

MORPHOLOGY

The morphology of the escarpment is spatially variable and related to both the primary platform depositional characteristics and to significant post-depositional events (Paull et al. 1990a, 1990b; Twichell et al. 1991). The northern part has not been signficantly eroded, and probably still retains much of its original depositional topography: coarse bioclastic limestones (typical of platform margins, not interiors) have been dredged from this area, and seismic data strongly indicate marginal-reef and back-reef lagoon facies (Locker and Buffler 1983; Corso et al. 1988; Paull et al. 1990a).

The central and southern parts of the escarpment are erosional, and estimates of retreat range from less than 1 km to about 26 km (Freeman-Lynde 1983; Paull et al. 1990a, 1990b). The primary evidence for erosion is (1) the irregular trend of the escarpment (large embayments and canyons extending into the platform); (2) the steep, vertical, or overhanging lower parts of the escarpment; (3) fractures and fresh, angular blocks (observed from the submersible DSRV *Alvin*); and (4) platform-interior wackestones and packstones directly exposed in the escarpment (Freeman-Lynde 1983; Mullins et al. 1986; Paull et al. 1990a, 1991a). The ages of the exposed rocks range from Aptian at the base to late Cenomanian near the top (Paull et al. 1990a). Finally, Corso et al. (1988) have made projections based on the buried mid-

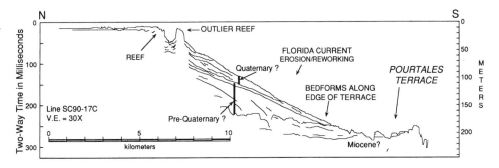

Figure 11.16. Interpretation of a high-resolution seismic line south of the Florida Keys, showing the Pourtales Terrace (unpublished data from S. R. Locker). The terrace is the exposed part of the eroded top of the southward-prograding early Neogene Florida Platform.

Cretaceous unconformity (MCU, MCSB) indicating that the escarpment has eroded back some 6 km (fig. 11.17).

The central part of the escarpment is characterized by large embayments, the largest of which is related to the multiple, large-scale, mid-Tertiary mass-wasting events discussed above (Mullins et al. 1986) (fig. 11.1). The southern part, with large box canyons that extend 26 km into the platform (figs. 11.1 and 11.18), is morphologically most complex. In addition, the escarpment is characterized by large terraces, as well as numerous smaller terraces that may be controlled by joints and bedding planes. The larger terraces are up to 4 km wide, but they are discontinuous and extend only 2 to 5 km along the escarpment. Terrace size and depth vary along the canyon headwall, suggesting spatial changes in erosional rates (Twichell et al. 1991). Smaller features include gullies, sediment chutes, sinkhole-like depressions, individual blocks, slumps, and basal rock-fall deposits.

PROCESS OF EROSION

Paull (1991a) suggested that the fresh appearance of the exposed rocks (due to collapse along joints) and the basal talus deposits indicate that the escarpment continues to erode, and that the entire feature should be considered an active unconformity—one that has been eroding since the mid-Cretaceous. The original, depositional slope resulted from differential sediment accumulation between the emerging platform and the adjacent basin. With continued subsidence, sedimentation on the platform kept pace with the relative sea-level rise as the basin became deeper; hence, platform relief and slope declivity increased (Schlager and Camber 1986). In time, the slope became less progradational, and then was characterized by sediment bypassing; ultimately, it became erosional, retreating by self-erosion (Schlager and Ginsburg 1981; McIreath and James 1984; Schlager and Camber 1986). However, self-erosion produces a concave-up profile that is not shown along the present West Florida Escarpment. Although initial erosion of the Lower Cretaceous may have occurred in this manner, discharge of corrosive fluids and cliff sapping are the primary causes of long-term retreat (Paull et al. 1990a, 1990b, 1991a; Twichell et al. 1991).

Seepage of brines from the Florida Platform through the West Florida Escarpment is indicated by (1) vent-type benthic communities, (2) black, iron-sulfide rich sediments, (3) high organic-carbon concentrations, (4) local heavily corroded limestone surfaces, and (5) sediment pore waters enriched in Cl^-, NH_4^+, and H_2S and depleted in SO_4^{2-} at the contact between the flat-lying Mississippi River fan deposits and the permeable lime-

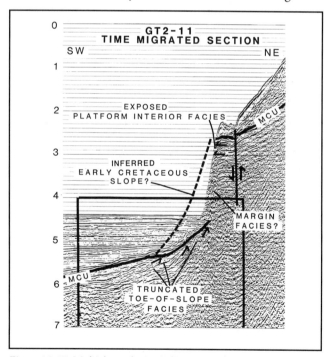

Figure 11.17. Multichannel seismic line across the West Florida Escarpment, illustrating the mid-Cretaceous unconformity (MCU) and the projected extent of erosional retreat of this part of the escarpment. Note that platform-interior facies are exposed on the upper escarpment (Corso et al. 1988).

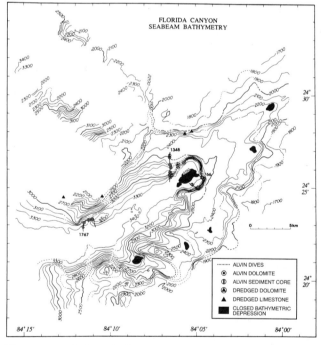

Figure 11.18. Seabeam bathymetric chart of the Florida Canyon area. The canyon has been eroded approximately 26 km into the Florida Platform. A semicircular 750 m high headwall (near site 1766) separates the lower U-shaped canyon from the upper V-shaped canyon. Note the angular trends of canyon walls (reflecting joint patterns?) and the closed circular sinkhole-sized depressions (shaded areas) (Paull et al. 1990b).

stone at the base of the escarpment (Paull et al. 1984; Paull and Neumann 1987; Commeau et al. 1987; Paull et al. 1990b; Chanton et al. 1991; Paull et al. 1991b). Presumably, dissolution of Mesozoic anhydrite deposits within the Florida Platform forms density-driven brine that flows through permeable zones extending to the West Florida Escarpment (Paull and Neumann 1987; Paull et al. 1991b) (fig. 11.19). It has been shown that high permeability and active fluid circulation are widespread within the Florida Platform, including horizontal exchange with the abyssal Gulf of Mexico (Kohout 1967; Kohout et al. 1988) (fig. 11.20).

When these brines mix with seawater, which seems to be concentrated within 2 m above the base of the escarpment in many areas (Paull and Neumann 1987), hydrogen sulfide is rapidly oxidized, creating sulfuric acid that can react with the escarpment carbonates. Corrosion occurs when water masses of different salinities are mixed (Plummer 1975; Hanshaw and Back 1980). Bacterially mediated oxidation of reduced inorganic compounds from the brines (e.g. HS^-, NH_4^+) provides the basis of the chemosynthetic food chain that supports dense populations of mussels (family Mytilidae), trochid gastropods, calyptogenid clams, stoloniferous corals, vestimentiferan worms *(Escarpia laminata)*, galatheid crabs, holothurians, and zoarcid fish (Paull et al. 1984; Paull and Neumann 1987; Commeau et al. 1987). Also, high concentrations of dissolved methane are found in sediment pore waters at seep sites; the methane is generated by microbial degradation of organic carbon in limestones as the dense brines migrate through the platform (Martens et al. 1991; Paull et al. 1991a, 1991b).

Ultimately, rock ledges are undercut, mass-wasting occurs along joints and fractures, and the escarpment as a whole is defaced and retreats. The lack of distinct debris

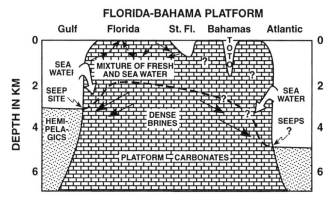

Figure 11.19. Schematic west to east cross-section of the entire Florida-Bahamas Platform, illustrating the concept of regional fluid flow. Denser brines will flow toward platform margins and exit at seep sites at or near sediment/water/rock boundary (Paull and Neumann 1987).

Figure 11.20. Schematic cross-section oriented west to east across south Florida from Naples to Ft. Lauderdale, illustrating the subsurface flow of fluids within the Florida Platform. Note the geothermally driven influx and efflux of seawater. Note also the sinkholes along the Miami Terrace where fluids exit the platform. Similar karst features may exist along the Pourtales Terrace to the south. The circulation described by Paull and Neumann (1987) and Paull et al. (1991a) probably represents the western part of this diagram (Kohout et al. 1988).

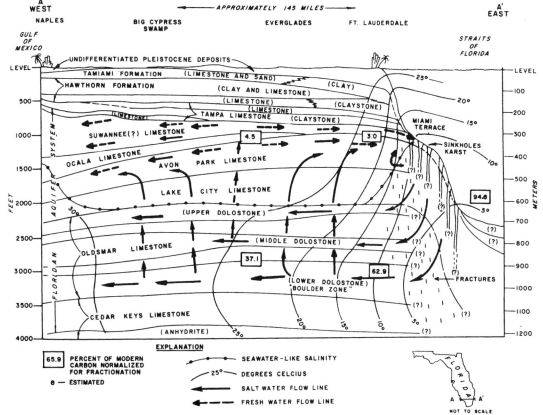

wedges at the base of the escarpment may be explained by widespread dissolution, the rapid sedimentation rate at the base of the escarpment on the onlapping distal Mississippi River fan, and the relative lack of bypassing sediments coming down the escarpment from above. Paull et al. (1990b) explained that the flat-floored box canyons represent hydrologic focusing at the canyon heads (fig. 11.21). Due to regional concentration of joints or other hydrologic conduits within the Florida Platform, groundwaters may be concentrated, forming areas of increased corrosion and cliff sapping. Through time, these areas have extended back into the platform 26 km or more, creating the canyons (figs. 11.18 and 11.21). To the south, in the Straits of Florida, the West Florida Escarpment disappears as a prominent bathymetric feature because the Lower Cretaceous carbonate platform is buried there (Corso et al. 1988).

Quaternary and Modern Shelf-Slope Systems

The bathymetric template of Quaternary and modern shelf-slope systems—modified by multiple sea-level fluctuations and estuarine, coastal oceanic, and open-ocean processes—has created a diverse mosaic of sedimentary facies. The shelf-slope Quaternary subsurface stratigraphic package beneath the surficial facies pattern is poorly known (Hine and Doyle 1991), except for a few detailed-study sites (Doyle and Holmes 1985; Mullins et al. 1988a, 1988b; and others).

WEST FLORIDA

The outstanding characteristics of the Quaternary and modern West Florida Shelf and West Florida Slope system are the breadth and low gradient inherited from the underlying carbonate ramp system. The average gradient of the shelf is 0.4 m/km, and the average gradient of the slope is 1.6 m/km (Ginsburg and James 1974); along parts of the inner shelf, the gradient is as low as 0.2 m/km (Hine and Belknap 1986). There is no precisely located shelf/slope break, although there is a slight steepening between the 100 and 200 m isobaths (fig. 11.1).

Karst and Antecedent Topographic Control Besides the broad, regional nature of the ramp, several important topographic features have controlled sedimentation. The western Florida coast and inner shelves are dominated by two large estuarine systems: Tampa Bay and Charlotte Harbor. Both of these drowned river-valley systems appear to occupy local structural depressions, perhaps resulting from concentrated dissolution of underlying limestones within the platform. Evans and Hine (1991) have shown prominent subsurface sinkhole development and fold structures beneath Charlotte Harbor (fig. 11.22). Other studies of the Florida Platform have shown that subsurface karst features are prominent (Snyder et al. 1989). Folds and sag structures are readily discernable in Paleogene strata beneath the east margin of the Florida Platform (fig. 11.8a). Macurda (1989) has shown the presence of extensive, buried Miocene karstification and prominent collapse features along the outer shelf off southwestern Florida. Perhaps low stands of sea level during the late Paleogene and early Neogene promoted accelerated fluid flow within the Florida Platform, caused extensive dissolution that produced the deformation seen in seismic profiles, and eventually created the shallow basins of Tampa Bay and Charlotte Harbor.

Concentration of surface runoff into large, distinct basins allows for transport of upland sediments out onto the shelf during sea-level low stands, where they are reworked during subsequent high stands. Seismic data reveal a 30-m-deep, 5-km-wide, buried channel that can be traced approximately 40 km seaward from the present-day coastline at Tampa Bay (Hebert 1985; A. C. Hine,

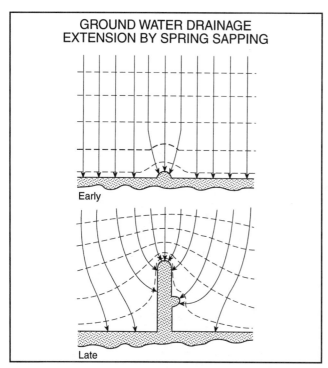

Figure 11.21. Postulated groundwater flow within the West Florida Platform. Dotted lines indicate a horizontal pressure gradient. Solid lines represent lines of flow perpendicular to the pressure gradient. The eroded platform edge is shown by a stippled pattern. In the early phase, flow captured by irregularities in the margin will concentrate the corrosive force of spring sapping at the head of an irregularity (small canyon). In the late phase, a large canyon is ultimately developed (Paull et al. 1990b).

unpublished data; Duncan 1992; Willis 1984; A. B. Tihansky, unpublished data) (fig. 11.23). This apparent westerly subaerial sediment transport accounts for the deposition of siliciclastic sediments (or reworked Miocene phosphorites) on more-distal (westerly) parts of the underlying carbonate ramp.

Where Tertiary limestones have been exposed, the karst topography has had a dominant effect on coastal morphology (Hine et al. 1988). The inner shelf reveals aligned sinkholes and linear depressions etched in bedrock, formed by springs discharging freshwater that migrated landward in response to sea-level rise (Hine and Belknap 1986).

Reefs and Antecedent Topographic Control Reef structures have been identified in the middle and outer shelf to upper slope in the areas of the Florida Middle Ground, Pulley Ridge, and Howell Hook (fig. 11.1). The Florida Middle Ground is a bathymetrically complex zone in approximately 40 m of water. Although there is no borehole information, the bathymetry and the carbonate-producing community on the present sea floor indicate that higher substrate areas have been built up by reef growth. Perhaps there is some antecedent regional topographic feature (subsurface karst?), on which the reef started, that is not seen in seismic profiles (Price 1954). Jordan (1952) suggested paleo-drainage control of initial reef development. Brooks and Doyle (1991) postulated that control may have been exerted by a combination of karst and an antecedent break in slope. Initial reef development may also have required Loop Current recruitment of tropical species. The reef structures were probably built up during numerous Pleistocene sea-level fluctuations (Brooks and Doyle 1991).

Video data (Brooks and Doyle 1991) reveal that the tops of the reefs are rugged and dominated by *Millepora,* and that broad areas are covered by carbonate sediments. Reef flanks support large blocks of reef material, as well as a living assemblage of coralline algae, *Millepora,* and sponges. Sediments are coarse and include *Madracis* fragments transported from above. Between the reefs are small, restricted basins, where trapped sediments are as much as 5 m thick. These sediments are carbonate sands, fine sands, and silty sands, and they support bedforms indicative of active winnowing and transport. The sediments associated with the Florida Middle Ground consist mostly of mollusks (36%), followed by barnacles, benthic foraminifers, annelids, bryozoans, and coralline

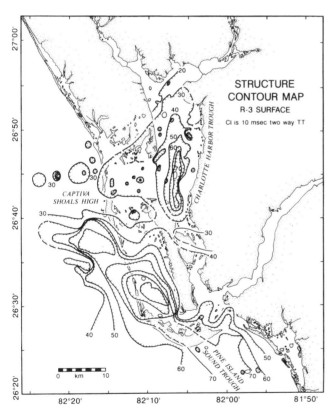

Figure 11.22. Structure-contour map of subsurface unconformity (R-3) in the Charlotte Harbor area, obtained from high-resolution seismic-reflection data. Contours represent seismic-travel times but can be roughly converted to depth below sea level (10 msec = 10 m). The map indicates subsurface highs and lows with as much as 50 m of relief. These structures probably were formed by deep-seated dissolution and collapse of underlying limestones during a low stand of sea level. They are probably similar to features seen in figure 11.8a (Evans and Hine 1991).

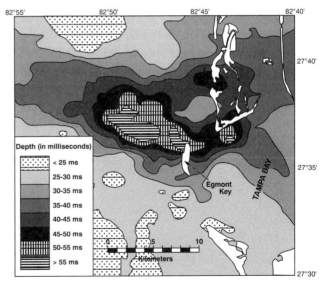

Figure 11.23. Structure-contour map of a seismic basement seen in a high-resolution seismic-reflection survey of lower Tampa Bay. The basin is part of a major east-west paleofluvial channel that extends from beneath modern Tampa Bay across the inner continental shelf (Duncan 1992).

algae (less than 10% each)—and a large unidentifiable fraction (38%) (Brooks and Doyle 1991).

The mix of sediment constituents is nearly indistinguishable from that of surrounding shelf carbonate sediments and does not resemble the mix from other modern coral-reef environments in the Caribbean. Most likely the carbonate-producing community that constructed the topographic highs differed significantly from the present community. Indeed, it is possible that the rocky substrates of the Florida Middle Ground are now undergoing significant bio-erosion. If carbonate-producing benthic communities structured like the present one did not build the Florida Middle Ground reefs, then either the marine environment changed from tropical (relative dominance of corals and algae) to subtropical or temperate as sea level was rising, or this area was more consistently bathed by warm Loop Current waters during past sea-level high stands (Brooks and Doyle 1991).

It is also possible that as depth increased due to sea-level rise, upwelling caused by the Loop Current could have increased, terminating the reef community by exposure to excess nutrients (Hallock and Schlager 1986). Another possibility is that the early reefs progressed in a start-up/keep-up mode, but as the rate of sea-level rise increased, they lagged behind, ultimately regressing into a give-up mode (Neumann and Macintyre 1985).

The Florida Middle Ground probably is an aborted analog to the active organic buildups seen on the Campeche Bank, a submerged extension of the Yucatan Peninsula (Logan et al. 1969). The Campeche Bank is also a drowned mid-Cretaceous carbonate bank capped by a modern carbonate ramp system that has probably undergone some of the same geologic history as the west Florida margin (Locker and Buffler 1983). However, the presence of active coral reefs in a middle- to outer-shelf setting on the Florida Platform has been a primary difference between these two carbonate systems during the Quaternary. Alacran Reef, rising from water depths of about 60 m on Campeche Bank, may be underlain by 38 m of unlithified Holocene carbonates (Purdy 1974). Coral reefs have persisted on the Campeche ramp, but not on the west Florida ramp, because (1) the Campeche ramp is at a lower latitude and overall has a warmer climate, (2) it experiences less-intense cold-air outbreaks associated with extra-tropical storms (Roberts et al. 1982), and (3) it has no known upwelling.

Farther out on the West Florida Shelf are a number of prominent, linear ridge systems, such as Pulley Ridge (70–90 m) and Howell Hook (150–190 m), that are presumed to be Quaternary reefs (fig. 11.1). There are multiple ridges within each system. Pulley Ridge may be controlled by an abrupt break in slope of an underlying Neogene surface (Holmes 1985; Brooks and Doyle 1991). Howell Hook is underlain by a Plio-Pleistocene reef system. Little is known about these features, except that they probably do not presently support Caribbean-type coral-reef communities, but rather coralline algae (Holmes 1985; Doyle and Holmes 1985).

Exposed hardbottoms at the shelf/slope break (i.e. at about 500 m) consist of either Miocene phosphatic crusts or Pliocene outcrops capped by Pleistocene coarse carbonate sand; they provide substrate and topographic control for inactive deep-water coral mounds dominated by ahermatypic corals (Newton et al. 1987). These hardbottoms represent extensive erosional surfaces, perhaps caused by more-intense Loop Current activity during glacial times.

Sediment Facies Because the West Florida Shelf and the West Florida Slope constitute a ramp system, facies changes are broader and more diffuse than on rimmed carbonate platforms (Reading 1978). However, unlike carbonate ramp models (Read 1985), sediment grain size remains relatively coarse well out onto the outer shelf and upper slope (at depths of 500 m). In fact, Blake and Doyle (1983) reported that sediment grain size is coarsest between depths of 75 and 100 m, becoming finer both landward and seaward. Muds and oozes are not encountered until water depths exceed 800 m (Mullins et al. 1988a). In addition, the West Florida Shelf is a mixed siliciclastic-carbonate system with a quartz-sand belt (fig. 11.24) that was introduced onto the Florida Platform after the late Paleogene closure of the Suwannee Strait. In general, facies boundaries trend parallel to the bathymetry (Doyle 1981). The quartz-sand facies is a second primary difference between the west Florida margin and the Campeche Bank. Siliciclastic sediments are being introduced to the upper slope off northwestern Florida by the Loop Current, which periodically carries muds from the Mississippi River (Walker 1984).

The quartz-sand belt at the coastline, which makes up the barrier-island system and underlies the marine-marsh system, is nowhere very thick (less than 10 m), and it abruptly thins seaward (Gould and Stewart 1956; Ginsburg and James 1974; Doyle and Sparks 1980; Blake and Doyle 1983; Hine and Mullins 1983; Sussko and Davis 1992). In some areas, the quartz/carbonate boundary lies within a few hundred meters of the beach. The quartz-sand facies pinches out where Tertiary limestone bedrock is exposed—indeed, much of the shelf has exposed hardbottom with no sedimentary veneer at all (NOAA 1985). Within the quartz-sand belt, the dominant carbonate constituents are mollusks (fig. 11.24).

Scattered corals can be found on exposed rocky surfaces in shallow water, and calcareous green algae (*Halimeda, Udotea, Penicillis*) are found in seagrass beds; however, none of these organisms produce an identifiable component in surrounding sediments. South of Cape Romano, quartz content drops from 80% to 2% on the inner shelf toward Florida Bay, a well-known carbonate sediment-producing environment that represents the bank-interior facies of the south Florida carbonate platform (Sussko and Davis 1992).

The middle shelf is characterized by a thin molluscan-sand sheet, about 1 m thick, where hardbottoms do not exist (fig. 11.24). The infauna here are mostly filter-feeders, whereas deposit feeders are dominant on the slope, and specific benthic-organism assemblages are largely restricted to specific sedimentary regimes (Blake and Doyle 1983). Molluscan sands can be found beyond the outer shelf; however, coralline-algal sands and ooids form identifiable facies belts within the outer shelf. The ooids are in water depths ranging from 80 to 100 m, but they were formed in shallow waters 2 to 5 m deep, indicating that they are a relict, autochthonous deposit formed when sea level was lower (Kump and Hine 1986). These coated grains formed in shallow-water, wave-dominated environments during the last sea-level low stand and during early phases of the following rise.

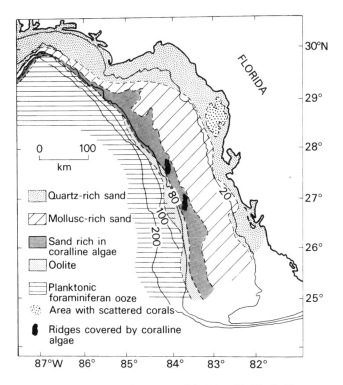

Figure 11.24. Sediment facies map of the West Florida Shelf, showing inner quartz sand belt and seaward carbonate belts, each dominated by a different carbonate sediment type. Facies belts are parallel to bathymetry (Reading 1978).

Both the molluscan sands and the coralline-algal sands are probably younger than the ooids, having formed after sea level had risen and created open-shelf conditions (Reading 1978). The algal sands probably dominate in areas where hardbottoms and/or rocky highs (relict reefs?) are more abundant.

The region of the outer shelf, shelf margin, and upper slope (200–600 m, 1° to 2° gradient) is a broad area consisting of two facies belts: hardgrounds with algal ridges (200–400 m) and winnowed sands (400–600 m). A bioturbated pelagic-ooze facies extends from 600 m to the top of the West Florida Escarpment at 2,000 m (Mullins et al. 1988a) (fig. 11.14). Three types of Quaternary hardgrounds have been discovered by dredging: (1) heavily bored, intraclastic, foraminiferal grainstones cemented by magnesian calcite, (2) deep-water coral framestones, and (3) gravel-sized rhodolith rudstones consisting of red-algae encrustations of skeletal fragments and intraclasts (Mullins et al. 1988a). The winnowed sands contain calcite (from planktic foraminifers), aragonite (from bivalves and gastropods, including pteropods), and magnesian calcite (from red algae, benthic foraminifers, echinoderms, and intergranular cement). Piston cores recovered from the shelf margin reveal that winnowed sands and hardgrounds overlie oozes (Mullins et al. 1988b).

The slope consists of winnowed foraminiferal sands, hardgrounds and nodules, planktic-foraminifer ooze, and nannofossil ooze (Mullins et al. 1988a). The carbonate oozes are products of Loop Current upwelling. Argillaceous ooze from the Mississippi River and organics make up the noncarbonate component. Microfossil abundance, sediment texture, mineralogy, and isotope data from the slope cores indicate cyclicity that has been related to climate changes in the Milankovitch frequency band (Gardulski et al. 1986, 1990; Roof et al. 1991). During glacial episodes, there was an increase in upwelling that stimulated pteropod (low-Sr aragonite), dolomite, and organic-sediment production. Also, more terrigenous muds were introduced from the Mississippi River during sea-level low stands. During high stands the sediments were more calcitic and coarser because of foraminifer production.

Roof et al. (1991) have identified evidence of different sediment-cycle frequencies occurring during the late Neogene and Quaternary on the West Florida Slope (fig. 11.25). They identified a time of variable climate during the late Miocene to Pliocene (their interval 3) that caused variations in Loop Current position, producing both large- and small-amplitude sediment cycles. Their interval 2 occurred during the late Pliocene to early Pleistocene and featured high-frequency (30,000- to

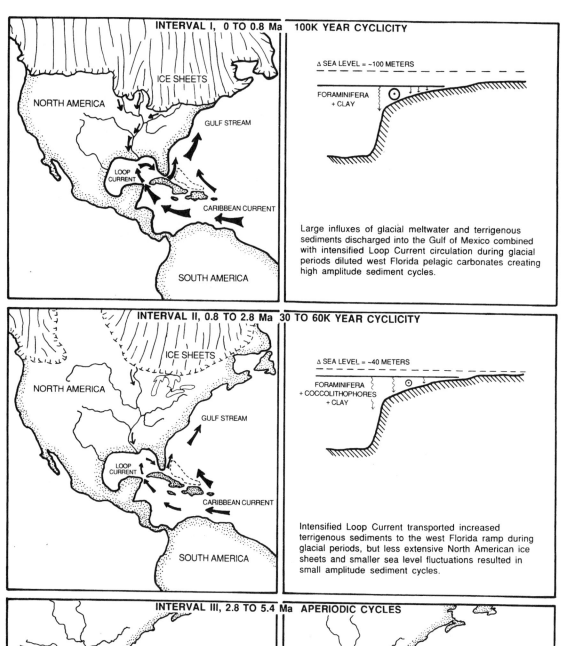
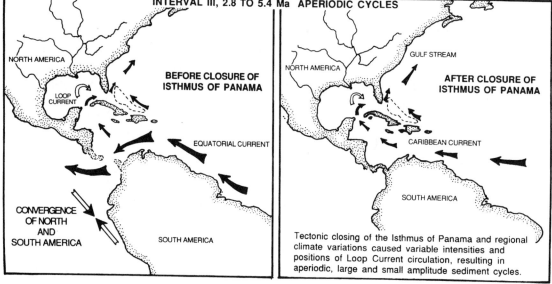

Figure 11.25. Major controls on sedimentation on the West Florida Platform for the past 5.4 million years, during three intervals of different climate-cycle frequencies (Roof et al. 1991).

60,000-year), low-amplitude sea-level cycles and thinner sediment cycles. Finally, during the last 0.8 million years (interval 1), sea-level cyclicity increased to 100,000 years; this, coupled with a large influx of meltwater and sediments into the Gulf of Mexico from the Mississippi River (due to larger ice sheets than during interval 2), produced larger-amplitude (thicker) sediment cycles.

Holocene turbidites and slumps are rare along the modern West Florida Slope. Seismic data indicate incipient pull-apart structures demonstrating potential downslope mass movement of sediment.

SOUTH FLORIDA

Along the south Florida non-rimmed margin (west of the rimmed margin of the Florida Keys), Holmes (1985), Brooks and Holmes (1989), and Locker et al. (1991, 1992) have demonstrated that the slope is actively prograding due to vigorous off-platform sediment transport (fig. 11.26). Large, south-oriented (off-bank) sand waves exposed on the sea floor provide additional evidence for this transport pathway (Holmes 1985). This off-bank transport is driven primarily by strong northerly winds generated by extra-tropical storms. To a lesser extent, the south-flowing Loop Current may have an effect (NOAA 1985).

Brooks and Holmes (1989) and Locker et al. (1991, 1992) have identified a number of seismic sequences, each probably deposited during a single sea-level fluctuation. These sequences contain current-controlled facies (sediment drifts) and sea level-generated features such as low-stand slope erosion, transgressive unconformities, and paleo-shorelines (Locker et al. 1992). The Florida Current has transported (and is now transporting) sediments alongslope, burying the Miocene Pourtales Terrace from the west (fig. 11.26). Overall, the slope sediments are a mixture of deep- and shallow-water components, comprising equal parts of sand, silt, and clay. The sediments are mostly carbonate (86%), and the insoluble minerals are quartz, kaolinite, and smectite. The dominant carbonate constituents are planktic foraminifers, pteropods, sponge spicules, benthic mollusks, benthic foraminifers, algae, and echinoids (Brooks and Holmes 1989).

More recently, drowned shorelines (beaches, and possibly eolian dunes) have been discovered in 60 to 100 m of water; the deposits consist of cemented shallow-water grainstones comprising tropical to subtropical skeletal components and some ooids and aggregate grains (Locker et al. 1992; Toscano and Hine 1992). These sea-level indicators were formed since the last glacial maximum, and suggest that the rate of sea-level rise has not been constant, but probably has accelerated and decelerated during the past 18,000 years. These low-relief shorelines may have been constructed when the rate of rise was relatively slow. The ability of carbonate sediments to rapidly lithify through cementation prevented these shorelines from disappearing completely during ensuing flooding.

Another indicator of past sea levels are outlier reefs constructed seaward of the main, shallow-water reefs fronting the Keys (Lidz et al. 1991) (fig. 11.27). These exposed Pleistocene reefs formed on terraces that may

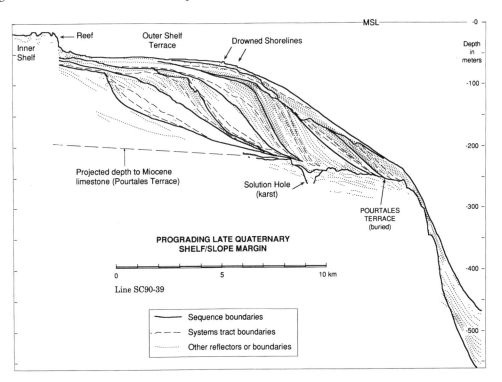

Figure 11.26. Interpretation of high-resolution seismic line across the northern boundary of the Florida Platform, showing sequences, sequence boundaries, and systems-tract boundaries and suggesting margin progradation via multiple sea-level fluctuations. Outer sequences downlap on the buried Pourtales Terrace. Note small bottom features, which are drowned shorelines (unpublished data from S. D. Locker).

also have supported beaches and dunes. Through time, multiple sea-level fluctuations may have caused reef-margin progradation seaward as a result of outlier-reef formation and back-filling of the basin that lies between the outlier and the main reef tract.

In general, the progradational slope off the south Florida Platform contrasts sharply with the west Florida aggradational ramp. The primary differences are these: (1) The south Florida Platform is an elevated, non-rimmed margin (probably drowned), but not a true ramp. (2) There is vigorous off-bank sediment transport, particularly during moderately high to high sea levels. (3) The more-tropical environment (as compared to subtropical) produces different grain constituents.

EAST FLORIDA

The shelf-slope system of the east Florida margin (fig. 11.1), like its southern and western counterparts, lies atop an older carbonate platform. In this case the carbonate platform persisted into the Eocene, and perhaps the Early Oligocene (figs. 11.7 and 11.8). Furthermore, the modern east-coast shelf is dominated by siliciclastic sediments (Duane and Meisburger 1969b; Meisburger and Duane 1971; Field and Duane 1974; Meisburger and Field 1975).

Background and Regional Setting Overall, the stratigraphy beneath the modern shelf consists of gently east-dipping Tertiary strata overlain by thin Quaternary sands. Eocene limestones correlative to the Ocala Limestone are overlain by the Miocene Hawthorn Group, which comprises a distinctive mixture of marls, clays, and phosphorites. Undifferentiated sands and clays represent the Late Miocene to Pliocene. The Quaternary is characterized by shelly quartz sands, and the coastal Anastasia calcarenite is prominent in some areas (Meisburger and Field 1975). The Anastasia probably is correlative with the Miami and Key Largo limestones to the south (Puri and Vernon 1964; Enos and Perkins 1977). Although the Anastasia Formation crops out along small sectors of the northeast barrier-island coastline and is locally the dominant geologic feature, its seaward extent out onto the shelf is unknown. However, cemented-carbonate facies (probably Anastasia-equivalent) are commonly encountered by shallow coring into the subsurface (Meisburger and Duane 1971).

Modern Shelf Sectors The modern East Florida Shelf is the southern extension of the large, contiguous feature that lies off the entire U.S. and Canadian east coasts (Emery 1969; Milliman et al. 1972). This shelf is generally a wave-dominated, low-gradient feature that has a reasonably well defined shelf/slope break. The East Florida Shelf displays significant north-to-south variations in morphology, width, sediment type, environmental parameters, and Quaternary geologic history. For example, the shelf off Jacksonville is 125 km wide, whereas to the south, off Lake Worth, it is only 2 km wide (fig. 11.1). Tidal range decreases to the south, as does wave-energy flux (Nummedal et al. 1977). The modern sedimentary cover is quartz-dominated in the north, and becomes more enriched in carbonate to the south. The carbonate sediments change from a mollusk-dominated temperate suite on the north to a more tropical, coral-algal suite on the south. Inner-shelf topography is characterized by shoreface sand ridges on the north, and by relict reefs on the south. Finally,

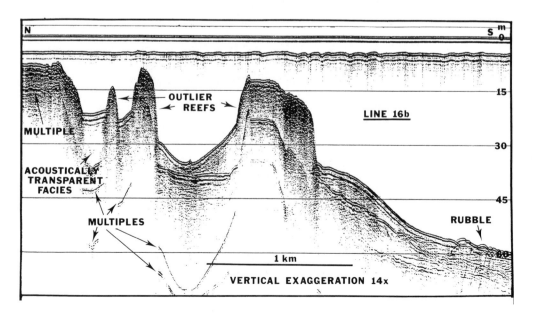

Figure 11.27. Seismic line off the western Florida Keys, illustrating outlier reefs. These reefs lie seaward of the main reef tract (far left) and have been built up on a deep terrace in 38 m of water. The reefs are mostly Pleistocene and probably were built during a number of sea-level high stands (Lidz et al. 1991).

buried Quaternary fluvial channels are more common on the north than on the south.

Siliciclastic Shelf Sector Most of the East Florida Shelf from the Florida/Georgia border to Lake Worth—where carbonate sediments and relict reefs dominate facies and topography, respectively—is covered by sediments containing more than 50% quartz sand (Duane and Meisburger 1969b). The dominant topographic feature of this shelf sector is the Cape Canaveral cuspate foreland and its associated offshore bathymetry (fig. 11.28). Here, the shelf supports a complex bathymetric mosaic of attached to isolated linear shoals, broad depressions and highs, and two cape-retreat (shoal-retreat) massifs—one off Cape Canveral and one off False Cape (Field and Duane 1974).

According to Swift et al. (1972), cape- and shoal-retreat massifs form on shelves as cuspate forelands migrate landward in response to sea-level rise. The track followed by the retreating cape is marked by a broad, shelf-normal sand deposit that may be reworked and re-shaped. Maximum Holocene sand thickness of the Cape Canaveral cape- and shoal-retreat massif is about 12 m. Sands contain approximately equal proportions of siliciclastic and carbonate components. The carbonate components are generally skeletal, comprising bivalve, gastropod, barnacle, bryozoan, foraminifer, and echinoid fragments (Field and Duane 1974).

Beyond the influence of the Cape Canaveral cape- and shoal-retreat massif to the north and south, the inner shelf is dominated by shoreface attached/detached sand ridges (Duane et al. 1972). These sand bodies are some 15 km in length, with 4 to 10 m of relief, and they form a 20° to 24° angle with the coastline, extending offshore toward the northeast. McBride and Moslow (1991) proposed this model for ridge formation: Ebb-tidal deltas are formed at inlets along barrier-island coasts. As the barrier-island coast retreats in response to sea-level rise, the ebb-tidal deltas are reworked into linear, northeast-trending ridges.

The most common surface sediment along the northern part of the East Florida Shelf is a fine- to

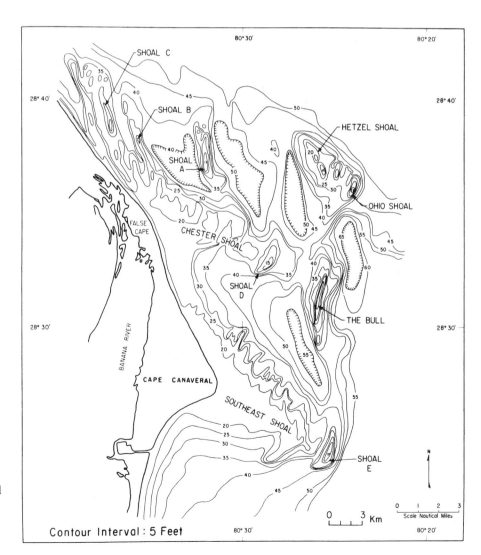

Figure 11.28. Bathymetry off Cape Canaveral, showing a large sand-shoal complex called a cape-retreat massif. Sand bodies like this commonly are associated with major cuspate forelands (Field and Duane 1974).

medium-grained, moderately sorted to well-sorted quartz sand having about 15% carbonate that is mostly bivalve fragments (Meisburger and Field 1975). Because these sediments are generally thin (1–2 m), underlying pre-Holocene stratigraphic units readily supply a variety of sediments to the modern environment, producing a patchwork mosaic of different surface-sediment facies.

Carbonate-Shelf Sector   The shelf from Miami to the Lake Worth area generally is only a few kilometers wide, and is distinguished by three (or sometimes four) linear rock ridges, each of which marks the seaward edge of a bedrock terrace (fig. 11.29). These rock-ridge and terrace couplets are progressively deeper (stepwise) in the seaward direction. The rock ridges themselves have less than 8 m of relief. A relatively thin veneer of sediment (0.5–4 m) covers the bedrock terraces betweeen the ridges. These ridges support a modern benthic community of alcyonarians, sponges, and scattered coral heads (Lighty et al. 1978).

The rock ridges were *Acropora palmata*-dominated barrier reefs that flourished during the early Holocene (Lighty et al. 1978). Radiocarbon dating indicates that these reef tracts terminated about 7,000 years ago. Lighty et al. (1978) postulated that exposed soil horizons were eroded as sea level rose, creating turbidity levels high enough to stress the reef community. Broad lagoons formed behind the reefs. The flow of cold water over the reef during winter frontal passages also contributed to reef demise (Lighty et al. 1978; Roberts et al. 1982; Wilson and Roberts 1992). The addition of nutrient material from soil erosion landward and aperiodic upwelling seaward probably played a role in reef termination (Hallock and Schlager 1986). The northernmost extent

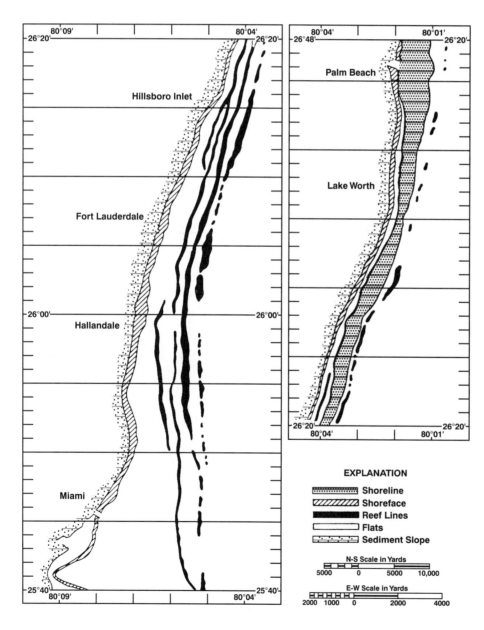

Figure 11.29. Extensive relict reefs lying offshore of the carbonate shelf off the southeast Florida coast. Four reef tracts exist in some places. Reefs extend northward beyond Lake Worth. These features were terminated in the Holocene (Duane and Meisburger 1969a).

of these reefs may be represented by the now-buried reef mass lying off Fort Pierce (Meisburger and Duane 1971).

Surface sediments generally consist of fragments of *Halimeda,* mollusks, benthic foraminifers, bryozoans, and corals. Very near the shoreface, in 3 to 10 m of water, there are sporadic, active sabellariid-worm reefs (Kirtley and Tanner 1968), which locally produce worm-tube debris. Nonskeletal carbonate components are pellets and ooids. Lithoclasts may be locally dominant, and are derived from the underlying Pleistocene Miami and Anastasia formations (Duane and Meisburger 1969b). This carbonate content contrasts with that of the siliciclastic shelf sector to the north, in that this southern sector has a more tropical assemblage (*Halimeda* and coral fragments), a lesser molluscan contribution, and no barnacle fragments.

Slope Below the shelf/slope break off eastern Florida is a depositional (progradational) surface known as the Florida-Hatteras Slope that extends to the floor of the Straits of Florida (Malloy and Hurley 1970; Popenoe et al. 1984). The Florida-Hatteras Slope is not significantly erosional south of Cape Hatteras; whereas, to the north, large submarine canyons formed as a result of massive late Pleistocene glacial-meltwater runoff and a large sediment influx that resulted in extensive mass wasting (Twichell et al. 1977; Farre et al. 1983; Riggs and Belknap 1988). As mentioned above, the most significant erosional feature of the East Florida Slope is the Miami Terrace (Mullins and Neumann 1979), which extends about 40 km north from Miami (fig. 11.1).

In the center of the northern Straits of Florida, Neumann et al. (1977) discovered enormous, hydrodynamically streamlined ahermatypic-coral reefs (fig. 11.30) that they called "lithoherms"—a name inspired by the pervasive and widespread submarine cementation. The lithoherms occur in water depths of 450 to 700 m, in an area 200 km long and 10 to 15 km wide, making this area one of the largest coral-reef tracts in the western Atlantic (Mullins 1983b), yet one completely hidden from normal view. Individual lithoherms are a few hundred meters long, 50 m high, and about 50 to 75 m wide. They are oriented into the strong (50–60 cm/s), north-flowing Florida Current. Living *Lophelia* sp. and *Enallopsammia* sp. occupy the tops of these mounds. As Newton et al. (1987) have indicated, these features were more actively accreting during the Pleistocene, for reasons not well understood.

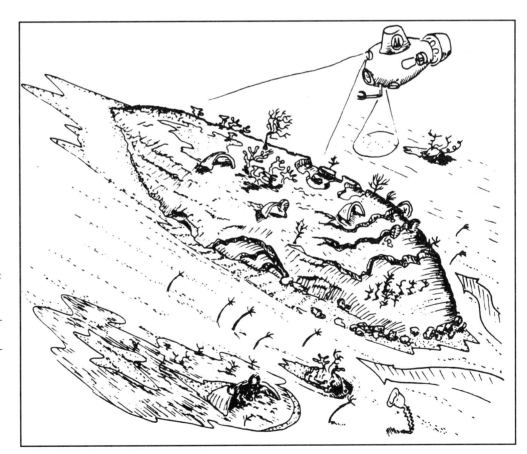

Figure 11.30. Fully developed lithoherm and small, incipient lithoherms in the Straits of Florida. These mounds have been built up from skeletal debris (mostly ahermatypic corals) and by widespread submarine cementation. Collectively, these features form a huge reef tract on the floor of this seaway in 450 to 700 m of water (Neumann et al. 1977).

Conclusion

The Florida Platform has experienced a dynamic geologic history. This history has been complicated by the platform's unique setting, separating two large water bodies, and it has been influenced by various tectonic, paleoceanographic, and paleogeographic factors: a changing major ocean-current system, the large landmass to the north, among others—all in the context of multiple eustatic sea-level fluctuations and global climate change.

As much as has been learned about the Florida Platform in the past decade, a great deal of Florida's geologic history is still hidden by and contained within this vast and thick carbonate rock mass. Only through acquisition of more regional seismic data, closely tied to well-studied boreholes, will there be substantial expansion of our understanding of Florida's geologic history.

12

Origin and Paleoceanographic Significance of Florida's Phosphorite Deposits

John S. Compton

The element phosphorus (P) is ubiquitous in the Earth's crust, but most sedimentary rocks contain small amounts, and total P contents rarely exceed 1% by weight. Large, economic phosphorite deposits that contain upwards of 37% P_2O_5 are unusual, and were preferentially formed during certain periods of Earth history (Cook and McElhinny 1979; Sheldon 1980; Arthur and Jenkyns 1981) (fig. 12.1). The Miocene phosphorite deposits of Florida and the southeastern U.S. continental margin together constitute one of these phosphorite "giants," with an estimated 10 billion tons of economic phosphorite (5×10^{13} moles P; Stowasser 1977). Non-economic phosphorite is more abundant than economic phosphorite (Riggs and Manheim 1988), and on the basis of studies of numerous cores from northern and southern Florida, total phosphorite is estimated to be at least an order of magnitude more abundant than economic phosphorite.

Understanding the origin of these unusually large phosphorite deposits is important, because P is a biolimiting nutrient, and changes in P cycling would have a large impact on the size, evolution, and productivity of the biosphere. In addition, the P and carbon (C) cycles are linked, such that changes in biologic productivity may have influenced atmospheric pCO_2, an important greenhouse gas. Formation and preservation of large phosphorite deposits are made possible by a relatively rare combination of tectonic setting, climate, sea level, and oceanic circulation. A better understanding of the origin of Florida's enormous phosphorite deposits will provide insight into the Miocene Epoch, a time of major climatic, tectonic, and oceanographic changes (Kennett 1982).

Phosphorite deposits largely similar to those in Florida occur throughout the Miocene of the southeastern U.S. continental margin in Georgia, South Carolina, and North Carolina (Riggs 1984; Scott 1988b) (fig. 12.2). The size and economic importance of the Miocene phosphorite deposits in Florida and North Carolina are shown by the fact that they supply nearly all of the U.S. domestic demand for P. These deposits contribute about 87% of total U.S. production, which has varied between 40 and 55 million tons annually over the last decade (Herring and Stowasser 1991). The processed phosphorite is primarily used in the manufacture of fertilizer products to increase agricultural yields. The phosphorite deposits of the southeastern U.S. are the largest known Miocene deposits, but smaller Miocene phosphorite deposits also occur on the California and Peru margins, Chatham Rise, South African shelf, Cuba, and elsewhere (Burnett and Riggs 1990; and others).

Large, economic phosphorite deposits are unusual in the rock record because formation of these deposits requires a sufficient flux of organic matter to the sediment to supply the P necessary to form phosphorite, and a mechanism to concentrate the phosphorite after it has formed. Phosphorite is used here to indicate a sedimentary rock that contains an economic amount of P (generally more than 15% P_2O_5 by weight). As is true of most sedimentary phosphorite, the bulk of the P in Florida's phosphorite deposits occurs as the carbonate-fluorapatite mineral francolite. Francolite is a diagenetic mineral that most commonly precipitates as a cryptocrystalline pore-filling cement during the very early stages of sediment burial, typically within tens of centimeters of the sediment/seawater interface. Francolite has the general formula $Ca_5(PO_4)_{3-x}(CO_3)_xF$, where $x \leq 0.75$; francolite also contains minor amounts of elements that substitute for Ca^{2+}, PO_4^{3-}, and F^-. Minor amounts of additional P can occur in association with organic matter, and as detrital apatite grains weathered from igneous, metamorphic, and sedimentary rocks. Other P-bearing minerals can occur in phosphorite deposits, the two most common in Florida being the

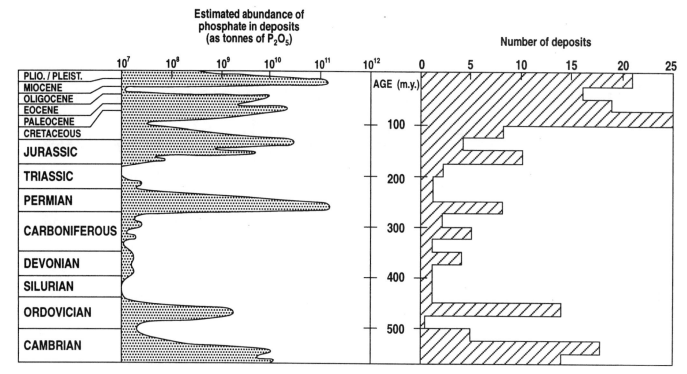

Figure 12.1. Estimated global abundance of P (metric tons of P_2O_5) and the number of phosphate deposits (averaged over 25-million-year intervals) throughout the Phanerozoic (Cook and McElhinny 1979; reprinted from *Economic Geology* 74 [1979]: 316).

hydrated aluminophosphate minerals crandallite and wavellite, but these minerals contribute a minor fraction of the total P, and represent alteration products of intensely weathered francolite.

A large flux of organic matter to sediment is limited to regions of upwelling ocean water, because P is very rapidly recycled. The P taken up by primary producers in the surface waters of the ocean is regenerated and released to the ocean as phosphate ion during its descent through the water column, or soon after it arrives on the ocean floor. To form phosphorite, the organic matter must make it to the sea floor, and the phosphate ion produced during regeneration of the organic P must precipitate as francolite in the form of intergranular cement or crusts of phosphorite. The phosphorite then must be concentrated by physical processes—such as bottom currents or shoreline migration across the shelf—in order to form enriched phosphorite deposits by removal of fine-grained sediment and organic matter. Phosphorite is currently forming in organic-rich deposits under areas of intense upwelling, as along the Namibia and Peru continental margins (Baturin et al. 1972; Veeh et al. 1973; Burnett 1977). These deposits provide a useful modern analog to phosphorite formation; however, these phosphate-bearing sediments would require further concentration by reworking to be considered economic deposits.

Given the importance of organic matter as a source of P in phosphorite formation, a conspicuous feature of large, economic phosphorite deposits is their general

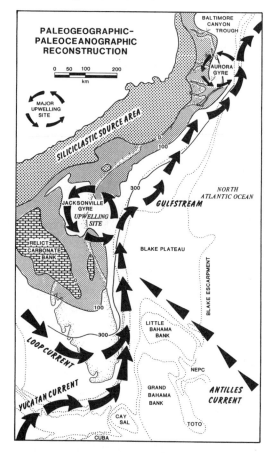

Figure 12.2. Paleogeographic and paleoceanographic reconstruction of the Miocene continental margin of the southeastern United States, showing currents and major upwelling sites (from Snyder et al., 1990). The fine stippled pattern represents areas of phosphorite deposition.

lack of organic matter, which results from its removal by oxidation during reworking. Phosphorites may therefore provide a proxy of organic-rich sediments on shallow-water continental shelves. Organic matter deposited on the shelf is highly susceptible to later oxidation or off-shelf transport during sediment reworking, particularly during marine regressions.

One of the major obstacles to understanding the complex history of reworked phosphorite deposits is the paucity of ages determined by biostratigraphy or magnetostratigraphy. The development of strontium-isotope chronostratigraphy over the last decade now makes it possible to determine the age of unaltered, authigenic Sr-bearing minerals with a precision of ±0.5 million years (Burke et al. 1982; Depaolo and Ingram 1985; Hodell et al. 1991). Besides the Sr that occurs in biogenic carbonates such as mollusk shells and foraminifer tests, Sr is incorporated into the minerals francolite and dolomite during their precipitation in the early burial diagenesis of sediments. The $^{87}Sr/^{86}Sr$ age of phosphorite and dolomite should be the same as the age of the sediment in which they originally formed, regardless of the extent of physical reworking—as long as the Sr-isotope ratio has not been altered since formation. Therefore, Sr isotopes provide an opportunity to date phosphogenic episodes and the timing of organic-rich sedimentation on the Florida Platform (Compton et al. 1993).

The timing of phosphogenesis and organic-carbon burial on the Florida Platform has global paleoceanographic significance, because the burial and subsequent removal of organic C from continental shelves globally may significantly impact the C cycle and, in particular, the CO_2 content of the atmosphere (Compton et al. 1990). The Miocene was an important epoch of Earth history because it was a time of large-scale buildup of ice on Antarctica that resulted in large and rapid changes in sea level (Kennett 1982). Sea-level fluctuations change the area of submerged continental shelf, which may in turn influence the rate of burial of organic matter. During warm periods with low ice volume, sea level and

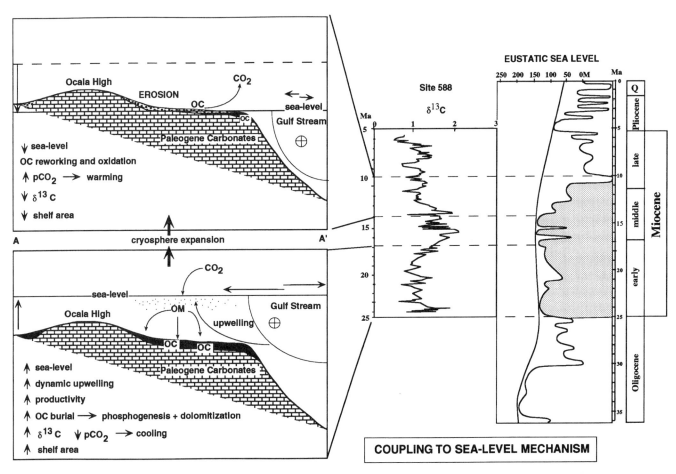

Figure 12.3. Schematic representation of postulated relationships among organic-C burial on the Florida Platform (continental margins in general), the $\delta^{13}C$ record of ODP Site 588 (Kennett 1986), and eustatic sea level (Haq et al. 1987a). The model indicates that the Early to Middle Miocene was a period of enhanced organic-C burial corresponding to a positive $\delta^{13}C$ shift. The overall rise in sea level increased meandering of the Gulf Stream, which resulted in more dynamic upwelling and a greater supply of nutrients to support an increase in primary productivity. According to the model, the Middle to Late Miocene was a period of enhanced organic-C oxidation associated with a decrease in $\delta^{13}C$ as sea level fell and shelf sediments were reworked.

submerged shelf area were relatively high, and CO_2 was biologically removed from the atmosphere and buried as organic matter in shelf sequences (fig. 12.3). An increased burial rate of isotopically light organic C would also result in a positive shift in the C-isotope composition of seawater (positive $\delta^{13}C$ shift). During cold, glacial periods, the buildup of ice caused sea level to fall, and organic-rich shelf sediments were exposed to reworking. Weathering of these sediments oxidized the organic C, and CO_2 was returned to the atmosphere. The increased oxidation of isotopically light organic C would result in a negative shift in the C-isotope composition of seawater. The rapid transfer of C between CO_2 in the atmosphere and organic matter in shelf sequences may have influenced global climate by varying the amount of atmospheric CO_2, a greenhouse gas. If this scenario is correct, then phosphorite formation should coincide with periods of rising or maximum sea level, and with a positive shift in the $\delta^{13}C$ of seawater. Conversely, sediment exposure and reworking should coincide with periods of falling or low sea level, and with a negative shift in the $\delta^{13}C$ of seawater. The integration of Sr-isotope data with lithostratigraphic and geochemical data on the phosphorite deposits of Florida will provide a test of this hypothesis by addressing two key questions: (1) Did the phosphorite form during the early diagenesis of organic-rich sediments? (2) Does phosphogenesis coincide with rising or maximum sea level, and with positive global $\delta^{13}C$ shifts?

Phosphorite in the Hawthorn Group of Florida

Most phosphorite-bearing sediments of Florida belong to the Hawthorn Group. The composition of the Hawthorn Group is highly variable, reflecting different depositional environments and repeated reworking of the sediment in response to sea-level fluctuations on the expansive shallow-water shelf (Riggs 1979a, 1979b, 1984; Scott 1988b; see chapter 5). The lithologies of the Hawthorn Group include moldic limestone; dolomite; friable, clayey dolosilt and quartz sand; and clay. Compositional end-members are rare; most lithologies contain a wide range of grain sizes and mineralogies, but most lithologies contain some amount of phosphorite. Depositional environments ranged from fluvial and lacustrine on shore, to estuarine, lagoonal, beach, and shallow-water shelf seaward. The biota and energy of these environments varied considerably. Complex lithologies resulted from mixing of different types of sediment as depositional environments shifted laterally in response to changes in sea level. Marine transgressions and regressions are marked by changes in lithofacies, erosional unconformities, and condensed intervals.

The Hawthorn Group extends into Georgia and South Carolina and is largely contemporaneous with the Pungo River Formation of eastern North Carolina (fig. 12.4). Phosphorite deposits also occur on the present-day shelf of the southeastern Atlantic Bight and the adjacent Blake Plateau. The Hawthorn Group of northern Florida is divided into several formations (Scott 1988b). The basal Penney Farms Formation lies unconformably on Oligocene or Eocene carbonates and consists of interbedded phosphatic dolomite, quartz sand, and clay. The overlying Marks Head and Coosawhatchie formations are a complex mixture of phosphatic carbonates, sand, and clay. The Statenville Formation contains economic deposits of phosphatic sand and clay, and fewer carbonate beds. The Hawthorn in central and southern Florida is generally less complex than in northern Florida and is divided into a lower carbonate unit and an upper clastic unit (Riggs 1979b). The carbonate-rich Arcadia Formation lies unconformably on Oligocene or Eocene limestones; it is predominantly limestone and dolomite, with occasional, laterally discontinuous clastic-rich beds. The carbonate beds contain variable amounts of quartz and phosphorite. Chert occurs in places. The unconformably overlying clastic unit (Peace River Formation) consists of quartz sand, clay, and highly variable amounts of carbonate and phosphorite. The extent of phosphorite deposits on the eastern and western Florida shelf is unknown.

The age of the Hawthorn Group has been difficult to determine, because much of it has been reworked, and because of its paucity of age-diagnostic fossils. In northern Florida, the age is tentatively considered to range from early to mid-Miocene on the basis of limited faunal evidence and correlations with more-fossiliferous sediments in southeastern Georgia (Scott 1988b). In southern Florida, the age of the Arcadia Formation is thought to range from earliest Miocene to no younger than mid-Burdigalian (late Early Miocene) on the basis of limited faunal evidence and physical correlations. Foraminifers, diatoms, mollusks, and vertebrate faunas suggest an age for the Peace River Formation of early Middle Miocene (early Langhian) to latest Miocene or early Pliocene (Webb and Crissinger 1983; Scott 1988b).

Sediment distribution within the Hawthorn Group in Florida is related to the structural framework of the Florida Platform (Riggs 1979a, 1979b). Several positive features—such as the Ocala High, the Sanford High, and the St. Johns, Central Florida, and Brevard platforms

EASTERN NORTH CAROLINA
EASTERN SOUTH CAROLINA
SE AND E GEORGIA
NW. FLA. AND SW GA.
NORTHERN FLORIDA
SOUTHERN FLORIDA
PLIOCENE
YORK TOWN FM.
RAYSOR / YORK TOWN FMS.
CYPRESSHEAD FM. / DUPLIN
MICCOSUKEE FM. / CITRONELLE FM.
CYPRESSHEAD FM. / NASHUA FM.
TAMIAMI FM.
PLIOCENE
MIOCENE
UPPER
MIDDLE
LOWER
REWORKED SEDIMENT
PUNGO RIVER FM.
COOSAW- HATCHIE FM.
COOSAW- HATCHIE FM.
HAWTHORN GROUP
STATENVILLE FM.
COOSAW- HATCHIE FM.
BONE VALLEY MBR.
PEACE RIVER FM.
MARKS HEAD FM.
MARKS HEAD FM.
TORREYA FM.
MARKS HEAD FM.
ARCADIA FM.
NOCATEE MBR.
TAMPA MBR.
PARACHUCLA FM.
PARACHUCLA FM.
CHATTA- HOOCHEE AND ST. MARKS FM.
PENNY FARMS FM.
UPPER
MIDDLE
LOWER
MIOCENE
OLIGOCENE
RIVER BEND FM.
COOPER FM.
SUWANNEE LS.
SUWANNEE LS.
SUWANNEE LS.
SUWANNEE LS.
OLIGOCENE
EOCENE
COOPER FM.
OCALA LS.
OCALA LS.
OCALA LS.
OCALA LS.
UPPER
CASTLE HAYNE
SANTEE LS. FM.
SANTEE LS.
AVON PARK FM.
AVON PARK FM.
AVON PARK FM.
AVON PARK FM.
MIDDLE

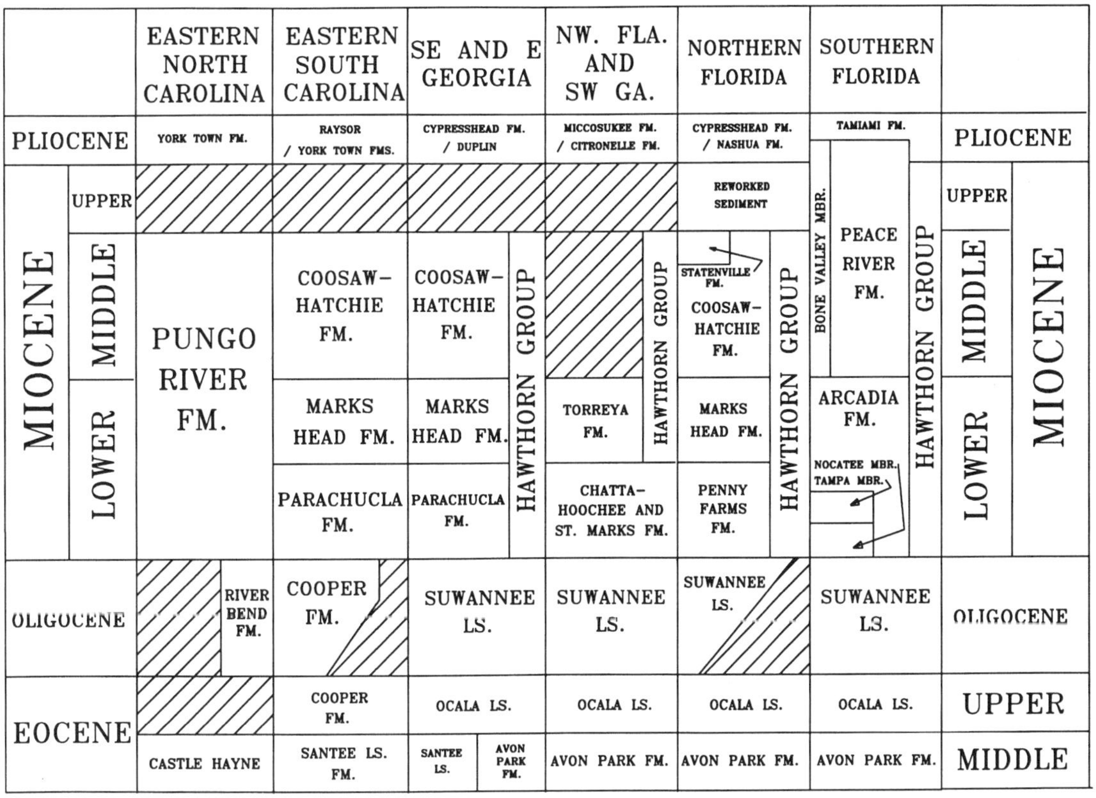

Figure 12.4. Stratigraphic correlations of the southeastern U.S. coastal plain (Scott 1988b).

(fig. 12.5)—influenced this distribution. Most of the economic phosphorite deposits currently being mined are located on the Central Florida Platform. The deepest Miocene depocenters were the Jacksonville and Okeechobee basins. Geologic conditions in Florida changed dramatically at the end of the Oligocene, when the previously isolated carbonate platform was inundated with terrigenous sediment from the North American continent after filling of the Gulf Trough (see chapter 4). The amount of terrigenous material is greatest in the north, and tends to decrease toward the southern end of the platform. The Kissimmee Saddle may have restricted transport of terrigenous sediment into the southern region (Riggs 1979a, 1979b). The amount of carbonate tends to increase to the south.

The mineralogy of the Hawthorn Group is related to grain size. Pebble-sized grains (>4 mm) are composed primarily of cryptocrystalline to microcrystalline francolite and dolomite, and contain variable amounts of included quartz, feldspar, pyrite, clay minerals, and organic matter. Pebbles occur in sandy gravels or in dolomite-cemented beds. Dolomitic pebbles and dolomite-cemented beds commonly are bored. Sand-sized grains (.064–4 mm) are predominantly quartz with variable amounts of phosphorite, calcite shell fragments or foraminifers, and feldspar. The silt-sized fraction (.002–.064 mm) is composed of dolomite, calcite, quartz, and, in places, the zeolite

SOUTHEAST GEORGIA EMBAYMENT
JACKSONVILLE BASIN
122
90
ST. JOHNS PLATFORM
BLAKE PLATEAU
OCALA HIGH
SANFORD HIGH
BREVARD PLATFORM
N
W
E
S
FLORIDA PLATFORM
90
122
180
OKEECHOBEE BASIN
Central Florida Phosphate District
0 40 80
kilometers
Babcock Deep Core
90
200 meters

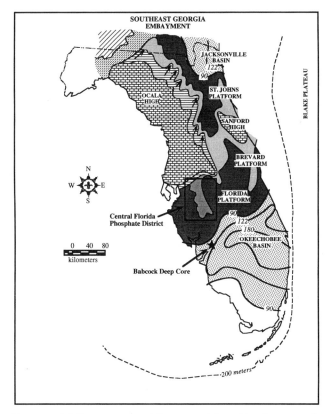

Figure 12.5. Miocene structural features and depositional pattern of the Florida Platform, showing thickness of Miocene sediments in meters (Riggs 1979a) and locations of the Central Florida Phosphate District and the Babcock Deep core (reprinted from *Economic Geology* 74 [1979]: 290).

mineral clinoptilolite. Biogenic calcite occurs as shell material, much of which has recrystallized to moldic limestone or dolomite. Carbonate in northern Florida is predominantly dolomite, whereas in southern Florida it consists of a mixture of calcite and dolomite (Prasad 1985). The clay-sized fraction (<.002 mm) is composed of variable amounts of mixed-layer illite/smectite, palygorskite, and sepiolite. Kaolinite and mica (discrete illite, muscovite, and biotite) are generally restricted to the uppermost part of the Hawthorn Group. Diatom frustules (opal-A silica) and diagenetic opal-CT and quartz chert occur in places.

Diagenetic minerals constitute a significant proportion of Hawthorn sediment. These minerals formed during early burial and remained at relatively low temperatures (<50°C) because they were never buried very deeply (several hundred meters maximum), and because of the low geothermal gradient in Florida. However, interpretation of the original sediment and its burial diagenesis is complicated by reworking and physical mixing of sediments from different environments. Sediment was influenced by later subaerial and supergene weathering, as much of the southeastern U.S. continental shelf was exposed during the low sea-level stand of the Middle to Late Miocene. Subaerial and supergene weathering can significantly alter the original composition of the sediment by oxidation of organic matter and pyrite, by removal of carbonate from francolite to form fluorapatite and (eventually) aluminum-phosphate minerals, by solution and recrystallization of the silica and carbonate minerals, and by conversion of palygorskite to smectite and smectite to koalinite in intensely weathered horizons (Altschuler et al. 1963; McClellan et al. 1985; Van Kauwenbergh et al. 1990). These weathering reactions occur at different rates: oxidation of organic matter and Fe-sulfides is most rapid, whereas alteration of francolite to fluorapatite and smectite to kaolinite occurs only in the most intensely weathered sections.

Origin of the Phosphorite

The variety of phosphorite grain types in the Hawthorn Group suggests that the phosphorite had several different origins. In the case of skeletal phosphorite grains, the precursor is obvious, and in the case of phosphorite fossil molds of gastropods and bivalves the locus of phosphorite formation is obvious. What is not obvious for these grains is the geochemical environment and timing of phosphorite formation. In addition, most phosphorite grains in the Hawthorn Group sediments are peloids. Peloids are defined here as rounded to subrounded grains that lack external or internal structures that might indicate their origin. The composition of the original sediment and the depositional environment in which the phosphorite formed are uncertain, because most phosphorite grains are no longer associated with the sediment in which they formed, but have been reworked into younger deposits. Therefore, determining the origins of the various phosphorite grains requires an integrated consideration of their texture, mineralogy, and elemental and isotopic compositions.

PHOSPHORITE GRAIN TYPES

The distribution of phosphorite grain types is highly variable, but in general phosphatic clasts and skeletal phosphorite tend to be most abundant near the Ocala High, whereas on the Sanford High and in the outer basins peloidal phosphorite is predominant (Riggs 1979a). For example, most phosphorite in cores from northeastern Florida (Jacksonville Basin) and southeastern Florida occurs as sand- to pebble-sized peloidal grains. Peloidal phosphorite is common near the Ocala High as well, but pebble- and cobble-sized clasts and skeletal bone material can be abundant. The peloidal grains are commonly rounded, with dark, polished surfaces (fig. 12.6). The interior of phosphorite peloids is black, gray, or tan, and consists of massive, cryptocrystalline francolite; however, euhedral to subhedral (hexagonal) francolite crystals as much as several micrometers in diameter can occur locally in the void spaces of some grains (fig. 12.7). The peloids lack nuclei, and have no internal structure. The phosphorite is tightly cemented and hard (apatite has a relative hardness of 5 on the Mohs scale). The phosphorite peloids also contain variable amounts of quartz, feldspar, dolomite, pyrite, organic matter, and the clay minerals palygorskite, sepiolite, and mixed-layer illite/smectite. Molds of diatom frustules (opal-A silica) have been observed in several peloidal phosphorite grains (fig. 12.7).

The origin of the peloidal grains is uncertain, but they may represent phosphatized fecal pellets, rounded rip-up clasts of phosphorite crusts, rounded foraminifer and bivalve molds, or pore-filling nodules. The mineralogic similarity of HCl-insoluble residues of the peloidal phosphorite grains (francolite readily dissolves in HCl) to the Hawthorn Group sediments in general suggests that some of the francolite precipitated as pore-filling cement that included surrounding sediment grains. The small amount of HCl-insoluble residue in the phosphorite (4–10%) is consistent with early cementation in high-porosity, near-surface sediments. The organic-matter content of HCl-insoluble residues of the peloids is

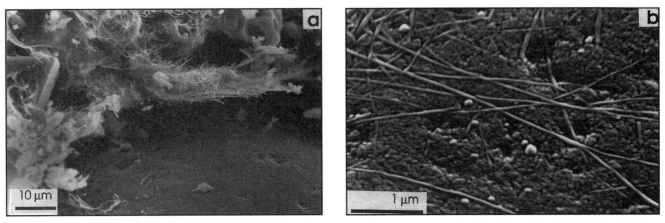

Figure 12.6. SEM photomicrographs of a phosphorite sample from the Babcock Deep core. A: Pitted surface of a polished peloidal phosphorite grain with surrounding fibrous Mg-rich clay and dolomite rhombs. B: Close-up of peloidal surface showing granular crystalline texture and fibrous clay.

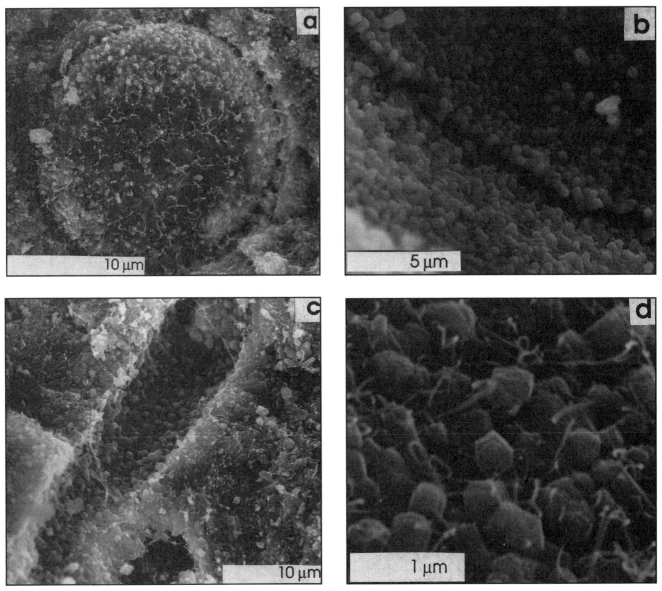

Figure 12.7. SEM photomicrographs of the interior of a phosphorite peloid from the Babcock Deep core. A: Diatom mold. B: Close-up showing small hexagonal francolite crystals. C: Void space in the phosphorite peloid. D: Close-up showing subhedral hexagonal francolite crystals partially coated with fibrous clays.

greater than that of the surrounding sediment, which indicates that the original organic-matter content of the host sediment was higher than that of the current host sediment, and that complete cementation by pore-filling francolite protects the organic matter from oxidation.

Other types of phosphorite grains include molds of gastropods and bivalves, skeletal grains (vertebrate teeth and bones), and phosphorite crusts. The abundance of phosphorite molds suggests that the microenvironment of shell interiors was particularly conducive to francolite precipitation, probably during early burial. Metastable aragonite and high-Mg calcite are highly susceptible to dissolution during early diagenesis (Rude and Aller 1991; and others). The dissolution of abundant metastable biogenic carbonate minerals should promote francolite precipitation by providing a source of calcium and carbonate ions (Ames 1959).

ASSOCIATED DIAGENETIC MINERALS

Diagenetic minerals commonly associated with francolite in phosphorite grains include pyrite, dolomite, and the fibrous clay minerals palygorskite and sepiolite (fig. 12.8). Pyrite, dolomite, and phosphorite appear to have formed in the same diagenetic environment. The inclusion of pyrite in many of the phosphorite grains supports the isotopic evidence (see below) for francolite precipitation within the sulfate-reduction zone. It is unlikely that the included pyrite is reworked from older sediment, because fine-grained pyrite would be rapidly oxidized during reworking. Laboratory experiments indicate that Mg-ion can inhibit precipitation of francolite (Martens and Harris 1970; Nathan and Lucas 1976). Aragonite dissolution may decrease the effects of Mg inhibition by decreasing the pore-water Mg/Ca ratio (Bentor 1980). In addition, pore-water Mg concentrations are lowered by precipitation of dolomite and Mg-rich clay minerals.

Several environments of formation of the palygorskite and sepiolite have been postulated (Weaver and Beck 1977; Strom and Upchurch 1985; McClellan and Van Kauwenbergh 1990), but the origin of these clay minerals in relation to phosphorite formation is uncertain. The early diagenetic environment indicated during deposition of the Hawthorn Group—namely, the presence of metastable biogenic carbonate, reduction of sulfate ion, and increased carbonate alkalinity—should promote dolomite precipitation (Baker and Kastner 1981; Compton 1988). The presence of dolomite in some phosphorite grains and the similar $^{87}Sr/^{86}Sr$ ratios of associated dolomite and francolite suggest that they formed at roughly the same time. Most of the dolomite within phosphorite grains appears to be in situ rather than derived from reworking of older deposits.

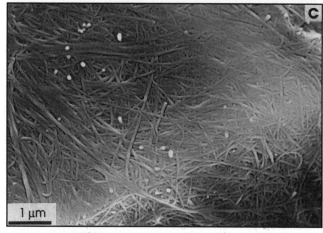

Figure 12.8. SEM photomicrographs of samples from the Babcock Deep core. A: Interior void space of phosphorite grain filled with euhedral dolomite crystals. B: Individual dolomite rhombs coated with fibrous clay; C: Close-up of matted texture of fibrous palygorskite/sepiolite clay minerals.

ISOTOPIC EVIDENCE

It is important to understand several key factors bearing on phosphogenesis: the original sedimentary environment, the subsequent physical-reworking history, and the extent of chemical alteration. In cores from northeastern and south-central Florida, most of the phosphorite appears to be reworked, and not in situ. Because the original sediment in which the phosphorite formed is generally not preserved on the Florida Platform, we must rely on the elemental and isotopic composition of the phosphorite to infer its sedimentary environment of formation. Substitution of carbonate for phosphate in the francolite structure is accompanied by substitutions of Na, Mg, and Sr for Ca, and excess amounts of F (Mc Clellan and Lehr 1969; McArthur 1985; McClellan and Van Kauwenbergh 1990). In addition, some sulfate can substitute for phosphate in the francolite structure. The amount of these substitutions and the C, O, S, and Sr-isotopic compositions can be used to infer the origin, age, and extent of alteration of the phosphorite (Longinelli and Nuti 1968; Kolodny and Kaplan 1970; McArthur 1985; McArthur et al. 1986, 1990).

Carbon and sulfur isotopes can be used to differentiate phosphorite formation among the successive oxic, suboxic, and anoxic (sulfate-reducing) burial environments (Nathan and Nielsen 1980; Benmore et al. 1983; McArthur et al. 1986). During the early burial degradation of organic matter by microorganisms the metabolism of isotopically light organic C to CO_2 results in an increasingly negative $\delta^{13}C$ of pore-water carbonate from the sediment/seawater interface down to the base of the sulfate-reduction zone. Sulfate-reducing bacteria in the anoxic zone preferentially utilize isotopically light sulfate, such that the remaining pore-water sulfate becomes increasingly heavy with depth in the anoxic zone. Depleted S-isotopic values can occur at the suboxic/anoxic interface, where isotopically light, reduced S, diffusing out of the underlying sulfate-reduction zone as H_2S, is oxidized to sulfate by bacteria. The depth to the suboxic and anoxic zones varies, but the sulfate-reduction zone commonly occurs at shallow burial depths (<1m) in organic-rich sediment (Martens et al. 1978; and others).

The C- and S-isotopic compositions of substituted carbonate and sulfate ions in the francolite indicate that the predominantly peloidal phosphorite grains formed during early diagenesis of organic-rich sediment, probably within a meter of the sediment/seawater interface (fig. 12.9). The negative C-isotopic composition of the carbonate in these francolites indicates that precipitation occurred between the sediment/seawater interface and the uppermost part of the sulfate-reduction zone (Claypool and Kaplan 1974). The sulfate in most of the francolite from the Babcock Deep core has a S-isotopic composition that is depleted relative to the average Miocene seawater value of +21.75‰ (Burdett et al. 1989). These depleted S-isotopic values suggest that the phosphorite formed at the suboxic/anoxic interface, where isotopically light, reduced S, diffusing out of the underlying sulfate-reduction zone, is oxidized to sulfate by bacteria. Mats of the S-oxidizing bacterium *Beggiatoa* often occur at the sediment/seawater interface in areas of intense upwelling, and are important to S recycling in the sediment (Reimers 1982; Jorgensen 1983). In addition, degradation of these bacterial mats should enhance phosphorite formation because of their relatively high P content (Reimers et al. 1990). It has been suggested that certain rod-shaped structures in Florida phosphorite are

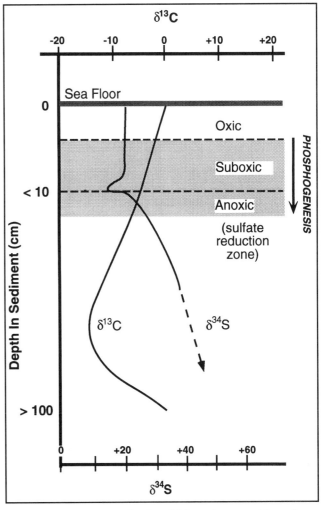

Figure 12.9. Variations in the C- and S-isotopic compositions of pore-water bicarbonate and sulfate ions with depth of burial (McArthur et al. 1986). The isotopic compositions of substituted carbonate and sulfate in francolite from phosphorite grains in the Hawthorn Group indicate that the francolite precipitated within the suboxic to anoxic zones of the sediment, near the sediment/seawater interface (shaded).

fossilized bacteria (Riggs 1979b). Similar depleted S-isotopic compositions are reported for phosphorite from the Cape Province of South Africa, and for the Namibian and South African shelf (McArthur et al. 1986).

The francolite sulfate in some phosphorite samples has an S-isotopic composition slightly more positive than that of average Miocene seawater. These phosphorite samples may have precipitated in the uppermost sulfate-reduction zone, where residual sulfate becomes heavier as lighter sulfate is removed preferentially by sulfate-reducing bacteria (Kaplan et al. 1963; Goldhaber and Kaplan 1974).

A comparison of the C- and O-isotopic compositions of Florida francolite with those of francolite from phosphorite deposits where the origin of the phosphorite is better established supports this evidence for an early burial origin (fig. 12.10). The C-isotopic composition is particularly useful for comparison, because it is less susceptible than the O-isotopic composition to later diagenetic alteration. The C- and O-isotopic compositions of francolite carbonate from the Babcock Deep core and the Jacksonville Basin are similar to values reported for onshore and offshore phosphorite deposits in Morocco (McArthur et al. 1986) and onshore deposits from Cape Province, South Africa (Birch 1979; Dingle et al. 1979) (fig. 12.10). The francolite carbonate has a similar negative C-isotopic composition and a lighter O-isotopic composition compared to phosphorite now forming in the uppermost meter of organic-rich sediment on the Peru-Chile and Namibian margins (Burnett 1977; Baturin 1982; Glenn et al. 1988). The lighter O-isotopic values of the Florida phosphorite may reflect formation in warmer waters, or minor meteoric alteration (see below). The C-isotopic compositions are also similar to those measured for in situ phosphorite from organic-rich rocks of the Miocene Monterey Formation (fig. 12.10). The lighter O-isotopic composition of the Monterey phosphorite may reflect its recrystallization at higher burial temperatures (significantly greater burial depths) (Mertz 1984; Kastner et al. 1990).

In summary, C- and S-isotopic analyses indicate that peloidal phosphorite grains in the Hawthorn Group formed during early diagenesis of organic-rich sediments, at or near the sediment/seawater interface, within the suboxic to anoxic zones. The isotopic values are similar to those of present-day analogs of phosphorite forming in similar environments on the Namibian and Peruvian margins. In addition to this isotopic evidence for an early diagenetic origin, significant francolite precipitation is generally considered to be limited to extremely shallow burial depths (uppermost tens of centimeters), because the pore-water phosphate-ion maximum usually occurs just below the sediment/seawater interface (Martens et al. 1978; and others), and the supply of fluoride ion can be limited by diffusion from the overlying seawater (Froelich et al. 1983).

PHYSICAL REWORKING OF THE PHOSPHORITE

The common association of rounded, sand-sized peloidal phosphorite grains in clayey, quartz sands, dolosilts, and moldic limestones with low organic matter contents suggests that much of the phosphorite that formed during early diagenesis in organic-rich sediment was reworked. Rounded, sand-sized phosphorite grains within phosphorite pebbles indicate that some phosphorite grains experienced multiple formation and reworking events. The uniform polish of spiral phosphorite gastropod molds suggests that the highly polished surfaces of most sand-sized phosphorite grains may reflect an original texture more than physical abrasion by detrital quartz grains in high-energy environments. Physical abrasion alone of a spiral-shaped phosphorite mold would be expected to polish only the exposed, outer surfaces, rather than uniformly polish all surfaces. In some samples, such as phosphorite crusts on dolomitic pebbles and beds, the phosphorite appears to be in situ, but many of these crusts occur on or as intraclasts. The bored and phosphorite-encrusted surfaces of many of the dolomite-cemented pebbles and beds indicate that they were exposed at the sediment surface. Dolomite clasts probably formed from

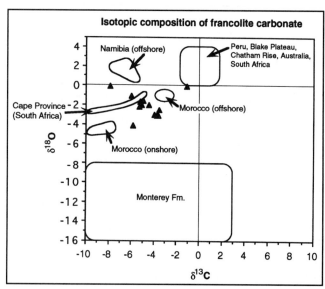

Figure 12.10. C- and O-isotopic compositions of francolite carbonate from the Hawthorn Group of northeast and south-central Florida (triangles) compared with compositions for other phosphorite samples (Mertz 1984; McArthur et al. 1986; Kastner et al. 1990).

boring of the upper and lower surfaces of dolomite beds that formed at shallow burial depths and were later exhumed by erosion of the overlying sediment. $^{87}Sr/^{86}Sr$ ages (see below) suggest that the amount of displacement during reworking varied considerably, and that it is related to grain size. Coarse-grained dolomite and phosphorite gravels usually have relatively young $^{87}Sr/^{86}Sr$ ages, suggesting that the gravel was reworked locally, whereas fine-grained phosphorite sands can be significantly older than the sediment in which they occur, suggesting that they were probably transported well beyond the site where they originally formed.

A schematic summary of the postulated early burial diagenesis of the Hawthorn Group and the results of sediment reworking are shown in figure 12.11. Deposition and early diagenesis of the original organic-rich Hawthorn sediment occurred during periods of rising or high sea level, when dissolved phosphate in upwelling waters supported high surface productivity that supplied organic matter to sediments. Bacterial degradation of the organic matter during early burial released phosphate to the pore water, where it diffused to sites of francolite precipitation in shell interiors and formed pore-filling peloids (fig. 12.11a). Sulfate diffused into the sediment in response to its removal by sulfate-reducing bacteria. The reduced sulfide combined with Fe to form monosulfides and pyrite, or it diffused to overlying *Begiattoa* mats and was re-oxidized to sulfate. Removal of sulfate, increase in carbonate alkalinity resulting from sulfate reduction, and dissolution of biogenic aragonite and high-Mg calcite promoted precipitation of dolomite at sediment depths below and somewhat overlapping the zone of phosphogenesis. The Mg required for dolomite precipitation was supplied by diffusion. Reworking of this sediment in migrating high-energy beach environments occurred during periods of falling or low sea level. The well-mixed, shallow water column supplied abundant oxygen to the sediment and supported an active infauna. The section was condensed and phosphorite was concentrated; most of the organic matter and pyrite were rapidly removed by oxidization; the cemented dolomite beds were exhumed, broken up, and heavily bored; and much of the unconsolidated, fine-grained dolosilt and clay were winnowed to deeper water.

CHEMICAL ALTERATION OF THE PHOSPHORITE

The extent of chemical alteration of the phosphorite can be determined from the elemental and isotopic

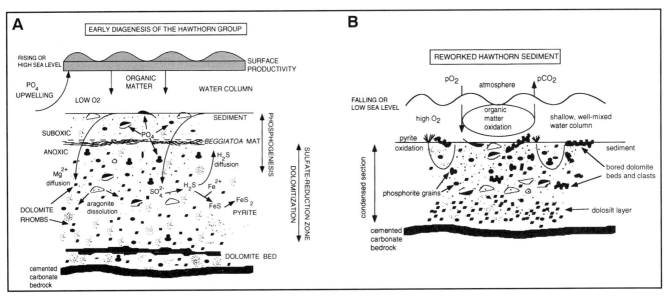

Figure 12.11. Schematic of early burial diagenesis and reworking of sediment on the Florida Shelf during the Miocene. A: Phosphogenesis occurs in organic-rich sediment deposited as a result of increased upwelling and surface productivity during periods of rising or high sea level. Early diagenesis results in the release of phosphate ion and precipitation of pore-filling francolite in the sediment to form nodular phosphorite grains or phosphorite interior molds of foraminiferal tests and bivalves. Sulfate reduction in the anoxic zone results in precipitation of pyrite and dolomite rhombs and cemented dolomite clasts and beds at depth in the sediment. The S-oxidizing *Beggiatoa* mat occurs at the suboxic/anoxic interface. B: After the above organic-rich sediment is reworked, much of the organic matter and pyrite have been removed by oxidation, and the higher bottom-water O concentrations support a burrowing and encrusting fauna that bore exhumed dolomite beds and clasts. Dolomite rhombs become concentrated into dolosilt layers, phosphorite grains become concentrated with sand grains, and much of the fine-grained clay is winnowed to deeper water.

compositions of the francolite. It is important to evaluate the amount of chemical alteration, because much of the phosphorite is reworked and occurs in near-surface deposits that may have been exposed to groundwaters. Chemical alteration during weathering of phosphorite by meteoric waters results in loss of carbonate, Sr, and other substituted ions (McArthur 1978, 1985). Whereas much of the phosphorite from surface mines is highly weathered, most of the dark phosphorite grains in core samples from the deeper basins do not appear to be significantly altered. For example, francolite of dark, peloidal phosphorite grains in the Babcock Deep core has a fairly uniform, highly substituted average composition of $(Ca_{4.61}Na_{0.22}Mg_{0.14}K_{0.05}Sr_{0.02})(PO_4)_{2.23}(CO_3)_{0.62}(SO_4)_{0.12}(F)_{1.04}$. The CO_3, SO_4, Na, Mg, and Sr contents of this francolite are similar to or exceed the maximum amounts reported for other sedimentary francolite (McClellan and Lehr 1969; McArthur 1978, 1985; McClellan 1980) and suggest that the francolite has undergone minimal chemical alteration. The polished, tightly cemented peloidal phosphorite grains are hard, and apparently resistant to chemical alteration or physical breakage during weathering and reworking. The color of phosphorite grains may provide an approximate guide to the extent of alteration. In surface mines, the uppermost phosphorite is commonly a dull, chalky white and the francolite has lost much of its carbonate in altering from a carbonate-fluorapatite to a fluorapatite. These dull white grains commonly contain the Al-phosphate minerals wavellite and crandallite. Surface mines also contain gray, tan, brown, and orange grains. Grains of these various colors are far less common in cores from the deeper basins.

The O-isotopic composition of the francolite carbonate may provide additional evidence of possible alteration by meteoric waters (McArthur et al. 1986). The somewhat depleted O-isotopic values of francolite CO_3 from the Babcock Deep core relative to Miocene seawater may indicate relatively high temperatures (possibly as high as 35°C; Kolodny et al. 1983), disequilibrium (Glenn et al. 1988), or meteoric alteration (McArthur et al. 1986). It is conceivable that ambient Miocene seawater temperatures may have approached 35°C; water temperatures as high as 32°C are common during the summer months in shallow-water environments along the west coast of Florida. Therefore, despite evidence of physical reworking, the large number of substituted ions—taken together with the only slightly negative O-isotopic values—indicates that meteoric alteration of these black to gray peloidal phosphorite grains was minor.

Age of the Phosphorite

Variations in the Sr-isotopic composition of seawater have been used to determine ages of Sr-bearing marine minerals (Burke et al. 1982; DePaolo and Ingram 1985). A detailed $^{87}Sr/^{86}Sr$ curve for Miocene seawater has been constructed from sediments obtained at Deep Sea Drilling Project Site 588 in the southwest Pacific (fig. 12.12); the Sr-isotopic composition of calcareous benthic foraminifers was measured, and the age of the fora-

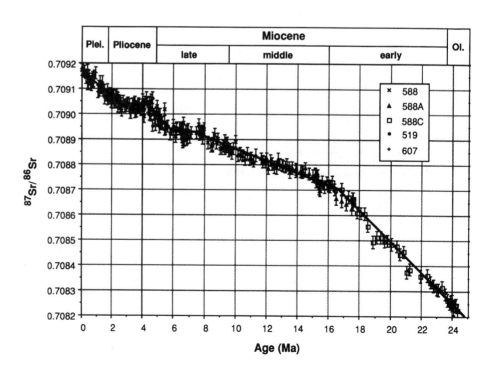

Figure 12.12. Age versus $^{87}Sr/^{86}Sr$ ratios of planktic foraminifers recovered from DSDP sites 588, 519, and 607 from 24 million years ago to the present (Hodell et al. 1991). Error bars represent 95% confidence limits.

minifers was determined independently by biostratigraphy and magnetostratigraphy (Hodell et al. 1991). The age of an Sr-bearing mineral that formed in equilibrium with seawater and whose initial $^{87}Sr/^{86}Sr$ ratio has not been altered can be determined by matching the ratio with that of seawater, obtained from constructed curves like that shown in figure 12.12; the resolution achievable with the $^{87}Sr/^{86}Sr$ ratio alone is highly variable, and depends on the steepness of the seawater curve. For example, during the Early to Middle Miocene (from 24 to 16 million years ago) the seawater $^{87}Sr/^{86}Sr$ ratio increased rapidly, and ages can be determined within ±0.74 million years, whereas from 8 to 6 million years ago the curve is essentially flat, and $^{87}Sr/^{86}Sr$ ages are indistinguishable over this 2-million-year period (Hodell et al. 1991).

$^{87}Sr/^{86}Sr$ AGES

Preliminary results indicate that $^{87}Sr/^{86}Sr$ ages can be extremely useful in working out the complex diagenetic and depositional history of the phosphorite deposits on the Florida Platform. However, several questions must be addressed before Sr-isotopic ratios can be unambiguously interpreted: Is it reasonable to assume that the measured $^{87}Sr/^{86}Sr$ ratio reflects that of seawater at the time of phosphorite formation? How is it determined that the original $^{87}Sr/^{86}Sr$ ratio has not been altered since phosphorite formation? What age resolution is possible with Sr isotopes? How much variation in age is there among the different phosphorite grain types and sizes from the same sample?

The Sr-isotopic ratio of the phosphorite is interpreted to represent the Sr-isotopic ratio of seawater at the time of phosphorite formation. It is assumed that the Sr incorporated in francolite was in isotopic equilibrium with pore-water Sr, as documented for Holocene francolite from the Peruvian and South African continental margins (McArthur et al. 1990). Pore-water Sr is most likely derived from seawater originally buried with the sediment and Sr released from dissolution of carbonate minerals. Aragonite, which has a high Sr content, is probably the largest source of pore-water Sr, because it is unstable and highly susceptible to dissolution. The isotopic compositions of Sr from pore waters and Sr from the dissolution of buried carbonate minerals should be the same and should reflect the seawater Sr value at the time of deposition. The amount of Sr from significantly older, reworked aragonite should be minor, because aragonite is highly soluble and unlikely to survive multiple reworking events. The Sr-isotopic composition of pore water should not change significantly during early burial diagenesis (McArthur et al. 1990).

The possibility that the $^{87}Sr/^{86}Sr$ ratio of the francolite has been altered during reworking and weathering can be evaluated by determining the elemental and isotopic compositions of the francolite. As discussed above, in the case of the dark peloidal phosphorite grains, the highly substituted composition of the francolite and the only slightly negative O-isotopic composition of its substituted carbonate indicate that the francolite has undergone minimal chemical alteration. The lack of chemical alteration strongly suggests that the $^{87}Sr/^{86}Sr$ ratio has not been altered since the francolite formed. Increased confidence in the $^{87}Sr/^{86}Sr$ ages stems from the fact that the phosphorite ages become increasingly younger up-section, as predicted by the Neogene seawater curve. The $^{87}Sr/^{86}Sr$ ages of the phosphorite are also consistent with the available biostratigraphic ages. None of the obtained ratios fall outside the Late Oligocene to Late Miocene seawater values. Weathering should result in loss of Sr, but the effect of weathering on the Sr-isotopic ratio is unknown. Leaching of Sr from the francolite should not affect the Sr-isotopic ratio (the slight difference in mass between ^{86}Sr and ^{87}Sr should not affect their rate of leaching from francolite). However, solution and reprecipitation of francolite should result in an Sr-isotopic ratio in equilibrium with the Sr-isotopic ratio of the solution during recrystallization. Recrystallization of francolite from groundwaters should result in markedly different and geologically unreasonable $^{87}Sr/^{86}Sr$ ages.

Caution must be exercized in analyzing the $^{87}Sr/^{86}Sr$ ratio of phosphorite, because Sr-bearing minerals other than francolite, such as dolomite and calcite, are present. Treatment of powdered phosphorite samples with Na acetate acetic acid buffer at a pH of 5 will preferentially remove most of the calcite at room temperature, and dolomite at 60°C. Strontium contributed by minor amounts of other HCl-soluble minerals should be insignificant because of the high Sr content and predominance of francolite. For example, francolite from the Babcock Deep core has Sr contents between 2,500 and 4,000 ppm—more than an order of magnitude higher than the Sr content of dolomite from the Hawthorn Group, which contains only a few hundred parts per million of Sr. Therefore, Sr contributed from minor amounts of calcite and dolomite in a sample predominantly composed of francolite will be insignificant; however, it is very important to remove all francolite before performing Sr analyses of dolomite or calcite samples.

The resolution achievable with Sr isotopes depends on the steepness of the seawater $^{87}Sr/^{86}Sr$ curve (fig. 12.12), and it is best to integrate $^{87}Sr/^{86}Sr$ age determinations with available biostratigraphy and magnetostratigraphy (DePaolo and Finger 1991). An advantage of using Sr isotopes to date samples from the Hawthorn Group is that the majority of the phosphorite formed during the Early to Middle Miocene, a period when the seawater $^{87}Sr/^{86}Sr$ ratio rapidly increased from 0.7082 to 0.7087 (Hodell et al. 1991). This allows fairly precise age determinations on the basis of Sr isotopes alone. In addition to DSDP Site 588, the Neogene seawater $^{87}Sr/^{86}Sr$ curve has been determined from sediment recovered at Ocean Drilling Project Site 747 in the southern Indian Ocean (Oslick et al. 1992). The two curves show good agreement between 22 and 16 million years ago; however, the curve for the Middle to Late Miocene becomes far less steep, and $^{87}Sr/^{86}Sr$ ages have a resolution of only ±1.4 million years. Between 8 and 5.5 million years ago, the $^{87}Sr/^{86}Sr$ curve is essentially flat, and sample ages cannot be distinguished by Sr isotopes. Fortunately, the amount of phosphorite on the Florida Platform younger than 16 million years is relatively small.

It is important to determine the age of a number of different grains within the same sample, because the phosphorite may have been reworked from a number of different source beds having different ages. The amount of age variation among different phosphorite grains appears to be related to the duration of the condensed section. In the Babcock Deep core, different peloidal phosphorite grains from most samples have the same age, within the resolution of the Sr technique. However, in some samples peloids of different grain size yield ages different by as much as 1 million years, and in a major condensed interval the $^{87}Sr/^{86}Sr$ ages of phosphorite pebbles vary by as much as 3.5 million years.

SEDIMENT AGE

Determining the age of Hawthorn Group sediments is difficult because of the repeated and extensive reworking that has occurred. Many of the fossils have been reworked from older deposits and therefore do not represent the age of final sediment deposition. The $^{87}Sr/^{86}Sr$ ages of calcareous benthic foraminifers are interpreted to represent final sediment deposition, because the predominant foraminifer, *Buliminella elegantissima,* is considered too fragile to survive reworking in high-energy environments (V. Waters, pers. comm. 1991). Therefore, in samples that contain both phosphorite and benthic foraminifers, it is possible to determine the age of the phosphorite (periods of organic rich sediment deposition) and the last reworking event. Unfortunately, foraminifers are not everywhere preserved. Where foraminifers are absent, the $^{87}Sr/^{86}Sr$ age of phosphorite or dolomite provides the maximum age of the sediment.

Recrystallization of the benthic foraminifers would decrease the Sr content, but might not necessarily alter the $^{87}Sr/^{86}Sr$ ratio. The assumption that the Sr-isotopic composition of the benthic foraminifers is unaltered is supported by the good agreement between the $^{87}Sr/^{86}Sr$ ages of the benthic foraminifers and independent biostratigraphic ages. For example, diatoms and nannofossils from a 3 m thick stratigraphic interval in the Babcock Deep core (32–35 m below sea level) indicate an age of latest Miocene to Pliocene (J. Barron, pers. comm. 1990), and the $^{87}Sr/^{86}Sr$ ages of benthic foraminifers from the same interval are around 5 million years. The Sr-isotopic ages of benthic foraminifers deeper in the core (83–86 m below sea level) are between 15 and 16.7 million years, in good agreement with Middle Miocene (CN3–CN4) nannofossil ages (M. Covington, pers. comm. 1992). In addition, the O-isotopic values of the benthic foraminifers from the Babcock Deep core range from 0 to –1‰ PDB, suggesting that they have not been significantly altered by meteoric waters. The light C-isotopic composition of the benthic foraminifers (–1‰ to –5‰ PDB) suggests that they lived in the near surface of organic-rich sediment and incorporated isotopically light carbonate produced by the bacterial degradation of organic matter.

$^{87}Sr/^{86}Sr$ AGES FROM THE BABCOCK DEEP CORE

$^{87}Sr/^{86}Sr$ ratios of separated peloidal phosphorite grains, benthic foraminifers, calcite shell fragments, dolomite, and shark teeth from the Babcock Deep core range from 0.708148 to 0.709005, and tend to increase up-section, as predicted by the seawater curve (Hodell et al. 1991) (fig. 12.13). $^{87}Sr/^{86}Sr$ ages of the francolite range from 25.6±0.7 million years near the base of the section to 9.2±1.4 million years at the top of the section (fig. 12.14). The age of sediment from the Babcock Deep core is taken to be the $^{87}Sr/^{86}Sr$ age of the contained benthic foraminifers (solid line in fig. 12.14; see above). The dashed line in figure 12.14 is the approximate age of the sediment obtained from the $^{87}Sr/^{86}Sr$ ages of two phosphatic and dolomitic beds that do not appear to have been significantly reworked. Calcareous benthic foraminifers are consistently younger than phosphorite from the same sample. Calcareous shell fragments tend to be older than benthic foraminifers, but younger than phosphorite. Dolomite is approximately the same age as associated phosphorite. Shark teeth from three samples are approximately the same age as, or older than, the associated phosphorite. These results suggest that skeletal phosphorite grains and calcite shell fragments can endure multiple reworking events, and are therefore of limited use in dating the sediment.

Paleoceanographic Significance of Phosphogenic Episodes on the Florida Platform

The formation of phosphorite ultimately depends on the enrichment of phosphorus by organic matter from upwelling seawater and the burial and regeneration of organic matter in the sediment (Bentor 1980). Although the phosphate concentration of seawater may have varied somewhat in the past, it is unlikely that the concentration was ever high enough during the Oligocene or Miocene to precipitate francolite directly from seawater. Most modern phosphorite formation occurs in organic-rich sediments near the sediment/seawater interface, in regions of upwelling and high productivity, as offshore Namibia (Baturin et al. 1972) and the Peru margin (Veeh et al. 1973). Phosphorite formation has also been documented on the eastern margin of Australia, where there is no upwelling (O'Brien and Veeh 1980). The Fe redox cycle effectively pumps phosphate ion into the sediment: phosphate ion is absorbed onto Fe-hydroxides, and then released when the Fe-hydroxide is reduced at depth in the sediment. However, large phosphorite deposits probably cannot form by this mechanism, because of the limited amount of Fe—especially where much of the reduced Fe is removed by pyrite formation.

The C-isotopic composition of marine organic matter is about 20‰ lighter than that of marine carbonates, because plants preferentially take up ^{12}C over ^{13}C during photosynthesis. It has been postulated that the C-isotopic composition ($\delta^{13}C$) of seawater should increase during marine transgressions, because of increased burial of isotopically light organic C in expanded shallow-water shelf areas (Tappan 1968; Broecker 1971, 1982; Fischer and Arthur 1977; Berger 1982). Increased removal of ^{12}C by burial of marine organic matter would require that the $\delta^{13}C$ of the oceans increase. Conversely, increased input of ^{12}C into the oceans by oxidation of organic matter during marine regressions would cause the $\delta^{13}C$ of the oceans to decrease. This idea is supported by the positive correlation found between Neogene sea-level fluctuations and variations in the $\delta^{13}C$ of seawater as recorded by benthic foraminifers (Loutit et al. 1983; Woodruff and Savin 1985, 1989).

The $\delta^{13}C$ records of benthic foraminifers from Miocene sediments recovered from several different ocean basins show a large positive excursion in the seawater C-isotopic composition (Woodruff and Savin 1985; Vincent and Berger 1985). The positive shift in $\delta^{13}C$ of +1‰ corresponds to the overall Early to mid-Miocene marine transgression, and the negative shift in $\delta^{13}C$ back to its former value corresponds to the Middle to Late Miocene marine regression (Haq et al. 1987a) (fig. 12.13).

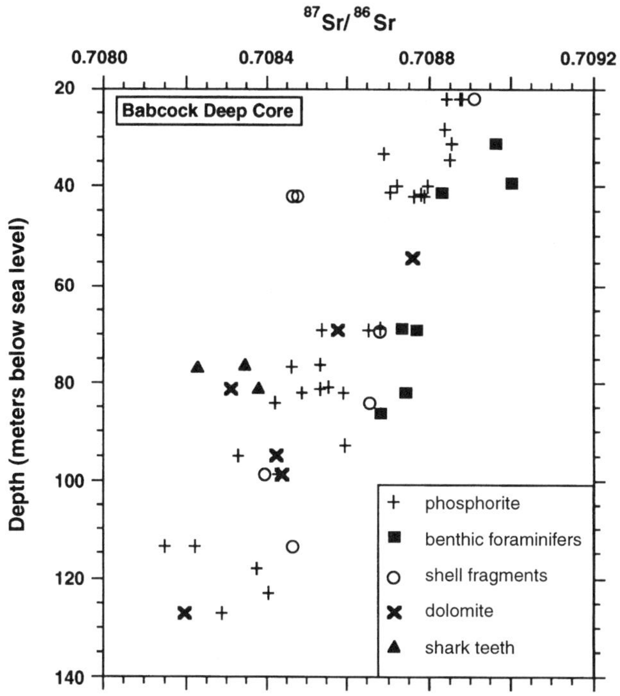

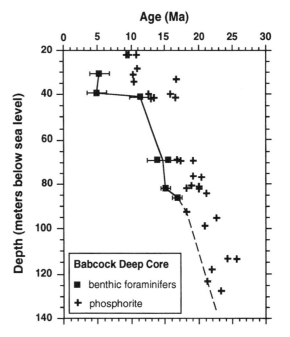

Figure 12.13. $^{87}Sr/^{86}Sr$ ratios of peloidal phosphorite grains, benthic foraminifers, shell fragments, dolomite, and shark teeth versus depth in the Babcock Deep core.

Figure 12.14. $^{87}Sr/^{86}Sr$ age of benthic foraminifers and peloidal phosphorite grains versus depth in the Babcock Deep core. Ages were calculated using the regression equations of Hodell et al. (1991). The solid line represents the age of the sediment, and the dashed line represents the maximum age of the sediment.

Compton et al. (1990) postulated that the positive $\delta^{13}C$ shift from the Early to Middle Miocene reflects an increase in the organic-C burial rate as transgressive seas flooded the continental margins and expanded the continental shelf area. Organic C may have been particularly abundant in shelf sequences of the southeastern U.S. during Early Miocene marine transgressions because of the increased surface productivity associated with upwelling caused by meandering of the Gulf Stream and its interaction with topographic highs (Riggs 1984; Snyder et al. 1990). The early diagenesis of these organic-rich sediments resulted in formation of phosphorite, dolomite, and pyrite. The present-day lack of abundant organic C in Miocene sediments of the southeastern U.S. is explained by its oxidation during marine regressions that occasioned sediment reworking and subaerial exposure. The reworked diagenetic phosphorite thereby proxies for previously deposited organic rich sediment (Compton et al. 1990).

If phosphorite in the Hawthorn Group formed during the early diagenesis of organic-rich sediment, then the age of the phosphorite should record periods of increased burial of organic C on the Florida Platform; furthermore, on the basis of this hypothesis and the short residence time of C in the ocean (10^5 years), the age of the phosphorite should correspond directly to marine transgressions and positive $\delta^{13}C$ shifts. The age of reworked sediment should correspond to marine regressions and negative $\delta^{13}C$ shifts. The relation of phosphogenesis and sedimentation as recorded in the Babcock Deep core to sea-level fluctuations and the C-isotopic record presented below provides an initial test of this hypothesis. The Babcock Deep core is emphasized here because it has provided the greatest amount of data. Other cores from widely different areas of the platform are presently being studied to evaluate the hypothesis more rigorously, and to determine how representative the Babcock Deep core is of other areas of the platform.

PHOSPHOGENESIS, SEDIMENTATION, AND SEA-LEVEL FLUCTUATIONS

The postulated cycle of organic rich sediment deposition, formation of phosphorite during early burial diagenesis, and sediment reworking probably was driven by high-frequency (100,000- to 300,000-year), high-amplitude (50–200 m) sea-level fluctuations, as documented in Miocene shelf sequences of the North Carolina continental margin (Snyder et al. 1990). Strontium isotopes cannot be used to resolve high-frequency, fourth-order sea-level fluctuations, but they can be used to correlate phosphogenic episodes with the longer, third-order eustatic cycles defined by Haq et al. (1987a). The hypothesis presented above predicts that phosphogenesis is favored during sea-level transgressions, or high stands, and that concentration of the phosphorite by reworking is most likely during sea-level regressions, or low stands. During marine transgressions, a larger part of the Florida Platform was submerged, and increased meandering of the Gulf Stream resulted in dynamic upwelling over the platform. Increased productivity and regeneration of organic matter buried in the sediment would have enhanced phosphorite formation on the shallow-water shelf. The extent of reworking probably was related to the amplitude and duration of sea-level low stands.

BABCOCK DEEP CORE

The relation of phosphogenesis and sedimentation as recorded in the Babcock Deep core to sea-level fluctuations is shown in fig. 12.15. The major phosphogenic episodes recorded in the Babcock Deep core occurred between 23 and 17 million years ago and correspond to the second-order Early to Middle Miocene sea-level transgression. This phosphorite occurs in cored sediment now lying 128 to 66 m below sea level that was deposited during the third-order transgression from 21 to 15 million years ago (TB2 cycles 2.1–2.4 of Haq et al. 1987a). The phosphorite from this interval can be roughly divided into two groups. Most of the phosphorite from 128 to 93 m formed between 23 and 20 million years ago and occurs in sediment that was deposited 21 to 18 million years ago; phosphorite from 93 to 66 m formed between 20 and 17 million years ago and occurs in sediment that was deposited 18 to 15 million years ago. The overall sea-level transgression from 21 to 15 million years ago was interrupted by several major, but brief, drops in sea level that may correspond to episodes of sediment reworking.

Concentrated phosphatic gravel and sand intervals at 41, 33, and 22 m appear to represent condensed intervals associated with sea-level low stands and regressions. The large range in age (13 to 16.5 million years) of the abundant phosphorite concentrated at the top of the essentially phosphorite-free underlying limestone (66–41 m) suggests that the phosphorite gravel at 41 m represents a particularly intense period of sediment reworking that corresponds to an overall sea-level low stand from 10.5 to 5.5 million years ago (TB3, 3.1–3.3) (fig. 12.15). The overlying diatomaceous mudstone from 41 to 22 m corresponds to the sea-level high stand 4 to 5 million years ago (TB3, 3.4–3.5), and contains phosphorite that formed 14 to 9 million years ago. The phosphorite pebbles at 22 m may represent a lag deposit associated with Pliocene sea-level low stands (TB3, 3.6–3.7). The overall trend of the phosphorite to become younger up-section reflects the decreasing probability of reworking older sediment as it became buried and further removed from erosional processes. The abundance of 15 to 17 million year old phosphorite in sediment as young as 5 million years may result from the greater probability of reworking upslope sediment deposited during the Middle Miocene sea-level maximum from 17 to 15 million years ago.

CORRELATION OF PHOSPHOGENESIS TO THE $\delta^{13}C$ RECORD OF DSDP SITE 588

The Sr-isotopic composition of phosphorite from the Babcock Deep core is directly compared to the C- and Sr-isotopic compositions of benthic foraminifers recovered at DSDP Site 588 in the southwest Pacific (Kennett 1986) (fig. 12.16) in order to test the hypothesis that the timing of phosphogenesis on the Florida Platform corresponds to the Early to Middle Miocene positive shift in the C-isotopic composition of seawater (Compton et al. 1990).

In addition to the age of the phosphorite, it is important to estimate the relative abundance of phosphorite that formed during specific time periods. The percentage of the total phosphorite in the core that formed during a given time is estimated by integrating the area of the gamma-ray log over the depth interval containing phosphorite with similar Sr-isotopic ratios (age) and dividing it by the total integrated area of the gamma-ray log. In general, phosphorite abundances determined by the gamma-ray log and X-ray diffraction (XRD) techniques are in good agreement (fig. 12.17). Phosphorite abundance was estimated by combining the results of the gamma-ray log and XRD techniques. Generally, the obtained phosphorite age/abundance relationship is probably correct, but it is somewhat crude because of the lim-

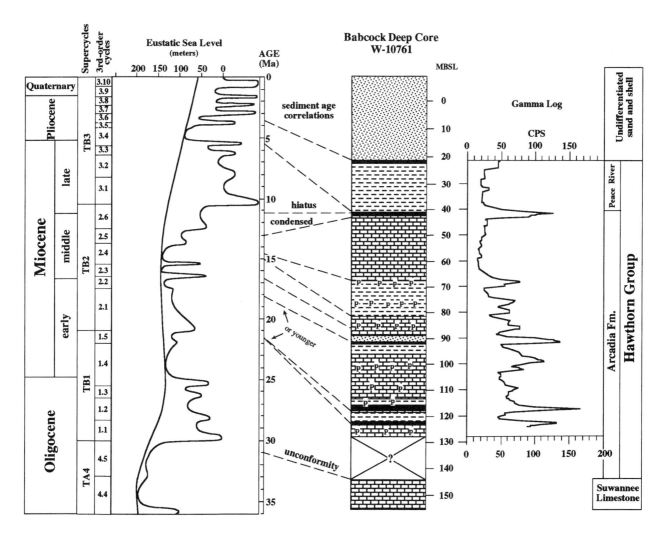

Figure 12.15. Comparison of eustatic sea-level fluctuations (from Haq et al. 1987a) with the stratigraphy and gamma-ray log of the Babcock Deep core. The dashed lines relate core sediments to the age scale and sea-level curve (see fig. 12.14). Dots: sand; dashes: clay; solid black: diatomaceous mudstone; brick: limestone; P: phosphate-rich.

ited number of dated samples and the uncertainty in calculating the relative phosphorite abundance in such highly variable sediments.

The positive $\delta^{13}C$ shift at site 588 between 340 and 315 m below sea floor, which corresponds to $^{87}Sr/^{86}Sr$ ratios of 0.7085 and 0.7087 (fig. 12.16), is considered correlative to the global Early to Middle Miocene positive $\delta^{13}C$ shift documented at other sites (Woodruff and Savin 1985; Vincent and Berger 1985). The two $\delta^{13}C$ maxima at Site 588 (315 and 275 m below sea floor) correspond to two (CM3 and CM6) of the six short-term global $\delta^{13}C$ maxima that are recognized within the overall $\delta^{13}C$ maximum between 17 and 14 million years (Woodruff and Savin 1991). The Sr-isotopic ratios of sediments from Florida and site 588 are comparable because the residence time of Sr is much longer than the mixing time of the oceans, and therefore the $^{87}Sr/^{86}Sr$ ratio of seawater should be the same throughout the oceans at a given time. The direct comparison of Sr-isotopic ratios is advantageous because it avoids any uncertainty involved in conversion of Sr- and C-isotopic ratios to ages.

Approximately three-quarters of the phosphorite in the Babcock Deep core formed between the Early Miocene $\delta^{13}C$ minimum of 0.6‰ at 370 m below sea floor and the Middle Miocene $\delta^{13}C$ maximum of 2.2‰ at 315 m. However, most of the phosphorite does not correlate with the steepest part of the positive $\delta^{13}C$ shift between 340 and 315 m. If the positive $\delta^{13}C$ shift represents increased burial of organic C on continental shelves, and if the amount of phosphorite that forms is proportional to the amount of organic C buried, then the greatest amount of phosphorite is expected to correlate with the steepest positive increase in $\delta^{13}C$. However, much of the phosphorite in the Babcock Deep core appears to have formed slightly before and during the initial increase in the positive $\delta^{13}C$ shift, and only about 30% of the total phosphorite correlates with the steep positive shift in $\delta^{13}C$.

That phosphogenesis does not coincide more directly with the major positive $\delta^{13}C$ shift can be explained in several ways. The age/abundance distribution of phos-

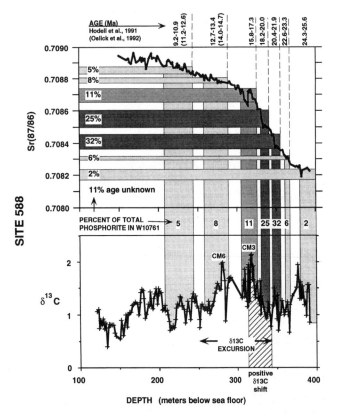

Figure 12.16. Comparison of $^{87}Sr/^{86}Sr$ ratios of phosphorite from the Babcock Deep core (W10761) to the Sr- and C-isotopic compositions of benthic foraminifers from DSDP Site 588 (Lord Howe Rise, southwest Pacific; Kennett 1986). The phosphorite samples were placed in six separate groups having similar $^{87}Sr/^{86}Sr$ values (shaded areas). The intersection of the range in Sr-isotopic values with the Sr-isotopic curve for Site 588 is projected onto the corresponding $\delta^{13}C$ record to show the correlation of phosphogenesis on the Florida Platform to the Early to Middle Miocene $\delta^{13}C$ positive excursion between 350 and 250 m below sea floor. The numbers in the shaded regions represent the percent of the phosphorite in the Babcock Deep core that has that range of Sr-isotopic values. CM3 and CM6 are two of the six $\delta^{13}C$ maxima that are recognized within the overall positive excursion by Woodruff and Savin (1991).

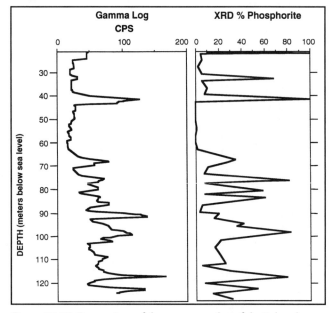

Figure 12.17. Comparison of the gamma-ray log of the Babcock Deep core (counts per second) and the percent phosphorite in bulk samples estimated from the integrated areas of francolite X-ray-diffraction peaks.

phorite from the Babcock Deep core may not be representative of phosphorite from other regions of the Florida Platform. The overall rise in sea level from 21 to 15 million years ago destabilized the Gulf Stream and caused it to migrate landward, which in turn caused increased upwelling, high surface productivity, and a parallel up-slope migration of organic-C deposition and phosphogenesis. To test this hypothesis, ages are being determined for phosphorite in cores from the Jacksonville Basin, and for phosphorite from the large economic deposits of the central Florida phosphate district, both of which are north of (up-slope from) the Babcock Deep core. Like phosphorite in the Babcock Deep core, most of the phosphorite in the Jacksonville Basin formed between 23 and 15 million years ago, and the detailed age/abundance distribution of these cores is currently being determined. Another contributing factor may be variations in the preservation of phosphorite. Phosphorite that formed around the maximum sea-level stand of the Middle Miocene has had the lowest preservation potential, because it has had the thinnest sediment cover and the longest exposure to subaerial weathering. Phosphorite formed during the earliest stages of the marine transgression has had the greatest potential for preservation in the rock record.

The earliest phosphogenic episode on the Florida Platform (25.6–24.3 million years ago) may correspond to an earlier global $\delta^{13}C$ excursion from the Late Oligocene to the Early Miocene; this positive excursion is apparent at the base of the section at site 588, between 400 and 360 m below sea floor (fig. 12.16). The amount of phosphorite from this earliest phosphogenic episode is probably greater than the estimated 2% of the total phosphorite in the Babcock Deep core, because the lowermost part of the core was not recovered. The oldest phosphorite from the Jacksonville Basin has a similar age, about 25 to 26 million years.

Phosphorite from the Babcock Deep core may correlate with other, short-term $\delta^{13}C$ variations recorded at Site 588. For example, the increase in $\delta^{13}C$ of +0.7‰ between 350 and 340 m below sea floor correlates with 32% of the phosphorite (fig. 12.16). The relation of phosphogenic episodes that occurred before and after the major $\delta^{13}C$ excursion to the C-isotopic record is uncertain. It is not known whether the low-amplitude, high-frequency variations in the C-isotopic record at Site 588 are truly global, or also reflect local changes in the average $\delta^{13}C$ of seawater from basin/basin fractionation, productivity, temperature, vital, and other effects (Kroopnick 1985; and others). Future high-resolution work may recognize low-amplitude, high-frequency global variations in the C-isotopic record, such as the six short-term maxima between 17 and 14 million years ago (Woodruff and Savin 1991).

CORRELATION TO OTHER MIOCENE PHOSPHORITE DEPOSITS

If sea-level fluctuations did in fact control organic-C deposition and removal from shelf sequences, then major phosphogenic episodes on other continental shelves should have been contemporaneous with those on the Florida Platform. $^{87}Sr/^{86}Sr$ ages for phosphorite from other regions of the southeastern U.S. continental margin can be compared to ages for the Florida Platform to determine if phosphogenic episodes occurred simultaneously throughout the southeastern U.S. For example, $^{87}Sr/^{86}Sr$ ages of phosphorite from the Texas Gulf phosphorite mine near Aurora, North Carolina, range from 22.9 to 12.1 million years (S. Riggs, pers. comm. 1993). Farther south, phosphorite samples from the Pungo River Formation in the offshore Onslow Bay area have $^{87}Sr/^{86}Sr$ ages that range from 19.3 to 17.0 million years (S. Riggs, pers. comm. 1992). These ages fall within the range determined for phosphorites from the Florida Platform.

The timing of phosphogenesis on the Florida Platform may not directly correspond to the maximum increase in the organic-C burial rate on a global scale. The positive global $\delta^{13}C$ shift of the Middle Miocene has been attributed to increased organic-C burial on continental shelves globally, the southeastern U.S. contributing a significant, but perhaps not dominant part. Less extensive Miocene continental-margin phosphorite deposits are known on the Blake Plateau, Peru, Cuba, California, and elsewhere (Burnett and Riggs 1990; McArthur et al. 1990; and others). The far greater amount of phosphorite in the southeastern U.S. indicates that conditions were particularly favorable for phosphorite formation and preservation.

Attributing the positive C shift to the deposition of organic-rich rocks in the Monterey Formation (Vincent and Berger 1985) appears unwarranted: although the phosphatic-marlstone facies of the Monterey Formation ranges in age from 23 to 7 million years, most of this organic-rich facies is 15.5 to 13.2 million years old on the basis of biostratigraphy (Garrison et al. 1987, 1990). In contrast to the Hawthorn Group, the Monterey Formation for the most part contains in-situ phosphorite that was not reworked into younger sediment. The majority of the phosphorite in deep, hemipelagic basins of the circum-Pacific rim appear to have formed after the major phosphogenic episodes on the continental shelves of the southeastern U.S. Organic C buried in the Monterey Formation may have contributed to the latter half of the overall mid-Miocene $\delta^{13}C$ maximum (CM6 at DSDP site 588), but most of its organic C was deposited after the steep positive C-isotopic shift prior to CM3 (fig. 12.16).

POSSIBLE RELATION OF PHOSPHOGENESIS TO LATE CENOZOIC GLOBAL COOLING EVENTS

Compton et al. (1990) speculated that changes in the organic-C burial rates (as reflected in $\delta^{13}C$ excursions) may have contributed to global cooling events during the Tertiary by altering pCO_2. The rapid uptake of atmospheric CO_2 during periods of high primary productivity and burial of isotopically light organic C should cause a decrease in pCO_2 during positive $\delta^{13}C$ shifts (fig. 12.3). A decrease in pCO_2 may have contributed to global cooling by the reverse greenhouse effect (Vincent and Berger 1985). Conversely, oxidation of reworked, organic-rich shelf sediment during sea-level low stands would cause an increase in pCO_2 during negative $\delta^{13}C$ shifts. However, the relation between $\delta^{13}C$ excursions and pCO_2 is uncertain, because additional factors can influence pCO_2 over the time scale of the $\delta^{13}C$ excursions (Lasaga et al. 1985; and others). Similar to the association of large phosphorite deposits in the southeastern U.S. with the Middle Miocene cooling event, the hypothesis presented here predicts that phosphogenesis most likely occurred during positive $\delta^{13}C$ shifts, and may have coincided with or preceded cooling events, as implied by increases in $\delta^{18}O$ in the Cenozoic stratigraphic record (fig. 12.18).

One way to increase rapidly the amount of CO_2 in the atmosphere is to oxidize organic C. This is currently being demonstrated by the accelerated burning of fossil fuels and the concomitant rise in atmospheric CO_2 from 290 to 350 ppm over the last 100 years (Keeling 1986). There is an abundance of oxygen in the atmosphere, such that the limit to the amount of organic C oxidized is the rate at which organic C is subaerially weathered. Rapid transfer of C between CO_2 in the atmosphere and organic matter in sediments can be accomplished either by depositing or reworking organic-rich shelf sediments according to fluctuations of sea level. Shelf area is important, because it represents a relatively large area where the ocean, atmosphere, and continents interface on the time scale of sea-level fluctuations. The ratio of submerged to exposed shelf area is highly sensitive to changes in sea level, because a relatively large part of the total continental surface area of the Earth is within a few hundred meters of sea level, above or below. Even relatively small changes in sea level can result in a large change in the area of submerged shelf.

The submerged-shelf area is also important as the locus of sediment accumulation and, hence, organic-matter preservation. Much of the primary productivity occurs over the continental margins, and high bulk-sediment accumulation rates preserve more of the deposited organic matter. On a global scale, most organic matter is buried on continental margins; organic-rich sediments from anoxic basins account for a relatively small amount of the total buried organic C (Berner 1982). Therefore, the shelf can alternately serve as a sink and a source of CO_2. The size of the C reservoir will depend on the amplitudes of sea-level fluctuations and the amount of sediment that is reworked during marine regressions. Organic C buried in the deep ocean (>200 m) likely will not be oxidized unless the sediment is reworked by bottom currents and exposed to dissolved O in bottom waters; otherwise, the organic C buried in deep-water sediments will not be returned to the atmosphere as CO_2 until it is uplifted or subducted over the generally much longer tectonic time scale of millions of years.

If sea-level fluctuations are a mechanism for changing organic-matter burial rates, then it is necessary to identify the products of organic-matter deposition. Organic matter of the continental shelves can be removed by reworking and exposure during sea-level low stands, and the more strikingly organic-rich deposits preserved in the rock record may not necessarily be significant in terms of the global organic-C burial rate. Accumulation of organic C in the terrestrial biosphere can also constitute a significant amount of the organic C buried globally. Probably, sea-level fluctuations strongly influence terrestrial organic-C accumulation as well, because peat deposits usually occur in low-lying coastal-plain settings. Unfortunately, diagenetic mineral proxies of peat or coal deposits are lacking, making it difficult to assess their importance.

Similar to the association of large phosphorite deposits in the southeastern U.S. with the Middle Miocene cooling event, phosphogenesis most likely occurs during positive $\delta^{13}C$ shifts, and should coincide with or precede cooling events implied from the $\delta^{18}O$ record. The sharp increase in $\delta^{18}O$ in the Early Eocene occurred between two large positive $\delta^{13}C$ excursions in the Late Paleocene and early Middle Eocene that correspond to enormous reworked phosphorite deposits in North Africa (fig. 12.18). The conceptual model also predicts that phosphorite deposits will have formed around the Eocene/Oligocene boundary because of a marine transgression-regression centered around a positive $\delta^{13}C$ excursion and a positive shift in $\delta^{18}O$ (fig. 12.18).

Preliminary work on the $^{87}Sr/^{86}Sr$ ages of phosphorite from the San Gregorio Formation of Baja California Sur, Mexico, suggests that phosphogenesis occurred from the latest Eocene to the middle Early Oligocene (41–36 million years ago), corresponding to a positive $\delta^{13}C$ shift and an overall marine transgression from 40 to 35 million years ago. The $^{87}Sr/^{86}Sr$ age of these largely reworked shelf deposits (Galli-Olivier et al. 1991; Grimm 1992) appears to be middle Early Oligocene (35 million years) to Late Oligocene, corresponding to a negative $\delta^{13}C$ shift and a marine regression from 35 to 30 million years ago. Therefore, these preliminary results suggest that phosphogenesis as recorded in the San Gregorio Formation may fit the phosphogenic episode predicted by the hypothesis presented here (fig. 12.18).

Conclusion

The predominantly peloidal phosphorite grains from the Hawthorn Group of Florida formed during early diagenesis in organic-rich sediments. The C- and S-isotopic compositions of carbonate and sulfate in the francolite indicate that the francolite precipitated in suboxic to

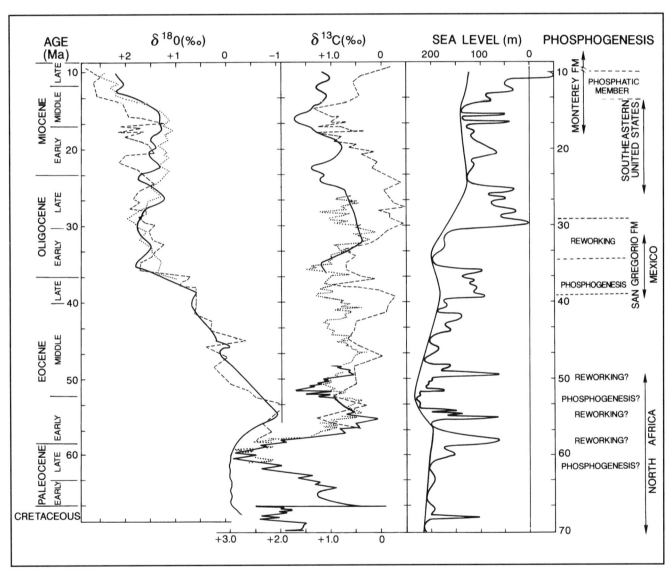

Figure 12.18. Cenozoic record of $\delta^{18}O$, $\delta^{13}C$, and sea level. $\delta^{18}O$ record of benthic foraminifers from the southwest Pacific (Shackleton and Kennett 1975; dashed line), Atlantic (solid line), and Pacific (dotted line) (Miller et al. 1987). $\delta^{13}C$ record of benthic foraminifers from the southwest Pacific (Shackleton and Kennett 1975; dashed line), Atlantic (Miller and Fairbanks 1985; solid line, Oligocene and Miocene), Weddell Sea (Kennett and Stott 1990; dotted line), and southeast Atlantic (Shackleton et al. 1984; solid line, Paleocene to Middle Eocene). Second- and third-order sea-level fluctuations from Haq et al. (1987a). The timing of phosphogenic episodes is plotted for the Monterey Formation phosphatic member and the southeastern U.S. continental margin. Phosphogenesis as recorded in the San Gregorio Formation, Baja California Sur, occurred from the Late Eocene to Early Oligocene, and reworking of the phosphorite during the Early Oligocene. Also shown is the postulated timing of phosphogenesis and reworking of the large Paleogene shelf phosphorite deposits of North Africa.

anoxic pore waters near the sediment/seawater interface, at burial depths corresponding to the upper sulfate-reduction zone. Precipitation of pore-filling francolite cement may have been enhanced by the presence of metastable biogenic carbonate minerals. The occurrence of hard, round peloidal phosphorite grains in unlithified quartz sands and the older $^{87}Sr/^{86}Sr$ ages of phosphorite relative to fragile benthic foraminifers indicate that much of the phosphorite has been extensively reworked. Most of the original organic matter and pyrite presumably were removed by oxidation during reworking, except where they were protected by tightly cemented francolite within peloidal phosphorite grains. The large amount of substituted elements in the francolite and the only slightly negative O-isotopic values of the francolite carbonate indicate that the phosphorite has had minimal chemical alteration, despite reworking and periodic exposure to subaerial weathering.

Srontium-isotopic compositions have proved to be extremely useful in working out the complex depositional and diagenetic history of the Hawthorn Group. Deposition corresponds to marine transgressions, and erosion and reworking correspond to marine regressions. The cycle of deposition of organic-rich sediments, phosphorite formation during early diagenesis, and organic-C removal and phosphorite concentration during sediment reworking was repeated as the shoreline migrated in response to sea-level fluctuations. Phosphogenesis occurred episodically from the Late Oligocene to the Late Miocene. The majority of the phosphorite formed between 22 and 16 million years ago and generally corresponds to the Early to Middle Miocene positive $\delta^{13}C$ shift of the ocean. Phosphorite may provide a useful proxy for organic-C deposition on continental shelves where the organic matter has since been removed by oxidation during reworking events. Sea-level fluctuations have a large impact on global shelf area, and may have served as a principal mechanism in the transfer of C between CO_2 in the atmosphere and organic matter in shelf sediments. The rapid burial or oxidation of shelf organic C may have influenced climate by altering atmospheric pCO_2.

13
Environmental Geology of Florida

Sam B. Upchurch and Anthony F. Randazzo

ENVIRONMENTAL GEOLOGY HAS BEEN defined as "the practical application of geologic principles in the solution of environmental problems" (Coates 1981 p. G4). It involves Earth processes that adversely affect human activity, including health, welfare, and living conditions (natural and anthropogenic). This rather broad definition dramatizes the scope of environmental geology to many geologists.

Florida statutes (FS492) regulate the practice of geology and require that persons who hold themselves out as geologists to the public, industry, or governmental agencies be licensed. The geologist must take a competency test that emphasizes those tasks that a geologist is likely to perform in Florida, and that may adversely affect the "health, welfare, and safety" of the public.

The issues of environmental geology in Florida can be subdivided into several categories: (1) natural hazards; (2) anthropogenic hazards; (3) water supply; and (4) airborne hazards derived from geologic materials.

Natural Hazards

Natural hazards include all of those hazards that arise independent of human activity. Examples of natural hazards that affect Florida include sinkhole collapse, landslides and slumps, storms (hurricanes) and floods, sea-level rise, and, less probably, earthquakes and tsunamis.

EARTHQUAKES

Earthquakes occur in Florida, although none has been hazardous or caused significant damage. Florida is one of the few low-risk areas for earthquakes in the coterminous United States (U.S. Geological Survey 1974). Earthquake activity in Florida has been summarized by Campbell (1943), Mott (1983), Smith and Randazzo (1989), and Lane (1991).

There is no evidence to suggest that Florida will ever suffer a major earthquake. Earthquakes are the result of slippage along faults. There is considerable controversy about the nature of faults in the state. Studies of the basement structure of Florida (see chapter 2) indicate that the region was faulted during the breakup of Pangaea. Vernon (1951) and others have mapped the traces of faults on the land surface, but these are minor, and there is little evidence of stress accumulation or slippage along these faults at present; however, earthquakes focused outside the state have caused minor damage in Florida.

Only twenty-seven "tremors" had been recorded in Florida as of 1991 (Lane 1991), and the foci of many of these were outside the state. The great earthquake of 31 August 1886 at Charleston, South Carolina, was felt as far south as Tampa. The vibrations caused church bells to ring in St. Augustine (Lane 1991). Many of the local earthquakes seem to be along the St. Johns River valley, which occupies a down-faulted trough formed during the rifting of Pangaea (Leve 1966; O'Connor 1984). Shocks along the St. Johns corridor may represent local stress release along that system. Figure 13.1 shows the intensities of Florida earthquakes based on the modified Mercalli intensity scale (table 13.1). Note that the worst tremor reported in Florida had a Mercalli intensity of VI, which means that many people noticed the event, but little damage was sustained.

TSUNAMIS

Tsunamis, or seismic sea waves, are giant waves created by submarine earthquakes, volcanic eruptions, and landslides. Submarine volcanic eruptions and earthquakes are most common along plate boundaries, especially plate-collision zones. The plate-collision zone nearest Florida is along the Antilles Archipelago, the boundary between the Caribbean and American plates. Cuba and the Bahamas Platform protect Florida from seismic sea waves generated in the Caribbean and Atlantic. The probability

Table 13.1. Modified Mercalli earthquake intensity scale of 1931

Scale	Criteria
I	Felt only by a few under especially favorable circumstances.
II	Felt only by a few persons at rest, especially on upper floors of buildings. Delicately suspended objects may swing.
III	Felt quite noticeably indoors, especially on upper floors of buildings, but many may not recognize it as being an earthquake. Standing automobiles may rock slightly. Vibration like passing of truck. Duration estimated.
IV	During the day felt indoors by many, outdoors by few. At night some are awakened. Dishes, windows, doors disturbed; walls make cracking sound. Sensation like heavy truck striking building. Standing automobiles rocked noticeably.
V	Felt by nearly everyone, many awakened. Some dishes, windows, etc. broken; a few instances of cracked plaster; unstable objects overturned. Disturbances of trees, poles, and other tall objects sometimes noticed. Pendulum clocks may stop.
VI	Felt by all, many frightened and run outdoors. Some heavy furniture moved; a few instances of fallen plaster or damaged chimneys. Damage slight.
VII	Everyone runs outdoors. Damage negligible in buildings of good design and construction; slight to moderate in well-built ordinary structures; considerable in poorly built or badly designed structures; some chimneys broken. Noticed by persons driving automobiles.
VIII	Damage slight in buildings of good design and construction; considerable in ordinary structures with partial collapse; great in poorly built structures. Panel walls thrown out of frame structures. Fall of chimneys, factory stacks, columns, monuments, walls. Heavy furniture overturned. Sand and mud ejected in small amounts. Changes in well water levels. Persons driving automobiles disturbed.
IX	Damage considerable in specially designed structures; well-designed frame structures thrown out of plumb; great in substantial buildings, with partial collapse. Buildings shifted off foundations. Ground conspicuously cracked. Underground pipes broken.
X	Some well-built wooden structures destroyed; most masonry and frame structures destroyed with foundations; ground badly cracked. Rails bent. Landslides from riverbanks and steep slopes considerable. Shifted sand and mud. Water splashed over banks.
XI	Few, if any, structures remain standing. Bridges destroyed. Broad fissures in ground. Underground pipelines completely out of service. Earth slumps and sand slips in soft ground. Rails bent greatly.
XII	Damage total. Practically all works of construction are damaged greatly or destroyed. Waves seen on ground surface. Lines of sight and level are destroyed. Objects thrown upward into air.

Source: Wood and Neumann 1931; cited in Coates 1981.

of an earthquake-generated tsunami striking the shores of Florida is, therefore, near zero.

Submarine slumps and landslides on the edge of the continental shelf may generate tsunamis, but there is little likelihood of a slump-generated tsunami in Florida, because there are no strong earthquakes that might trigger a submarine slide, and suitably thick siliciclastic sediments are absent. Submarine landslides are most common in thick, clay-rich, siliciclastic sediments, not found on the Florida continental margin.

An unusual event occurred about 11 P.M. on Friday, 3 July 1992: a "rogue wave" struck the Daytona Beach area. The wave was initially estimated to be about 5 m high and 45 km long (*Tampa Tribune* 1992). After further study, the wave was downgraded to an estimated 3 to 4.5 m in height and 32 km in length (Harger 1992). About 100 automobiles were damaged and 20 people received minor injuries. At the time of the event, it was suggested that the wave might have resulted from a submarine slide. Later speculations as to the origin included collapse of a sinkhole offshore (T. M. Scott, personal communication 1992) and emanation of a gas bubble from the subsurface. D. L. Smith and N. D. Opdyke of the University of Florida suggested that the wave may have been caused by a meteorite 1 to 3 m in diameter striking the water (Harger 1992); a meteor was reported by an observer shortly before the wave struck. The wave was not a true tsunami.

HURRICANES AND TROPICAL STORMS

Because of the peninsular configuration of Florida and its subtropical location, tropical storms and hurricanes are a yearly hazard. Tropical storms and hurricanes are

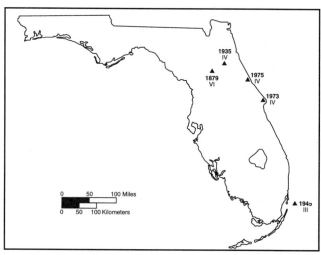

Figure 13.1. Historic seismic events in Florida, by year. Roman numerals represent intensity of the earthquake on the Modified Mercalli Intensity Scale (table 13.1) (Reagor et al. 1987).

spawned in the tropical Atlantic, Caribbean, and Gulf of Mexico. They are intense low-pressure cells caused by heat transfer from the warm ocean. The hurricane season extends from June through October.

Storm Classification Tropical weather disturbances are classified according to their maximum sustained wind speed. A tropical disturbance consists of unstable weather that may intensify into a tropical depression. A tropical depression consists of a disturbance with a well defined low-pressure region and wind speeds no greater than 61 km/hr. Tropical storms are more intense, with wind speeds of 62 to 117 km/hr. Tropical storms have sufficient energy to cause localized damage. A hurricane is characterized by well organized, counterclockwise (cyclonic) circulation and wind speeds in excess of 118 km/hr. Hurricanes are further classified according to the Saffir-Simpson hurricane scale (table 13.2). Most hurricanes rank in category I or II of the Saffir-Simpson scale. A storm of category IV is rare.

The short-term human and economic impacts of hurricanes and tropical storms on coastal and inland areas are the primary concerns. Hurricanes are accompanied by high winds, high rainfall, and storm surges. When they impinge on land areas, the waves and storm surges cause coastal and inland flooding, and the wind is likely to inflict significant damage.

Waves and Storm Surges The sustained high winds of tropical storms and hurricanes build up large waves. The heights of these waves vary with water depth, wind velocity, length of time the wind blows over the water, and distance over which the wind is in contact with the water. Waves in excess of 6 m are common, but nearshore wind-driven waves are seldom this large because they lose energy through contact with the sea floor in shallow water.

Wind-generated waves driven into the coastal environment can cause severe damage to coastal structures and shores. Beaches, coastal dunes, mangrove swamps, tidal marshes, and other natural features can absorb this energy. If a natural shoreline system is present (see chapter 10), absorption of the wave energy by the system should minimize damage in human terms. The problem arises when waves and water overreach the shoreline system. This occurs when the largest waves impact the coast, when there is a significant storm surge, where the coastal system is compromised by erosion, or where construction is too near the sea.

Storm surges are caused when sustained onshore winds pile the water up along the coastline, resulting in flooding and severe wave action. Flooding of coastal areas makes evacuation of low-lying areas difficult and causes significant property damage. Salt left in coastal soils may cause long-term loss of soil fertility and crop production. The saltwater that is piled up along the coast hinders runoff of the large amounts of rain typical of a hurricane, and flooding well inland can result. Historically, the primary cause of loss of life in a hurricane is drowning. If the storm surge coincides with a high tide, flooding can be significantly worse. Since the storm surge and waves are both a result of the wind, wave action becomes significantly more important if a storm surge is present. The water is deeper, so wave energy is not lost, and waves can move farther inland (fig. 13.2). The storm-surge heights for hurricanes given in table 13.2 are theoretical, and the actual height of a storm surge and the inland extent of flooding are determined by local topography, tide stage, wind direction and velocity, and other factors.

The highest storm surge recorded at Tampa was 4.2 m in the hurricane of 25 September 1848. Clearly, if an equivalent storm surge were to occur today, damage to downtown Tampa and St. Petersburg, and to vast areas of residential development, would be severe (fig. 13.3).

Table 13.2. Saffir-Simpson hurricane scale

Category	Wind speed		Probable storm surge	
	km/hr	mph	m	ft
I	119–53	74–95	1.2–1.5	4–5
II	154–177	96–110	1.8–2.4	6.8
III	178–209	111–30	2.7–3.6	9–12
IV	210–49	131–55	3.9–5.4	13–18
V	>249	>155	>5.4	>18

Figure 13.2. A storm surge and waves breaking on the coast.

Examples of Hurricane Damage Presented below are three examples of hurricane flood and wind damage in Florida, and some of the responsive measures taken to minimize future problems. The first example involves a pair of great hurricanes that caused significant loss of life in southern Florida in 1926 and 1928. The 1926 hurricane developed winds well in excess of 160 km/hr. The wind and associated waves caused a breach in a small dike on the southwest corner of Lake Okeechobee, and the lake surged out of its boundaries into neighboring communities. At least 250 persons perished during this storm, and the city of Moore Haven was virtually destroyed. The dike was rebuilt, but the second strong storm in 1928 again breached it. A storm surge estimated to be 3.6 m in height drowned some 1,849 to 2,300 persons—the largest recorded loss of life from a Florida natural disaster (Federal Writer's Project 1949; Carter 1974). The town of South Bay was largely destroyed. These surges, coupled with high rainfall throughout the area and poor drainage, resulted in a dreadful loss of life and property. The Hoover Dike, which now surrounds the lake, was built because of these two storms. Twenty years later the Central and Southern Florida Flood Control District was created to manage flooding and water supply in southern Florida.

The second example is the September 1935 storm, which crossed the Keys, causing widespread property damage and loss of life. A railroad to Key West was the only land link between the Keys and the mainland. Much of the railroad grade had been built with fill, which was mounded higher than surrounding islands. The storm struck with winds up to 320 km/hr and a surge of 3.6 m. The storm destroyed the railroad from Florida City to Key West, with a loss of life estimated by the Red Cross at 425 (the total death count finally rose to an estimated 800; Federal Writer's Project 1949). Towns were wiped out and railroad grades were washed away. U.S. Highway 1 now follows parts of the old railroad grade: some filled areas have been replaced by bridges, but the highway still occupies some of the original fill.

Finally, the most devastating storm in U.S. history in terms of property damage was Hurricane Andrew, which struck the south Miami-Homestead area in August 1992 (fig. 13.4), and then crossed to the Gulf Coast just south of Naples. Andrew was a category-IV storm, with winds gusting up to 260 km/hr. Andrew differed from the other storms in that there was little rainfall, and most of the damage was caused by the storm surge and wind. Damage was estimated at $7 to $10 billion! There was remarkably low loss of life (17 persons). Storm surges caused considerable damage along the coast, but the greatest (and most surprising) damage was caused by wind inland, near Florida City and Homestead. Poor or inadequate construction seems to have been a significant contributor to this damage.

These examples dramatize an important trend in the effects of hurricanes: as coastal development expands and communication improves, there has been a trend toward

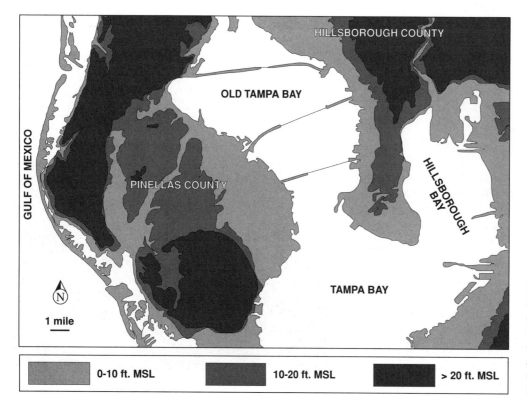

Figure 13.3. Flood-prone areas in the Tampa-St. Petersburg area. Areas prone to flooding by storm surges are zoned by elevation above mean sea level. After the National Oceanic and Atmospheric Administration (undated).

reduction in loss of life, and toward increases in property damage.

Hurricanes as Geologic Agents Hurricanes are geologic agents that affect the stratigraphic record and the configurations of coastlines. Some of the best examples of sediment deposition by a hurricane in Florida are a result of Hurricane Donna (September 1960). The 3 m storm surge of Hurricane Donna deposited 3 to 15 cm of carbonate mud in the Flamingo area of Everglades National Park, which caused significant mangrove mortality (Craighead 1971). Perkins and Enos (1968) and Enos (1977) reported up to 5 cm of mud deposited on Crane Key, and formation of emergent rubble mounds in the reef tract. Clearly, in a geologic sense, hurricanes are responsible for constructive deposition, as well as destruction.

Knowles and Davis (1983) and Davis et al. (1989a) found evidence of powerful hurricane activity in Holocene sediments along the coast of Sarasota County. Radiometric dates for these storm deposits, which consist of layers of shell transported over the barrier-island system into Sarasota Bay, show storms at 2,270, 1,320, and 240 years ago. Few historical storms have created deposits known to be as extensive.

Upchurch et al. (1992) have suggested that shell bars (termed "finger ridges" by archaeologists) on shell middens in Charlotte Harbor may represent storm ridges produced during hurricanes. These ridges extend laterally away from the exposed shores of coastal middens and suggest that the middens were subject to wave attack before development of the mangrove fringe.

Tidal inlets between barrier islands frequently open and close as a result of storms (see chapter 10). For example, Captiva Pass, the tidal inlet between La Costa and North Captiva islands, apparently developed between 1,300 and 600 years ago (Stapor et al. 1988). Redfish Pass, which separates North Captiva and Captiva islands, formed in 1921. A major cause of inlet development is breaching of barrier islands during major storms.

SEA-LEVEL RISE

One of the most important issues of the late twentieth century is "global change." Accumulating evidence suggests that increased CO_2 in the atmosphere resulting from combustion of fossil fuels is exacerbating the "greenhouse" effect. If more heat is trapped by an enhanced greenhouse effect, and if global temperatures rise, polar ice will melt and sea level will rise. The result will be a loss of coastal properties and structures, changes in local climate, and saltwater intrusion in Florida's aquifers.

Regardless of how Holocene sea-level curves (see chapter 14) and the Pleistocene geologic record are interpreted, the data are unequivocal: sea level has risen significantly over the last 10,000 years, and judging from past interglacial times, it will continue to rise. The greenhouse effect simply exacerbates the problem.

The map shown in figure 13.3, prepared to assist in development of storm evacuation plans, shows that a 3 m rise in sea level will flood significant areas of Tampa and St. Petersburg. As sea level rises, storm surges will penetrate farther inland, and wave damage will be increased. Patterns of longshore drift of coastal sediments will change. Ecosystems will be transformed. Property values and uses will be changed. Inland areas will suffer because saltwater will intrude aquifers, reducing the state's water resources. Regional climate and microclimates will be altered.

KARST

Dissolution of limestone and dolostone can create a landscape known as karst (see chapter 1). Sinkholes, the most widely known karst feature, are funnel-shaped depressions that form as a result of dissolution of underlying, fractured rock; they are typically characterized by internal drainage. Sinkhole formation commonly damages buildings, roads, and utilities, diminishing the usefulness of the affected land. Millions of dollars are lost in Florida each year to structural damage by sinkholes. Sinkholes result from a number of related processes associated with the development of karst and related to aquifer water chemistry. The examples given below illustrate sinkhole

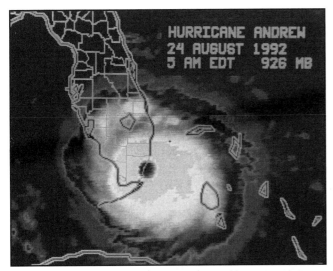

Figure 13.4. Radar image of Hurricane Andrew as it crossed south Florida (photo courtesy of the National Oceanic and Atmospheric Administration).

origins and the mechanisms involved in aquifer water chemistry.

Sources of Acidity The pH in aquifer systems is initially controlled by chemical reactions with carbon dioxide (CO_2) in the atmosphere and in soils. Equilibration of water with atmospheric CO_2, which has an average partial pressure (gas concentration) of $10^{-3.5}$ atmospheres, results in a minimum pH of rainfall of about 5.5, depending on the equilibrium between H_2O and CO_2. The equilibrium reaction is

$$(1) \quad CO_{2\,gas} + H_2O \Leftrightarrow H_2CO_{3\,carbonic\ acid}$$

Carbonic acid, with a pH of 5.5, is a weak acid, roughly equivalent to the acidity of soda pop. The acidity is derived from dissociation of carbonic acid by

$$(2) \quad H_2CO_3 \Leftrightarrow H^+ + HCO_3^-$$

where H^+ is the hydrogen ion and the source of acidity in most waters. HCO_3^- is bicarbonate ion, a common weathering product.

The pH of average rainwater in Florida is 4.77 (Upchurch, 1992), which is lower than predicted, because of "acid rain" problems. Acid rain is formed when sulfur and nitrogen oxides, produced by burning of fossil fuels, combine with atmospheric water to form sulfuric (H_2SO_4) and nitric (HNO_3) acids. Acid rain has increased with industrialization, and it is important as a modern weathering agent.

Many soils have high CO_2 concentrations produced by soil microbes as they metabolize humus (plant debris). As rainwater infiltrates soils, it reacts with soil CO_2 (eq. 1), and the pH drops even more. The partial pressure of CO_2 in the soil can be as high as 10^{-2} atmospheres, which is 10 to 50 times higher than CO_2 concentrations in the atmosphere. Dissolved organic acids are also a byproduct of the microbial decay of humus. Therefore, the pH of soil waters and shallow, surficial-aquifer waters is commonly in the range of 3 to 5. At this lower pH, groundwater is capable of dissolving many rock materials, especially carbonate minerals. An insoluble residue may form if the minerals being weathered contain aluminum or ferric iron, which are often relatively immobile in groundwaters and soil waters.

Equation (3) illustrates the reaction of acidic soil water with calcite to form karst features:

$$(3) \quad CaCO_{3\,calcite} + H_2CO_3 \Leftrightarrow Ca^{2+} + 2HCO_3^-$$

In this reaction, dissolved calcium and bicarbonate are produced. The consumption of hydrogen ions causes the pH to increase to approximately 7.0 to 7.5, depending on temperature and initial CO_2 concentrations.

Where and When Karst Forms Long-term dissolution can result in a general lowering of the land surface (denudation), but local underground dissolution features, such as caverns, may not be recognizable. The estimated denudation rate of north-central Florida by rock dissolution is 4 cm per 1,000 years (Brooks 1967); this estimate was based on analyses of water from the Suwannee and Waccasassa rivers, and the actual rate probably varies considerably with location and local geology (Opdyke et al. 1984).

Obvious karst features form when the flow of water is concentrated along well-defined conduits. Such conduits include joints or fractures, faults, and bedding planes in the rock, enlarged by rock dissolution. White (1988) discussed the preference for karst development along joints and fractures. Littlefield et al. (1984) showed that sinkholes in Hillsborough County are preferentially developed along fractures and features seen as photolineaments. Dissolution of limestone in Florida appears to occur preferentially in recharge areas and near the saltwater/freshwater coastal mixing zone; recharge areas are the more important of these two environments of sinkhole development (see chapter 6).

Shallow karst formation. Waters that have not equilibrated with carbonate minerals tend to be acidic owing to the presence of carbonic and organic acids. The water is, therefore, undersaturated with respect to calcite and dolomite, and dissolution of these minerals is predicted. First contact between the acidic water of the surficial aquifer system and carbonate minerals is in recharge areas. Once the water has reacted with the carbonates, the water becomes equilibrated, and additional dissolution is minimal. Therefore, the most dynamic region of rock dissolution is at the top of the limestone aquifer in recharge regions where there is a source of soil CO_2 and organic acids. Back and Hanshaw (1970) characterized the regional trends in equilibration of groundwater with respect to calcite and dolomite in the Upper Floridan aquifer system of central Florida. They showed that water is undersaturated near the Green Swamp recharge area (northern Polk County and southern Sumter and Lake counties), and saturated to oversaturated along flow paths away from the recharge area. Sprinkle (1989) and Katz (1992) suggested that undersaturated water is more widespread than predicted by Back and Hanshaw (1970), and that karstification is, therefore, more extensive.

Karst formation in mixing zones. Karst formation in Florida also occurs in the saltwater/freshwater coastal mixing zone. Sea-level fluctuations have altered the position of the freshwater lens in the past, and mixing-zone karst likely formed throughout the state (see chapter 4).

Runnels (1969) offered an explanation for this phenomenon, stating that mixing of two water masses that are saturated with respect to calcite will result in a new water mass that is undersaturated with respect to calcite. Badiozomani (1973), Plummer (1975), and Wigley and Plummer (1976) expanded the concept and showed that mixtures of calcium bicarbonate-rich water of the Floridan aquifer system and seawater—both of which may be saturated with respect to calcite and dolomite before mixing—become undersaturated with respect to calcite, in the approximate mixing range of 4% to 45% seawater at 25°C. The water is oversaturated with respect to dolomite in this same salinity range. Thus, in the landward "half" of the saltwater mixing zone, the equilibrium models predict that calcite would be either dissolved, thereby producing karstic porosity, or replaced by dolomite. Hanshaw and Back (1980) have documented calcite dissolution in the mixing zone in the Yucatan, but the possibility of dolomitization remains controversial. Thus, a belt of karstification that roughly follows the inner half of the saltwater/freshwater mixing zone is predicted. Near the coast, this zone should cross stratigraphic horizons and dip toward the interior of the state. Randazzo and Cook (1987) have presented data to support mixing-zone dolomitization in Florida.

Age of the karst. Strata of peninsular Florida have been subjected to extensive dissolution through many sea-level cycles (see chapter 4); therefore, the karst is multicyclic and complicated. Karst generally forms during sea-level stands low enough for freshwater flow systems to extend onto the continental shelf. During times of high sea level, karst formation is restricted.

Sea level has been rising for the last 10,000 to 15,000 years, and there are many examples of drowned karst offshore (see chapter 11). Many of the state's springs (e.g. Silver Springs, Homossassa Springs, Wakulla Springs) are sinkholes that have been converted to springs as rising sea level has forced the zone of coastal aquifer discharge inland.

A consequence of this multicyclicity is that sinkholes and other karst features are found over much of the Florida Platfrom. Karstic horizons associated with the modern saltwater/freshwater mixing zone or water-table position cannot be identified. They are superimposed and distributed in three dimensions to form a complex network throughout the carbonate rocks of the platform.

The karst system extends deep into the subsurface and constitutes a major part of the porosity and permeability of the Floridan aquifer system. Deeper parts of the cavernous system were in part formed during earlier low stands of sea level. For example, karst topography occurs in Miocene limestones at a water depth of 250 m on the Pourtales Terrace, in the southern Florida Straits off the Keys (Gomberg 1976). In southern Florida, a high-transmissivity zone (Boulder Zone) extends into Upper Cretaceous strata at depths up to 2,500 m below land surface (F. W. Meyer 1989). Even the most recent carbonate rocks contain karst features. Small sinkholes and solution channels are common in the Pleistocene (Sangamonian) Miami Limestone of southeastern Florida.

While carbonate strata of Late Cretaceous and younger ages contain cavernous zones, only fossils (and, perhaps, superposition) can be used as indications of the times when the caverns were formed. The fossil terrestrial faunas found in many sinkholes provide a minimum age of the karst. No terrestrial vertebrates are known from Early Oligocene or older rocks (Olsen 1959). Late Paleogene and Neogene terrestrial vertebrate faunas are known from sinkholes and caverns formed in older rocks (MacFadden and Webb 1982). There were large variations in sea level during the Late Oligocene and Miocene, and the sea-level low stands, especially the Messinian, appear to represent the times of extensive karst formation (see chapter 4).

Sinkhole Classification and Development The multicyclicity of karst development has resulted in a number of different types of sinkholes in Florida. Understanding these different forms of sinkholes and their modes of development is critical to detecting and mitigating damage to structures resulting from sinkhole development. Sinkhole classifications and distributions have been discussed by Sinclair et al. (1985), Beck and Sinclair (1986), and Lane (1986).

Collapse sinkholes. Figure 13.5 illustrates "before" and "after" cross-sections of a cavern system with a collapse sinkhole. The roof of the cavern is limestone. The clays and sands overlying the limestone are collectively called cover. If the cover is cohesive and thick enough, it may be able to bridge a cavern in the absence of a limestone roof. When the cover collapses, a large, funnel-shaped depression results. The size and depth of the funnel are determined by the thickness and angle of repose of the cover materials. Loose sands yield gentle slopes, while cohesive, clay-rich materials and rock result in near vertical sides. The thicker the cover, the greater will be the areal extent of the inverted cone.

The relationships of the water-table aquifer and the potentiometric surface of the Upper Floridan aquifer system are important in sinkhole development. Both surfaces represent water-pressure distributions in the aquifers. When they are separate and independent of each other, there is no significant transfer of water between the two aquifers. In the case illustrated in figure 13.5, the water table is at a higher elevation than the potentiometric surface, so there is a natural tendency for water to flow from the water-table aquifer into the confined, limestone aquifer, but the clay bed significantly retards this flow. If the two aquifers become interconnected along a joint or fracture, and if flow is from the surficial aquifer, the water-table surface would decline and form a cone of depression over the leaky area. The potentiometric surface would rise and form a mound under this depression as recharge occurs. If recharge of the limestone aquifer from the surficial aquifer is unimpeded, the water table and potentiometric surface will merge and coincide. Leakage of water from the surficial aquifer system into the lower, limestone aquifer is a significant cause of sinkhole development.

In figure 13.5a, dissolution has formed a cavern, which may or may not be growing, and the cavern roof is intact. In figure 13.5b the limestone roof of the cavern has collapsed, and the water table and potentiometric surface have combined, indicating drainage from the surficial aquifer into the limestone aquifer. This type of sinkhole in limestone or dolomite is called a rock-collapse sinkhole, whereas a sinkhole caused by loss of strength of the cover materials is called a cover-collapse sinkhole.

The formation of a collapse sinkhole is sudden and catastrophic. There may be no warning, and subsidence may be so swift that it is impossible to salvage any overlying structure or its contents (fig. 13.6). Collapse sinkholes have vertical or near vertical sides and are usually circular. Bedrock may be visible at depth in the sinkhole. There is a cone-shaped pile of rubble on the floor of the cavern below the collapse zone. Cover-collapse sinkholes are common. Many of the sinkholes that develop each year in Florida result from collapse over a piping failure in the cover, the pipe being a slowly growing tube caused by "raveling" of cover materials from the bottom up.

Although many sinkholes initially form by rock collapse, rock collapse is a relatively infrequent event. Cave divers report collapse features (rockfalls within caverns, rock piles at the throats of springs), suggesting that formation of rock-collapse sinkholes is important over geologic time, but most of the sinkholes that form each year apparently are secondary reactivation features, not primary rock collapses. Perhaps the best known example of a rock-collapse sinkhole is the Devil's Millhopper, a state geologic site near Gainesville, and one of the largest sinkholes in North America.

Increasing the load on a cavern roof through construction or water impoundment may hasten failure and promote formation of collapse sinkholes. Water is especially important with respect to cover-collapse sinkholes, because it lubricates cover materials and reduces cohesiveness. Water can be introduced from septic tanks and other domestic waste-disposal systems, leaky swimming pools, storm water retention ponds, waste- and cooling-

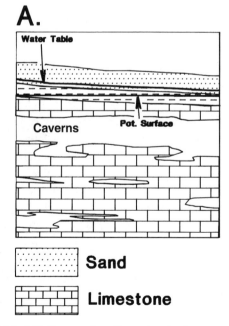
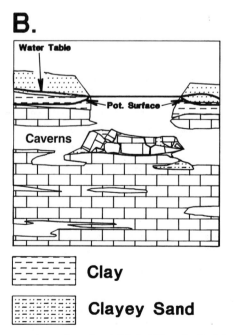

Figure 13.5. A-B: Cross-sections illustrating development of a collapse sinkhole. The potentiometric surface is labeled "Pot. Surface" (see text for explanation).

water ponds, reservoirs, and other structures. Also, vibrations created during construction or long-term land use may hasten collapse. Drilling may trigger an immediate collapse of overburden into a more deeply buried cavity (fig. 13.7).

Solution sinkholes. Sinkholes can also form slowly as a result of dissolution of rock. Figure 13.8 illustrates the formation of a solution sinkhole. The geologic framework is similar to that shown in figure 13.5, with the exception that a "swarm" of nearly vertical joints is present. Joints have been drawn in to illustrate the importance of fractures in the rocks to vertical flow of water. In figure 13.8a the confining clays are intact, and there is minimal vertical flow (note that the water table is above the potentiometric surface, indicating a potential for downward

Figure 13.7. Drilling rig that triggered a cover-collapse sinkhole by the vibrational forces it produced, Keystone Heights (photo courtesy of Thomas Scott).

Figure 13.6. Damage caused by a cover-collapse sinkhole in Odessa, Hillsborough County (photo courtesy of Marc J. Defant).

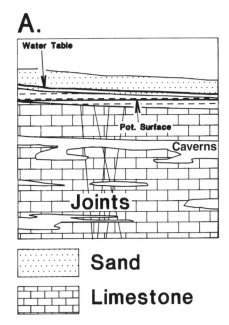
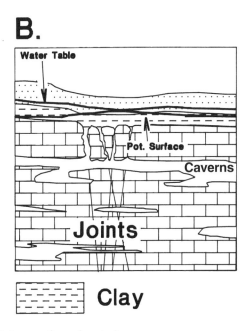

Figure 13.8. A-B: Cross-sections illustrating the development of a solution sinkhole (see text for explanation).

flow). Should water pass vertically along the joints, they will become enlarged by dissolution according to equation (3). The removal of rock mass by dissolution allows settling of rock and washing of overburden into the cavern ("raveling"). Consequently, slow subsidence occurs (fig. 13.8b).

Solution sinkholes develop gradually, and sudden, catastrophic damage to structures is infrequent. The most likely forms of structural damage are cracking and associated settling features. Insurance settlements are difficult to obtain where solution sinkholes are involved. Legal and engineering interpretations have leaned heavily on the concept of catastrophic collapse; in other words, if subsidence isn't sudden and observable, it isn't a sinkhole. Further, the standard sinkhole-identification technique is the drilling of test holes, but test holes likely will miss the small conduits associated with solution sinks. The result is that losses associated with solution sinkhole development may not be justly compensated. Raveling is an important factor in the development of solution sinkholes and has been recognized by the legal and engineering professions because of the use of surface geophysical testing (electrical resistivity and ground-penetrating radar).

It is unlikely that human activities substantially increase the rate of dissolution and the development of solution sinkholes; disposal of acidic wastes may do so locally, but is of little significance volumetrically. It is possible that installation of septic tanks, swimming pools, or other sources of mildly acidic waters may increase the rate of dissolution slightly. The most direct consequence of human activity is not increase in the rate of rock dissolution, but inducement of failure of already weakened rock through construction or by increasing the hydraulic head in the surficial aquifer. These activities could cause subsidence of rock materials and downward mechanical erosion of the cover (raveling).

Alluvial sinkholes. Because there is a long history of karstification in Florida, sinkholes vary in age and degree of development. Many of the older sinkholes have been partially filled by marine and wetland sediments and by lateral soil creep subsequent to sinkhole development (fig. 13.9). These older, partly to wholly filled sinkholes are called alluvial sinkholes. Where the water table is shallow, they form lakes and cypress "heads" or "domes" (fig. 13.10).

Care must be taken to avoid attributing all circular features to sinkholes. Circular depressions occur in marine sediments and sand dune fields as a result of localized erosion or sedimentation. If, however, there is a history of sinkhole development in the area, and if several of the depressions are aligned according to the principal joint orientations, then a sinkhole origin should be suspected. Often, unfortunately, the depression is so shallow as to be ignored.

Alluvial sinkholes are not easy to recognize. The home depicted in figure 13.11 was located in a gentle swale that was desirable because of the large trees nearby and the better soils in the depression. In this case, the water table and potentiometric surfaces are independent, and there is no indication that water is passing through the plug in the throat of the sinkhole. There was no indication to the homeowner that a paleo-sinkhole existed

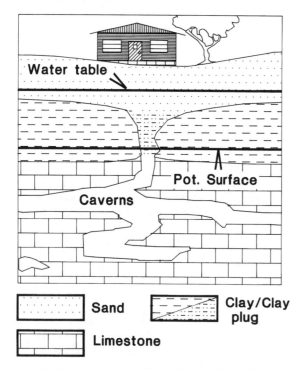

Figure 13.9. Cross section of an alluvial sinkhole. The sediment "plug" is intact, and there is little surface indication of the sinkhole. The water table and potentiometric surface do not indicate interconnection through the clay plug in the throat of the ancient sinkhole.

Figure 13.10. Cypress "dome" or "head" located over an alluvial sinkhole in Pasco County.

under the structure. There is a high probability that this is a stable configuration and that no subsidence will occur. The probability is even higher that any subsidence would be minor, and the only damage sustained by the home would be a few "settling cracks." This scenario applies to many Florida homes. Typically, the paleo-sinkhole could have been confirmed by drilling or shallow geophysical surveys. Most damage with respect to sinkholes in Florida results from construction over undetected alluvial sinkholes.

Raveling sinkholes. Rejuvenated alluvial sinkholes are sometimes called "raveling sinks." Alluvial sinkholes are reactivated by a number of different processes. The scenario shown in figure 13.11 is typical. Here a house has been built over a sediment-filled sinkhole, and a well has been constructed near the house (fig. 13.11a). In figure 13.11b the pump has been turned on and excessive pumping has caused turbulent flow in the cavernous limestone aquifer. This is a common occurrence in eastern Hillsborough County, where heavy pumping for agricultural freeze protection has reactivated a number of sinkholes (Metcalf and Hall 1984; Bengtsson 1987). Turbulent flow causes erosion of the plug (fig. 13.11c), and reduction of hydraulic head in the limestone aquifer reduces buoyancy and support of the plug. The erosion is initially accompanied by loss of cohesion and upward piping. Piping failures generally have small diameters (0.3–15 m at the land surface), but their vertical extent can be more than 30 m, depending on the lithology of the cover. Settling and cracking in the structure begins,

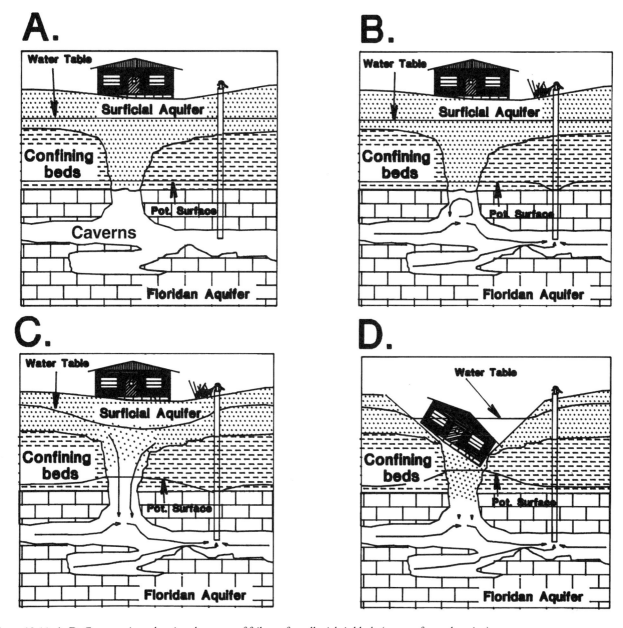

Figure 13.11. A–D: Cross-sections showing the stages of failure of an alluvial sinkhole (see text for explanation).

the failure works upward until the cover can no longer be supported, and finally rapid subsidence begins (fig. 13.11d). Other causes of rejuvenation of alluvial sinkholes include loading the fill with water, which reduces cohesion and lubricates sediment; vibrations; puncturing the plug with a well or other excavation; and loss of buoyancy due to pumping.

It is difficult to place a time frame on rejuvenation, because it is usually not known when the initial erosion of the plug began. The failures in eastern Hillsborough County occurred within hours of turning on the pumps. Other sinkholes may take years, depending on the dynamics of flow; cohesiveness, thickness, and diameter of the plug; and conditions of the overlying materials.

Because rejuvenated alluvial sinkholes are so difficult to predict, and because cover usually obscures bedrock, few have ever been studied. One detailed study (Bloomberg et al. 1988) used standard penetration tests and piezo cone-penetrometer profiles to characterize alluvial sinkholes at the karst-research area on the campus of the University of South Florida. They found that "stable" alluvial sinkholes include a central core, or conduit, characterized by sediment with low cohesion. Water can flow through the more-permeable core, thus recharging the Floridan aquifer. Vibrations caused by driving the drilling rig over a conduit induced a piping failure, which illustrates how unstable the sediment-filled conduit can be. Figure 13.12 is a cross-section through one of the alluvial sinkholes described by Bloomberg et al. (1988). Soil stratigraphy indicates two episodes of sinkhole activity (during and after deposition of units I and III), and a central conduit (loose sand filling a pipe).

This stratigraphy suggests a sequence of events—initial vertical transport of material during reactivation of the sinkhole, followed by slow lateral creep of soil into the central conduit. The lateral-creep phase may have lasted for years. The loose conduit fill suggests that downward transport of sand is continual.

Evaluating sinkhole risks. Construction of large commercial structures usually involves foundation testing by drilling, which may detect underground cavities and their potential for sinkhole development; however, foundation tests of this type may miss buried karst. If karst features are discovered, good engineering practice can avoid or correct hazards to the structure. Surface geophysical techniques, such as electrical resistivity (Smith and Randazzo 1975, 1986), provide an economical, comprehensive, and nondisruptive method to detect underground cavities that might form sinkholes.

Unfortunately, construction of small structures, including private homes, often does not involve foundation testing of any kind. Many homes are constructed each year over ancient karst features without the knowledge of the builder or homeowner. Each structure is potentially susceptible to damage. Therefore, much of the sinkhole damage in Florida is caused by subsidence of previously undetected karst features beneath structures.

As population density and investment in structures increase, the importance of sinkhole development to the public becomes magnified. Most modern sinkholes are small, and one or two homes may be affected. Prior to 1981, the state response to this scattered damage was to require insurance companies to provide a sinkhole-damage rider to insurance policies. In the spring of 1981

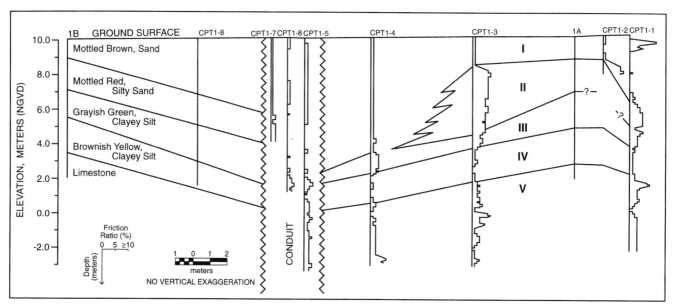

Figure 13.12. Cone-penetrometer-derived cross-section of an alluvial sinkhole on the University of South Florida campus, Hillsborough County (Bloomberg et al. 1988; see text for explanation).

everything changed—the famous Winter Park sinkhole, near Orlando, opened.

The Winter Park sink was either a cover-collapse or raveling alluvial sinkhole that formed in a deeply buried karst terrain. The sinkhole began to form in May 1981, and rapidly grew to 90 by 106 m (fig. 13.13). Water levels have fluctuated in the sinkhole, depending on potentials in the underlying aquifer. The Winter Park sink was small (about 0.8 hectare, or 2 acres) compared to some of the sinkholes (now lakes) in the Orlando area: some of the larger sinkhole lakes cover more than 3 km^2. Even so, this sinkhole electrified the news media because it formed in an urban area where damage was dramatic. Some of the damage and losses included a municipal swimming pool, a house and shed, a car dealership, several large trees, and sections of two streets. An estimated 110,000 m^3 of earth was lost down the hole (Lane 1986).

The Winter Park sink initiated a flurry of studies of sinkholes in Florida. Because of the damage from that one sinkhole, the Florida legislature created the Florida Sinkhole Research Institute, and researchers at the U.S. Geological Survey and several universities began intensive studies of sinkhole mechanics. The Winter Park sink clearly demonstrates the need for detailed understanding of sinkholes and development of risk models.

Although sinkholes are known throughout the state, most damage is localized in central and west-central Florida, around the flanks of the Ocala Platform. This suggests that bedrock geology—particularly, thickness of cover—exerts a significant control on sinkhole development. Sinclair and Stewart (1985) used this concept to develop a map of the state showing the expected types of sinkholes, based on cover thickness (fig. 13.14). Limestone-collapse and solution sinkholes are most likely to form in regions of bare or thinly covered karst. As cover thickens, rejuvenated alluvial and cover-collapse sinks become more probable. Finally, where cover is thickest, sinkhole formation is unlikely. This map can be used to predict general areas of sinkhole formation and the most likely mode of formation; however, it does not indicate the probability of occurrence of sinkholes in an area.

Identification of sinkhole-prone areas will minimize damage, and will allow for development of sound actuarial data at the local level. Site-specific sinkhole predictions can be made with geophysical data, and areas that are at highest risk can be delineated. This will require use of data on historical and ancient sinkhole occurrences.

Upchurch and Littlefield (1988) compared ancient sinks, identified from topography, and historical sinkholes reported from 1964 to 1985. They found, with some exceptions, that the distribution of ancient sinkholes can be used to predict modern ones. Through the efforts of several government agencies, a sinkhole-risk map has been produced, which shows the possibility of sinkhole development in Florida (fig. 13.15).

Other environmental impacts of sinkholes. In addition to structural damage, sinkholes can operate as pathways of local or regional groundwater contamination. In the past, sinkholes were considered to be convenient, low-cost waste-disposal sites. Many rural areas were dotted with sinkholes that had been filled with everything imaginable, including construction debris, domestic solid wastes (fig. 13.16a), and industrial wastes. Owing to the high costs of lift stations, many municipalities in Florida direct treated waste water (fig. 13.16b) and storm runoff (fig. 13.16c) to sinkholes rather than pump it through a hilly karst terrain to central disposal facilities. Finally, because karst areas typically have internal drainage, agricultural wastes, lawn fertilizers, and other sources of nutrients are inadvertently washed into the sinkholes. This nutrient loading may cause enrichment of surface waters (plate 9) and groundwater.

Live Oak, in Suwannee County, is typical of Florida

Figure 13.13. A-B: The Winter Park sink near Orlando (photos courtesy of the Florida Sinkhole Research Institute).

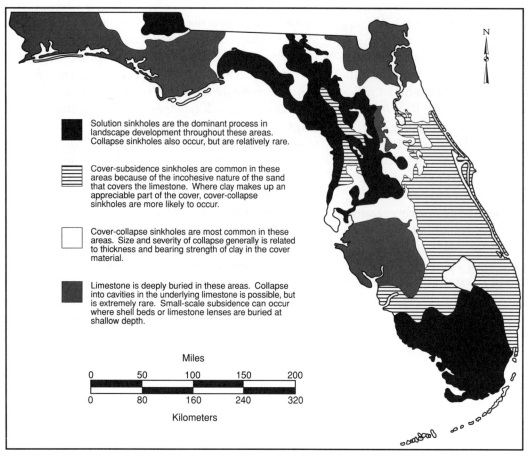

Figure 13.14. Predicted sinkhole types in Florida (Sinclair and Stewart 1985).

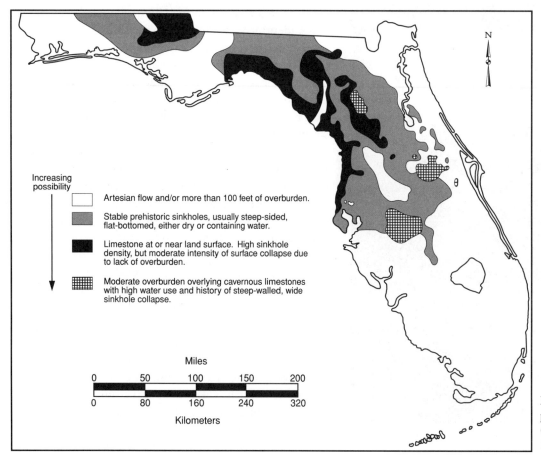

Figure 13.15. Sinkhole possibility map of Florida (Sinclair and Stewart 1985).

Figure 13.16. Examples of sinkholes used for waste disposal. A: Sinkhole filled with domestic wastes, Marion County. B: Sinkhole receiving treated municipal waste water in north-central Florida. C: Stormwater retention pond with a drainage well in Live Oak, Suwannee County.

cities in a karst terrain where it is impractical to construct an integrated storm water disposal system. Instead, the city uses drainage wells in sinkholes (fig. 13.16c). These drainage wells, numerous septic tanks, rural animal wastes, and other sources have combined to produce a large area characterized by relatively high nitrate concentrations in groundwater. Orlando (Kimrey and Fayard 1982), Tampa, Lake City, and many other cities are similarly drained. Many runoff-detention ponds along Florida highways serve the same function.

Brandon, in Hillsborough County, provides another example of nitrate contamination resulting from various land uses in a karst terrain. While the area has many residential and commercial developments today, nitrogen isotopes and other indicators suggest that nitrates in the limestone aquifer came from fertilizers used on citrus and other crops that have not been grown since the freezes of the 1980s (Jones and Upchurch 1993). Springs along the Alafia River are contaminated by nitrates, and the Alafia River is a major source of nitrates in Tampa Bay. Although the primary source of nitrates no longer operates, Brandon's groundwater is contaminated, and the springs likely will continue to discharge nitrates for many years to come. Brandon is now dotted with residential subdivisions, many of which use septic tanks or other on-site waste-treatment facilities; these waste-disposal facilities are relatively new (post-1980), and probably will constitute a new source of nitrate in groundwater.

SHRINKING AND SWELLING CLAYS AND ORGANIC MATTER

Damage to structures from shrinking and swelling of clays and from oxidation of organic matter in sediments is a problem worldwide. Structural damage in excess of $40 million per year occurs in the U.S. as a result of these processes. These problems are very important in Florida, where both kinds of materials are common components of the surficial sediments (Millan and Townsend 1981).

Certain clays, notably the smectites, have the capability of shrinking and swelling, because their crystal structure comprises sheets that are weakly bound together, with water and exchangeable cations in the interlayer spaces (Grim 1968). Swelling is caused by wetting and by introduction of large ions, such as Na⁺. Dehydration and exchange of larger ions for smaller ions, such as Ca^{2+}, cause shrinking. The amount of shrinking or swelling ranges between 2 to 3 angstroms, and is reversible (Grim 1968). Expansion of these clays can result in significant volume changes which can effect overlying structures if the clay is concentrated within 1 m of the ground surface and has some lateral continuity.

Humic substances are common components of soils. As long as the soils remain wet and net chemical reduction operates, these organic materials retain their volume; however, if organic aggregation (flocculation) or oxidation is induced, the organic materials may shrink or be destroyed. Oxidation results from lowering of the water table or changing the reduction-oxidation potential of the soil or aquifer water. Aggregation is caused by increasing the ionic strength (total dissolved solids) of the water. In either case, there is a net volume reduction, and subsidence may result.

Smectites are widespread in Florida, and are characteristic of sediments of the Hawthorn Group. Thus, wherever the Hawthorn is near the land surface, there is a potential for shrinking or swelling clays. Expansive clays occur to a lesser extent in other strata in many areas of the state. Organic materials in soils are even more widespread, occurring in Pliocene to Holocene sediments throughout the state. They are a common pedogenic feature, forming organic hardpans and lenses in many of the state's soils.

Consequences of Ion Exchange If a Ca- or Mg-saturated smectite is bathed in a Na-rich solution, ion exchange will cause the clay to swell. Inland, most of the smectitic soils in Florida are Ca- or Mg-saturated; near the coasts, they may be Na-saturated. When Na-rich swimming-pool water, septic-tank effluent, treated municipal waste water, or landfill leachate is brought into contact with a clay-rich soil, the ionic composition of the water bathing clays in the soil is changed, and the soil can be expected to swell. Na-saturated clays that are bathed in Ca- or Mg-rich water can shrink. This happens when Na-saturated clays in siliciclastic soils and aquifers are flooded with Ca-rich water from the Upper Floridan aquifer—as might occur when lawns or fields are irrigated, or septic-tank effluent is introduced. Shrinking or swelling of clays can cause failures of landfill liners, foundations, and other structures. It can also dramatically change the permeability of the soil or aquifer and reduce its effectiveness as a water-supply or waste-disposal medium.

The ability of clays to shrink or swell is predicted by the sodium-absorption ratio (SAR):

$$(4) \quad SAR = \frac{Na_{meq}}{\sqrt{\frac{Ca_{meq} + Mg_{meq}}{2}}}$$

where Na_{meq}, Ca_{meq}, and Mg_{meq} are concentrations of these elements in milliequivalents per liter. If this ratio is greater than 8 to 10, smectities can be expected to swell (Bouwer 1978).

Subsidence Many areas where clays and organic materials are especially abundant are also areas of sinkhole development. Clays and organic materials often fill alluvial sinkholes and near-surface karst features. There can be considerable confusion about the cause of structural damage where sinkholes and either clays or organic materials coexist. Typically, processing of a sinkhole-damage insurance claim involves standard penetration tests, and, more recently, surface geophysical testing (e.g., electrical resistivity, ground-penetrating radar) on the site. A standard penetration test is very localized (at most, an area of a few square centimeters to a few square meters is tested), but when used in conjunction with surface geophysical tests, definitive conclusions can be reached. Structural damage caused by a sinkhole is compensable under existing insurance laws, but damage caused by shrinking or swelling of clays or by oxidation of organic materials is not.

The combination of standard penetration tests and surface geophysical testing is often adequate to resolve the problem definitively. However, where two or more causes of settling are involved at the test site, results may be inconclusive. For example, several residential subdivisions in the city of Dunedin, in Pinellas County, are plagued by structural damage. Numerous geotechnical and geologic firms have been hired by homeowners and insurance companies to determine the causes of the damage. The results have been an inconsistent pattern of determinations (fig. 13.17), because of the presence of cavities undergoing some degree of collapse, and the action of shrinking/swelling clays and organic soils.

To date, there is no single definitive set of criteria that allows consistent differentiation of sinkhole damage from that caused by shrinking/swelling clays or oxidation of organic materials, nor is there a single satisfactory standard practice for identification of the processes responsible for structural damage where clays, organic materials, and karst coexist. Frank and Beck (1991) have compiled a list of criteria for distinguishing sinkhole activity from the action of expansive soils (table 13.3). Shrinking/swelling clays and organic materials are commonly associated with alluvial sinkholes and may be concentrated within sinkholes. Better detection practices, site-specific coordination and interaction between engineers and geologists, and detailed field and laboratory testing are needed to resolve this problem.

SLOPE-STABILITY PROBLEMS

Surficial materials move wherever there are slopes on the land surface. When this movement is slow, and there is no evidence of short-term dislocation or damage, the motion is called soil creep. When the motion is rapid, and displacement is evident, the resultant features are called landslides or slumps.

Landslides and slumps occur where slopes are steep and the regolith lacks cohesion. Landslides and slumps are not a serious problem in Florida, because slopes are relatively gradual and there is little relief; they occur most often at construction sites, where slopes are oversteepened and cohesiveness is lost through human activity. They also occur in hilly terrains, such as the Tallahassee Hills, Cody Escarpment, Brooksville Ridge, Lake Wales Ridge, and other areas. The Pitt landslide, which occurred on 2 April 1948 in Gadsden County (Rupert 1990), is a rare example.

Soil creep is not considered a hazard, because of the slow rate of movement and low potential for catastrophic damage; however, many structures develop "settling cracks" as a result of soil creep. These features can often be recognized as cracks in driveways and foundations that parallel the contours of the land and indicate

Table 13.3. Criteria for differentiation of sinkhole and expansive clay damage

Sinkhole Activity	Expensive Clay Activity
Presence of a buried karst surface	Expansive clay at different depth below structure than adjacent
Loose zones or voids encountered by standard penetration testing and/or geophysical surveys	Evidence of moisture changes at time of damage
Jumbled or broken strata above limestone	Evidence of differential drying or wetting at time of damage
Evidence of raveling zones or subsurface erosion	Lateral continuity of clay across structure and absence of evidence of disruption or sagging into sinkholes
Evidence of overlying strata slumping or sagging into deeper karstic features	Cracks that open and close in response to rainfall or water-table elevation
Locally depressed water table	
Organic muck at depth due to sinkhole infilling	
Nearby sinkhole activity in same time frame	
Intermixing of surficial sands with clays at depth	
Deep solution basins in limestone surface	
Absence or disruption of residual clays over a depression or limestone surface	
Cracks which do not open or close with moisture changes	
Presence of expansive clay in subsurface	

Source: Frank and Beck (1991).

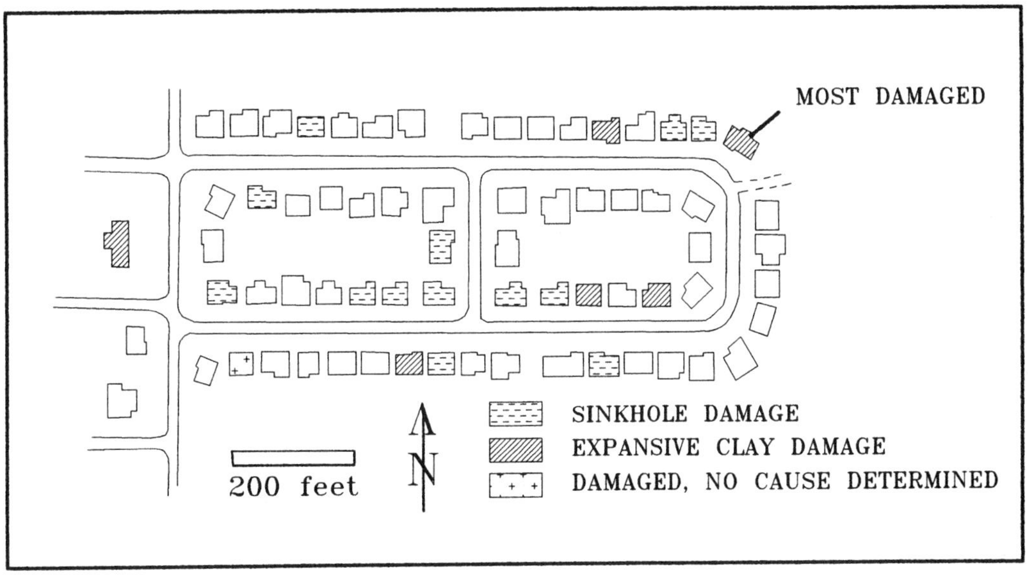

Figure 13.17. Pattern of damage reports in a Dunedin subdivision. Different engineering firms diagnosed adjacent homes as being damaged by shrinking/swelling clays, sinkholes, and other causes, notably oxidation of organic materials (modified from Frank and Beck 1991, their fig. 9, updated to 1992).

Table 13.4. Summary of the chemical composition of precipitation from selected sites in Florida

Stat.	Ca (mg/L)	Mg (mg/L)	K (mg/L)	Na (mg/L)	NH_4 (mg/L)	NO_3 (mg/L)	Cl (mg/L)	SO_4 (mg/L)	PO_4 (mg/L)	pH (field)	Conductivity (field)	Na/Cl (mole ratio)	Ratio dev. seawater
Quincy, Gadsden County													
x	0.14	0.07	0.06	0.44	0.15	1.05	0.75	1.77	0.02	4.68	17.7	0.90	0.05
s	0.17	0.08	0.18	0.53	0.24	1.05	0.89	1.84	0.09	0.41	15.9	0.22	0.22
n	179	179	179	179	179	179	179	179	179	160	162	179	179
Min.	0.01	0.01	0.00	0.03	0.00	0.00	0.06	0.10	0.00	3.57	4.0	0.40	−0.45
Max.	1.15	0.53	2.12	3.23	1.53	6.19	5.58	12.78	0.78	5.90	132.5	2.90	2.05
Austin-Cary Forest, Alachua County													
x	0.28	0.09	0.09	0.78	0.15	1.00	0.96	2.03	0.00	4.79	17.87	1.39	0.54
s	0.28	0.07	0.16	0.85	0.85	0.88	0.99	1.47	0.00	0.53	12.09	1.01	1.01
n	92	92	92	92	92	92	92	92	92	89	92	92	92
Min.	0.02	0.00	0.00	0.02	0.02	0.07	0.10	0.10	0.00	3.49	3.70	0.13	−0.73
Max.	1.33	0.39	0.95	4.87	4.87	4.70	8.88	8.88	0.00	6.70	66.70	6.04	5.19
Bradford Forest, Bradford County													
x	0.26	0.11	0.07	0.80	0.16	1.04	1.19	1.96	0.01	4.70	16.77	1.05	0.20
s	0.44	0.26	0.22	2.13	0.24	0.95	3.16	2.13	0.07	0.46	12.82	0.64	0.64
n	367	367	367	367	367	367	367	367	367	340	337	366	366
Min.	0.01	0.00	0.00	0.04	0.00	0.00	0.00	0.00	0.00	3.22	2.20	0.31	−0.54
Max.	5.40	4.31	3.81	29.30	1.92	6.60	52.62	22.80	1.19	6.60	99.00	6.17	5.32
Kennedy Space Center, Brevard County													
x	0.28	0.20	0.09	1.58	0.10	1.05	2.81	1.93	0.02	4.92	23.15	0.87	0.02
s	0.34	0.25	0.10	2.03	0.17	1.13	3.64	1.71	0.07	0.37	14.99	0.14	0.14
n	229	229	229	229	229	229	229	229	229	208	201–0.80	229	229
Min.	0.01	0.00	0.00	0.09	0.00	0.00	0.15	0.20	0.00	3.73	85.80	0.37	−0.48
Max.	3.28	1.70	0.80	13.82	1.18	10.12	12.81	12.81	0.58	5.72		1.91	1.06
Verna Well Field, Sarasota County													
x	0.31	0.14	0.23	0.81	0.21	1.00	1.39	1.53	0.05	4.85	15.02	0.90	0.05
s	0.42	0.25	1.40	1.50	0.63	1.15	2.62	1.47	0.39	0.54	10.70	0.22	0.22
n	202	202	202	202	202	202	202	202	202	159	177	202	202
Min.	0.02	0.01	0.00	0.07	0.00	0.00	0.14	0.15	0.00	3.39	3.00	0.51	−0.34
Max.	3.49	1.93	17.40	13.31	7.30	10.32	24.53	13.64	4.98	7.30	85.30	2.31	1.46
Everglades National Park, Dade County													
x	0.36	0.20	0.20	13.2	0.22	0.73	2.31	1.41	0.07	4.98	15.98	0.88	0.03
s	0.73	0.31	1.03	1.89	1.12	0.85	3.21	1.62	0.63	0.57	13.90	0.13	0.13
n	304	304	304	304	304	304	304	304	304	2.61	269	304	304
Min.	0.01	0.01	0.00	0.05	0.00	0.00	0.12	0.00	0.00	3.00	3.50	0.56	−0.30
Max.	7.68	2.66	12.80	15.91	17.12	8.37	26.89	15.42	9.98	7.27	141.50	2.22	1.37
Statewide													
x	0.28	0.14	0.13	1.00	0.17	0.97	1.66	1.75	0.03	4.77	17.58	0.96	0.11
s	0.48	0.25	0.74	1.80	0.61	1.01	2.98	1.80	0.34	0.50	13.80	0.47	0.47
n	1373	1373	1373	1373	1373	1373	1373	1373	1373	1217	1231	1373	1372
Min.	0.01	0.00	0.00	0.02	0.00	0.00	0.00	0.00	0.00	3.0	0.80	0.13	−0.73
Max.	7.68	4.31	17.4	29.3	17.2	10.32	52.62	22.8	9.98	7.3	141.5	6.17	5.32

Source: From Upchurch 1992 and based on data from the National Atmospheric Deposition Program, National Trends Network.
Note: x = arithmetic median; s = standard deviation; n = number of samples.

down-slope movement. Leaking swimming pools, septic tanks, and runoff-detention ponds can exacerbate creep through soil wetting and loss of cohesion.

NATURAL GROUNDWATER QUALITY

The natural quality of groundwater can represent a hazard in the sense that water use is limited where it is too salty for consumption or irrigation. Natural water quality may also contribute to such problems as boiler scale, water hardness (which inhibits soapsuds formation), unpleasant taste and odor, and turbidity.

Rainfall Chemistry A major factor that affects groundwater is the chemical composition of rainfall. Natural rainfall is affected by reactions with atmospheric gases and particulates. The sea also has a profound effect on rainfall composition, because sea spray is transported inland as an aerosol; this aerosol is mixed with precipitation, so that Florida rainfall is a very dilute mixture that has the ionic proportions—but not the concentrations—of seawater.

Table 13.4 summarizes the chemical quality of rainfall in Florida. Note that small amounts of all the major groundwater chemicals are present, as well as some nitrate and sulfate (probably being introduced as acid rain; Upchurch 1992). As a general rule, newly fallen precipitation, uncontaminated surface runoff, and uncontaminated soil waters in Florida have the compositions of dilute seawater. Whereas total dissolved solids are low, the ratios of dissolved metals (especially Na) to chloride are nearly constant and reveal an origin in marine aerosols. As this dilute seawater solution evaporates from the land surface and subsoils, or is transpired by plants, the compounds dissolved in the water are concentrated. The increase in dissolved solids by evapotranspiration is an important starting point in the evolution of groundwaters.

Soil and Aquifer Mineralogy Water that has passed through the soil is characteristically acidic. Reactions with calcite (equation 3) cause dissolution. Reaction of acidic meteoric water with potassium feldspar results in the formation of the common soil clay mineral kaolinite. Silicon is mobilized as silicic acid, and potassium is released as a dissolved cation (K^+). Gibbsite can also form.

Iron and aluminum oxyhydroxides, sulfide and sulfate minerals, and a few other minerals are weathered by different processes, which are discussed in Upchurch (1992). Clay minerals and some of the oxyhydroxides are also important as sites for ion exchange, which may also affect groundwater quality.

Table 13.5 lists the compositions of common minerals found in Florida and their dissolved weathering products. Most of these minerals are weathered slowly, so an important factor in determining how much of the weathering product enters the groundwater is the length of time the water is in contact with a particular rock type (i.e. the residence time).

Table 13.6 lists the most common minerals found in Florida aquifer systems and confining beds. The surficial aquifer system is composed predominantly of chemically inert quartz sand, but contains varying amounts of calcite and aragonite in coastal areas. This surficial aquifer system contains highly variable amounts of clay, oxyhydroxides, and humic material (table 13.6), all of which are reactive and have the ability to adsorb ions. In all, the sand-rich surficial aquifer system has the ability to adsorb low to moderate amounts of cations and anions. Carbonate-rich parts of the aquifer system have the ability to buffer acidity.

The intermediate aquifer system and confining beds (Hawthorn Group) comprise a complex array of materials (table 13.6), with important consequences for groundwater quality. Weathering of clays releases Fe, Mg, and silica, and produces kaolinite as a residue (table 13.5). Clays have very high absorption capacities and can effectively bind most metals. Reactions of groundwater with carbonate minerals buffer the acidity and release Ca and Mg into the system (table 13.5). The primary phosphorite deposits are composed of carbonate-fluorapatite, while weathered and reprecipitated deposits contain carbonate-hydroxylapatite (table 13.5). Carbonate-fluorapatite contains F, P, and small amounts of U (Altschuler et al. 1958), which are released upon weathering. Also present are ubiquitous but small quantities of pyrite, which releases Fe and S as sulfate or sulfide to groundwater. The sulfate may be in the form of sulfuric acid; the sulfide may be present as hydrogen sulfide, which imparts a "rotten egg" odor to water. The Hawthorn also contains gypsum, a source of sulfate (table 13.5), at scattered localities (Upchurch 1992).

The mineral assemblage of the Upper Floridan aquifer system is less complex than that of the other aquifer systems, and the predominant minerals are calcite and dolomite (table 13.6); therefore, water that has equilibrated with the Upper Floridan aquifer system includes Ca^{2+}, Mg^{2+}, and HCO_3^- as dominant chemical species. The pH is buffered by dissolution of carbonate minerals and generally ranges from 7 to 8. The base of the aquifer system is characterized by reduced permeability, partly because of intergranular gypsum and anhydrite (tables 13.5 and 13.6; see chapter 6), and sulfate is a major constituent of water that has come in contact with the base of the Upper Floridan aquifer system.

The length, depth, and tortuosity of the flow path

Table 13.5. Common minerals in Florida and their dissolved weathering products

Mineral	Composition	Dissolved weathering products
anhydrite	$CaSO_4$	Ca^{2+}, SO_4^{2-}
aragonite	$CaCO_3$	Ca^{2+}, HCO_3^-
calcite	$CaCO_3$	Ca^{2+}, HCO_3^-
carbonate-hydroxylapatite	$Ca_5(PO_4CO_3)_3(OH)$	Ca^{2+}, PO_4^{3-}, HCO_3^-, plus trace $U^{4+, 6+}$
carbonate-fluorapatite	$Ca_5(PO_4CO_3)_3F$	Ca^{2+}, PO_4^{3-}, HCO_3^-, plus trace $U^{4+, 6+}$
dolomite	$CaMg(CO_3)_2$	Ca^{2+}, Mg^{2+}, HCO_3^-
ferric hydroxide	$Fe(OH)_3$	$Fe^{2+, 3+}$
gibbsite (G), boehmite and diaspore (BD)	G: $Al(OH)_3$ or $Al_2O_3 \cdot 3H_2O$ BD: $AlO(OH)$	Al^{3+}
goethite	$FeO(OH)$	$Fe^{2+, 3+}$
gypsum	$CaSO_4 \cdot 2H_2O$	Ca^{2+}, SO_4^{2-}
hematite	Fe_2O_3	$Fe^{2+, 3+}$
K-feldspar	$KAlSi_3O_8$	K^+, H_4SiO_4*
K-mica	$KAl_2(Si_3Al)O_{10}(OH,F)_2$	K^+, F^-, H_4SiO_4*
kaolinite	$Al_2Si_2O_5(OH)_4$	Al^{3+}, H_4SiO_4*
Opal (–A, –CT)	$SiO_2 \cdot nH_2O$	H_4SiO_4
palygorskite	$(Mg,Al)_2Si_4O_{10}(OH) \cdot 4H_2O$	Mg^{2+}, Al^{3+}, H_4SiO_4*
pyrite	FeS_2	$Fe^{2+, 3+}$, SO_4^{2-}
quartz	SiO_2	Essentially inert
sepiolite	$Mg_4Si_5O_{15}(OH)_2 \cdot 6H_2O$	Mg^{2+}, H_4SiO_4*
smectite, var. montmorillonite	$(Na,Ca)_{0.3}(Al,Mg)_2Si_4O_{10}(OH)_2 \cdot nH_2O$	Na^+, Ca^{2+}, Mg^{2+}, Al^{3+}, H_4SiO_4*
smectite, var. nontronite	$Na_{0.3}Fe_2(Si,Al)_4O_{10}(OH)_2 \cdot nH_2O$	Na^+, Fe^{2+}, Al^{3+}, H_4SiO_4*

Sources: Upchurch 1992; mineral formulae from Fleischer 1987.
*Plus residual solids.

Table 13.6. Common minerals in Florida aquifer systems and confining beds

Mineral fraction	Surficial aquifer system	Intermediate aquifer system	Floridan aquifer system
silicate	**quartz** **potassium feldspar** **potassium mica** kaolinite chlorite smectite	**quartz** **potassium feldspar** **potassium mica** palygorskite sepiolite **smectite** kaolinite	quartz
carbonate	**calcite** **aragonite**	**dolomite** calcite	**calcite** **dolomite** aragonite (?)
oxyhydroxides, sulfides	**ferric hydroxide** **Goethite** Gibbsite, boehmite, diaspore, pyrite	pyrite ferric hydroxide Goethite	pyrite ferric hydroxide Goethite
other	humic substances[a]	carbonate-fluorapatite carbonate-hydroxylapatite opal-A opal-CT gypsum	gypsum anhydrite

Source: Upchurch 1992.
Note: See table 13.5 for mineral compositions and weathering products. Volumetrically or chemically important minerals indicated in bold.
[a] Refers to particulate organics, including organics concentrated in peats and mucks, and disseminated in other sediments.

water follows in an aquifer system profoundly affect the quality of water and soil/aquifer-rock interaction. In general, shallow, short flow paths—which are characteristic of the surficial aquifer system—result in low residence times, and chemical reactions may not go to completion. Consequently, total dissolved solids are lower than those anticipated for systems with a longer flow path. By the same reasoning, systems with a short flow path are more vulnerable to contamination.

If the flow path is long (on the order of tens of kilometers), reactions between rock and water become more probable, and the total dissolved solids of the water increase as a result of continued rock weathering. Flow paths of the Upper Floridan aquifer system in central Florida are characteristically long, and changes in composition along the flow paths reflect chemical maturation as reactions occur.

Typical residence times range from days to thousands of years, depending on the nature of the flow system. Residence times in the surficial aquifer system range from days to perhaps hundreds of years, and the flow systems are short (Tóth 1962, 1963); primary discharge is into local wetlands, lakes, streams, and canals. Residence times in the intermediate aquifer system are variable because of the complexity of pathways through the lithologically diverse Hawthorn Group. Water that passes through sinkholes, fractures, and karst conduits that penetrate the Hawthorn may have short residence times. Conversely, water that passes along the tortuous flow paths in the clay-rich horizons may have residence times of thousands to tens of thousands of years. Residence times in the Upper Floridan aquifer system also span a wide range. Short times have been recorded for some sinking streams and resurgences in northern and central Florida (Ceryak 1977). These short residence times are associated with conduit flow over distances of a few kilometers. In contrast, Hanshaw et al. (1965) used ^{14}C dating methods to approximate residence times in excess of 30,000 years for flow from northern Polk County to coastal Sarasota County. This long residence time is associated with groundwater velocities as high as 2 to 8 m per year.

All aquifers contain microbes (bacteria, fungi, and other organisms) that play important roles in the chemistry of the water. To survive, these microbes require sources of organic C, N, P, S, Fe, and other nutrients. Because these nutrients are normally introduced into the aquifer systems near the land surface, the microbes are most abundant there, and they decline in abundance with depth. The most important roles microbes play in maintaining aquifer chemistry involve transformations of N, Fe, and S species. For example, sulfate-reducing bacteria transform sulfate to hydrogen sulfide according to

(5) $\quad SO_4^{2-} + 2C_{organic} + 2H_2O \stackrel[activity]{microbial}{\Leftrightarrow} H_2S + 2HCO_3^-$

By this reaction, organic C and sulfate are consumed and hydrogen sulfide and bicarbonate are produced. Since bicarbonate is also a product of weathering of calcite, addition of bicarbonate through sulfate reduction may induce calcite precipitation. Thus, microbial activity may result in a complex array of "spinoff" reactions that affect major- and minor-element compositions of groundwater and aquifer cements.

Saltwater Intrusion The coastal saltwater/freshwater transition zone can be steep and narrow, or shallow and broad, depending on the hydraulic gradient on the freshwater lens. In northern and central Florida, this zone is usually steep and narrow; in southern Florida, it is usually shallow and broad. As deep water from the Upper Floridan aquifer wells up along the transition zone, it brings a calcium-magnesium-sulfate-bicarbonate signature. This water mixes with seawater to produce a series of water-quality facies that parallel the coast (fig. 13.18). The outermost belt of the transition zone has the composition of seawater. The innermost belt is calcium-magnesium-bicarbonate water. In between, there is often a sulfate-rich belt, and gradations in water quality. Upwelling brings water from the Upper Floridan aquifer into contact with waters of the intermediate and surficial aquifer systems, where mixing also occurs.

Figure 13.19 illustrates Stiff diagrams showing water quality in the principal aquifers of coastal Lee County (Upchurch 1986). Note that the shallow aquifers show an increasing influence of seawater toward the coast. Surficial-aquifer water has low total ionic strength and is a calcium-bicarbonate water, even near the coast. The intermediate aquifer has transitional compositions, with an increasing influence of seawater mixed with calcium-bicarbonate water. The Upper Floridan aquifer is predominantly a sodium-chloride solution with strong Mg and sulfate components. Water from well L2527 was incorrectly analyzed, and Na is abnormally low. Water in well L1983 is a sodium-bicarbonate water because of exchange of Ca^{2+} for Na^+. This pattern is typical of waters upwelling and mixing in aquifers along the coasts of Florida.

Saltwater intrusion is a problem throughout the state. Heavily urbanized areas, such as Tampa–St. Petersburg and Miami, have used so much water that draconian measures are required to limit additional intrusion. These measures include salinity dams on rivers and canals to limit tidal intrusion and maximize freshwater head, restrictions on water use, and limitation of pumping in threatened areas. The chloride-abundance map of Katz (1992) clearly illustrates regions where wedges of saltwater intrusion have occurred (fig. 13.18B). Inland, pumping and reduction of freshwater head leads to upward migration of saline water (Upchurch 1992). This water may resemble seawater in composition, but it is usually enriched in sulfate relative to seawater.

Maturation Patterns in the Aquifers As water flows through the near-surface aquifer systems and then into the Floridan aquifer at depth, water quality changes through a process called maturation. In general, the groundwater becomes more alkaline as it matures, and there are increases in sulfate content and total dissolved solids.

Back and Hanshaw (1970) tracked the evolution of water quality in the Upper Floridan aquifer from recharge to discharge. Combining their data with water-quality data from the surficial and intermediate aquifer systems (Upchurch 1992) reveals a three-step process of chemical maturation in the Floridan aquifer system (fig. 13.20). Step 1 involves initial equilibration of the acidic, sodium chloride-rich water of the surficial aquifer with

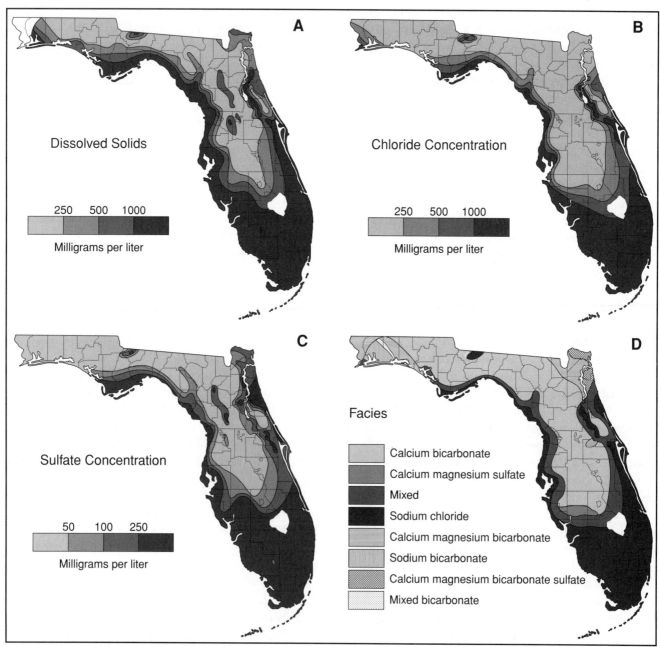

Figure 13.18. Abundance of selected chemical constituents in the waters of the Floridan aquifer system. A: Total dissolved solids. B: Chloride. C: Sulfate. D: Chemical facies (maps modified from Katz 1992).

carbonates; this step occurs upon recharge to the Upper Floridan aquifer and along the downward-directed segment of the flow path. Step 2 involves dissolution of gypsum and anhydrite and mixing with the resulting sulfate-rich water; this step occurs along both the downward- and upward-directed segments of the flow path, but the maximum increase in sulfate occurs near the base of the aquifer. Step 3 involves dissolution of carbonate rocks and mixing seawater; it occurs along the upward-directed segment of the flow path, as the water skims along the saltwater/freshwater transition zone. Total dissolved solids increase throughout the three-step process. Thus, groundwater continuously changes in chemical composition as it matures in the upper Floridan aquifer.

HAZARDS ASSOCIATED WITH SURFACE WATER

Florida's maritime climate is prone to drought and flood. The last twenty years have seen several extended periods of drought. Major floods have occurred in 1960 (central Florida, from Hurricane Donna), 1964 (northern Florida, from Hurricane Dora), 1969 (Ochlockonee River basin, from about 60 cm of precipitation in 24 hours), 1973 (northern Florida), 1975 (northwestern Florida, 50 cm of rain), and so on (U.S. Geological Survey 1989).

Flooding from hurricanes of the 1920s and demands for water in southern Florida and the Everglades led to creation of the Central and Southern Florida Flood Control District in the 1940s. The district, now the South Florida Water Management District, is responsible for one of the most elaborate flood-control systems in the world (fig. 13.21). Drainage projects began in southern Florida as early as 1881, when dredging was undertaken to connect the Caloosahatchee River and Lake Okeechobee. Drainage of the Everglades and Kissimmee River valley began in 1883 and continued into the 1980s. The floods associated with Hurricane Donna in 1960 led to expansion of the concept of the flood-control district, and the four other water-

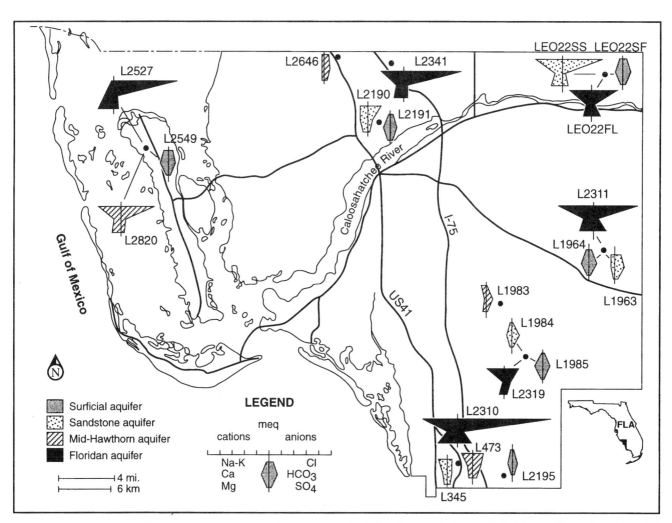

Figure 13.19. Stiff diagram of groundwater compositions for various aquifers in Lee County (Upchurch 1986). Water wells are identified by "L" and four number designations.

management districts were created by the legislature over the next ten years.

Flooding is still a problem. Many areas are low-lying and not amenable to adequate drainage. Others have been protected by flood-diversion systems or impoundments. Today, controversies have developed as to the efficacy of flood control relative to environmental consequences. For example, by authorization of Congress in 1954 the Kissimmee River was straightened by the Corps of Engineers to enhance drainage and navigation. By the 1970s increased nutrient loading in Lake Okeechobee occasioned a number of studies that concluded that the Kissimmee River Project had been detrimental to the lake. Today, plans are being made to restore the river to its original course, which will decrease drainage capacity, but increase the quality of river and lake water.

Anthropogenic Hazards

Human activity can cause environmental deterioration and hazards through many pathways. Emphasis in this section will be on those hazards that affect the largest segments of Florida's population.

HAZARDS ASSOCIATED WITH RESOURCE EXTRACTION

Mining practices in Florida can cause local environmental damage. Each type of mine has different problems, and it is not possible to discuss all of them. With the exception of phosphate mining, few mining operations have received attention relative to water quality and off-site environmental impacts.

Sand and Gravel Mining   Sand and gravel are mined throughout the state. Sand and gravel mines are subject to such problems as lowering of the water table or potentiometric surface, disposal of clay wastes, and local water contamination by hydrocarbons from machinery. Most sand and gravel mines have remarkably good records of minimal damage to the environment outside the mine itself. The fate of the mined-out pit may be of more concern than the mining operation itself, because in the past these pits have been used for disposal of wastes, posing a threat to groundwater quality. Today, only construction debris and similar "clean" wastes can be placed in these pits—and then only with government permits.

Limestone Mining   Like sand and gravel mines, limestone mines have good records of controlling environmental damage. The potential sources of damage are the same: adverse impacts on aquifer hydrology, disposal of fine-grained materials in slurries, and local hydrocarbon contamination from machinery operation. Crushing of limestone increases its solubility and rate of

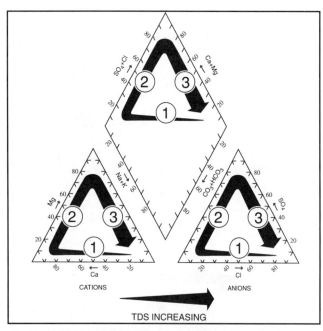

Figure 13.20. Three-step process of maturation of water quality in the Floridan aquifer system. Step 1: equilibration of water with carbonate rock; step 2: mixing with sulfate derived from gypsum and anhydrite; step 3: mixing with sea water (modified from Back and Hanshaw 1970 and Upchurch 1992).

Figure 13.21. "Structure 33," a gated spillway used to regulate water discharge from a flood-control canal. It also serves to prevent saltwater intrusion in the Broward County area. The spillway is part of the flood-control and water-conservation system of the South Florida Water Management District (photo courtesy of the South Florida Water Management District).

dissolution, so there may be increases in water hardness downstream from quarries. Lawrence and Upchurch (1982) observed a plume of calcium-bicarbonate water from a quarry that penetrated the Upper Floridan aquifer near Live Oak, in Suwannee County. The plume had traveled several kilometers, which suggests that there is some potential for aquifer contamination.

Phosphorite Mining Perhaps that largest single focus of environmental concern in Florida has been the phosphate industry—a highly visible and important industry in Florida (U.S. Environmental Protection Agency 1978a, 1978b) (see chapter 9). There are several areas of concern, including the effects of mining, reclamation, and agrichemical plants. The mining and processing industries have been accused of creating numerous water-related problems, but extensive monitoring and surveillance by regulatory agencies have greatly limited provable damage. These industries have also received much attention because of the possibility of hazards from radiation.

Mining is done by dragline, and the overburden is cast aside in the pit for later reclamation (fig. 13.22a). Modern practice emphasizes reburial of any radioactive overburden. The pit is dewatered at the time of mining. Dewatering offers two advantages: (1) greater control on mining, so that clays and other non-phosphate, low-profit strata can be avoided, and (2) increased slope stability, so that the dragline and longwall are stable during mining. Mining concerns include disposal of water during dewatering, possible contamination of aquifers exposed during mining, and disruption of the landscape.

These concerns are compounded by other aspects of the industry (Gordon Palm and Associates 1983; U.S. Environmental Protection Agency 1978a, 1978b; Upchurch 1989). Reclamation, which is required by Florida law, is highly controlled, and hazards from radioactivity and other sources are minimized. But this has not always been the case. As figure 13.22b shows, some parts of the mined-out phosphate land were not reclaimed in the past. These lands display rugged topography characterized by linear mounds surrounded by serpentine lakes, many of which ultimately develop into scenic and biotically diverse woodlands that are utilized

Figure 13.22. Phosphorite mining processes. A: A dragline is used to mine the ore. Note the overburden cast to the side in the dewatered pit. B: Reclaimed land. Note that reclamation leaves the land recontoured with numerous lakes. C: Typical clay-waste disposal pond. These "slimes" ponds are typically above grade and are contained by dams. D: An agrichemical plant with associated gypsum-disposal facility ("gypstack").

as parks and other natural areas. In some areas, these lands were converted into residential areas with great natural amenities. Unfortunately, the early wastes were not disposed of in a way to minimize radiation emanation from included uranium progeny, and some of these subdivisions have developed radon problems as a consequence.

Failure to reclaim mined land limits land use and the tax base. Typical modern reclamation goals include production of usable and safe lands and viable wetlands (fig. 13.22b). Overburden and waste products from phosphorite-ore processing (beneficiation) are used in reclamation.

Beneficiation takes place near the mine. The ore is transported to the beneficiation plant in a water slurry. At the plant, the ore is separated into three fractions: phosphate concentrate, sand, and clay waste. The phosphate is transported to the agrichemical plant. The sand is transported back to the mine, where it is used to reclaim mined-out land and clay-waste ponds. Disposal of the clay is one of the biggest problems the industry faces. The clay wastes, or "slimes," are transported to slimes ponds, where settling and dewatering occur. Since the clay wastes occupy a larger volume than the material removed from the ground, mine cuts are insufficient for waste disposal, and the clay wastes are disposed of in large, above-grade ponds surrounded by dams (figure 13.22c). In the past, dam failures have been a problem. One such failure occurred in December 1971, when two billion gallons of clay wastes entered the Peace River (Harris et al. 1972). The river was turbid all the way to Charlotte Harbor, there was a massive fish kill, and as much as 15 cm of slime was deposited in the river and upper estuary. Grace (1977) searched for the clay wastes in the Peace River estuary; he found that bioturbation had removed physical evidence of the spill, but P and Fe contents of the sediments indicated that the effects of the spill had not completely disappeared. Modern regulations governing the construction of dams have all but eliminated this risk in new ponds. The clays appear to seal the bottoms of the ponds, so groundwater contamination is minimal. Most of the present concerns involve the long-term fate of the ponds. Settling of the clay wastes is slow, and it takes tens of years for a pond to settle to the point where it will bear traffic. Construction on the dewatered ponds likely is not feasible, owing to the long-term high water contents of the clays, so use is limited to some forms of agriculture.

The phosphorite concentrate is shipped to an agrichemical plant for processing (fig. 13.22d). The typical plant treats the concentrate with sulfuric acid, which produces two products: (1) phosphoric acid, from which fertilizers are produced, and from which F and U are recovered, and (2) gypsum in an aqueous solution. There is moderate concern about the phosphoric acid product, because spills occasionally occur. On the other hand, the gypsum is of great concern, because it is slightly radioactive and contains radium, numerous trace metals, and rare-earth elements. It has little commercial value. The gypsum is disposed of in above-grade ponds (locally called "gypstacks") and the waste water is recirculated into the plant. The ponds typically grow to elevations in excess of 30 m, which gives them considerable hydraulic head. Gypstacks, therefore, are potential sources of acidic runoff and groundwater contamination. They may also act as sources of airborne dust, radon, and other contaminants. Studies by Miller and Sutcliffe (1982, 1984) and others have documented localized groundwater contamination and other impacts, which are typically limited to areas immediately adjacent to the gypstack. In the absence of a better disposal technique and lack of evidence of regional contamination—and given the importance of the industry to the economy of Florida and the nation—gypstacks are an undesirable byproduct whose utilization awaits future technologies. One of the major aims of the Florida Institute for Phosphate Research in Bartow is to find uses for the gypsum. Some possible applications being explored include use of gypsum as a soil amendment and in road and building construction materials. To date, signfiicant applications and markets have not developed.

Oil and Gas Extraction Florida is not known as an oil-producing state, yet in 1978 it ranked ninth in oil and gas production in the U.S. (see chapter 9). This production is from two widely separated regions of the state. The oldest production is from several small fields in the northwestern part of the panhandle, in northern Santa Rosa County. Oil is also produced in southern Florida. The southern Florida fields extend from western Lee County, through southwestern Hendry and north-central Collier County, to northwestern Dade County (Campbell 1986).

The onshore petroleum industry has a good environmental record, including close cooperation with state agencies and environmental groups in the environmentally sensitive Big Cypress Swamp and northern Everglades. Damage to overlying aquifers is minimized by active wellhead protection and monitoring. Development of oil fields involves protection of wetlands and minimization of impact at the land surface.

Controversy is focused on offshore oil and gas exploration and production in Florida. Offshore exploration in the panhandle and off Lee County has not been en-

Table 13.7. Proportions of groundwater samples exceeding state and federal contaminant guidelines or standards as determined in assessment of background water quality under the Florida Ground-Water Quality Monitoring Program (in %)

Aquifer	Synthetic organics	Pesticides	Mercury	Lead	Nitrate
Surficial aquifer system	7	3	2	7	<1
Biscayne aquifer	10	0	0	1	<1
Sand-and-gravel aquifer	3	1	11	14	<1
Surficial aquifer	7	4	2	10	<1
Intermediate aquifer system	1	7	3	8	0
Floridan aquifer system	3	5	1	9	1

Source: Maddox et al. 1992.

couraging, but interest continues for potential development, especially in western Florida. The state and many environmental groups have actively tried to block offshore exploration, especially in environmentally sensitive regions such as the Keys. At present, there is a moratorium on exploration, and attempts are being made to indemnify leaseholders and make the ban permanent.

Environmental damage from oil and gas exploration, and from field development and production, has been limited to the site itself. The only long-term damage has resulted from poorly abandoned test wells drilled in the early days of exploration. Many of these wells are free-flowing and are inducing saltwater intrusion in overlying aquifers. This practice is no longer allowed, and old, free-flowing wells are being plugged by the water-management districts.

HAZARDS ASSOCIATED WITH WASTE DISPOSAL AND CONTAMINANTS

Waste disposal includes intentional disposal of wastes in landfills, septic-tank systems, and other designated systems, and accidental losses, as from agriculture, leaky underground storage tanks, and equipment maintenance on industrial sites.

The Florida Department of Environmental Regulation (now called the Department of Environmental Protection) has assessed the background quality of groundwater in Florida (Maddox et al. 1992). As part of this assessment, a number of indicators of pollution from wastes and pest control were investigated. Tests were made for synthetic organics, pesticides, mercury, lead, and nitrate (table 13.7). Heavily contaminated urban and industrial sites were avoided, so that the assessment reflects general water quality—not water quality in regions likely to be contaminated. Also, because of the potentials for sample contamination or bad analytical procedures, the authors of the study recommended that resampling and analysis be undertaken where there might be concerns about the results. The results are not

Table 13.8. Synthetic organic compounds found in Florida groundwater

acrylonitrile	1,2 dichloroethane
benzene	trans-1,2 dichloroethane
bromodichloromethane	1,2 dichloropropane
bromoform	ethylbenzene
chlorobenzene	hexachlorobenzene
chloroform	methylene chloride
chloromethane	PCB-1016
dibromochloromethane	1,1,2,2 tetrachloroethane
1,2 dibromomethane	1,1,1 trichloroethane
1,2 dichlorobenzene	tetrachloroethane
1,3 dichlorobenzene	toluene
1,4 dichlorobenzene	trichloroethane
dichlorodifluoromethane	trichlorofluoromethane
1,1 dichloroethane	vinyl chloride

Source: Florida Department of Environmental Regulation 1989.

weighted by area or aquifer, so a region with a small number of samples may show a higher proportion of exceeding values than one with a large sample size. Finally, table 13.7 lists only the proportion of samples in which the analysis exceeded the state groundwater-quality guidance criterion (Florida Department of Environmental Regulation 1989). Many samples had detections that did not exceed standards.

Synthetic Organics The study of synthetic organics (table 13.7) in Florida aquifers included 144 compounds, many of which are industrial solvents, such as benzene, xylene, toluene, and trichloroethylene. In general, the level of contamination was found to be low. The detected compounds are listed in table 13.8.

At most of the sites where a synthetic organic compound was detected, the aquifer is unconfined and therefore vulnerable to contamination. The proportion of samples possibly containing synthetic organics is low. The Biscayne aquifer contained the highest number of positive samples, but many of the samples are suspect, and resampling to validate the results is under way.

Pesticides Pesticides include 109 herbicides, insecticides, and other chemicals that are primarily utilized, or have been previously utilized, in the agricultural industries. Most are synthetic organics, such as DDT, chlordane, ethylene dibromide (EDB), aldicarb, and 2,4-D. Arsenic and arsenic-based compounds were also included (Upchurch 1992). Again, the proportion is low (table 13.7), the greatest number of possible positive samples being from the intermediate aquifer and the Upper Floridan aquifer. Intensely developed agricultural areas, such as large citrus groves and truck farms, were avoided in the background study. The detected chemicals include Aldrin, arsenic, a-BHC, b-BHC, 2,4-D, 4,4'-DDE, 4,4'-DDT, Dieldrin, Endrin, Methoxychlor, and Mirex. Many of these detections reflect past agricultural practices. For examples, DDT has been banned from use in the U.S. for almost twenty years. Arsenic was widely used at the turn of the century as a livestock dip to control ticks.

Several pesticide problems have been identified in the last few years in these intensely utilized areas. It was discovered in 1982 that Aldicarb (Temik), a widely used pesticide, had contaminated a number of drinking-water wells in Hillsborough, Martin, Polk, St. Johns, Seminole, and Volusia counties. Migration had proceeded as far as 90 m from the point of application (Florida Department of Environmental Regulation 1984). Aldicarb use was temporarily suspended in 1983, until solutions to the contamination problem could be found. Contamination by EDB, which was used to control nematodes in soils, has been found in more than 1,000 water wells since 1983. Fifty public-supply wells were found to be contaminated (U.S. Geological Survey 1986a). Highest incidences of contamination were found in Jackson, Lake, Highlands, and Polk counties. The Florida Department of Agriculture took swift action to assess the extent of contamination and hazards of these pesticides, and to modify regulations concerning their use to further protect the public.

Mercury (Hg) has been used in pesticides and a number of industrial applications. It is also a natural constituent of fossil fuels. There is much concern about Hg in Florida, because it has been found in fish in many areas of the state. As table 13.7 indicates, small amounts of Hg were found in groundwaters throughout the state. The high proportion of occurrences in the sand-and-gravel aquifer is misleading: eight of seventy-five samples contained Hg, and several of these are suspect. Lead (Pb) is also found in small quantities throughout the state (table 13.7). Lead was a constituent of some gasoline products until the 1980s. Use of leaded gasoline, and use of Pb in solders and as weights in monitor wells, has produced the results shown. Again, although the proportion of samples with Pb in excess of the standard is higher than for other analytes, some of the samples are suspect. Finally, nitrate, introduced by application of animal wastes and fertilizers, is a widespread problem (table 13.7). Groundwater contamination by nitrates is most common in regions with dairies and feedlots.

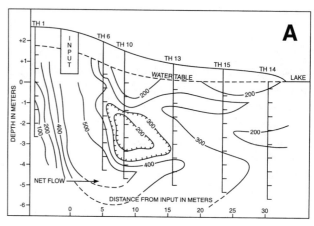

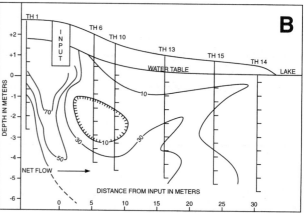

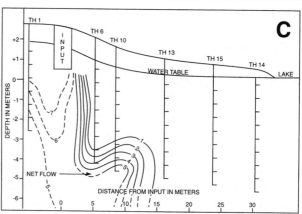

Figure 13.23. Example of chloride (A), nitrate (B), and phosphate (C) plumes from a septic tank in Hillsborough County. Chloride, the most rapidly moving constituent, has reached the vicinity of the lake (chloride concentrations are in milligrams per liter) (Rea and Upchurch 1980).

Contamination in Suburban, Urban, and Industrial Contexts Sources of pollution vary widely, but they can be assigned to several categories based on their importance in Florida, and ranked in order of decreasing relative incidence: septic tanks and on-site sewage disposal; leaky underground storage tanks; industrial sites, feedlots, dairies, food-processing plants; landfills; and others.

Septic-tank systems. More than 40% of Florida's households (1.4 million; Swihart et al. 1984) rely on septic-tank systems. The effluent from septic tanks contains N, P, and other constituents that can threaten surface waters and groundwaters. This effluent is especially hazardous in densely developed areas where many septic tanks are placed in a small area. In order to reduce the loading of septic-tank wastes, local ordinances limit the number of septic tanks allowed per acre, the setbacks allowed from nearby surface waters and domestic wells, and the design of the system relative to local soil conditions and size of the structure the system serves. Most of these systems function properly, although high water levels and failure to remove sludge (septage) from the tank periodically become problems.

Figure 13.23 shows a typical effluent plume from a Florida septic tank. Note that the plume is complex and that chloride (A) has moved the greatest distance, while phosphate (C) has moved least. Chloride is conservative and does not react with regolith particles or microbes, so it moves at the velocity of the water. Nitrate does not react with regolith materials, but soil microbes and plants may remove it. Phosphate is least mobile, because it is strongly absorbed on ferric hydroxides in the soil. Many of Florida's soils have low ability to absorb phosphates and other waste products. High water tables tend to inhibit aerobic bacteria that convert organic N and ammonium into nitrates. Use of garbage disposals and other appliances produces particles that plug the pore throats in the soil and reduce permeability. The septic tank diagrammed in figure 13.23 suffers from all of these problems, and is a likely candidate for failure.

There is a growing realization that septic-tank systems are not necessarily the solution for all domestic wastewater problems. Many local governments are now reviewing their ordinances, and there is much pressure to reduce the allowable density of septic-tank systems.

Leaky underground fuel-storage tanks. Leaky underground fuel-storage tanks have national environmental priority because they are so abundant and the escaping

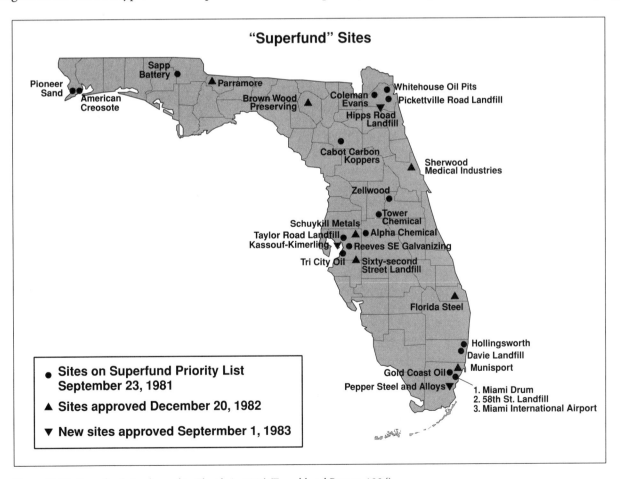

Figure 13.24. Superfund sites located in Florida in 1984 (Fernald and Patton, 1984).

fuel products are toxic and/or carcinogenic. The Florida Department of Environmental Regulation (1986) identified more than 400 sites where gasoline-station storage tanks were contaminating groundwater. The greatest number of incidents was reported in Dade County (77), with Broward (57) and Palm Beach (25) counties second and third. Leakage into the Biscayne aquifer is especially serious, because this aquifer is a major water source. One leaky tank lost over 10,000 gallons of gasoline between October 1979 and March 1980 and contaminated the public supply of Belleview, in Marion County, to the extent that the well field was shut down (U.S. Geological Survey 1986a). The state has a trust fund and assistance program to encourage remediation of underground storage tank leaks.

Hazardous-waste sites. Disposal of municipal and industrial wastes is a continuing problem. Alternatives, such as incineration and recycling, are growing in importance, but there will always be a need for places to dispose of solid and liquid wastes that may be hazardous. Hazardous wastes include any materials known or suspected to be toxic, carcinogenic, mutagenic, or teratogenic in humans or plants and animals. The U.S. Environmental Protection Agency maintains a list of materials that qualify a disposal site as a hazardous-waste site.

The Comprehensive Environmental Response Compensation and Liability Act of 1980 ("superfund" act) required the state and the Environmental Protection Agency to identify hazardous-waste sites and remediate those deemed a threat to the public (fig. 13.24). These superfund sites include landfills that contain hazardous wastes, industrial sites, and other locations where hazardous materials were disposed of in an unsuitable environment. The Florida Department of Environmental Regulation developed a list of 286 possible sites in November 1983. Of these sites, 111 were found to be free of hazardous materials; groundwater contamination was initially identified at 89 sites. By 1986 the list had grown to 413 potential sites, and 70 sites were undergoing remediation (U.S. Geological Survey 1986a).

Municipal-waste sites. Florida has more than 300 active, and 500 inactive, landfills. Most are unlined (U.S. Geological Survey 1986a) and are leaking leachate into the state's aquifers. Many landfills have contaminated the aquifers to the extent that municipal well fields (as in Dade County) have been closed. In some instances domestic wells have been abandoned, necessitating installation of public water-supply systems (as in Hillsborough County).

Injection wells. Injection wells are used to dispose of unwanted water from many sources in Florida. Drainage wells are used to dispose of storm-water runoff and to dewater the surficial aquifer. There are approximately 9,600 drainage wells in Florida (U.S. Geological Survey 1986a). Most of these drainage wells are associated with storm-water disposal systems in karst terrains. The storm water contains trace metals, synthetic organics, solids, and nutrients swept from paved surfaces. The impacts of injection of storm runoff through drainage wells have been documented by Hull and Yurewicz (1979) and Kimrey and Fayard (1982).

Dewatering wells (recharge on inter-aquifer connector wells) were once widely used in the phosphate industry, but strict regulations governing groundwater-quality monitoring have made their use too expensive, so they are being phased out (Upchurch et al. 1991). These wells were used to transfer surficial-aquifer water to the Upper Floridan aquifer through gravity flow. They made mining more cost-effective and mitigated withdrawals from the Upper Floridan aquifer by the industry.

Deep injection wells are used to dispose of treated municipal and industrial effluent (Hickey and Veccioli 1986). The injection target zones are in saline waters of the aquifer system, well below and isolated from potable-water zones. Permitting of deep injection wells requires that the zone be able to physically and chemically accept the waste water and that there be an adequate confining zone above the injection zone to isolate it from potable water. Several wells in Pinellas and Brevard counties are thought to be failing, as waste water has migrated from the injection zone. Concern is rising that deep injection may not be an adequate disposal solution.

Water Supply

Florida has abundant supplies of potable water, especially groundwater. Because groundwater is so readily available and requires less treatment than surface water, it is the source of choice. Surface-water impoundments are difficult to design because the topography is not suitable for large reservoirs. Notable exceptions to the general use of groundwater include the city of Tampa, which uses water from a small reservoir on the Hillsborough River, and Manatee County, which maintains a reservoir on the Manatee River. A few water systems utilize above-grade impoundments (see below), and mined-out land has been converted into below-grade reservoirs.

SURFACE WATER SUPPLIES

Public water systems in southeastern Florida utilize surface water indirectly. The Biscayne aquifer is so thoroughly interconnected to surface water that surface-water management is used to maintain the aquifer. The South

Florida Water Management District maintains a complex system of drainage canals and large, above-groundwater impoundments (conservation areas). During periods of high discharge, water can be pumped into Lake Okeechobee and the conservation areas and retained until needed; therefore, the conservation areas serve as valuable flood-control tools, as well as reservoirs. The head developed by impoundment in these reservoirs assists in recharge of the Biscayne aquifer and serves agriculture in the northern Everglades. Water is released through canals during periods of drought, and to maintain discharge through agricultural areas and Everglades National Park. Canal water is then utilized by means of groundwater pumping, owing to the high permeability of the Biscayne aquifer.

Other municipalities are developing facilities to treat river water during periods of high discharge, and then store the water in aquifer storage-and-retrieval facilities. The Peace River/Manasota Regional Water Supply Authority operates a treatment plant for the City of Port Charlotte, with a capacity of 12 million gallons per day. The facility has a 1,920-acre-foot reservoir and an aquifer storage-and-retrieval facility. Raw water is obtained from the Peace River; this water is treated and injected into permeable zones with fresh to slightly brackish water in the Upper Floridan aquifer. The injected water is recovered when needed, then retreated and distributed (Aikin and Pyne 1991).

GROUNDWATER SUPPLIES

Groundwater is the principal source of public and domestic water in Florida (see chapter 6). Withdrawal of groundwater by industry, agriculture, and municipalities has caused a number of problems in the state, including a decline in the potentiometric surface, saltwater intrusion, and competition among users. These problems arise in part because people and industries are located near the coast. Pumping near the coast has resulted in saltwater intrusion and loss of well fields. Population growth has increased demand. New technologies have been developed to recycle water and artificially recharge the aquifer in addressing the demand and distribution factors impacting regional water supplies.

WATER-SUPPLY ALTERNATIVES

Water supply is a problem in coastal environments and in southern Florida because pumping tends to induce intrusion of saline water. Many communities have been forced to seek alternative water supplies, such as importation of water from sources outside the jurisdiction, aquifer storage and retrieval, and desalination of brackish water.

Importation of water is a growing option—and a concern. Several large, inland well fields have been developed, with pipeline systems that extend into the interior of the state. The most extensive is the system of the West Coast Regional Water Supply Authority, which maintains a number of interconnected well fields serving much of Pasco, Pinellas, and Hillsborough counties. The authority was created to minimize competition between counties for water. "Water wars" have often been a result of this competition. Today, the water wars have diminished, largely through cooperative agreements and regional water-supply authorities.

A number of smaller communities have opted for reverse-osmosis treatment of brackish, transition-zone waters near the coast. Although reverse osmosis is expensive, many communities are finding that it is the most cost-effective means of providing water in regions where potential for importation is limited. Reverse osmosis is most commonly used on islands, as in the Keys and on Sanibel Island.

Airborne Hazards and Radioactivity Problems

The only major airborne hazard of a geologic nature in Florida comes from radioactive radon (Rn) gas emanating from water, rocks and soils. Radium (Ra) and polonium (Po), radioactive elements contained in rock and soil minerals, also create problems in the state. Most problems associated with radioactivity are natural; human activities simply place people in environments where exposure to the radioactivity is hazardous. For summaries of the chemistry of uranium (U) and its progeny, see Upchurch et al. (1991).

ORIGIN OF RADIOACTIVITY

The ultimate source of most radioactivity in Florida is U incorporated in carbonate-fluorapatite (table 13.5) at the time of formation (Altschuler et al. 1958). Other possible sources include rare-earth minerals in the heavy-mineral deposits on Trail Ridge (see chapter 9) that may contain other ^{238}U-^{234}Th (thorium) decay products.

Uranium, which is predominantly ^{238}U in carbonate-fluorapatite, is released upon weathering. Uranium concentration and the activity ratio of ^{234}U to ^{238}U have been used as tracers in determining the origins of Florida groundwaters and their relative ages (Osmond et al. 1974; Osmond and Cowart 1977; Cowart et al. 1978).

Uranium-238 undergoes a series of sequential decay events (fig. 13.25). Each step results in a new isotope. Some isotopes have short half-lives or decay by emission of beta particles. These isotopes are of less consequence than the isotopes with longer half-lives that emit alpha particles. In the ^{238}U decay scheme, five such progeny are ^{234}U, ^{230}Th, ^{226}Ra, ^{222}Rn, and ^{210}Po. The latter three are most important as hazardous substances because of their later occurrence in the decay series.

Radium-226 has a half-life of 1,620 years. The Ra cation is divalent, and chemically similar to Ca^{2+}; it substitutes for Ca in bone, and can cause leukemia and other carcinomas. The standard for Ra in water (^{226}Ra plus ^{228}Ra) is 5 picoCuries per liter (pCi/L).

Radon-222 is the decay product of ^{226}Ra (fig. 13.25). It has a half-life of 3.8 days and decays to ^{218}Po, then ^{214}Pb (lead), ^{214}Bi (bismuth), and then ^{214}Po. With the exception of ^{214}Bi, each of these isotopes emits an alpha particle and has a short half-life. Therefore, when Rn decays, not one, but three alpha particles are quickly emitted, making Rn triply hazardous. Rn is an inert gas, so it can travel through regolith and atmosphere. Radon is a gas with a short half-life, and the primary hazard from Rn is lung cancer as a result of inhalation.

There is no standard for Rn in water, although there is an information guideline that 10,000 pCi/L Rn in water will result in exposure equivalent to 1 pCi/L in the atmosphere. The standard for indoor, atmospheric Rn was adopted in 1986 by rule (10D-91 Florida Administrative Code) at 0.02 Working Levels; based on a series of assumptions (Nagda et al. 1987), this standard equates to 4 pCi/L.

Polonium-210 (fig. 13.25) is produced by decay of ^{210}Bi. Polonium is a multivalent ion that behaves chemically like Al or Fe (Upchurch et al. 1991); it is mobilized in acidic groundwaters and it is fixed as a Po-hydroxide radiocolloid in neutral and alkaline solutions. Polonium-210 is an alpha-particle emitter with a half-life of 138 days. There is no standard for Po, but it is covered in the guidance criterion of 15 pCi/L for gross-alpha radiation.

Polonium-210 was not considered a problem until recently. Outside of Florida, natural ^{210}Po in water has still not been identified as a major problem; however, in Florida it has been found in quantities sufficiently high to cause violation of the 15-pCi/L standard for gross-alpha radioactivity in the surficial aquifer (Burnett et al. 1987). The distribution of ^{210}Po is related to the distribution of phosphatic sediments and is a result of migration and concentration of Rn gas in the aquifer system. Apparently, the highest activities yet found anywhere occur at a homesite in Hillsborough County: the highest ^{210}Po activity found in a filtered sample was 566.8 pCi/L, and in an unfiltered sample it was 2569.4 pCi/L.

RADIOACTIVITY IN GROUNDWATER

Radon-222 activities in surficial-aquifer waters are almost always less than 10,000 pCi/L, but scattered analyses have shown activities in excess of 40,000 pCi/L. Where present, ^{222}Rn progeny often contribute more than 95% of the total alpha radiation in surficial-aquifer

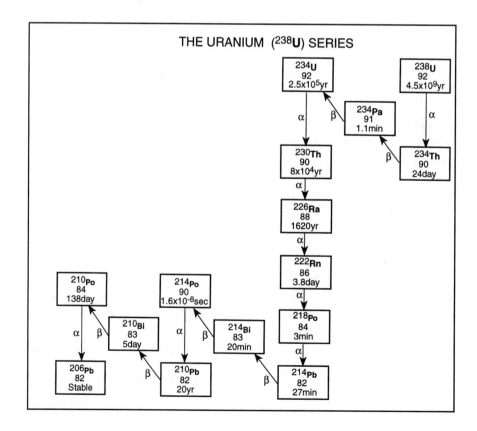

Figure 13.25. Decay products of ^{238}U.

waters. Radon may be present in high concentrations in the Upper Floridan aquifer, but holding times in municipal water systems generally result in loss of Rn by decay (Oural et al. 1989).

Radium-226 exceeds the 5-pCi/L standard in coastal groundwaters of the surficial and Upper Floridan aquifers south of the Tampa area (Sarasota, Manatee, and possibly southern Hillsborough counties) (Kaufmann and Bliss 1977). This Ra originates in weathering of phosphorites and transport in the aquifer systems, but the reasons for its concentration in that area are unknown. Water of the Upper Floridan aquifer that comes to the coast in these counties is exposed to the extensively weathered Hawthorn Group sediments. It also contains significant sulfate, which can transport Ra as a chemical complex. Finally, saltwater intrusion may cause ion exchange and release Ra to groundwater. Radium content occasionally is high in deep wells of central Florida, where sulfate-rich waters occur in nearly stagnant, lower-permeability zones. The nature and extent of any problems with Po are unknown; it has been found in coastal areas and in the interior of the state, wherever Rn is present (Burnett et al. 1987; Upchurch et al. 1991).

RADIOACTIVITY IN THE ATMOSPHERE

Radon problems arise when the gas enters structures. A recent statewide survey of homes (Nagda et al. 1987), which included 3,050 homes on a population basis and 3,106 homes on a land basis, found average indoor Rn activities of 1.0 pCi/L (population base) and 0.7 pCi/L (land base). Only 3.8% of homes in the land-based study exceeded the 4-pCi/L action level, and 2.6% exceeded the level in the population-based study. Soil Rn measurements were taken at 2896 homes in the land-based study; the average Rn activity was 155 pCi/L, and 4.2% exceeded 630 pCi/L, which is the action level recommended by Swedjemark (1986).

It is difficult to predict indoor Rn concentrations from soil concentrations accurately (Nagda et al. 1987). Many houses on highly radioactive regolith contain little or no Rn. Others with high indoor Rn have been built on relatively low-Rn materials. The reason for this poor correlation is construction practice. Well ventilated structures or structures with effective vapor barriers are likely to contain low Rn, regardless of soil Rn content. Tight, poorly ventilated structures may accumulate Rn, even when soil Rn concentration is low. In spite of the poor correlation between soil and indoor Rn, it is intuitively obvious that higher risks of indoor Rn must exist in structures built on high-Rn ground. Swedjemark (1986) has recommended that if soil Rn exceeds 630 pCi/L construction practice should include armoring the structure against Rn emanation. To minimize exposure of the public, the 1988 Florida Legislature authorized the Board of Regents to develop a building code for Rn-resistant structures.

Radon is highly soluble in water, so where the water table is high, Rn in water is high, and emanation into homes is inhibited. Where the water table is low and a significant vadose zone exists in or above phosphatic sediments, the risk of Rn entry into structures is great.

Summary

Florida does not have environmental problems associated with high-profile hazards such as earthquakes, landslides, and volcanoes. Its significant geologic hazards involve sinkholes, hurricanes, floods, water pollution, radon, and coastal erosion. A rapidly growing population and economically important agricultural and mining industries complicate the natural responses of Florida's fragile ecosystem. Through the prudent application of geologic principles, natural and anthropogenic hazards can be effectively managed to preserve the quality of the environment.

14

Geology of the Florida Keys

Anthony F. Randazzo and Robert B. Halley

The Florida Keys are an arcuate chain of Pleistocene reef and oolitic-limestone islands extending from Miami to the Dry Tortugas and separating Florida Bay from the Straits of Florida (fig. 14.1). These islands, together with the modern carbonate sediments and reef fringe in surrounding waters, are the surficial layers of one of the great carbonate platforms of the world. More than 6 km thick, the southern Florida carbonate platform has been accumulating shallow-water carbonate sediments since the Early Cretaceous—about 140 million years. Today, the Keys are the subaerial parts of the Pleistocene limestones that are the underpinnings of modern carbonate sedimentation in southern Florida.

Modern carbonate sediments around the Keys are similar in many ways to the sediments composing the Keys themselves, but with several important differences. These islands and the modern marine environments of the area provide an outstanding example of the application of uniformitarianism and comparative carbonate sedimentology. Several nearby modern environments of deposition have ancient counterparts in the sediments of the islands. The area has intermittently supported reef growth for hundreds of thousands of years. Fossil coral species in the coralline limestones of the Keys are identical to those of many living patch reefs. On the other hand, ooid sand shoals that were widespread in the Keys during the last interglacial interval have not redeveloped during the Holocene; the reasons for their absence are not entirely clear.

Numerous studies of descriptive and genetic aspects of the sedimentology and geology of the region have been published, including noteworthy contributions by Ginsburg (1956, 1964), Taft and Harbaugh (1964), Stanley (1966), Hoffmeister and Multer (1968), Hoffmeister (1974), Enos and Perkins (1977, 1979), Multer (1977), Halley and Evans (1983), Shinn (1988), Shinn et al. (1989a), and Lidz and Shinn (1991).

Although the general stratigraphic setting and surficial sedimentology of the Florida Keys are well known, many details—such as the ages and subdivision of Pleistocene units—are still sketchy, because outcrops are scattered and low in elevation, because so much of the area is under water, and because well preserved material for high-resolution paleontologic or radiometric dating is rare.

Geology of the Keys

The term "key" is derived from the Spanish word *cayo*, which means shoal, or reef. During the 1700s, English-speaking Caribbeans used this word for islands and pronounced it "key," but spelled it "kay." Later, the spelling became phonetic: key. The islands are the result of interplay among climate, seawater chemistry, biology, sea-level change, tidal interactions, antecedent topography, and tectonic stability. The uninhabited mud islands of Florida Bay—although Florida Keys in name—are geologically different from the Keys facing the Atlantic. Florida Bay islands, islands west of Key West (Marqueses Keys, Dry Tortugas), and scattered and ephemeral reef-rubble islands along the reef tract (e.g. Sand Key, Sambo Key, Pelican Shoal) are almost all accumulations of modern sediment, carbonate mud, sand, and mangrove peat—not limestone (the exception being Lignumvitae Key, a Pleistocene key in Florida Bay). The limestone islands are the populated Florida Keys—those with elevations as high as 6 m that people travel across on the road between Miami and Key West.

The two principal surficial stratigraphic units of the Florida Keys are the Key Largo Limestone and the Miami Limestone (Hoffmeister and Multer 1968). Although dissolution has modified details of surface features (Dodd and Siemers 1971), the general geometry and surficial geology of the Keys reflect the depositional

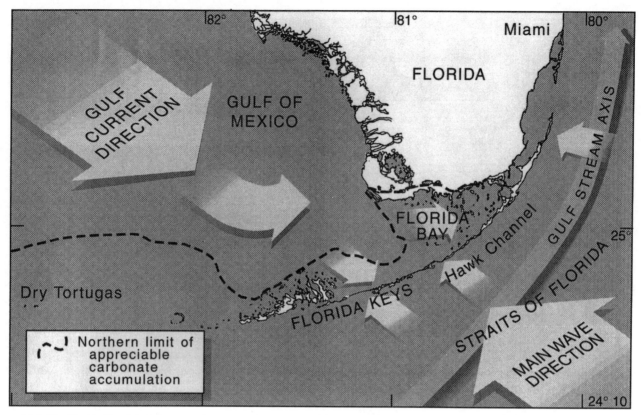

Figure 14.1. The Florida Keys in relation to various water movements and the principal area of carbonate-sediment accumulations.

distribution of these two formations (Hoffmeister 1974). The Upper Keys are generally reefs of the Key Largo Limestone oriented parallel to the shelf edge. The Lower Keys are elongated perpendicular to the shelf edge and are composed of oolite of the Miami Limestone.

The Key Largo Limestone (fig. 14.2) was named by Sanford (1909), and described in detail as a raised coral reef (Sanford 1913). It stretches in the subsurface at least from Miami to the Dry Tortugas, and its thickness, although varying widely, is greater than 60 m (Hoffmeister and Multer 1968). The Key Largo Limestone in the Keys generally comprises frame-building hermatypic corals (table 14.1) with intra- and interbedded calcarenites. The upper part of the Key Largo can be viewed in many cuts and quarries in the Upper Keys. Coral geometries vary with phyletic groups and conditions of preservation. In general, the formation contains a surprising amount of coral in growth position (Hoffmeister et al. 1967). Coral distribution appears to be in large, elongate patches—similar to the geometry of the islands (Stanley 1966; Hoffmeister and Multer 1968).

The Miami Limestone (fig. 14.2) was named by Hoffmeister et al. (1967) after earlier studies had described facies of this unit as separate stratigraphic formations (Sanford 1909; Cooke and Mossom 1929). The Miami Limestone is comprised of a bryozoan facies and an oolitic facies (Hoffmeister and Multer 1968); its thickness is variable, reaching a maximum of 15 m. Sanford (1909) recognized two oolitic facies in southern Florida, which he named the Key West Oolite (Lower Keys) and the Miami Oolite (Miami area). Mitchell-Tapping (1980) renamed the Miami Limestone the Fort

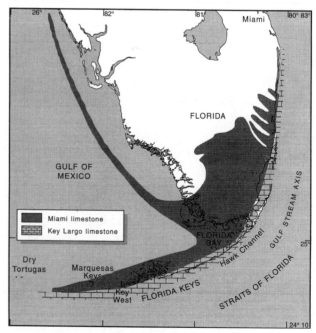

Figure 14.2. Geologic map of Pleistocene limestones of the Florida Keys (from Mitchell-Tapping 1980).

Dallas Oolite, and included the bryozoan facies of Hoffmeister and Multer (1968), but this subdivision has not been followed by subsequent workers (Kindinger 1986; Merriam 1988; Shinn et al. 1989a; Scott and Allmon 1992).

The oolitic facies of the Miami Limestone consists of well-sorted ooids, with varying amounts of skeletal material (corals, echinoids, molluscs, algae) and some quartz sand. Hoffmeister et al. (1967) and Perkins (1977) interpreted the oolitic facies of the Miami Limestone as a subtidal marine ooid-shoal complex formed during a sea-level high stand. Oolite in the Lower Keys has abundant marine fossils and less quartz sand (Weisbord 1974), and also represents a marine oolitic bank or bar system. Overall, the Lower Keys have the configuration of a tidal bar with flooding of interbar low areas—similar to, but generally lower than, the topography of the Miami area (Halley et al. 1977).

Limestones beneath Florida Bay are known only from scattered drill cores (Perkins 1977; Merriam et al. 1989; Shinn et al. 1989a). Surficial Pleistocene limestones beneath the bay comprise a variety of restricted platform-interior facies. The bryozoan facies of the mainland extends southward beneath much of Florida Bay, but it has not been mapped in detail. These limestones are generally micritic and peletal, with a variety of foraminifer and mollusc skeletal debris, characteristic of low-energy, lagoonal settings. The oolitic facies of the Lower Keys extends offshore northward and eastward for a kilometer or more, but does not extend to the oolite in the Miami area.

Hoffmeister and Multer (1964) found the Key Largo Limestone to be more than 60 m thick in test borings in the Florida Keys (fig. 14.3). At Big Pine Key they found the oolitic facies of the Miami Limestone to be less than 10 m thick, and to occur as a facies equivalent to the uppermost Key Largo Limestone. From these stratigraphic relations it was determined that the Miami Limestone is equivalent to the upper Key Largo Limestone and is underlain by older parts of the Key Largo Limestone. Perkins (1977) recognized regional subaerial-discontinuity surfaces, which he used to subdivide the Pleistocene of southern Florida into five "Q" units. In the Florida Keys he assigned the Miami Limestone and equivalent parts of the Key Largo Limestone to units Q4 and Q5, and he assigned the older, underlying Key Largo Limestone to units Q1, Q2, and Q3. The facies transition between the Key Largo Limestone and Miami Limestone in unit Q5 occurs laterally over a few hundred meters on southern Big Pine Key (Kindinger 1986).

Using stratigraphic information from closely spaced, shallow core borings on Key Largo, Harrison et al. (1984) and Harrison and Coniglio (1985) confirmed Perkins's units Q5, Q4, and Q3 among 10 borings on the island; they noted a quartz sandstone layer about 10 m below the surface, just above the boundary between Q3 and Q4. The upper Key Largo is comprised of boundstones, wackestones, and packstones (terminology

Table 14.1. Coral species of the Florida Keys and Dry Tortugas

Species	Dry Tortugas	Long Key Reef	Key Largo	Carpsfort Reef
Acropora palmata	p	—	p	p
Acropora cervicornis	p	p	p	p
Mycetophyllia lamarackan	p	p	p	p
Mycetophyllia ferox	p	p	p	p
Mycetophyllia danana	p	p	p	p
Mycetophyllia aliciae	p	p	p	p
Solenastrea hyades	p	p	p	p[a]
Agaricia agaricites	p	p	p	p
Agaricia lamarcki	p	p	p	p
Agaricia fragilis	p	p	p	p
Helioseris cucullata	p	p	p	p
Colpophyllia natans	p	p	p	p
Colpophyllia breviserialis	p	p	p	p
Scolymia cubensis	p	p	p	p
Scolymia lacera	p	p	p	p
Mussa angulosa	p	p	p	p
Montastraea annularis	p	p	p	p
Montastraea cavernosa	p	p	p	p
Dendrogyra cylindrus	—	—	p	—
Manicina areolata	p	p	p	p
Favia fragum	p	p	p	p
Favia conferta	p	—	—	—
Siderastrea radians	p	p	p	p
Siderastrea siderea	p	p	p	p
Dichocoenia stokesii	p	p	p	p
Dichocoenia stellaris	p	p	p	p
Stephanocoenia michelini	p	p	p	p
Diploria strigosa	p	—	p	—
Diploria clivosa	—	—	p	p
Diploria labyrinthiformis	p	—	p	p
Isophyllia sinuosa	p	p	p	p
Isophyllastraea rigida	p	p	p	—
Porites porites	p	p	p	p
Porites astreoides	p	p	p	p
Porites furcata	—	—	p	p
Madracis spp.	p	p	p	p
Madracis decactis	p	p	p	p
Millepora alcicornis	p	p	p	p
Millepora complanata	p	—	p	p
Eusmilia fastigiata	p	p	p	p
Oculina diffusa	p	p	p	—
Cladocora spp.	p	—	—	—
Meandrina meandrites	p	p	p	p

Source: Dustan 1985.
Note: p = present; — = absent.
a. Outer terrace only.

of Dunham 1962) at Key Largo (Harrison et al. 1984), and similar lithologies at Big Pine Key (Shinn et al. 1989a) (fig. 14.4). Harrison and Coniglio (1985) demonstrated a topographic high beneath Key Largo that persisted through several sea-level high stands; this high was a locus of reef growth, and explains in part the location of the present Florida Keys. Late Tertiary siliciclastic sediments underlie the Quaternary carbonate rocks. It has been suggested that the distribution and nature of these siliciclastic sediments controlled the position and arcuate shape of the modern shelf and slope of southern Florida (E. R. Warzeski, personal communication 1993).

The Miami Limestone and Key Largo Limestone are interpreted as subtidal marine deposits by almost all recent authors. Whereas the marine versus subaerial origin of the Keys limestones has been debated since the last century (Agassiz 1880), most geologists subscribe to the marine origin, largely because of distinctive marine burrows in the oolite (Shinn 1963; Halley and Evans 1983) and the significant proportion of coral in growth position in the Key Largo Limestone (Stanley 1966). Mitchell-Tapping (1980) considered the oolite (which he termed the Fort Dallas) to be an eolianite. He believed that steeply dipping cross-beds and mangrove roots to have formed as windblown dunes during a low sea-level stand. Reversals of strong tidal currents caused by interactions of Atlantic and Gulf waters could also have been responsible for the cross-bedding (Shinn 1988). During deposition of the Key Largo Limestone, sea level is estimated to have been about 6 to 8 m higher than today, and the entire southern Florida region was flooded (Perkins 1977). The absence of the reef-building coral *Acropora palmata* from the Key Largo Limestone is attributed to the absence of a barrier separating Key Largo reefs from the shelf waters of the platform interior (Harrison and Coniglio 1985): this species is one of the most sensitive reef corals, and it now grows where it is protected from shelf water (Ginsburg and Shinn 1964).

Radiometric dating (^{230}Th/^{234}U) has been attempted several times, using corals and oolite from the Q5 unit of Perkins (1977). Osmond et al. (1965) and Broecker and Thurber (1965) reported ages of 95,000 to 140,000 years for the Key Largo and Miami limestones. More recent dating of the Key Largo Limestone reported by Harmon et al. (1979) and Muhs et al. (1992) yielded radiometric ages of about 140,000 years. A single aragonitic coral from unit Q4 on Key Largo was also dated (Szabo and Halley 1988), yielding an age of 320,000 years. Unfortunately, many of the samples used in deter-

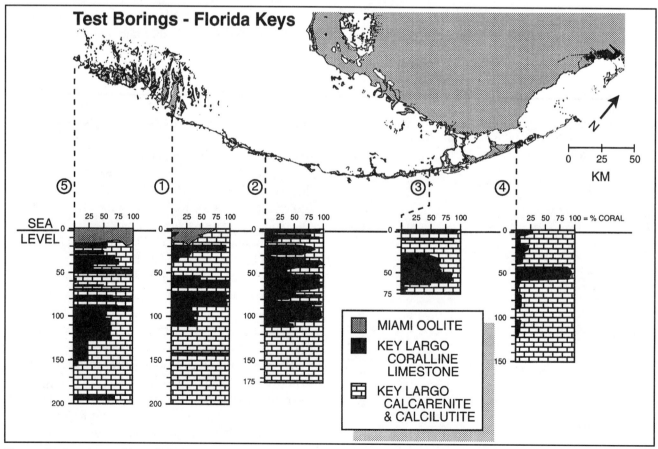

Figure 14.3. Locations and logs of selected test borings on the Florida Keys (Hoffmeister and Multer 1964).

mining these ages were not pristine, or did not meet strict criteria for closed-system conditions during the time since deposition. Because samples lost some primary U, or assimilated secondary U during diagenesis, these reported radiometric ages are considered estimates.

Holocene Reefs and Other Modern Sediments

Much attention has been given to coral reefs, but it should be realized that the bulk of carbonate sediment in the Florida Keys area is sand and mud that has accumulated on Pleistocene "bedrock" during the last 7,000 years. Ginsburg (1956), Swinchatt (1965), Ball et al. (1967), Enos (1977), Enos and Perkins (1979), Lidz et al. (1985), and Shinn et al. (1989a) have provided comprehensive investigations of the nature and distribution of modern carbonate sediments on this classic example of a rimmed shelf (Ginsburg and James 1974). Enos (1977) treated this area as a shelf margin with four major divisions: outer shelf margin, shallow slope margin (shelf break), slightly restricted inner shelf margin (Hawk Channel), and restricted inner shelf (Florida Bay).

Reef-tract sediments have accumulated to a thickness of as much as 12 m; these sediments are predominantly carbonate sand and coral rubble, with scattered coral-framework deposits on the outer shelf margin. The Caribbean reef-crest coral *Acropora palmata* forms buttresses or spurs between troughlike grooves on the seaward edge of the best-developed Keys reefs (Shinn 1963). Wave and current action and preferential coral growth on the spurs accentuate this configuration of the reef structure (plate 10; plate 11). Settlement of coral larvae is prevented by the sedimentological processes in the grooves (Shinn 1988). The reef platform generally is comprised of fore-reef, reef-tract, and back-reef zones (Ginsburg 1956) (fig. 14.5). Core-boring transects across reefs (Shinn et al. 1989a) show that modern reefs have formed on the topographic relief provided by Pleistocene reefs. Reefs have not built seaward, but have grown in place or migrated landward a short distance (less than 100 m) over their back-reef sand and rubble deposits. The reef framework and rubble are characterized by the coral genera *Acropora, Montastrea, Diploria, Siderastrea, Porites,* and *Colpophyllia.*

Halimeda, a green alga, produces much of the carbonate sand in the region; after death, its disarticulated coarse plates accumulate as sand. Other algae, foraminifers, molluscs, echinoderms, bryozoans, and corals also contribute skeletal material to the sand deposits. A particularly large *Halimeda* sand deposit, as much as 12 m thick, is found between the Dry Tortugas and Key West, where coral reefs are poorly developed. These sand deposits—known as "the quicksands"

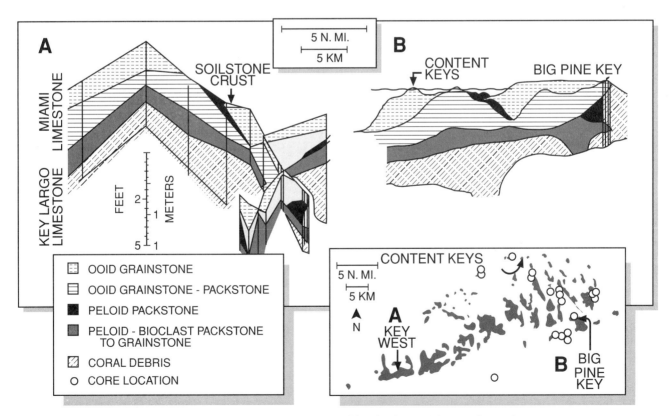

Figure 14.4. Panel diagrams of carbonate facies of Pleistocene limestones of the Florida Keys (Shinn et al. 1989a).

because of their shifting nature—have wave forms with amplitudes as high as 3 m (Shinn et al. 1989a). Other outstanding examples of carbonate-sand deposits are White Bank (particularly east of Key Largo), the Dry Tortugas, and the molluscan beaches of Sandy Key and Cape Sable (Davis et al. 1992).

Lime mud, consisting of aragonite needles 1 to 2 μm long, is principally derived from algae such as *Penicillus*, and from mechanically and biologically abraded skeletal debris (Stockman et al. 1967; Neumann and Land 1975). Suspended carbonate muds known as "whitings" are common (Shinn et al. 1989b), and may represent inorganic precipitation of aragonite. Muds generally settle and accumulate in low-energy environments such as Florida Bay and Hawk Channel, but they can be transported and accumulated within the framework of reefs as well. Turtle-grass (*Thalassia*) beds are particularly effective in trapping and accumulating muddy sediments. Turtle-grass is partly responsible for the accumulation of mud banks (Wanless and Tagett 1989; Enos 1989). Additionally, epibionts on *Thalassia* are believed capable of producing significant quantities of lime mud (Nelsen and Ginsburg 1986). Florida Bay islands have formed as the result of colonization of the mudbanks by mangroves, but some islands have long histories, and may have evolved from erosional remnants of shorelines during the Holocene transgression (Enos and Perkins 1979).

It is interesting to note that some of the earliest reported Holocene dolomite occurs on Sugarloaf Key (Shinn 1964). The dolomite crystals are embedded in calcareous supratidal crusts and have ^{14}C ages of less than 1,000 years; the crystals are typically poorly ordered, nonstoichiometric, and less than 4 μm in diameter. The supratidal environment of Sugarloaf Key has made it popular for study of carbonate sediment and dolomite formation. Steinen et al. (1977) and Swart et al. (1989) reported modern dolomite on several of the Florida Bay islands, where it is associated with hypersaline brines.

Evolution of the Florida Keys

The ancient reefs fossilized in the Keys and surrounding subsurface were a flourishing reef system sometime between 90,000 and 145,000 years ago. Sea level was some 6 to 8 m higher, and the reef system extended to the edge of the Gulf Stream. Waters of the Gulf of Mexico and the Atlantic Ocean interchanged across a much broader area then (fig. 14.1). Lacking *Acropora palmata* (the elkhorn, or moosehorn, coral; plate 12), the present-day Florida Keys are a series of coalescing patch reefs (Hoffmeister and Multer 1968). These reefs grew on preexisting topographic relief on an otherwise open shelf (Harrison and Coniglio 1985). There is a zone of older, dead reefs near the outer shelf margin (Shinn et al. 1989a; Lidz et al. 1991); these "outlier reefs" indicate multiple reef tracts and may be of Pleistocene age.

The general arcuate pattern of the Keys is related to the bathymetry of the shelf edge and the action of the Florida Current, which control many of the environmental parameters affecting reef growth. Growth of patch reefs is controlled by nutrient availability (Hallock and Schlager 1986) and substrate topography. Hoffmeister and Multer (1968) believed that there may have been two similarly elongate and parallel patch reefs behind an ancient reef tract (fig. 14.6). The interpreta-

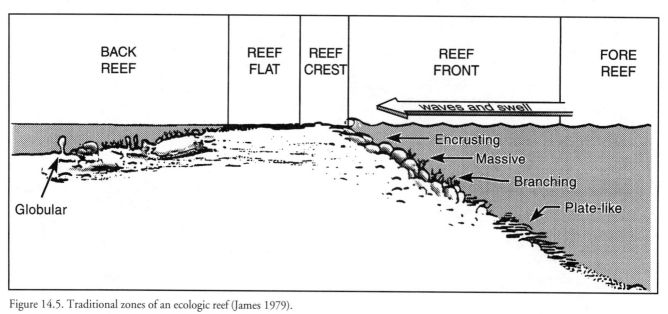

Figure 14.5. Traditional zones of an ecologic reef (James 1979).

tion that the Florida Keys are not an ancient reef tract proper is based on the absence of *Acropora palmata,* and the presence of corals typical of the lower-energy environments of patch reefs. *Acropora palmata* generally requires a high-energy environment associated with a reef tract or platform edge.

Harrison and Coniglio (1985) and Shinn (1988) proposed another explanation for the absence of *Acropora palmata,* based on water temperature; an influx of colder waters from the Gulf of Mexico could limit the temperature-sensitive *A. palmata* to an area near the platform edge, within the influence of the warmer Gulf Stream. Shinn hypothesized that if sea level remained high for a longer time, the slower coral growth would have led to an eventual connection of patch reefs. Cold-water stress on modern corals is well documented (Roberts et al. 1982).

Shinn (1988) has outlined the control of reef distribution by eustatic sea-level changes and topography. A sea-level fall that began about 100,000 years ago is believed responsible for the eventual distribution of a large part of the Pleistocene coral reef system and the exposure of the Florida Keys. Sea level may have fallen more than 100 m below the present level (Fairbanks 1989) (fig. 14.7). During this exposure, laminated calcareous crusts formed (Multer and Hoffmeister 1968) (fig. 14.8). The exposure surface formed the substrate colonized by more-recent corals when sea level began to rise some 18,000 years ago, eventually flooding the South Florida Shelf. Peat deposits formed in bogs that were later drowned, and ^{14}C dates of these materials have helped to recognize geologic events over the last 10,000 years (Robbin and Stipp 1979).

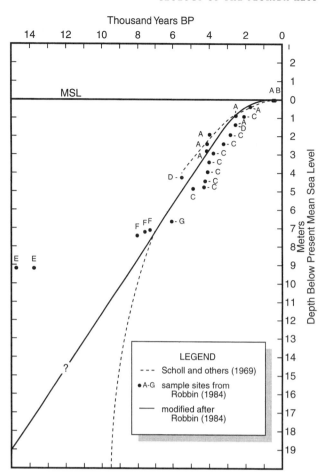

Figure 14.7. Curve showing sea-level changes during the past 15,000 years (Robbin 1984; Lidz and Shinn 1991).

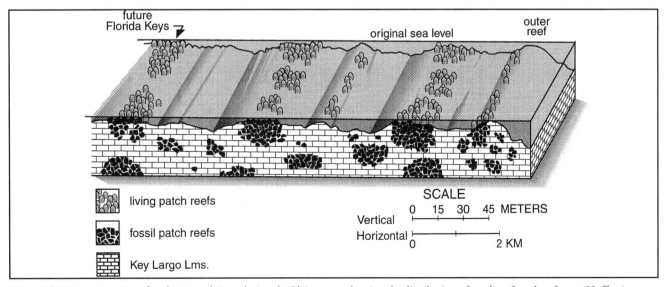

Figure 14.6. Reconstruction of geologic conditions during the Pleistocene, showing the distribution of patch reefs and reef tract (Hoffmeister and Multer 1968).

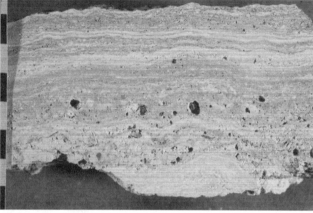

Figure 14.8. Examples of laminated crusts representing subaerial exposure and unconformities in the Florida Keys. Scales in centimeters (Multer and Hoffmeister 1968; photos courtesy of D. F. McNeill).

Sea level has steadily risen over the past 15,000 years (fig. 14.8). It has brought about the drowning of what is now Florida Bay and Hawk Channel (fig. 14.1) and renewed cross-shelf water interchanges between the Gulf of Mexico and the Atlantic Ocean. Waters of the Gulf of Mexico and Florida Bay provided a combination of higher salinity during the summer and colder temperatures during the winter that significantly inhibited the ability of reef-building corals to recolonize; this was especially true for *Acropora palmata*. The distribution of living coral reefs is directly linked to the presence or absence of tidal passes which serve as conduits for Gulf of Mexico water (Ginsburg and Shinn 1964).

Localized coral-reef colonization appears to be controlled by isolated topographic conditions that provided substrate highs. Grecian Rocks (Shinn 1980), Looe Key Reef (Shinn et al. 1981), and the reef mounds of Rodriguez Key (Turmel and Swanson 1976) and Tavernier Key (Bosence et al. 1985) are examples of such local development.

Ecologic zones appear to have expanded upward and landward in response to a rise in sea level, a process referred to as "backstepping" (Shinn 1988). Coral debris distributed on the landward side of reefs provides the hard substrate required for colonization during periods of rising sea level. Storm action can produce erosion on the seaward side of reefs, causing reef retreat. The backstepping process can thus produce "keep-up" reefs that

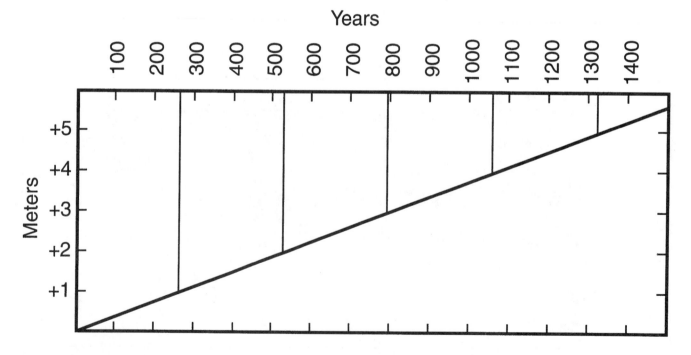

Figure 14.9. Possible future sea-level changes for Key West, Florida, assuming zero subsidence and maintenance of the current rate of sea-level rise of 38 cm per 100 years (Lidz and Shinn 1991).

are migrating inland, and "give-up" reefs that are being eroded (Neumann and MacIntyre 1985; Shinn 1988).

Growth rates of corals vary considerably. *Acropora cervicornis,* the staghorn coral, grows fast, averaging about 10 mm per year (Shinn 1966; Buddemeier and Smith 1988). Massive corals grow more slowly, but can also "keep up" if sea level rises slowly. Shinn (1988) reported that sea level has risen about 1 m per 1000 years over the past 10,000 years in southern Florida. This rate is certainly slow enough to allow all of the coral forms to keep up; their inability to do so can be attributed to storm action, disease, seasonal salinity and temperature changes, and variations in nutrient supplies (Shinn 1988; Shinn et al. 1989a).

Future Changes in the Florida Keys

The future of the Florida Keys will be controlled by sea level. Florida's present reef communities have developed during only some 7000 years, while sea level rose about 10 m (Robbin 1984). Shinn (1988) hypothesized that if sea level remained relatively stable over a period of only 1 million years, Florida Bay and the back-reef lagoons such as Hawk Channel would be filled with sediment. Backstepping would cease, and reef growth would be progradational (toward the Straits of Florida).

Continual rise of sea level would allow backstepping to progress, and the Keys would be reoccupied as places for coral growth. Waters from the Gulf of Mexico might limit the areal distribution of corals, particularly temperature-sensitive forms such as *Acropora palmata.* Lidz and Shinn (1991) presented a series of paleogeographic maps reconstructing shoreline positions of the past 8,000 years; they also projected future sea-level rises of 1 to 5m (fig. 14.9), and corresponding shoreline positions.

Rates and frequency of sea level changes are equally important. Numerous sea level changes have been recorded in the rock record, including at least five (Perkins 1977) over the past 125,000 years. Wanless (1989) has documented a significant relative sea-level rise over the past 58 years in southern Florida. A rise of 4 to 5 m would drown almost all of the Keys. Future distributions of reef zones and the size and configuration of carbonate-rock islands will reflect the dynamic processes of climate and water-body interchanges, with antecedent influences of substrate topography (fig. 4.6).

References

Abbott, R. T. 1974. American seashells. 2nd edition. New York: Van Nostrand. 663 p.

Addy, S. K., and R. T. Buffler. 1984. Seismic stratigraphy of shelf and slope, northeastern Gulf of Mexico. Bulletin of the American Association of Petroleum Geologists 68:1782–89.

Agassiz, A. 1880. Report on the Florida reefs by Louis Agassiz. Memoirs of the Museum of Comparative Zoology 7(1). 61 p.

Agassiz, L. 1851. Annual report of the superintendent of the coast survey, appendix no. 10. Executive Documents of the Senate of the United States, 32nd Cong., 1st sess., vol. 5, no. 3, 145–60.

Ahearn, M. E. 1981. A revision of the North American Hydrochoeridae. Master's thesis, University of Florida, Gainesville.

Ahearn, M. E., and J. F. Lance. 1980. A new species of *Neochoerus* (Rodentia: Hydrochoeridae) from the Blancan (late Pliocene) of North America. Proceedings of the Biological Society of Washington 92:435–42.

Ahr, W. M. 1973. The carbonate ramp: an alternative to the shelf model. Transactions of the Gulf Coast Association of Geological Societies 23:221–25.

Aikin, A. R., and R. D. G. Pyne. 1991. Aquifer storage and recovery. In Hydrology and hydrogeology in the '90s, edited by S. B. Upchurch, J. E. Moore, and A. Zaporozec, 218–26. Proceedings of the 10th Annual Meeting of the American Institute of Hydrology. Dubuque, Iowa: Kendall-Hunt.

Allegre, C., and D. Turcotte. 1985. Geodynamic mixing in the mesosphere boundary layer and the origin of ocean islands. Geophysical Research Letters 12:207–10.

Allen, J. H. 1846. Some facts respecting the geology of Tampa Bay, Florida. American Journal of Science 51:38–42.

Allen, J. R. 1991. Beach dynamics and dune retreat at the north end of Canaveral National Seashore. National Park Service Final Report. 105 p.

Allmon, W. D. 1992. Whence southern Florida's Plio-Pleistocene shell beds? In Plio-Pleistocene stratigraphy and paleontology of southern Florida, edited by T. M. Scott and W. D. Allmon, 1–20. Florida Geological Survey Special Publication no. 36.

———. 1993. Age, environment, and mode of deposition of the densely fossiliferous Pinecrest Sand (Pliocene of Florida): Implications for the role of biological productivity in shell bed formation. Palaios 8:183–201.

Allmon, W. D., G. Rosenberg, R. W. Portell, and K. S. Schindler. 1993. Diversity of Atlantic coastal plain mollusks since the Pliocene. Science 260:1626–29.

Alt, D., and H. K. Brooks. 1965. Age of the Florida marine terraces. Journal of Geology 73:406–11.

Altschuler, Z. S., R. S. Clarke, and E. J. Young. 1958. Geochemistry of uranium in apatite and phosphorite. U.S. Geological Survey Professional Paper no. 314D, 45–90.

Altschuler, Z. S., E. J. Dwornik, and H. Kramer. 1963. Transformation of montmorillonite to kaolinite during weathering. Science 141:148–52.

Altschuler, Z. S., E. B. Jaffe, and F. Cuttitta. 1956. The Florida aluminum phosphate zone of the Bone Valley Formation, Florida, and its uranium deposits. U.S. Geological Survey Professional Paper no. 300, 495–504.

Altschuler, Z. S., and E. J. Young. 1960. Residual origin of the Pleistocene sand mantle in Central Florida Uplands and its bearing on marine terraces and Cenozoic uplift. U.S Geological Survey Professional Paper no. 400B, B202–7.

Ames, L. L. 1959. The genesis of carbonate apatites. Economic Geology 54:829–41.

Ampian, S. G. 1985. Clays. Bulletin of the U.S. Bureau of Mines 675:157–69.

Andress, N. E., F. H. Cramer, and R. F. Goldstein. 1969. Ordovician chitinozoans from Florida well samples. Transactions of the Gulf Coast Association of Geological Societies 19:369–75.

Angstadt, D. M., J. A. Austin Jr., and R. T. Buffler. 1985. Early Late Cretaceous to Holocene seismic stratigraphy and geologic history of the southeastern Gulf of Mexico. Bulletin of the American Association of Petroleum Geologists 69:977–95.

Antoine, J. W., and J. L. Harding. 1965. Structure beneath the continental shelf, northeastern Gulf of Mexico. Bulletin of the American Association of Petroleum Geologists 49:157–71.

Applegate, A. V., G. O. Winston, and J. G. Palacas. 1981. Subdivision and regional stratigraphy of the pre–Punta Gorda rocks (lowermost Cretaceous–Jurassic?) in south Florida. Transactions of the Gulf Coast Association of Geological Societies 31 (supp.):447–53.

Applin, E. R., and L. Jordan. 1945. Diagnostic foraminifera from subsurface formations in Florida. Journal of Paleontology 19:129–48.

Applin, P. L. 1951. Preliminary report on buried pre-Mesozoic rocks in Florida and adjacent states. U.S. Geological Survey Circular no. 91. 28 p.

Applin, P. L., and E. R. Applin. 1944. Regional subsurface stratigraphy and structure of Florida and southern

REFERENCES

Georgia. Bulletin of the American Association of Petroleum Geologists 28:1673–1753.

———. 1965. The Commanche series and associated rocks in central and south Florida. U.S. Geological Survey Professional Paper no. 447. 84 p.

———. 1967. The Gulf series in the subsurface in northern Florida and southern Georgia. U.S. Geological Survey Professional Paper no. 524G. 35 p.

Arden, D. D. 1974. Geology of the Suwannee Bsin interpreted from geophysical profiles. Transactions of the Gulf Coast Association of Geological Societies 24:223–30.

Arthur, J. D. 1988. Petrogenesis of early Mesozoic tholeiite in the Florida basement and an overview of Florida basement geology. Florida Geological Survey Report of Investigations no. 97.

Arthur, M. A., and H. C. Jenkyns. 1981. Phosphorites and paleoceanography. Oceanologica Acta 4 (supp.):83–96.

Arthur, M. A., and S. O. Schlanger. 1979. Cretaceous "oceanic anoxic events" as causal factors in reef-reservoired oil fields. Bulletin of the American Association of Petroleum Geologists 69:870–85.

Austin, J. A. Jr., P. L. Stoffa, J. D. Phillips, J. Oh, D. S. Sawyer, G. M. Purdy, E. Reiter, and J. Makris. 1990. Crustal structure of the Southeast Georgia Embayment–Carolina Trough: Preliminary results of a composite seismic image of a continental suture (?) and a volcanic passive margin. Geology 18:1023–27.

Ayuso, R. A. 1986. Lead-isotopic evidence for distinct sources of granite and for distinct basements in the northern Appalachians, Maine. Geology 14:322–25.

Back, W. B., and B. B. Hanshaw. 1970. Comparison of chemical hydrogeology of the carbonate peninsulas of Florida and Yucatan. Hydrology 10:330–68.

Back, W. B., B. B. Hanshaw, J. S. Herman, and J. N. Van Driel. 1986. Differential dissolution of a Pleistocene reef in the ground-water mixing zone of coastal Yucatan, Mexico. Geology 14:137–40.

Bader, R. S. 1956. A quantitative study of the Equidae of the Thomas farm Miocene. Bulletin of the Harvard Museum of Comparative Zoology 115:49–78.

Badiozomani, K. 1973. The dorag dolomitization model. Journal of Sedimentary Petrology 43:965–84.

Bailey, J. W. 1850. Discovery of an infusorial stratum in Florida. American Journal of Science 60:282.

———. 1851. Miscellaneous notes. American Journal of Science 61:85–86.

Baker, P. A., and M. Kastner. 1981. Constraints on the formation of sedimentary dolomite. Science 213:214–16.

Ball, M. M., R. M. Martin, W. D. Bock, R. E. Sylvester, R. M. Bowles, D. E. Taylor, L. Coward, J. E. Dodd, and L. Gilbert. 1985. Seismic structure and stratigraphy of northern edge of Bahaman-Cuban collision zone. Bulletin of the American Association of Petroleum Geologists 69:1275–94.

Ball, M. M., R. M. Martin, R. Q. Foote, and A. V. Applegate. 1988. Structure and stratigraphy of the western Florida shelf. U.S. Geological Survey Open-File Report no. 88-439. 60 p.

Ball, M. M., E. A. Shinn, and K. W. Stockman. 1967. The geologic effects of Hurricane Donna in south Florida. Journal of Geology 75:583–97.

Banks, J. E. 1978. Southern Florida subsurface features related to oil exploration. Transactions of the Gulf Coast Association of Geological Societies 28:25–30.

Baria, L. R., D. L. Stoudt, P. M. Harris, and P. D. Crevello. 1982. Upper Jurassic reefs of Smackover Formation, United States Gulf Coast. Bulletin of the American Association of Petroleum Geologists 66:1449–82.

Barnett, R. S. 1975. Basement structure of Florida and its tectonic implications. Transactions of the Gulf Coast Association of Geological Societies 25:122–42.

Barr, D. E., A. Maristany, and T. Kwader. 1981. Water resources of southern Okaloosa and Walton Counties, northwest Florida. Northwest Florida Water Management District Water Resources Assessment no. 81-1. 41 p.

Barraclough, J. T. 1962. Ground water resources of Seminole County, Florida. Florida Geological Survey Report of Investigations no. 27. 91 p.

Basaltic Volcanism Study Project. 1981. Basaltic volcanism on the terrestrial planets. New York: Pergamon Press. 1286 p.

Baskin, J. A. 1980. Evolutionary reversal in *Mylagaulus* [sic] (Mammalia, Rodentia) from the late Miocene of Florida. American Midland Naturalist 104:155–62.

———. 1981. *Barbourofelis* (Nimravidae) and *Nimravides* (Felidae), with a description of two new species from the late Miocene of Florida. Journal of Mammalogy 62:122–39.

———. 1982. Tertiary Procyoninae (Mammalia: Carnivora) of North America. Journal of Vertebrate Paleontology 2:71–93.

Bass, M. N. 1969. Petrography and ages of crystalline basement rocks of Florida. American Association of Petroleum Geologists Memoir no. 11, 283–310.

Bassot, J. P., and M. Caen-Vachette. 1983. Données nouvelles sur l'âge du massif de granitoide du Niokola-Koba (Senegal oriental): Implication sur l'âge du stade precoce la chaine Mauritanides. African Earth Sciences Journal 1:159–65.

Bates, R. L., and J. A. Jackson, eds. 1980. Glossary of geology. 2nd edition. Falls Church, Va.: American Geological Institute. 751 p.

Baturin, G. N. 1982. Phosphorites on the sea floor. Amsterdam: Elsevier. 343 p.

Baturin, G. N., K. I. Merkulova, and P. I. Chavlov. 1972. Radiometric evidence for recent formation of phosphatic nodules in marine shelf sediments. Marine Geology 13:37–41.

Baum, G. R., and P. R. Vail. 1988. Sequence stratigraphic concepts applied to Paleogene outcrops, Gulf and Atlantic basins. In Sea-level changes: An integrated approach, edited by C. K. Wilgus, B. S. Hastings, C. G. St.

Kendall, H. W. Posamentier, C. A. Ross, and J. C. Van Wagoner, 309–27. Society of Economic Paleontologists and Mineralogists Special Publication no. 42.

Beck, B. F., and W. C. Sinclair. 1986. Sinkholes in Florida. Orlando: Florida Sinkhole Research Institute, University of Central Florida, Report no. 85-86-4. 17 p.

Bengtsson, T. O. 1987. The hydrologic effects from intense ground-water pumpage in east-central Hillsborough County, Florida. In Karst hydrogeology: Engineering and environmental applications, edited by B. F. Beck and W. L. Wilson, 109–14. Rotterdam: A. A. Balkema.

Benmore, R. A., M. L. Coleman, and J. M. McArthur. 1983. Origin of sedimentary francolite from its sulphur and carbon isotope composition. Nature 302:516.

Bennett, K. C., and D. L. Smith. 1979. A residual magnetic anomaly map for Alachua County, Florida. Florida Scientist 42:236–42.

Bentor, Y. K. 1980. Phosphorites. In Marine phosphorites: Geochemistry, occurrence, genesis, edited by Y. K. Bentor, 3–18. Society of Economic Mineralogists and Paleontologists Special Publication no. 29.

Berdan, J. M. 1964. Stratigraphy and faunas of subsurface lower Paleozoic rocks, Florida and adjacent states. Geological Society of America Special Paper no. 82. 10 p.

Berger, W. H. 1982. Deep-sea stratigraphy. In Cyclic and event stratigraphy, edited by G. Einsele and A. Seilacher, 121–57. Berlin: Springer-Verlag.

Berggren, W. A., and C. D. Hollister. 1974. Paleogeography, paleobiology, and history of circulation in the Atlantic Ocean. In Studies in paleoceanography, edited by W. W. Hay, 126–86. Society of Economic Mineralogists and Paleontologists Special Publication no. 20.

Bernasconi, A. 1987. The major Precambrian terranes of eastern South America. Precambrian Research 37:107–24.

Berner, R. A. 1982. Burial of organic carbon and pyrite sulfur in the modern ocean. American Journal of Science 282:451–73.

Berner, R. A., A. C. Lasaga, and R. M. Garrels. 1983. The carbonate-silicate geochemical cycle and its effect on atmospheric CO_2 over the past 100 million years. American Journal of Science 283:641–83.

Berta, A. 1981. The Plio-Pleistocene hyena *Chasmoporthetes ossifragus* from Florida. Journal of Vertebrate Paleontology 1:341–56.

———. 1987. The sabercat *Smilodon gracilis* from Florida and a discussion of its relationships (Mammalia, Felidae, Smilodontini). Bulletin of the Florida State Museum 31:1–63.

Birch, G. F. 1979. Phosphorite pellets and rock from the western continental margin and adjacent coastal terrace of South Africa. Marine Geology 33:91–116.

Bischoff, R. J., W. C. Shanks III, and R. J. Rosenbauer. 1994. Karstification without carbonic acid: Bedrock dissolution by gypsum-driven dolomitization. Geology 22:995–98.

Black, W. W. 1980. Chemical characteristics of metavolcanics in the Carolina slate belt. Proceedings, the Caledonides in the U.S.A., Virginia Polytechnic Institute and State University Memoir 2:271–78.

Blake, N. J., and L. J. Doyle. 1983. Infaunal-sediment relationships at the shelf-slope break. In The shelfbreak: Critical interface on continental margins, edited by D. J. Stanley and G. T. Moore, 381–89. Society of Economic Paleontologists and Mineralogists Special Publication no. 33.

Bland, A. E., and W. H. Blackburn. 1980. Geochemical studies on the greenstones of the Atlantic seaboard volcanic province, south-central Appalachians. Proceedings, the Caledonides in the U.S.A., Virginia Polytechnic Institute and State University Memoir 2:263–70.

Bloomberg, D., S. B. Upchurch, M. L. Hayden, and R. C. Williams. 1988. Cone-penetrometer exploration of sinkholes. Environmental Geology and Water Science 12:99–105.

Bollinger, G. A. 1973. Seismicity of the southeastern United States. Bulletin of the Seismological Society of America 63:1785–1808.

Bond, G., P. Nickerson, and M. Kominz. 1984a. Breakup of a supercontinent between 635 Ma and 555 Ma: New evidence and implications for continental histories. Earth and Planetary Science Letters 70:325–45.

Bond, P., K. M. Campbell, and T. M. Scott. 1984b. An overview of peat in Florida and related issues. Florida Bureau of Geology Open-File Report no. 4. 228 p.

Bond, P., L. Smith, and W. F. Tanner. 1981. Structural patterns in South Florida. Transactions of the Gulf Coast Association of Geological Societies 31:239–42.

Bosence, D. W. J., J. R. Royston, and M. L. Quine. 1985. Sedimentology and budget of a recent carbonate mound, Florida Keys. Sedimentology 32:317–43.

Bouma, A., et al. 1986. Initial Reports of the Deep Sea Drilling Project no. 14, 28–31.

Bouwer, H. 1978. Groundwater hydrology. New York: McGraw-Hill. 480 p.

Boyle, J. R., and W. Schmidt. 1992. Florida minerals. In Minerals yearbook—1990. U.S. Bureau of Mines. 11 p.

Braunstein, J., P. Huddlestun, and R. Biel. 1988. Gulf Coast region, correlation of stratigraphic units of North America (COSUNA) project. American Association of Petroleum Geologists. Chart.

Brewster-Wingard, G. L., Scott, T. M., Edwards, L. E., and Weedman, S. D. In press. Reinterpretation of the peninsular Florida Oligocene: A multidisciplinary view. Sedimentary Geology.

Bridge, J., and J. M. Berdan. 1952. Preliminary correlation of the Paleozoic rocks from test wells in Florida and adjacent parts of Georgia and Alabama. Guidebook for the 44th Annual Meeting of the Association of American State Geologists, April 18–19, pp. 29–38.

REFERENCES

Briggs, J. C. 1974. Marine zoogeography. New York: McGraw-Hill. 475 p.

Brito Neves, B., and G. Cordani. 1991. Tectonic evolution of South America during the late Proterozoic. Precambrian Research 53:23–40.

Broecker, W. S. 1971. A kinetic model for the chemical composition of seawater. Quaternary Research 1:188–207.

———. 1982. Ocean chemistry during glacial time. Geochimica et Cosmochimica Acta 46:1689–1705.

Broecker, W. S., and D. L. Thurber. 1965. Uranium-series dating of corals and oolites from Bahaman and Florida Keys limestones. Science 149:58–60.

Brooks, G. R., and L. J. Doyle. 1991. Geologic development and depositional history of the Florida Middle Ground: A mid-shelf, temperate-zone reef system in the northeastern Gulf of Mexico. In From shoreline to abyss: Contributions in marine geology in honor of Francis Parker Shepard, edited by R. H. Osborne, 189–203. Society for Sedimentary Geology Special Publication no. 46.

Brooks, G. R., L. J. Doyle, and J. I. McNeillie. 1986. A massive carbonate gravity flow deposit intercalated in the lower Mississippi fan. Initial Reports of the Deep Sea Drilling Project no. 98, 541–46.

Brooks, G. R., and C. W. Holmes. 1989. Recent carbonate slope sediments and sedimentary processes bordering a non-rimmed platform: southwest Florida continental margin. In Controls on carbonate platform and basin development, edited by P. D. Crevello, J. L. Wilson, J. F. Sang, and J. F. Reed, 259–72. Society of Economic Paleontologists and Mineralogists Special Publication no. 44.

Brooks, H. K. 1966. Stops 5–8. Geological history of the Suwannee River. In Geology of the Miocene and Pliocene series in the North Florida–South Georgia area, 37–45. Guidebook for the 12th Annual Field Conference of the Southeastern Geological Society.

———. 1967. Rate of solution of limestone in the karst terrane of Florida. Gainesville: Water Resources Research Center, University of Florida. 67 p.

———. 1968. The Plio-Pleistocene of Florida with special reference to the strata outcropping on the Caloosahatchee River. In Late Cenozoic stratigraphy of south Florida, compiled by R. D. Perkins, 3–42. Guidebook for the 2nd Annual Field Trip of the Miami Geological Society.

———. 1981. Physiographic divisions of Florida. Gainesville: Institute of Food and Agricultural Sciences, University of Florida. 1 sheet.

———. 1982. Geologic map of Florida. Gainesville: Institute of Food and Agricultural Science, University of Florida. 1 sheet.

Brown, P. M., and R. Van der Voo. 1983. A paleomagnetic study of Piedmont metamorphic rocks in southern Delaware. Bulletin of the Geological Society of America 94:815–22.

Brown, R. C. 1988. Florida's fossils. Sarasota, Fla.: Pineapple Press. 208 p.

Bryan, J. R. 1991. Stratigraphic and paleontologic studies of Paleocene and Oligocene carbonate facies of the eastern Gulf Coastal Plain. Ph.D. diss., University of Tennessee, Knoxville. 324 p.

———. 1993. Late Eocene and early Oligocene carbonate facies and paleoenvironments of the eastern Gulf Coastal Plain. In Geologic field studies of the coastal plain in Alabama, Georgia, and Florida, edited by S. A. Kish, 23–47. Southeastern Geological Society Guidebook no. 33.

Bryan, J. R., and P. F. Huddlestun. 1991. Correlation and age of the Bridgeboro Limestone, a coralgal limestone from southwestern Georgia. Journal of Paleontology 65:864–68.

Bryant, J. D. 1991a. Age-frequency profiles of micromammals and population dynamics of *Proheteromys floridanus* (Rodentia) from the early Miocene Thomas farm site, Florida (U.S.A.). Palaeogeography, Palaeoclimatology, Palaeoecology 85:1–14.

———. 1991b. New early Barstovian (middle Miocene) vertebrates from the upper Torreya Formation, eastern Florida panhandle. Journal of Vertebrate Paleontology 11:472–89.

Bryant, J. D., B. J. MacFadden, and P. A. Mueller. 1992. Improved chronologic resolution of the Hawthorn and Alum Bluff groups in northern Florida. Bulletin of the Geological Society of America 104:208–18.

Bryant, W. R., A. A. Meyerhoff, N. K. Brown, M. A. Furrer, T. E. Pyle, and J. W. Antoine. 1969. Escarpments, reef trends, and diapiric structures, eastern Gulf of Mexico. Bulletin of the American Association of Petroleum Geologists 53:2506–42.

Budd, A. F., T. A. Stemann, and R. H. Stewart. 1992. Eocene Caribbean reef corals. Journal of Paleontology 66:570–94.

Buddemeier, R. W., and S. V. Smith. 1988. Coral reef growth in an era of rapidly rising sea level. Coral Reefs 7:51–56.

Buffler, R. T. 1991. Seismic stratigraphy of the deep Gulf of Mexico basin and adjacent margins. In The Gulf of Mexico basin, edited by A. Salvador, 353–87. The Geology of North America, vol. J. Geological Society of America.

Buffler, R. T., and D. S. Sawyer. 1985. Distribution of crust and early history, Gulf of Mexico basin. Transactions of the Gulf Coast Association of Geological Societies 35:333–44.

Buffler, R. T., J. S. Watkins, F. J. Schaub, and J. L. Worzel. 1980. Structure and early geologic history of the deep central Gulf of Mexico basin. In The origin of the Gulf of Mexico and the early opening of the central North Atlantic Ocean, edited by R. H. Pilger, 3–26. Baton Rouge: Louisiana State University Press.

Bullard, E., J. E. Everett, and A. G. Smith. 1965. The fit of the continents around the Atlantic. Philosophical

Transactions of the Royal Society of London 258A:941–66.

Bullen, R. P., S. D. Webb, and B. I. Waller. 1970. A worked mammoth bone from Florida. American Antiquity 35:203–5.

Burdett, J. W., M. A. Arthur, and M. Richardson. 1989. A Neogene seawater sulfur isotope age curve from calcareous pelagic microfossils. Earth and Planetary Science Letters 94:189–98.

Burke, K., C. Cooper, J. F. Dewey, P. Mann, and J. L. Pindell. 1984. Caribbean tectonics and relative plate motions. In The Caribbean–South American plate boundary and regional tectonics, edited by W. E. Bonini, R. B. Hargraves, and R. Shagam, 31–63. Geological Society of America Memoir no. 162.

Burke, W. H., R. E. Denison, E. A. Heatherington, R. B. Koepnick, H. F. Nelson, and J. B. Otto. 1982. Variations of seawater $^{87}Sr/^{86}Sr$ throughout Phanerozoic time. Geology 10:516–19.

Burnett, W. C. 1977. Geochemistry and origin of phosphorite deposits from off Peru and Chile. Bulletin of the Geological Society of America 88:813–23.

Burnett, W. C., J. B. Cowart, and P. A. Chin. 1987. Polonium in the surficial aquifer of west central Florida. In Radon, radium, and other radioactivity in groundwater, edited by B. Graves, 251–69. Chelsea, Mich.: Lewis.

Burnett, W. C., and S. R. Riggs, eds. 1990. Phosphate deposits of the world, vol. 3: Neogene to modern phosphorites. Cambridge and New York: Cambridge University Press. 464 p.

Bush, P. W., G. L. Barr, J. S. Clarke, and R. H. Johnston. 1987. Potentiometric surface of the Upper Floridan aquifer in Florida and in parts of Georgia, South Carolina, and Alabama. U.S. Geological Survey Water-Resources Investigations, Report no. 86-4316. 1 sheet.

Bush, P. W., and R. H. Johnston. 1988. Ground-water hydraulics, regional flow, and ground-water development of the Floridan aquifer system in Florida and in parts of Georgia, South Carolina, and Alabama. U.S. Geological Survey Professional Paper no. 1403C. 80 p.

Caby, R. 1989. Precambrian terranes of Benin-Niger and northeast Brazil and the late Proterozoic South Atlantic fit. Geological Society of America Special Paper no. 230, 145–58.

Cahen, L., and N. Snelling. 1984. The geochronology and evolution of Africa. Oxford: Clarendon. 512 p.

Caldeira, K., M. A. Arthur, R. A. Berner, and A. C. Lasaga. 1993. Cenozoic cooling. Nature 361:1–23.

Campbell, K. M. 1986. The industrial minerals of Florida. Florida Bureau of Geology Information Circular no. 102. 94 p.

Campbell, L., S. Campbell, D. Colquhoun, and J. Ernisse. 1975. Plio-Pleistocene faunas of the central Carolina coastal plain. Geologic Notes 19(3):50–124.

Campbell, R. B. 1939a. Paleozoic under Florida? Bulletin of the American Association of Petroleum Geologists 23:1712–13.

———. 1939b. Deep test in Florida Everglades. Bulletin of the American Association of Petroleum Geologists 23:1713–14.

———. 1943. Earthquakes in Florida. Proceedings of the Florida Academy of Sciences 6:1–4.

Cander, H. S. 1991. Dolomitization and water-rock interaction in the Middle Eocene Avon Park Formation, Floridan aquifer. Ph.D. diss., University of Texas, Austin. 172 p.

———. 1994. An example of mixing-zone dolomite, Middle Eocene Avon Park Formation, Floridan aquifer system. Journal of Sedimentary Research A64:615–29.

Cantrell, R. L. 1992. Peat. Annual Report, U.S. Bureau of Mines. 30 p.

Carpenter, J. H., J. C. Detweiler, J. L. Gillson, E. C. Weichel Jr., and J. P. Wood. 1964. Mining and concentration of ilmenite and associated minerals at Trail Ridge, Florida. Mining Engineering 5:789–95.

Carroll, D. 1963. Petrography of some sandstones and shales of Paleozoic age from borings in Florida. U.S. Geological Survey Professional Paper no. 454A. 15 p.

Carter, B. D., and R. E. Hammack. 1989. Stratigraphic distribution of Jacksonian (Priabonian) echinoids in Georgia. Palaios 4:86–91.

Carter, B. D., and M. L. McKinney. 1992. Eocene echinoids. Paleobiology 18:299–325.

Carter, L. J. 1974. The Florida experience. Baltimore: Johns Hopkins University Press. 355 p.

Carver, R. E., G. A. Brook, and R. A. Hyatt. 1986. Trail Ridge and Okefenokee Swamp. In Geological Society of America Centennial Field Guide, vol. 6: Southern Section, edited by T. L. Neathery, 331–34. Florida Geological Survey.

Cathcart, J. B. 1962. Economic geology of the Keysville Quadrangle, Florida. U.S. Geological Survey Bulletin no. 1128. 82 p.

Causaras, C. R. 1985. Geology of the surficial aquifer system, Broward County, Florida. U.S. Geological Survey Water-Resources Investigations, Report no. 84-4068. 167 p.

———. 1987. Geology of the surficial aquifer system, Dade County, Florida. U.S. Geological Survey Water-Resources Investigations, Report no. 86-4126. 240 p.

Ceryak, R. 1977. Alapaha River Basin. Suwannee River Water Management District Information Circular no. 5. 20 p.

Chaki, S., and W. R. Oglesby. 1972. Bouguer anomaly map of northwest Florida. Florida Bureau of Geology Map Series no. 52.

Channell, J. E. T., C. McCabe, T. H. Torsvik, A. Trench, and N. H. Woodcock. 1992. Paleozoic paleomagnetic studies in the Welsh basin—recent advances. Geologic Magazine 129:533–42.

Chanton, J. P., C. S. Martens, and C. K. Paull. 1991. Control of pore water chemistry at the base of the Florida Escarpment by processes within the platform. Nature 349:229–331.

Cheetham, A. H. 1963. Late Eocene zoogeography of the

eastern Gulf Coast region. Geological Society of America Memoir no. 91. 113 p.

Chen, C. S. 1965. The regional lithostratigraphic analysis of Paleocene and Eocene rocks of Florida. Florida Geological Survey Bulletin no. 45.

Chowns, T. M., and C. T. Williams. 1983. Pre-Cretaceous rocks beneath the Georgia coastal plain—regional implications. In Studies related to the Charleston, South Carolina, earthquake of 1886—tectonics and seismicity, edited by G. S. Gohn, L1–L42. U.S. Geological Survey Professional Paper no. 1313.

Claypool, G. E., and I. R. Kaplan. 1974. The origin and distribution of methane in marine sediments. In Natural gases in marine sediments, edited by I. R. Kaplan, 99–140. New York: Plenum Press.

Cloetingh, S. 1988. Integrated stresses: A tectonic cause for third-order cycles in apparent sea level? In Sea level changes: An integrated approach, edited by C. K. Wilgus, B. S. Hastings, C. G. St. Kendall, H. W. Posamentier, C. A. Ross, and J. C. Van Wagoner, 19–29. Society of Economic Paleontologists and Mineralogists Special Publication no. 42.

Coates, A. G., J. B. C. Jackson, L. S. Collins, T. M. Cronin, H. J. Dowsett, L. M. Bybell, P. Jung, and J. A. Obando. 1992. Closure of the Isthmus of Panama. Bulletin of the Geological Society of America 104:814–28.

Coates, D. R. 1981. Environmental Geology. New York: John Wiley and Sons. 791 p.

Cole, W. S., and G. M. Ponton. 1930. The foraminifera of the Marianna Limestone of Florida. Florida Geological Survey Bulletin no. 5, 19–69.

Colquhoun, D. J., S. M. Herrick, and H. G. Richards. 1968. A fossil assemblage from the Wicomico Formation in Berkeley County, South Carolina. Bulletin of the Geological Society of America 79:1211–20.

Colton, R. C. 1978. The subsurface geology of Hamilton County, Florida, with emphasis on the Oligocene age Suwannee Limestone. Master's thesis, Florida State University, Tallhassee. 185 p.

Commeau, R., C. K. Paull, J. A. Commeau, and L. J. Poppe. 1987. Chemistry and mineralogy of pyrite-enriched sediments at a passive margin sulfide brine seep. Earth and Planetary Science Letters 82:62–74.

Compton, J. S. 1988. Degree of supersaturation and precipitation of organogenic dolomite. Geology 16:318–21.

Compton, J. S., D. A. Hodell, J. R. Garrido, and D. J. Mallison. 1993. Origin and age of phosphorite from the south-central Florida platform: Relation of phosphogenesis to sea level fluctuations and $\delta^{13}C$ excursions. Geochimica et Cosmochimica Acta 57:131–46.

Compton, J. S., S. W. Snyder, and D. A. Hodell. 1990. Phosphogenesis and weathering of shelf sediments from the southeastern United States: Implications for Miocene $\delta^{13}C$ excursions and global cooling. Geology 18:1227–30.

Conover, C. S., J. J. Geraghty, and G. G. Parker. 1984. Ground water. In Water resources atlas of Florida, edited by E. A. Fernald and D. J. Patton, 36–53. Tallahassee: Florida State University.

Conrad, T. A. 1846a. Observations on the geology of a part of East Florida with a catalogue of Recent shells of the coast. American Journal of Science 52:36–48.

———. 1846b. Descriptions of new species of organic remains from the Upper Eocene limestone of Tampa Bay. American Journal of Science 52:399–400.

Cook, P. J., and M. W. McElhinny. 1979. A reevaluation of the spatial and temporal distribution of sedimentary phosphate deposits in the light of plate tectonics. Economic Geology 74:315–30.

Cooke, C. W. 1930. Correlation of coastal terraces. Journal of Geology 38:577–89.

———. 1939. Scenery of Florida, interpreted by a geologist. Florida Geological Survey Bulletin no. 17. 118 p.

———. 1945. Geology of Florida. Florida Geological Survey Bulletin no. 29.

Cooke, C. W., and S. Mossom. 1929. Geology of Florida. In Twentieth Annual Report of the Florida Geological Survey, 29–228.

Cooper, H. H. Jr. 1959. A hypothesis concerning the dynamic balance of freshwater and saltwater in a coastal aquifer. Journal of Geophysical Research 64:461–67.

Corso, W. 1987. Development of the Early Cretaceous northwest Florida carbonate platform. Ph.D. diss., University of Texas, Austin. 136 p.

Corso, W., J. A. Austin Jr., and R. T. Buffler. 1989. The Early Cretaceous platform off northwest Florida: Controls on morphologic development of carbonate margins. Marine Geology 86:1–14.

Corso, W., R. T. Buffler, and J. A. Austin Jr. 1988. Erosion of the southern Florida Escarpment. In Atlas of seismic stratigraphy, edited by A. W. Bally, 149–55. American Association of Petroleum Geologists Studies in Geology no. 27(2).

Covington, J. M. 1992. Neogene nannofossils of Florida. In Proceedings of the Third Bald Head Island Conference on Coastal Plains Geology—the Florida Neogene. University of North Carolina, Wilmington. 52 p.

Cowart, J. B., M. I. Kaufman, and J. K. Osmond. 1978. Uranium-isotope variations in groundwaters of the Floridan aquifer and boulder zone of south Florida. Journal of Hydrology 36:161–72.

Craighead, F. C. Sr. 1971. The Trees of South Florida. Coral Gables, Fla.: University of Miami Press. 212 p.

Crain, L. J., G. H. Hughes, and L. J. Shell. 1975. Water resources of Indian River County, Florida. Florida Bureau of Geology Report of Investigations no. 80. 75 p.

Cramer, F. H. 1971. Position of the north Florida lower Paleozoic block in Silurian time—phytoplankton evidence. Journal of Geophysical Research 67:4754–57.

———. 1973. Middle and Upper Silurian chitinozoan succession in Florida subsurface. Journal of Paleontology 47:278–88.

Croft, M., and G. D. Shaak. 1985. Ecology and stratigraphy of the echinoids of the Ocala Limestone (Late Eocene).

Tulane Studies in Geology and Paleontology 18:127–43.

Cronin, T. M. 1981. Rates and possible causes of neotectonic vertical crustal movements of the emerged southeastern United States Atlantic Coastal Plain. Bulletin of the Geological Society of America 92:812–33.

———. 1988. Evolution of marine climates of the U.S. Atlantic coast during the past four million years. Philosophical Transactions of the Royal Society of London 318(B):661–78.

———. 1990. Evolution of Neogene and Quaternary marine Ostracoda, United States Atlantic Coastal Plain: Evolution and speciation in Ostracoda, IV. U.S. Geological Survey Professional Paper no. 1367C. 43 p.

Cummins, L. E., J. D. Arthur, and P. C. Ragland. 1992. Classification and tectonic implications for Early Mesozoic magma types of the circum-Atlantic. Geological Society of America Special Paper no. 268, 119–35.

Cushman, J. A., and G. M. Ponton. 1932. The foraminifera of the upper, middle, and part of the lower Miocene of Florida. Florida Geological Survey Bulletin no. 9. 147 p.

Cushman-Roison, M., and B. J. Franks. 1982. Sand-and-gravel aquifer. In Principal aquifers in Florida, edited by B. J. Franks, U.S. Geological Survey Water-Resources Investigations Open-File Report no. 82-255. 4 sheets.

Dall, W. H. 1887. Notes on the geology of Florida. American Journal of Science 134:161–70.

———. 1890–1903. Contributions to the Tertiary fauna of Florida, with especial reference to the Miocene silex-beds of Tampa and the Pliocene beds of the Caloosahatchie River. Transactions of the Wagner Free Institute of Science 3(I–V):1–1654.

———. 1915. A monograph of the molluscan fauna of the *Orthaulax pugnax* zone of the Oligocene of Tampa, Florida. Bulletin of the U.S. National Museum 90. 173 p.

Dall, W. H., and G. D. Harris. 1892. Correlation papers—Neocene. U.S. Geological Survey Professional Paper no. 84. 349 p.

Dallmeyer, R. D. 1987. ^{40}Ar/^{39}Ar age of detrital muscovite within Lower Ordovician sandstone in the coastal plain basement of Florida—implications for West African terrane linkages. Geology 15:998–1001.

———. 1988. A tectonic linkage between the Rockelide orogen (Sierra Leone) and the St. Lucie metamorphic complex in the Florida subsurface. Journal of Geology 89:183–95.

———. 1989a. Contrasting accreted terranes in the southern Appalachian orogen and Atlantic-Gulf coastal plains and their correlations with West African sequences. In Terranes in the circum-Atlantic Paleozoic orogens, edited by R. D. Dallmeyer, 247–68. Geological Society of America Special Paper no. 230.

———. 1989b. ^{40}Ar/^{39}Ar ages from subsurface crystalline basement of the Wiggins uplift and southwesternmost Appalachian Piedmont: Implications for late Paleozoic terrane accretion during assembly of Pangaea. American Journal of Science 289:812–33.

———. 1991. Exotic terranes in the central-southern Appalachian orogen. In The West African orogens and circum-Atlantic correlatives, edited by R. D. Dallmeyer and J. P. Lecorche, 335–71. New York: Springer-Verlag.

Dallmeyer, R. D., M. Caen-Vachette, and M. Villeneuve. 1987. Emplacement age of post-tectonic granites in southern Guinea (West Africa) and the peninsular Florida subsurface: Implications for origins of southern Appalachian exotic terranes. Bulletin of the Geological Society of America 99:87–93.

Dallmeyer, R. D., and M. Villeneuve. 1987. ^{40}Ar/^{39}Ar mineral age record of polyphase tectonothermal evolution in the southern Mauritanide orogen, southeastern Senegal. Bulletin of the Geological Society of America 98:602–11.

Dalziel, I. W. D. 1991. Pacific margins of Laurentia and East Antarctica–Australia as a conjugate rift pair: Evidence and implications for an Eocambrian supercontinent. Geology 19, 598–601.

Daniels, D. L., I. Zeitz, and P. Popenoe. 1983. Distribution of subsurface lower Mesozoic rocks in the southeastern United States, as interpreted from regional aeromagnetic and gravity maps. In Studies related to the Charleston, South Carolina, earthquake of 1886—tectonics and seismicity, edited by G. S. Gohn, K1–K23. U.S. Geological Survey Professional Paper no. 1313.

Davis, D. G. 1980. Cave development in the Guadalupe Mountains: A critical review of recent hypotheses. Bulletin of the National Speleological Society 42:42–48.

Davis, J. H. 1943. The natural features of southern Florida. Florida Geological Survey Bulletin no. 25. 311 p.

———. 1946. The peat deposits of Florida—their occurrence, development and uses. Florida Geological Survey Bulletin no. 30. 250 p.

Davis, R. A. 1989. Morphodynamics of the west-central Florida barrier system: The delicate balance between wave- and tide-domination. Proceedings of KNGMG Symposium, Coastal Lowlands, Geology, and Geotechnology, 225–35. Dordrecht, Neth.: Kluwer Academic Publishers.

———. 1994. Barriers of the Florida Gulf peninsula. In Geology of Holocene barrier island systems, edited by R. A. Davis, 167–206. Heidelberg: Springer-Verlag.

Davis, R. A., and W. T. Fox. 1981. Interaction between wave- and tide-generated processes at the mouth of a microtidal estuary: Matanzas Inlet, Florida (U.S.A.). Marine Geology 40:49–68.

Davis, R. A., and M. O. Hayes. 1984. What is a wave-dominated coast? Marine Geology 60:313–29.

Davis, R. A., and A. C. Hine. 1989. Quaternary geology and sedimentology of the barrier island and marshy coast, west-central Florida, U.S.A. 28th International

REFERENCES

Geology Congress, Field Trip Guidebook no. T375. Washington, D.C.: American Geophysical Union. 38 p.

Davis, R. A., A. C. Hine, and D. F. Belknap. 1985. Geology of the barrier island and marsh-dominated coast, west-central Florida. Guidebook of the Annual Meeting of the Geological Society of America, Orlando. 119 p.

Davis, R. A., A. C. Hine, and E. A. Shinn. 1992. Holocene coastal development on the Florida Peninsula. In Quaternary coasts of the United States: Marine and lacustrine systems, edited by C. H. Fletcher III and J. F. Wehmille, 193–212. Society of Economic Paleontologists and Mineralogists Special Publication no. 48.

Davis, R. A., and J. M. Klay. 1989. Origin and development of Quaternary terrigenous inner shelf sequences, southwest Florida. Transactions of the Gulf Coast Association of Geological Societies 39:341–47.

Davis, R. A., S. C. Knowles, and M. P. Bland. 1989a. Role of hurricanes in the Holocene stratigraphy of estuaries: Examples from the Gulf Coast of Florida. Journal of Sedimentary Petrology 59:1052–61.

Davis, R. A., and B. J. Kuhn. 1985. Origin and development of Anclote Key, west-peninsular Florida. Marine Geology 63:153–71.

Davis, R. A., P. Newell, and R. J. Sussko. 1989b. Inner continental shelf off southwest Florida. Society of Economic Paleontologists and Mineralogists Foundations, Gulf Coast Section, 7th Annual Research Conference, 53–61.

Dean, R. G., and M. P. O'Brien. 1987. Florida's east coast inlets, shoreline effects, and recommendations for action. Report no. 87-17, Coastal and Ocean Engineering Department, University of Florida, Gainesville. 65 p.

DeBalko, D. A., and R. T. Buffler. 1992. Seismic stratigraphy and geologic history of Middle Jurassic through Lower Cretaceous rocks, deep eastern Gulf of Mexico. Transactions of the Gulf Coast Association of Geological Societies 42:89–105.

den Hartog, C. 1970. The seagrasses of the world. Amsterdam: North Holland. 275 p.

Denny, W. M. III, J. A. Austin Jr., and R. T. Buffler. 1994. Seismic stratigraphy and geologic history of middle-Cretaceous through Cenozoic rocks, southern Straits of Florida. Bulletin of the American Association of Petroleum Geologists 78:461–87.

DePaolo, D. J. 1986. Detailed record of the Neogene Sr isotopic evolution of seawater from DSDP Site 590B. Geology 14:103–106.

———. 1988. Neodymium isotope geochemistry. New York: Springer-Verlag. 187 p.

DePaolo, D. J., and K. L. Finger. 1991. High-resolution strontium-isotope and biostratigraphy of the Miocene Monterey Formation. Bulletin of the Geological Society of America 103:112–24.

DePaolo, D. J., and B. L. Ingram. 1985. High-resolution stratigraphy with strontium isotopes. Science 227:938–41.

Diblin, M. C., A. F. Randazzo, and D. S. Jones. 1991. *Lithoplaision ocalae:* a new trace fossil from the Ocala Limestone (Eocene), Florida. Ichnos 1:255–60.

Dietz, R. S., and J. C. Holden. 1970. Reconstruction of Pangaea: Breakup and dispersion of continents, Permian to present. Journal of Geophysical Research 75:4939–56.

Dillon, W. P., C. K. Paull, and L. E. Gilbert. 1985. History of the Atlantic continental margin off Florida: The Blake Plateau basin. In Geologic evolution of the United States Atlantic margin, edited by C. W. Poag, 189–215. New York: Van Nostrand Reinhold.

Dillon, W. P., and P. Popenoe. 1988. The Blake Plateau and Carolina Trough. In The Atlantic continental margin, U.S., edited by R. E. Sheridan and J. A. Grow, 291–328. The geology of North America, vol. I-2. Geological Society of America.

Dillon, W. P., and J. M. A. Sougy. 1974. Geology of West African and Canary and Cape Verde islands. In The North Atlantic, edited by A. E. M. Nairn and F. G. Stehli, 315–90. Ocean basins and margins, vol 2. New York: Plenum Press.

Dillon, W. P., A. M. Trehu, P. G. Valentine, and M. M. Ball. 1988. Eroded carbonate platform margin—the Blake Escarpment off southeastern United States. In Atlas of seismic stratigraphy, edited by A. W. Bally, 140–47. Studies in Geology, vol. 2(27). American Association of Petroleum Geologists.

Dingle, R. V., A. R. Lord, and Q. B. Hendey. 1979. New sections in the Varswater Formation (Neogene) of Langebaan Road, South Western Cape, South Africa. Annals of the South African Museum 78. 81 p.

Dobson, L. M. 1990. Seismic stratigraphy and geologic history of Jurassic rocks, northeastern Gulf of Mexico. Master's thesis, University of Texas, Austin. 165 p.

Dobson, L. M., and R. T. Buffler. 1991. Basement rocks and structure, northeast Gulf of Mexico. Transactions of the Gulf Coast Association of Geological Societies 41:191–206.

Dodd, J. F., and C. T. Siemers. 1971. Effect of Late Pleistocene karst topography on Holocene sedimentation and biota, lower Florida Keys. Bulletin of the Geological Society of America 82:211–18.

Domning, D. P. 1988. Fossil Sirenia of the west Atlantic and Caribbean region. I. *Metaxytherium floridanum* Hay, 1922. Journal of Vertebrate Paleontology 8:395–426.

Domning, D. P., G. S. Morgan, and C. E. Ray. 1982. North American Eocene sea cows (Mammalia: Sirenia). Smithsonian Contributions to Paleobiology 52:1–69.

Donoghue, J. F. 1991. St. Vincent Island. In Field guidebook, Research Conference on Quaternary Coastal Evolution, edited by J. F. Donoghue et al., 76–85. Society of Economic Paleontologists and Mineralogists.

Donovan, A. D., G. R. Baum, G. L. Bleckschmidt, T. S. Loutit, C. E. Pflum, and P. R. Vail. 1988. Sequence stratigraphic setting of the Cretaceous-Tertiary boundary in central Alabama. In Sea level changes: An integrated approach, edited by C. K. Wilgus, B. S. Hastings, C. G. St. Kendall, H. W. Posamentier, C. A. Ross, and J. C. Van Wagoner, 299–307. Society of Economic Paleontologists and Mineralogists Special Publication no. 42.

Douglas, M. S. 1947. The Everglades: River of grass. New York: Rinehart. 406 p.

Doyle, L. J. 1981. Depositional systems of the continental margin of the eastern Gulf of Mexico west of peninsular Florida: A possible modern analog to some depositional models for the Permian Delaware basin. Transactions of the Gulf Coast Association of Geological Societies 31:279–82.

Doyle, L. J., and C. W. Holmes. 1985. Shallow structure, stratigraphy, and carbonate sedimentary processes of west Florida upper continental slope. Bulletin of the American Association of Petroleum Geologists 69:1133–44.

Doyle, L. J., and T. N. Sparks. 1980. Sediments of the Mississippi, Alabama, and Florida (MAFLA) continental shelf. Journal of Sedimentary Petrology 50:905–16.

Droxler, A., A. D. Cunningham, A. C. Hine, P. Hallock, D. Duncan, E. Rosencrantz, R. T. Buffler, and E. Robinson. 1992. Late Middle (?) Miocene segmentation of an Eocene–early Miocene carbonate megabank on the northern Nicaraguan Rise: Response to tectonic activity along the North America/Caribbean plate boundary zone. Abstracts of the American Geophysical Union 299.

Duane, D. B., M. E. Field, E. P. Meisburger, D. J. P. Swift, and S. J. Williams. 1972. Linear shoals on the Atlantic inner continental shelf, Florida to Long Island. In Shelf sediment transport: Process and pattern, edited by D. J. P. Swift, D. B. Duane, and O. H. Pilkey, 447–99. Stroudsberg, Penn.: Dowden, Hutchinson, and Ross.

Duane, D. B., and E. P. Meisburger. 1969a. Geomorphology and sediments of the inner continental shelf, Palm Beach to Cape Kennedy, Florida. U.S. Army Corps of Engineers Coastal Engineering Research Center Technical Memorandum no. 34. 82 p.

———. 1969b. Geomorphology and sediments of the nearshore continental shelf, Miami to Palm Beach, Florida. U.S. Army Corps of Engineers Coastal Engineering Research Center Technical Memorandum no. 29. 47 p.

DuBar, J. R. 1958a. Neogene stratigraphy of southwestern Florida. Transactions of the Gulf Coast Association of Geological Societies 8:129–55.

———. 1958b. Stratigraphy and paleontology of the late Neogene strata of the Caloosahatchee River area of southern Florida. Florida Geological Survey Bulletin no. 40. 267 p.

———. 1962. Neogene biostratigraphy of the Charlotte Harbor area in southwestern Florida. Florida Geological Survey Bulletin no. 43. 83 p.

———. 1974. Summary of Neogene stratigraphy of southern Florida. In Post-Miocene stratigraphy, central and southern Atlantic Coastal Plain, edited by R. Q. Oaks and J. R. DuBar, 206–31. Logan: Utah State University Press.

———. 1991. Quaternary geology of the Gulf of Mexico Coastal Plain–Florida Peninsula. In Quaternary nonglacial geology, edited by R. B. Morrison, 595–604. The Geology of North America, vol. K-2. Geological Society of America.

Dudley, J. P. 1987. Wildlife and landscapes of late Pleistocene Florida: The paleoecology of Rancholabrean mammoth-mastodont faunas in southeastern North America. Master's thesis, University of Florida, Gainesville. 232 p.

Duerr, A. D., and R. M. Wolansky. 1986. Hydrogeology of the surficial and intermediate aquifers of central Sarasota County, Florida. U.S. Geological Survey Water-Resources Investigations Report no. 86-4068. 48 p.

Duncan, D. S. 1992. Neogene to Recent seismic stratigraphy of the lower Tampa Bay estuary, west-central Florida. Master's thesis, University of South Florida, St. Petersburg. 101 p.

Duncan, J. G., W. L. Evans III, and K. L. Taylor. 1994. Geologic framework of the lower Floridan aquifer system, Brevard County, Florida. Florida Geological Survey Bulletin no. 64. 122 p.

Dunham, R. J. 1962. Classification of carbonate rocks according to depositional texture. In Classification of carbonate rocks, edited by W. E. Ham, 108–21. American Association of Petroleum Geologists Memoir no. 1.

Dupre, B., and C. Allegre. 1983. Pb-Sr isotope variation in Indian Ocean basalts and mixing phenomena. Nature 303:142–46.

Dustan, P. 1985. Community structure of reef-building corals in the Florida Keys: Carysfort Reef Key, Key Largo, and Long Reef, Dry Tortugas. Atoll Research Bulletin 288:1–27.

Eberli, G. P., and R. N. Ginsburg. 1987. Segmentation and coalescence of Cenozoic carbonate platforms, northwestern Great Bahama Bank. Geology 15:75–79.

———. 1989. Cenozoic progradation of northwestern Great Bahama Bank, a record of lateral platform growth and sea-level fluctuations. In Controls on carbonate platform and basin development, edited by P. D. Crevello, J. L. Wilson, J. F. Sarg, and J. F. Reed, 339–51. Society of Economic Paleontologists and Mineralogists Special Publication no. 44.

Edgerton, C. D. 1974. Effects of urbanization on the availability of construction materials in south Florida. U.S. Bureau of Mines Information Circular no. 8664.

Edmund, A. G. 1985. The armor of fossil giant armadillos (Pampatheriidae, Xenarthra, Mammalia). Texas Memorial Museum Pearce-Sellards Series no. 40.

———. 1987. Evolution of the genus *Holmesina* (Pampatheriidae, Mammalia) in Florida, with remarks

REFERENCES

on taxonomy and distribution. Texas Memorial Museum Pearce-Sellards Series no. 45.

Egemeier, S. J. 1988. A theory for the origin of Carlsbad Caverns. Bulletin of the National Speleological Society 49:73–76.

Elsner, H. 1992. Granulometry and mineralogy of some northeastern Florida placers: A consequence of heavy mineral concentration in nearshore bars. Sedimentary Geology 76:233–55.

Emery, K. O. 1969. The continental shelves. Scientific American 221:106–21.

Emry, R. J., L. S. Russell, and P. J. Bjork. 1987. The Chadronian, Orellan, and Whitneyan North American land mammal ages. In Cenozoic mammals of North America: Geochronology and biostratigraphy, edited by M. O. Woodburne, 118–52. Berkeley: University of California Press.

Enos, P. 1977. Holocene sediment accumulations of the south Florida shelf margin. In Quaternary sedimentation in south Florida, edited by P. Enos and R. D. Perkins, 1–130. Geological Society of America Memoir no. 147.

———. 1989. Islands in the bay—a key habitat of Florida Bay. Bulletin of Marine Science 44:365–86.

Enos, P., and R. D. Perkins. 1977. Quaternary sedimentation in south Florida. Geological Society of America Memoir no. 147. 198 p.

———. 1979. Evolution of Florida Bay from island stratigraphy. Bulletin of the Geological Society of America 90:59–83.

Evans, M. W., and A. C. Hine. 1991. Late Neogene sequence stratigraphy of a carbonate-siliciclastic transition: Southwest Florida. Bulletin of the Geological Society of America 103:679–99.

Evans, M. W., A. C. Hine, and D. F. Belknap. 1989. Quaternary stratigraphy of Charlotte Harbor estuary-lagoon system, southwest Florida: Implications of the carbonate-siliciclastic transition. Marine Geology 88:319–48.

Evans, M. W., A. C. Hine, D. F. Belknap, and R. A. Davis. 1985. Bedrock controls on barrier island development: West-central Florida coast. Marine Geology 63:263–83.

Fairbanks, R. B. 1989. A 17,000-year glacio-eustatic sea level record: Influence of glacial melting rates on the Younger Dryas event and deep-ocean circulation. Nature 342:637–42.

Fairbridge, R. W. 1961. Eustatic changes in sea level. In Physics and chemistry of the Earth, vol. 4, edited by L. H. Ahrens et al., 99–185. New York: Pergamon Press.

Farre, J. A., B. A. McGregor, W. B. Ryan, and J. M. Robb. 1983. Breaching the shelfbreak: Passage from youthful to mature phase in submarine canyon evolution. In The shelfbreak: Critical interface on continental margins, edited by D. J. Stanley and G. T. Moore, 25–39. Society of Economic Paleontologists and Mineralogists Special Publication no. 33.

Faust, M. J. 1986. Seismic stratigraphy of Middle Cretaceous Unconformity (MCU) in central Gulf of Mexico basin (abstract). Bulletin of the American Association of Petroleum Geologists 70:588.

———. 1990. Seismic stratigraphy of the mid-Cretaceous unconformity (MCU) in the central Gulf of Mexico basin Geophysics 55:868–84.

Federal Writer's Project. 1949. Florida—a guide to the southernmost state. American Guide Series. New York: Oxford University Press. 600 p.

Fenneman, N. M. 1938. Physiography of eastern United States. New York: McGraw-Hill. 691 p.

Fernald, E. A., and D. J. Patton, eds. 1984. Water resources atlas of Florida. Tallahassee: Florida State University. 291 p.

Field, M. E. 1974. Buried strandline deposits on the central Florida inner shelf. Bulletin of the Geological Society of America 85:57–60.

Field, M. E., and D. B. Duane. 1974. Geomorphology and sediments of the inner continental shelf, Cape Canaveral, Florida. U.S. Army Corps of Engineers Coastal Engineer Research Center Technical Memorandum no. 42. 87 p.

Fischer, A. G. 1951. The echinoid fauna of the Inglis Member, Moodys Branch Formation. Florida Geological Survey Bulletin no 34(2), 45–101.

Fischer, A. G., and M. A. Arthur. 1977. Secular variations in the pelagic realm. In Deep-water carbonate environments, edited by H. E. Cook and P. Enos. Society of Economic Paleontologists and Mineralogists Special Publication no. 25.

Fischer, A. G., and D. J. Bottjer. 1991. Orbital forcing and sedimentary sequences. Journal of Sedimentary Petrology 61:1063–69.

Fish, J. E. 1988. Hydrogeology, aquifer characteristics, and ground-water flow of the surficial aquifer system, Broward County, Florida. U.S. Geological Survey Water-Resources Investigations Report no. 87-4034. 92 p.

Fleischer, M. 1987. Glossary of mineral species, 5th edition. Mineralogical Record. 234 p.

Florida Bureau of Geology. 1986. Hydrogeological units of Florida. Florida Bureau of Geology Special Publication no. 28. 9 p.

Florida Department of Environmental Regulation. 1984. Summary of aldicarb studies. Pesticide Review Section, Florida Department of Environmental Regulation. 87 p.

———. 1986. Florida sites list, petroleum contamination incidents. Bureau of Operations, Florida Department of Environmental Regulation. 82 p.

———. 1989. Florida ground water guidance concentrations. Florida Department of Environmental Regulation. 14 p.

Folk, R. L. 1965. Some aspects of recrystallization in ancient limestones. In Dolomitization and limestone diagenesis, edited by L. C. Pray and R. C. Murray, 14–48.

Society of Economic Paleontologists and Mineralogists Special Publication no. 13.

———. 1974. The natural history of crystalline calcium carbonate: Effect of magnesium content and salinity. Journal of Sedimentary Petrology 44:40–53.

Frailey, D. 1978. An early Miocene (Arikareean) fauna from north-central Florida (the SB-1A local fauna). Occasional Papers of the Museum of Natural History no. 75, 1–20. Lawrence: University Press of Kansas.

———. 1979. The large mammals of the Buda local fauna (Arikareean: Alachua County, Florida). Bulletin of the Florida State Museum 24:123–73.

Frank, E. F., and B. F. Beck. 1991. An analysis of the cause of subsidence damage in the Dunedin, Florida, area—1990/1991. Florida Sinkhole Research Institute Report no. 90-91-2. 64 p.

Frazier, M. K. 1981. A revision of the fossil Erethizontidae of North America. Bulletin of the Florida State Museum 27:1–76.

Freeman-Lynde, R. P. 1983. Cretaceous and Tertiary samples dredged from the Florida Escarpment, eastern Gulf of Mexico. Transactions of the Gulf Coast Association of Geological Societies 33:91–99.

Froelich, P. N., K. H. Kim, R. Jahnke, W. C. Burnett, A. Soutar, and M. Deakin. 1983. Pore water fluoride in Peru continental margin sediments: Uptake from seawater. Geochimica et Cosmochimica Acta 47:1605–12.

Frost, S. H., and R. L. Langenheim. 1974. Cenozoic reef biofacies. DeKalb: Northern Illinois University Press. 388 p.

Fuller, W. R. 1976. Heat flow reconnaissance of Florida. Master's thesis, University of Florida, Gainesville. 78 p.

Galli-Oliver, C., G. Garduno, and J. Gamino. 1991. Phosphorite deposits in the upper Oligocene, San Gregorio Formation at San Juan de la Costa, Baja California Sur, Mexico. In Phosphate deposits of the world, vol. 3, edited by W. C. Burnett and S. R. Riggs, 22–126. Cambridge: Cambridge University Press.

Gallivan, L. B. and R. A. Davis. 1981. Sediment transport in a microtidal estuary: Matanzas River, Florida (U.S.A.). Marine Geology 40:69–84.

Galloway, W. E., D. G. Bebout, W. L. Fisher, J. B. Dunlap Jr., R. Cabrera-Castro, J. E. Lugo-Rivera, and T. M. Scott. 1991. Cenozoic. In The Gulf of Mexico basin, edited by A. Salvador, 245–324. The geology of North America, vol. J. Boulder, Colo.: Geological Society of America.

Gardner, J. 1926–50. The molluscan fauna of the Alum Bluff Group of Florida. U.S. Geological Survey Professional Paper no. 142. 656 p.

Gardulski, A. F., M. H. Gowen, A. Milsark, S. D. Weiterman, S. W. Wise Jr., and H. T. Mullins. 1991. Evolution of a deep-water carbonate platform: Upper Cretaceous to Pleistocene sedimentary environments on the west Florida margin. Marine Geology 101:163–79.

Gardulski, A. F., H. T. Mullins, B. Oldfield, J. Applegate, and S. W. Wise Jr. 1986. Carbonate mineral cycles in ramp slope sediment, eastern Gulf of Mexico. Paleoceanography 1:555–65.

Gardulski, A. F., H. T. Mullins, and S. D. Weiterman. 1990. Carbonate mineral cycles generated by foraminiferal and pteropod response to Pleistocene climate—west Florida ramp slope. Sedimentology 37:727–43.

Garrison, R. E., M. Kastner, and Y. Kolodny. 1987. Phosphorites and phosphatic rocks in the Monterey Formation and related Miocene units, coastal California. In Cenozoic basin development in coastal California, edited by R. V. Ingersoll and W. G. Ernst, 6:348–81. Engelwood Cliffs, N.J.: Prentice-Hall.

Garrison, R. E., M. Kastner, and C. E. Reimers. 1990. Miocene phosphogenesis. In Phosphate deposits of the world, vol. 3, edited by W. C. Burnett and S. R. Riggs, 285–99. Cambridge: Cambridge University Press.

Gebert, W. A., D. J. Graczyk, and W. R. Krug. 1987. Average annual runoff in the United States, 1951–80. U.S. Geological Survey Hydrologic Atlas HA-710, scale 1:7,500,000, 1 sheet.

Gelbaum, C. S. 1978. The geology and ground water of the Gulf trough. Bulletin of the Georgia Geological Survey 93:38–47.

Gelbaum, C. S., and J. E. Howell. 1982. The geohydrology of the Gulf trough. In Second symposium on the geology of the southeastern coastal plain, edited by D. D. Arden, B. F. Beck, and E. Morrow, 140–53. Georgia Geological Survey Information Circular no. 53.

Gibeaut, J. C., and R. A. Davis. 1993. Statistical geomorphic classification of ebb-tidal deltas along the west-central Florida coast. Journal of Coastal Research 18:165–84.

Gillson, J. V. 1959. Sand deposits of titanium minerals. Mining Engineering 11:421–29.

Ginsburg, R. N. 1956. Environmental relationships of grain size and constituent particles in some south Florida carbonate sediments. Bulletin of the American Association of Petroleum Geologists 40:2384–2427.

———. 1964. South Florida carbonate sediments. Field Trip Guide for the Annual Meeting of the Geological Society of America, Miami Beach. 72 p.

Ginsburg, R. N., and N. P. James. 1974. Holocene carbonate sediments of continental shelves. In The geology of continental margins, edited by C. A. Burk and C. L. Drake, 137–56. New York: Springer-Verlag.

Ginsburg, R. N., and E. A. Shinn, 1964. Distribution of the reef-building community in Florida and the Bahamas (abstract). Bulletin of the American Association of Petroleum Geologists 48:527.

Givens, C. R. 1989. First record of the Tethyan genus *Volutilithes* (Gastropoda: Volutidae) in the Paleogene of the Gulf Coastal Plain, with a discussion of Tethyan molluscan assemblages in the Gulf Coastal Plain and Florida. Journal of Paleontology 63:852–56.

Glaser, K. S., and A. W. Droxler. 1991. High production and highstand shedding from deeply submerged carbonate banks, northern Nicaraguan Rise. Journal of Sedimentary Petrology 61:128–42.

REFERENCES

Glenn, C. R., M. A. Arthur, H. W. Yeh, and W. C. Burnett. 1988. Carbon isotopic composition and lattice-bound carbonate of Peru-Chile margin phosphorites. Marine Geology 80:287–307.

Goldhaber, M. B., and I. R. Kaplan. 1974. The sulfur cycle. In The sea, edited by E. D. Goldberg, 5(17):569–656. New York: Wiley Interscience.

Goldstein, R. F., F. H. Cramer, and N. E. Andress. 1969. Silurian chitinozoans from Florida well samples. Transactions of the Gulf Coast Association of Geological Societies 19:377–84.

Gomberg, D. N. 1974. Geology of the Pourtales Terrace (abstract). Florida Scientist 37 (supp. 1):15.

———. 1976. Geology of the Pourtales Terrace, Straits of Florida. Ph.D diss., University of Miami (Florida). 371 p.

Gordon Palm and Associates. 1983. Water data acquisition—surface and ground water. Unpublished report submitted to the Florida Phosphate Council, Lakeland, Fla. 1: 151 p.; 2: 211 p.

Gough, D. I. 1967. Magnetic anomalies and crustal structure in eastern Gulf of Mexico. Bulletin of the American Association of Petroleum Geologists 51:200–211.

Gould, H. R. and R. H. Stewart. 1956. Continental terrace sediments in the northeastern Gulf of Mexico. Society of Economic Paleontologists and Mineralogists Special Publication no. 3.

Gould, J. 1989. Gulf of Mexico update: May, 1988–July, 1989. U.S. Department of the Interior Minerals Management Service, Office of Contractor Service, Information Report no. MMS 89-0079. 51 p.

Grace, S. R. 1977. Sedimentary phosphorus in the Myakka and Peace River estuaries, Charlotte Harbor, Florida. Master's thesis, University of South Florida, Tampa. 74 p.

Griffin, G. M., D. A. Reel, and R. W. Pratt. 1977. Heat flow in Florida oil test holes and indications of oceanic crust beneath the southern Florida-Bahamas Platform. In The geothermal nature of the Floridan Plateau, edited by D. L. Smith and G. W. Griffin, 43–65. Florida Bureau of Geology Special Publication no. 21.

Griffin, G. M., P. A. Tedrick, D. A. Reel, and J. P. Manker. 1969. Geothermal gradients in Florida and southern Georgia. Transactions of the Gulf Coast Association of Geological Societies 19:189–93.

Griffin, G. M., C. C. Weiland, L. Q. Hood, R. W. Goode III, R. K. Sawyer, and D. F. McNeil. 1982. Assessment of the peat resources of Florida with a detailed survey of the northern Everglades. Tallahassee, Fla.: Governor's Energy Office. 190 p.

Griffith, L. S., M. G. Pitcher, and G. W. Rice. 1969. Quantitative environmental analysis of a Lower Cretaceous reef complex. In Depositional environments in carbonate rocks, edited by G. M. Friedman, 120–38. Society of Economic Paleontologists and Mineralogists Special Publication no. 14.

Grigg, R. W. 1982. Darwin Point: A threshold for atoll formation. Coral Reefs 1:29–54.

Grim, R. E. 1968. Clay mineralogy. New York: McGraw-Hill. 596 p.

Grimm, K. A. 1992. Regional stratigraphic and paleoceanographic significance of the Oligocene-Miocene San Gregorio Formation, Baja California Sur Mexico. Ph.D. diss., University of California, Santa Cruz. 431 p.

Gromet, L. P. 1989. Avalonian terranes and late Paleozoic tectonism in southeastern New England: Constraints and problems. In Terranes in the circum-Atlantic Paleozoic orogen, edited by R. D. Dallmeyer, 193–212. Geological Society of America Special Paper no. 230.

Grow, J. A., R. E. Mattick, and J. S. Schlee. 1979. Multichannel seismic depth sections and interval velocities over outer continental shelf and upper continental slope between Cape Hatteras and Cape Cod. In Geological and geophysical investigations of continental margins, edited by J. S. Watkins, L. Montadert, and P. W. Dickerson, 65–83. American Association of Petroleum Geologists Memoir no. 29.

Gunter, H. 1928. Basement rocks encountered in a well in Florida. Bulletin of the American Association of Petroleum Geologists 12:1107–8.

———. 1934. Florida's disappearing lakes. Florida Conservator 5–6.

Gut, H. J. 1959. A Pleistocene vampire bat from Florida. Journal of Mammalogy 40:534–38.

Guthrie, G. M., and D. E. Raymond. 1992. Pre–Middle Jurassic rocks beneath the Alabama Gulf Coastal Plain. Alabama Geological Survey Bulletin no. 150. 155 p.

Hall, D. J. 1990. Gulf Coast–East Coast magnetic anomaly I: Root of the main décollement for the Appalachian-Ouachita orogen. Geology 18:862–65.

Halley, R. B., and C. C. Evans. 1983. The Miami Limestone: A guide to selected outcrops and their interpretation (with a discussion of diagenesis in the formation). Miami, Fla.: Miami Geological Society. 65 p.

Halley, R. B., E. A. Shinn, J. H. Hudson, and B. H. Lidz. 1977. Pleistocene barrier bar seaward of ooid shoal complex near Miami, Florida. Bulletin of the American Association of Petroleum Geologists 61:519–26.

Hallock, P., and W. Schlager. 1986. Nutrient excess and the demise of coral reefs and carbonate platforms. Palaios 1:389–98.

Hampson, P. S. 1984. Wetlands in Florida. Florida Bureau of Geology Map Series no. 109, scale 1:2,000,000. 1 sheet.

Handford, C. R., and R. G. Loucks. 1993. Carbonate depositional sequences and system tracts—responses of carbonate platforms to relative sea-level changes. In Carbonate sequence stratigraphy, edited by R. G. Loucks and J. F. Sarg, 3–42. American Association of Petroleum Geologists Memoir no. 57.

Hanshaw, B. B., and W. Back. 1980. Chemical mass-wasting of the northern Yucatan Peninsula by groundwater dissolution. Geology 8:222–24.

Hanshaw, B. B., W. Back, and R. G. Deike. 1971. A geochemical hypothesis for dolomitization by groundwater. Economic Geology 66:710–24.

Hanshaw, B. B., W. Back, and M. Rubin. 1965. Radiocarbon determinations for estimating ground-water flow velocities in central Florida. Science 148:494–95.

Haq, B. U., J. Hardenbol, and P. R. Vail. 1987a. Chronology of fluctuating sea levels since the Triassic. Science 235:1156–67.

———. 1987b. The new chronostratigraphic basis of Cenozoic and Mesozoic sea level cycles. In Timing and depositional history of eustatic sequences: Constraints on seismic stratigraphy, edited by C. A. Ross and D. Hama, 7–13. Cushman Foundation for Foraminiferal Research Special Publication no. 24.

———. 1988. Mesozoic and Cenozoic chronostratigraphy and cycles of sea-level change. In Sea-level: An integrated approach, edited by C. K. Wilgus, B. S. Hastings, C. G. St. Kendall, H. W. Posamentier, C. A. Ross, and J. C. Van Wagoner, 71–108. Society of Economic Paleontologists and Mineralogists Special Publication no. 42.

Harger, C. 1992. Freak wave linked to meteorite. Tampa Tribune, Florida/Metro section, August 4, p. 1.

Harmon, R. S., T. L. Ku, R. K. Matthews, and P. L. Smart. 1979. Limits of U-series analysis. Phase I of the Uranium-series intercomparison project. Geology 7:405–9.

Harper, R. M. 1914. Geography and vegetation of northern Florida. 6th Annual Report of the Florida Geological Survey, 63–437.

———. 1921. Geography of central Florida. 13th Annual Report of the Florida Geological Survey, 71–307.

Harris, G. D. 1951. Preliminary notes on Ocala bivalves. Bulletins of American Paleontology 33(138). 55 p.

Harris, R. C., H. Mattraw, J. Alberts, and A. R. Hanke. 1972. Effect of pollution on the marine environment—a case study. In Proceedings of the Coastal Zone Managment Symposium, U.S. Environmental Protection Agency, 41–57.

Harrison, R. S., and M. Coniglio. 1985. Origin of the Key Largo Limestone, Florida Keys. Bulletin of Canadian Petroleum Geology 33:350–58.

Harrison, R. S., L. D. Cooper, and M. Coniglio. 1984. Late Pleistocene of the Florida Keys. Canadian Society of Petroleum Geologists Core Conference, Calgary, Alberta, October 18–19, 1–14.

Hart, S. R. 1984. A large-scale anomaly in the southern hemisphere mantle. Nature 309:753–57.

———. 1988. Heterogeneous mantle domains: Signatures, genesis, and mixing chronologies. Earth and Planetary Science Letters 90:273–96.

Hatcher, R. D. 1989. Tectonic synthesis of the U.S. Appalachians. In The Appalachian-Ouachita orogen in the United States, edited by R. D. Hatcher, W. A. Thomas, and G. W. Viele, 511–36. The Geology of North America, vol. F-2. Geological Society of America.

Hay, O. P. 1901. Bibliography and catalogue of the fossil Vertebrata of North America. U.S. Geological Survey Bulletin no. 179. 868 p.

Hayes, L. R., and D. E. Barr. 1983. Hydrology of the sand-and-gravel aquifer, southern Okaloosa and Walton Counties, Florida. U.S. Geological Survey Water-Resources Investigations Report no. 82-4110. 41 p.

Hayes, M. O. 1975. Morphology and sand accumulation in estuaries. In Estuarine research, vol. 2, edited by L. E. Cronin, 3–22. New York: Academic Press.

———. 1979. Barrier island morphology as a function of tidal and wave regime. In Barrier islands from the Gulf of St. Lawrence to the Gulf of Mexico, edited by S. L. Leatherman, 1–27. New York: Academic Press.

———. 1994. The Georgia Bight barrier system. In Geology of Holocene barrier island systems, edited by R. A. Davis, 223–304. New York: Springer-Verlag.

Hazel, J. E. 1968. Pleistocene ostracode zoogeography in Atlantic Coast submarine canyons. Journal of Paleontology 42:26–171.

———. 1983. Age and correlation of the Yorktown (Pliocene) and Croatan (Pliocene and Pleistocene) formations at the Lee Creek mine. In Geology and paleontology of the Lee Creek Mine, North Carolina, edited by C. E. Ray, 81–200. Smithsonian Contributions to Paleobiology 553.

Healy, H. G. 1975. Terraces and shorelines of Florida. Florida Bureau of Geology Map Series no. 71. 1 sheet.

———. 1982. Surficial and intermediate aquifers. In Principal aquifers in Florida, edited by B. J. Franks, 82–255. U.S. Geological Survey Water-Resources Investigations Open-File Report. 4 sheets.

Heatherington, A. L., and P. A. Mueller. 1990. Mesozoic igneous suites of the Florida basement: Implications for mantle sources and tectonic reconstructions. Geological Society of America annual meeting. Abstracts with Programs 22(7):A162.

———. 1991. Geochemical evidence for Triassic rifting in southwestern Florida. Tectonophysics 188:291–302.

Heatherington, A. L., P. A. Mueller, and K. A. D'Arcy. 1988. Exotic terranes of Florida: Nd and Sr isotopic evidence. EOS, Transactions of the American Geophysical Union 69:1500.

———. 1989. Geochemical constraints on the evolution of the Florida basement. Geological Society of America annual meeting. Abstracts with Programs 21(6):A164.

Heatherington, A. L., P. A. Mueller, and K. Hoyle. 1992. Geochemistry and geochronology of Suwannee terrane volcanic rocks and proposed correlative sequences in West Africa. Geological Society of America annual meeting. Abstracts with Programs 24(2):21.

Hebert, J. A. 1985. High resolution seismic stratigraphy of the inner west Florida shelf west of Tampa Bay. Master's thesis, University of South Florida, St. Petersburg. 52 p.

Heilprin, A. 1887. Explorations of the west coast of Florida and in the Okeechobee wilderness. Transactions of the Wagner Free Institute of Science 1:1–134.

Hendry, C. W., and J. W. Yon. 1958. Geology of the area in and around the Jim Woodruff Reservoir. Florida Geological Survey Report of Investigations no. 16, part 1.

Henry, H. R., and F. A. Kohout. 1972. Circulation patterns of saline groundwater affected by geothermal heating—as related to waste disposal. American Association of Petroleum Geologists Memoir no. 18, 202–21.

Hermanson, J. W., and B. J. MacFadden. 1992. Evolutionary and functional morphology of the shoulder region and stay apparatus in fossil and extant horses (Equidae). Journal of Vertebrate Paleontology 1:377–86.

Herrick, S. M., and R. C. Vorhis. 1963. Subsurface geology of the Georgia coastal plain. Georgia Geological Survey Information Circular no. 25. 78 p.

Herring, J. R., and W. F. Stowasser. 1991. Phosphate—our nation's most important agricultural mineral commodity and its uncertain future. Geological Society of America annual meeting. Abstracts with Programs 23(5):A299.

Hickey, J. J., and J. Veccioli. 1986. Subsurface injection of liquid waste with emphasis on injection practices in Florida. U.S. Geological Survey Water-Supply Paper no. 2281. 25 p.

Higgins, M., and I. Zietz. 1983. Geological interpretation of geophysical maps of the pre-Cretaceous basement beneath the coastal plain of the southeastern United States. Geological Society of America Memoir no. 158, 125–30.

Hill, C. A. 1990. Sulfuric acid speleogenesis of Carlsbad Cavern and its relationship to hydrocarbons, Delaware Basin, New Mexico and Texas. Bulletin of the American Association of Petroleum Geologists 74:1685–94.

Hine, A. C., and D. F. Belknap. 1986. Recent geological history and modern sedimentary processes of the Pasco, Hernando, and Citrus county coastline: West-central Florida. Florida Sea Grant College Publication no. 79. 160 p.

Hine, A. C., D. F. Belknap, J. G. Hutton, E. B. Osking, and M. W. Evans. 1988. Recent geological history and modern sedimentary processes along an incipient, low-energy, epicontinental-sea coastline: Northwest Florida. Journal of Sedimentary Petrology 58:567–79.

Hine, A. C., and L. J. Doyle. 1991. Quaternary development of the west Florida continental shelf. Gulf Coast Section, Society of Economic Mineralogists and Paleontologists Foundation, 12th annual research conference program and abstracts. 93–94.

Hine, A. C., M. W. Evans, R. A. Davis, and D. F. Belknap. 1987a. Depositional response to seagrass mortality along a low-energy, barrier island coast: West-central Florida. Journal of Sedimentary Petrology 57:431–39.

Hine, A. C., M. W. Evans, D. L. Mearns, and D. F. Belknap. 1987b. Effect of Hurricane Elena on Florida's marsh-dominated coast: Pasco, Hernando, and Citrus counties. Florida Sea Grant College final report. 33 p.

Hine, A. C., and H. T. Mullins. 1983. Modern carbonate shelf-slope breaks. In The shelfbreak: Critical interface on continental margins, edited by D. J. Stanley and G. T. Moore, 169–88. Society of Economic Paleontologists and Mineralogists Special Publication no. 33.

Hine, A. C., and J. C. Steinmetz. 1984. Cay Sal Bank, Bahamas—a partially drowned carbonate platform. Marine Geology 59:135–64.

Hirschfeld, S. E., and S. D. Webb. 1968. Plio-Pleistocene megolonychid sloths of North America. Bulletin of the Florida State Museum 12:213–96.

Hodell, D. A., P. A. Mueller, and J. R. Garrido. 1991. Variations in the strontium isotopic composition of seawater during the Neogene. Geology 19:24–27.

Hoerle, S. E. 1970. Mollusca of the 'Glades' unit of southern Florida: Part II—list of molluscan species of the Belle Glade rock pit, Palm Beach County, Florida. Tulane Studies in Geology and Paleontology 8:56–68.

Hoffmeister, J. E. 1974. Land from the sea—the geologic story of south Florida. Coral Gables, Fla.: University of Miami.

Hoffmeister, J. E., and H. G. Multer. 1964. Pleistocene limestones of the Florida Keys. In Guidebook for Field Trip no. 1, edited by R. N. Ginsburg, 57–61. Annual Meeting of the Geological Society of America, Miami, Florida.

———. 1968. Geology and origin of the Florida Keys. Bulletin of the Geological Society of America 79:1487–1502.

Hoffmeister, J. E., K. W. Stockman, and H. G. Multer. 1967. Miami Limestone of Florida and its Recent Bahamian counterpart. Bulletin of the Geological Society of America 78:175–90.

Hoganson, J. W. 1972. The *Spirulaea vernoni* Zone of the Crystal River Formation in peninsular Florida. Master's thesis, University of Florida, Gainesville. 70 p.

Holmes, C. W. 1985. Accretion of south Florida platform—late Quaternary development. Bulletin of the American Association of Petroleum Geologists 69:149–60.

Horton, J. W. Jr., A. A. Drake Jr., and D. W. Rankin. 1989. Tectonostratigraphic terranes and their Paleozoic boundaries in the central and southern Appalachians. In Terranes in the circum-Atlantic Paleozoic orogens, edited by R. D. Dallmeyer, 213–46. Geological Society of America Special Paper no. 230.

Howell, B. F., and H. G. Richards. 1949. New Paleozoic linguloid brachiopod from Florida. Bulletin of the Wagner Free Institute of Science 24:35–36.

Hoyt, J. H. 1969. Late Cenozoic structural movement, northern Florida. Transactions of the Gulf Coast Association of Geological Societies 19:1–9.

Huddlestun, P .F. 1984. The Neogene stratigraphy of the central Florida panhandle. Ph.D. diss., Florida State University, Tallahassee. 210 p.

———. 1988. A revision of the lithostratigraphic units of the coastal plain of Georgia: The Miocene through Holocene. Georgia Geologic Survey Bulletin no. 104. 162 p.

———. 1993. Revision of the lithostratigraphic units of the coastal plain of Georgia—the Oligocene. Georgia Geologic Survey Bulletin no. 105. 152 p.

Huddlestun, P. F., M. E. Hunter, and B. C. Carter. 1988. Suwannee Strait as a faunal province boundary. Annual meeting of the Southeastern Section of the Geological Society of America. Abstracts with Programs 20(4):271.

Hughes, J. L. 1979. Saltwater barrier line in Florida—concepts, considerations, and site examples. U.S. Geological Survey Water-Resources Investigations Report no. 79-75. 29 p.

Hulbert, R. C. Jr. 1982. Population dynamics of the three-toed horse Neohipparion from the late Miocene of Florida. Paleobiology 8:159–67.

———. 1984. Paleoecology and population dynamics of the early Miocene (Hemingfordian) horse *Parahippus leonensis* from the Thomas farm site, Florida. Journal of Vertebrate Paleontology 4:547–58.

———. 1992. A checklist of the fossil vertebrates of Florida. Papers in Florida Paleontology 6:1–35.

Hulbert, R. C. Jr., and B. J. MacFadden. 1991. Morphologic transformation and cladogenesis at the base of the adaptive radiation of Miocene horses. American Museum Novitates 3,000:1–61.

Hulbert, R. C. Jr., and G. S. Morgan. 1989. Stratigraphy, paleoecology, and vertebrate fauna of the Leisey shell pit local fauna, early Pleistocene (Irvingtonian) of southwestern Florida. Papers in Florida Paleontology 2:1–19.

———. 1993. Quantitative and qualitative evolution in the giant armadillo *Holmesina* (Edentata: Pampatheriidae) in Florida. In Morphological change in Quaternary mammals of North America, edited by R. R. Martin and A. Barhosky, 134–77. Cambridge: Cambridge University Press.

Hull, J. P. D. Jr. 1962. Cretaceous Suwanee Strait, Georgia, and Florida. Bulletin of the American Association of Petroleum Geologists 46:118–22.

Hull, R. W., and M. C. Yurewicz. 1979. Quality of storm runoff to drainage wells in Live Oak, Florida, April 4, 1979. U.S. Geological Survey Open-File Report no. 79-1073. 58 p.

Humphrey, S. R. 1974. Zoogeography of the nine-banded armadillo (*Dasypus novemcintus*) in the United States. Bioscience 24:457–62.

Hunter, M. E. 1968. Molluscan guide fossils in late Miocene sediments in southern Florida. Transactions of the Gulf Coast Association of Geological Societies 18:439–50.

———. 1976. Biostratigraphy. In Tertiary carbonates, Citrus, Levy, Marion counties, west central Florida, edited by M. E. Hunter, 66–87. Guidebook for the 20th annual field trip of the Southeastern Geological Society.

———. 1981. Stop no. 4: Crystal River quarry, Citrus County, Florida. In Survey of central Florida geology, edited by J. F. Meeder, D. R. Moore, and P. Harlem, 22–26. Miami Geological Society field trip guidebook, 1981.

Hutchinson, D. R., J. A. Grow, K. A. Klitgord, and B. A. Swift. 1983. Deep structure and evolution of the Carolina trough. In Studies in continental marine geology, edited by J. S. Watkins and C. L. Drake, 129–54. American Association of Petroleum Geologists Memoir no. 34.

Hutchinson, D. R., K. A. Klitgord, and A. M. Trehu. 1990. Integration of COCORP deep reflection and magnetic anomaly analysis in the southeastern United States: Implications for origin of the Brunswick and East Coast magnetic anomalies: Alternative interpretation and reply. Bulletin of the Geological Society of America 102:271–74.

Huwyler, S. 1983. Uranium recovery from phosphates and phosphoric acid. 2nd ed. Switzerland: Federal Institute for Reactor Research. 72 p.

Ivany, L. C., R. W. Portell, and D. S. Jones. 1990. Animal–plant relationships and paleobiogeography of an Eocene seagrass community from Florida. Palaios 5:244–58.

Jackson, J. B. C., P. Jung, A. G. Coates, and L. S. Collins. 1993. Diversity and extinction of tropical American mollusks and emergence of the Isthmus of Panama. Science 260:1624–26.

Jakes, P., and A. J. R. White. 1972. Major and trace element abundance in volcanic rocks of orogenic areas. Bulletin of the Geological Society of America 83:29–40.

James, N. P. 1979. Reefs. In Facies models, edited by R. G. Walker, 121–33. Geoscience Canada Reprint Series no. 1.

Jansa, L. F. 1981. Mesozoic carbonate platforms and banks of eastern North American margin. Marine Geology 44:97–117.

Jarrett, M. L., W. B. LaFrenz, and D. L. Smith. 1984. A seismic refraction profile of the north Florida crust. Florida Scientist 47:153–60.

Jee, J. L. 1993. Seismic stratigraphy of the western Florida carbonate platform and the history of Eocene strata. Ph.D. diss., University of Florida, Gainesville. 215 p.

Jenkyns, H. C. 1991. Impact of Cretaceous sea-level rise and anoxic events on the Mesozoic carbonate platform of Yugoslavia. Bulletin of the American Association of Petroleum Geologists 75:1007–17.

Joeckel, R. M. 1990. A functional interpretation of the masticatory system and paleoecology of entelodonts. Paleobiology 16:459–82.

Johnson, J. H., and B. J. Ferris. 1948. Eocene algae from Florida. Journal of Paleontology 22:762–66.

Jones, D. S. 1982. Some considerations of the late Eocene faunas of northwest peninsular Florida. In Cenozoic vertebrate and invertebrate paleontology of north Florida, edited by D. L. Smith, 14–32. Guidebook for the 24th annual field trip of the Southeastern Geological Society.

———. 1992. Integrated stratigraphic approach to geochronology of marine–nonmarine sites in the Plio-Pleistocene of Florida. In Plio-Pleistocene stratigraphy and paleontology of southern Florida, edited by T. M. Scott and W. D. Allmon, 41–50. Florida Geological Survey Special Publication no. 36.

Jones, D. S., and P. F. Hasson. 1985. History and development of the marine invertebrate faunas separated by the Central American isthmus. In The great American interchange, edited by F. G. Stehli and S. D. Webb, 325–55. New York: Plenum Press.

Jones, D. S., B. J. MacFadden, S. D. Webb, P. A. Mueller, D. A. Hodell, and T. M. Cronin. 1991. Integrated geochronology of a classic Pliocene fossil site in Florida: Linking marine and terrestrial biochronologies. Journal of Geology 99:637–48.

Jones, D. S., P. A. Mueller, J. R. Bryan, J. P. Dobson, J. E. T. Channell, J. C. Zachos, and M. A. Arthur. 1987. Biotic, geochemical, and paleomagnetic changes across the Cretaceous/Tertiary boundary at Braggs, Alabama. Geology 15:311–15.

Jones, D. S., and D. Nicol. 1986. Origination, survivorship, and extinction of rudist taxa. Journal of Paleontology 60:107–15.

———. 1989. Eocene clavagellids (Mollusca: Pelecypoda) from Florida: The first documented occurrence in the Cenozoic of the western hemisphere. Journal of Paleontology 63:320–23.

Jones, D. S., and R. W. Portell. 1988. Fossil invertebrates from Brooks Sink, Bradford County, Florida. In Heavy mineral mining in northeast Florida and an examination of the Hawthorne Formation and post-Hawthorne clastic sediments, edited by F. L. Pirkle and J. G. Reynolds, 41–52. Guidebook for the 29th annual field trip of the Southeastern Geological Society.

Jones, G. W., and S. B. Upchurch. 1993. Origin of nutrients in ground water discharging from Lithia and Buckhorn springs. Ambient Ground-Water Quality Monitoring Program, Southwest Florida Water Management District. 209 p.

Jordan, G. F. 1952. Reef formation in the Gulf of Mexico off Apalachicola Bay, Florida. Bulletin of the Geological Society of America 63:741–44.

Jorgensen, B. B. 1983. The microbial sulfur cycle. In Microbial geochemistry, edited by W. E. Krumbein, 91–124. Oxford: Blackwell.

Kane, B. C. 1984. Origin of the Grandin (Plio-Pleistocene) sands, western Putnam County, Florida. Master's thesis, University of Florida, Gainesville. 86 p.

Kaplan, I. R., K. O. Emery, and S. C. Rittenberg. 1963. The distribution and isotopic abundance of sulfur in recent marine sediments off Southern California. Geochimica et Cosmochimica Acta 27:297–331.

Kastner, M., R. E. Garrison, Y. Kolodny, C. E. Reimers, and A. Shemesh. 1990. Coupled changes of oxygen isotopes in PO_3^- and CO_3^{2-} in apatite, with emphasis on the Monterey Formation, California. In Neogene to modern phosphorites, edited by W. C. Burnett and S. R. Riggs, 312–24. Phosphate deposits of the world, vol. 3. Cambridge: Cambridge University Press.

Katz, B. G. 1992. Hydrochemistry of the upper Floridan aquifer, Florida. U.S. Geological Survey Water-Resources Investigations Report no. 91-4196. 37 p.

Kaufmann, R. F., and J. D. Bliss. 1977. Effects of the phosphate industry on radium-226 in ground water of central Florida. U.S. Environmental Protection Agency Office of Radiation Programs Publication no. EPA/520-6-77-010. 111 p.

Keeling, C. D. 1986. Atmospheric CO_2 concentrations—Mauna Loa Observatory, Hawaii, 1958–1986. Carbon Dioxide Information Center Publication no. NDP-001/R1. Oak Ridge, Tenn.: Oak Ridge National Laboratory. 1064 p.

Keen, A. M. 1971. Seashells of tropical west America. Palo Alto: Stanford University Press.

Keigwin, L. 1978. Pliocene closing of the Isthmus of Panama, based on biostratigraphic evidence from nearby Pacific Ocean and Caribbean Sea cores. Geology 6:630–34.

Kennett, J. P. 1982. Marine Geology. Englewood Cliffs, N.J.: Prentice-Hall. 813 p.

———. 1986. Miocene to early Pliocene oxygen and carbon isotope stratigraphy in the southwest Pacific, Deep Sea Drilling Project Leg 90. Initial Reports of the Deep Sea Drilling Project 90:1383–1411.

Kennett, J. P., and L. D. Stott. 1990. Proteus and protooceanus: Ancestral Paleogene oceans as revealed from Antarctic stable isotope results, ODP Leg 113. Proceedings of the Ocean Drilling Program, Scientific Results 113:865–80.

Keppie, J. D. 1989. Northern Appalachian terranes and their accretionary history. In Terranes in the circum-Atlantic Paleozoic orogens, edited by R. D. Dallmeyer, 159–92. Geological Society of America Special Paper no. 230.

Keppie, J. D., R. D. Nance, J. B. Murphy, and J. Dostal. 1991. Northern Appalachians: Avalon and Meguma terranes. In The West African orogens and circum-Atlantic correlatives, edited by R. D. Dallmeyer and J. P. Lecorche, 315–34. New York: Springer-Verlag.

Kerrick, D. M., and K. Caldeira. 1994. Metamorphic CO_2 degassing and early Cenozoic paleoclimate. GSA Today 4(3):57, 62–65.

Ketcher, K. M. 1992. Stratigraphy and environment of bed 11 of the Pinecrest Beds at Sarasota, Florida. In Plio-Pleistocene stratigraphy and paleontology of southern Florida, edited by T. M. Scott and W. D. Allmon, 167–78. Florida Geological Survey Special Publication no. 36.

Kier, P. M. 1963. Tertiary echinoids from the Caloosahatchee and Tamiami formations of Florida. Smithsonian Miscellaneous Collections no. 145. 63 p.

Kimrey, J. O., and L. D. Fayard. 1982. Geohydrologic reconnaissance of drainage wells in Florida—an interim report. U.S. Geological Survey Open-File Report no. 82-860. 59 p.

Kindinger, J. L. 1986. Geomorphology and tidal-bar depositional model of lower Florida Keys (abstract). Bulletin of the American Association of Petroleum Geologists 70:607.

King, C. A. M. 1972. Beaches and coasts. New York: St. Martin's. 570 p.

King, D. T. Jr., and M. C. Skotnicki. 1990. Upper Cretaceous stratigraphy and relative sea-level changes, Gulf Coastal Plain, Alabama. In Sequence stratigraphy as an exploration tool—concepts and practices in the Gulf Coast section, 199–212. Society of Economic Paleontologists and Mineralogists eleventh annual research conference, Houston, Texas.

King, E. R. 1959. Regional magnetic map of Florida. Bulletin of the American Association of Petroleum Geologists 43:2844–54.

King, P. B. 1961. The subsurface Ouachita structural belt east of the Ouachita mountains. In The Ouachita system, edited by P. T. Flawn, A. Goldstein Jr., P. B. King, and C. E. Weaver, 83–98. Austin: University of Texas Press.

King, W., and G. Simmons. 1972. Heat flow near Orlando, Florida, and Uvalde, Texas, determined from well cuttings. Geothermics 1:133–39.

Kirkland, B. L., R. G. Lighty, R. Rezak, and T. T. Tieh. 1987. Lower Cretaceous barrier reef and outer shelf facies, Sligo Formation, south Texas. Transactions of the Gulf Coast Association of Geological Societies 37:371–82.

Kirtley, D. W., and W. F. Tanner. 1968. Sabellariid worms: Builders of a major reef type. Journal of Sedimentary Petrology 38:73–79.

Kjellesvig-Waering, E. N. 1950. A new Silurian eurypterid from Florida. Journal of Paleontology 24:229–31.

———. 1955. A new phyllocarid and eurypterid from the Silurian of Florida. Journal of Paleontology 29:295–97.

Klein, H., J. T. Armbruster, B. F. McPherson, and H. J. Freiberger. 1975. Water and the south Florida environment. U.S. Geological Survey Water-Resources Investigations Report no. 24-75. 165 p.

Klein, H., and C. R. Causaras. 1982. Biscayne aquifer, southeast Florida, and contiguous surficial aquifer to the north. In Principal aquifers in Florida, edited by B. J. Franks, 82–255. U.S. Geological Survey Water-Resources Investigations Open-File Report. 4 sheets.

Klein, H., and J. E. Hull. 1978. Biscayne aquifer, southeast Florida. U.S. Geological Survey Water-Resources Investigations Report 78-107. 52 p.

Klitgord, K. D., W. P. Dillon, and P. Popenoe. 1983. Mesozoic tectonics of the southeastern United States coastal plain and continental margin. In Studies related to the Charleston, South Carolina, earthquake of 1866—tectonics and seismicity, edited by G. S. Gohn, 1–15. U.S. Geological Survey Professional Paper no. 1313.

Klitgord, K. D., D. R. Hutchinson, and H. Schouten. 1988. U.S. Atlantic continental margin—structural and tectonic framework. In The Atlantic continental margin, U.S., edited by R. E. Sheridan and J. A. Grow, 19–55. The geology of North America, vol. I-2. Geological Society of America.

Klitgord, K. D., P. Popenoe, and H. Schouten. 1984. Florida: A Jurassic transform plate boundary. Journal of Geophysical Research 89:7753–72.

Klitgord, K. D., and H. Schouten. 1986. Plate kinematics of the central Atlantic. In The western North Atlantic region, edited by P. R. Vogt and B. E. Tucholke, 351–78. The geology of North America, vol. I-2. Geological Society of America.

Knapp, M. S. 1977. The northern Brooksville Ridge, a case for topographic inversion. Florida Scientist 40:25 (supp.).

Knowles, S. C., and R. A. Davis Jr. 1983. Hurricane influence on Holocene sediment accumulation in Sarasota Bay, Florida. Bulletin of the American Association of Petroleum Geologists 67:496.

Koepnick, R. B., W. H. Burke, R. E. Denison, E. A. Heatherington, H. F. Nelson, J. B. Otto, and L. E. Waite. 1985. Construction of the seawater $^{87}Sr/^{86}Sr$ curve for the Cenozoic and Cretaceous: supporting data. Chemical Geology 58:55–81.

Kohout, F. A. 1964. Flow of freshwater and saltwater in the Biscayne aquifer of the Miami area, Florida. U.S. Geological Survey Water-Supply Paper no. 1613-C, C12-C32.

———. 1965. A hypothesis concerning cyclic flow of salt water related to geothermal heating in the Floridan aquifer. Transactions of the New York Academy of Sciences, series 2 28:249–71.

———. 1967. Ground-water flow and the geothermal regime of the Floridan Plateau. Transactions of the Gulf Coast Association of Geological Societies 17:339–54.

Kohout, F. A., H. R. Henry, and J. E. Banks. 1977. Hydrogeology related to geothermal conditions of the Floridan Plateau. In The geothermal nature of the Florida Plateau, edited by D. L. Smith and G. M. Griffin, 1–41. Florida Geological Survey Special Publication no. 21.

Kohout, F. A., H. Meisler, F. W. Meyer, G. W. Meyer, and R. F. Waite. 1988. Hydrology of Atlantic continental margins. In The Atlantic continental margin, U.S., edited by R. E. Sheridan and J. A. Grow, 463–80. The Geology of North America, vol. I-2. Geological Society of America.

Kolodny, Y., and I. R. Kaplan. 1970. Carbon and oxygen isotopes in apatite—CO_2 and co-existing calcite from sedimentary phosphates. Journal of Sedimentary Petrology 40:954–59.

Kolodny, Y., B. Luz, and O. Navon. 1983. Oxygen isotope variations in phosphate of biogenic apatite. I. Fish bone apatite—rechecking the rules of the game. Earth and Planetary Science Letters 64:398–404.

Kramer, D. A. 1990. Titanium. In 22nd Annual Report of the U.S. Bureau of Mines. 22 p.

Krantz, D. E. 1990. Mollusk-isotope record of Plio-Pleistocene marine paleoclimate, U.S. Middle Atlantic Coastal Plain. Palaios 5:317–34.

REFERENCES

Krause, R., E. and R. B. Randolph. 1989. Hydrology of the Floridan aquifer system in southeast Georgia and adjacent parts of Florida and South Carolina. U.S. Geological Survey Professional Paper no. 1403D. 65 p.

Krivoy, H. L., and T. E. Pyle. 1972. Anomalous crust beneath west Florida shelf. Bulletin of the American Association of Petroleum Geologists 56:107–113.

Kroopnick, P. 1985. The distribution of ^{13}C of TCO_2 in the world oceans. Deep-Sea Research 32:57–84.

Krumm, D. K., and D. S. Jones. 1993. New coral–bivalve association (*Actinastrea–Lithophaga*) from the Eocene of Florida. Journal of Paleontology 67:945–51.

Kump, L. R., and A. C. Hine. 1986. Ooids as sea-level indicators. In Sea-level research: A manual for the collection and evaluation of data, edited by O. van de Plassche, 175–93. Norwich, U.K.: Geo Books.

Kurten, B. 1965. The Pleistocene Felidae of Florida. Bulletin of the Florida State Museum 9:215–73.

———. 1966. Pleistocene bears of North America. 1. Genus *Tremarctos,* spectacled bears. Acta Zoologica Fennica 11:1–120.

———. 1967. Pleistocene bears of North America. 2. Genus *Arctodus,* short-faced bears. Acta Zoologica Fennica 117:1–60.

Kurten, B., and E. Anderson. 1980. Pleistocene mammals of North America. New York: Columbia University Press. 442 p.

Ladd, J. W., and R. E. Sheridan. 1987. Seismic stratigraphy of the Bahamas. Bulletin of the American Association of Petroleum Geologists 71:719–36.

Lahann, R. W. 1978. A chemical model for calcite crystal growth and morphology control. Journal of Sedimentary Petrology 481:337–44.

Lambert, W. D. 1990. Rediagnosis of the genus *Amebelodon* (Mammalia, Proboscidea, Gomphotheriidae), with a new subgenus and species, *Amebelodon* (*Konobelodon*) *britti.* Journal of Paleontology 64:1032–40.

———. 1992. The feeding habits of the shovel-tusked gomphotheres: evidence from tusk and wear patterns. Paleobiology 18:132–47.

Lane, E. 1986. Karst in Florida. Florida Geological Survey Special Report no. 29. 100 p.

———. 1991. Earthquakes and seismic history of Florida. Florida Geological Survey Open-File Report no. 40. 11 p.

Langer, W. H. 1988. Natural aggregates of the conterminous United States. U.S. Geological Survey Bulletin no. 1594. 33 p.

Larson, R. L. 1991. Latest impulse of the Earth: Evidence for mid-Cretaceous super plume. Geology 19:547–50.

Lasaga, A. C., R. A. Berner and R. M. Garrels. 1985. An improved geochemical model of atmospheric CO_2 fluctuations over the past 100 million years. In The carbon cycle and atmospheric CO_2: Natural variations Archean to present, edited by E. T. Sundquist and W. S. Broecker, 397–411. American Geophysical Union Monograph no. 32.

Lawrence, B. 1943. Miocene bat remains from Florida, with notes on the generic characters of the humerus of bats. Journal of Mammology 24:356–69.

Lawrence, F. W., and S. B. Upchurch. 1982. Identification of recharge areas using geochemical factor analysis. Ground Water 20:680–87.

LeConte, J. 1857. On the agency of the Gulf Stream in the formation of the peninsula and keys of Florida. American Journal of Science 73:46–60.

———. 1883. The reefs, keys, and peninsula of Florida. Science 2:764.

Lee, F. W., J. H. Schwartz, and S. J. Hemberger. 1945. Magnetic survey of the Florida peninsula. U.S. Bureau of Mines Report no. 3810. 49 p.

Leg 101 Scientific Party. 1988. Leg 101—an overview. Proceedings of the Ocean Drilling Program, Scientific Results 101:455–72.

Leidy, J., and F. Lucas. 1896. Fossil vertebrates from the Alachua clays. Philadelphia: Wagner Free Institute of Science. 61 p.

Le Pichon, X., and P. J. Fox. 1971. Marginal offsets, fracture zones, and the early opening of the North Atlantic. Journal of Geophysical Research 76:6294–6308.

Leve, G. W. 1966. Ground water in Duval and Nassau counties, Florida. Florida Geological Survey Report of Investigations no. 43. 91 p.

Lewelling, B. R. 1988. Potentiometric surface of the intermediate aquifer system, west-central Florida, May 1987. U.S. Geological Survey Open-File Report no. 87-705. Scale 1:1,000,000, 1 sheet.

Lichtler, W. F., W. Anderson, and B. F. Joyner. 1968. Water resources of Orange County, Florida. Florida Geological Survey Report of Investigations no. 50. 150 p.

Lidz, B. H., A. C. Hine, E. A. Shinn, and J. L. Kindinger. 1991. Multiple outer-reef tracts along the south Florida bank margin: Outlier reefs, a new windward-margin model. Geology 19:115–18.

Lidz, B. H., D. M. Robbin, and E. A. Shinn. 1985. Holocene carbonate sedimentary petrology and facies accumulation, Looe Key National Marine Sanctuary, Florida. Bulletin of Marine Science 36:672–700.

Lidz, B. H., and E. A. Shinn. 1991. Paleoshorelines, reefs, and a rising sea: South Florida, U.S.A. Journal of Coastal Research 7:203–29.

Lighty, R. G., I. G. Macintyre, and R. Stuckenrath. 1978. Submerged early Holocene barrier reef, southeast Florida shelf. Nature 276:59–60.

Littlefield, J. R., M. A. Culbreth, S. B. Upchurch, and M. T. Stewart. 1984. Relationship of modern sinkhole development to large-scale photolinear features. In Sinkholes: Their geology, engineering, and environmental impact, edited by B. F. Beck, 189–95. Rotterdam: A. A. Balkema.

Lloyd, J. M. 1985. Annotated bibliography of Florida basement geology and related regional and tectonic studies. Florida Geological Survey Information Circular no. 98. 72 p.

———. 1991. 1988 and 1989 Florida petroleum production and exploration. Florida Geological Survey Information Circular no. 107, part 1. 62 p.

Locker, S. D., and R. T. Buffler. 1983. Comparison of Lower Cretaceous carbonate shelf margins, northern Campeche Escarpment and northern Florida Escarpment, Gulf of Mexico. In Seismic expression of structural styles, edited by A. W. Bally, 2.2.3-123–2.2.3-128. American Association of Petroleum Geologists Studies in Geology 15(2).

Locker, S. D., A. C. Hine, and E. A. Shinn. 1991. Sea level and geostrophic current control on carbonate shelf–slope depositional sequences and erosional patterns, south Florida Platform margin. Official Program for the Annual Convention of the American Association of Petroleum Geologists, Dallas.

———. 1992. High resolution sequence stratigraphic framework of carbonate deposition controlled by sea level and geostrophic bottom currents, south Florida platform margin. Annual Meeting of the Geological Society of America. Abstracts with Programs 24(7):A83.

Logan, B. W., J. L. Harding, W. M. Ahr, J. D. Williams, and R. G. Snead. 1969. Late Quaternary carbonate sediments of Yucatan shelf, Mexico. American Association of Petroleum Geologists Memoir no. 11, 5–128.

Long, L. T. 1974. The Florida earthquake of October 27, 1973. Earthquake Notes 45:37–42.

Long, L. T., and R. P. Lowell. 1973. Thermal model for some continental margin sedimentary basins and uplift zones. Geology 1:87–88.

Longinelli, A., and S. Nuti. 1968. Oxygen isotopic compositions of phosphorites from marine formations. Earth and Planetary Science Letters 5:13–16.

Longman, M. W. 1980. Carbonate diagenetic textures from near-surface diagenetic environments. Bulletin of the American Association of Petroleum Geologists 64:461–87.

Lord, K. M. 1993. The basement structure, tectonic history, seismicity, and seismic hazard potential of the Floridan Plateau region. Ph.D. diss., University of Florida, Gainesville. 221 p.

Lord, K. M., and D. L. Smith. 1991. The Floridan Plateau: Evidence for a uniquely stable basement. Geological Society of America, Southeastern Section, Annual Meeting. Abstracts with Programs 23(4):60.

Loucks, R. G., and J. F. Sarg. 1993. Carbonate sequence stratigraphy. Memoir no. 57. American Association of Petroleum Geologists. 545 p.

Loutit, T. S., J. Hardenbol, P. R. Vail, and G. R. Baum. 1988. Condensed sections: The key to age determination and correlation of continental margin sequences. In Sea level changes: An integrated approach, edited by C. K. Wilgus, B. S. Hastings, C. G. St. Kendall, H. W. Posamentier, C. A. Ross, and J. C. Van Wagoner, 183–213. Special Publication no. 42. Tulsa, Okla.: Society of Economic Paleontologists and Mineralogists.

Loutit, T. S., N. G. Pisias, and J. P. Kennett. 1983. Pacific Miocene carbon isotope stratigraphy using benthic foraminifera. Earth and Planetary Science Letters 66:48–62.

Lowell, R. P., and L. T. Long. 1977. Thermal model for the Florida crust. In The geothermal nature of the Florida Plateau, edited by D. L. Smith and G. M. Griffin, 149–161. Special Publication no. 21. Tallassee: Florida Geological Survey.

Lyell, C. 1830. Principles of geology, vol. 1. London: John Murray. 511 p.

Lyons, P. L. 1950. A gravity map of the United States. Tulsa Geological Science Digest 33:33–43.

Lyons, W. G. 1991. Post-Miocene species of *Latirus* Montfort, 1810 (Mollusca: Fasciolariidae) of southern Florida with a review of regional marine biostratigraphy. Bulletin of the Florida Museum of Natural History, Biological Sciences 35(3):131–208.

———. 1992. Caloosahatchee-age and younger molluscan assemblages at APAC mine, Sarasota County, Florida. In Plio-Pleistocene stratigraphy and paleontology of southern Florida, edited by T. M. Scott and W. D. Allmon, 133–160. Special Publication no. 36. Tallahassee: Florida Geological Survey.

MacFadden, B. J. 1979. Fossil rhinoceroses of Florida. The Plaster Jacket 30:16.

———. 1980. An early Miocene land mammal (Oreodonta) from a marine limestone in northern Florida. Journal of Paleontology 54:93–101.

———. 1982. New species of primitive three-toed browsing horse from the Miocene phosphate mining district of central Florida. Florida Scientist 45:117–25.

———. 1986a. Fossil horses from "Eohippus" *Hyracotherium* to *Equus:* Scaling, Cope's Law, and the evolution of body size. Paleobiology 12:355–69.

———. 1986b. Late Hemphillian monodactyl horses (Mammalia, Equidae) from the Bone Valley Formation of central Florida. Journal of Paleontology 60:466–75.

———. 1992. Fossil horses: Systematics, paleobiology, and evolution of the Family Equidae. New York: Cambridge University Press. 369 p.

MacFadden, B. J., J. D. Bryant, and P. A. Mueller. 1991. Sr-isotopic, paleomagnetic, and biostratigraphic calibration of horse evolution: Evidence from the Miocene of Florida. Geology 19:242–45.

MacFadden, B. J., and H. Galiano. 1981. Late Hemphillian cat (Mammalia, Felidae) from the Bone Valley Formation of central Florida. Journal of Paleontology 55:218–26.

MacFadden, B. J., and R. C. Hulbert Jr. 1988. Explosive speciation at the base of the adaptive radiation of Miocene grazing horses. Nature 336:466–68.

———. 1990. Body size estimates and size distribution of ungulate mammals from the late Miocene Love bone bed of Florida. In Body size in mammalian paleobiology: Estimation and biological implications, edited by J. Damuth and B. J. MacFadden, 337–63. New York: Cambridge University Press.

MacFadden, B. J., and S. D. Webb. 1982. The succession of Miocene (Arikareean through Hemphillian) terrestrial mammalian localities and faunas in Florida. In Miocene of the southeastern United States, edited by T. M. Scott and S. B. Upchurch, 186–99. Florida Geological Survey Special Publication no. 25.

MacNeil, F. S. 1949. Pleistocene shore lines in Florida and Georgia. U.S. Geological Survey Professional Paper no. 221-F, 95–106.

Macurda, D. B. Jr. 1989. Seismic stratigraphy of carbonate platform sediments, southwest Florida. In Atlas of seismic stratigraphy, edited by A. W. Bally, 159–61. Studies in Geology no. 2(27). American Association of Petroleum Geologists.

Maddox, G. L., J. M. Lloyd, T. M. Scott, S. B. Upchurch, and R. Copeland, eds. 1992. Florida ground water quality monitoring program—background hydrogeochemistry. Florida Geological Survey Special Publication no. 34. 347 p.

Maglio, V. J. 1966. A revision of the fossil selenodont artiodactyls from the middle Miocene Thomas farm, Gilchrist County, Florida. Breviora of the Museum of Comparative Zoology 255. 27 p.

Mallinson, D. J., J. Compton, and D. Hodell. 1994. Isotope chronostratigraphy, lithostratigraphy, and sequence stratigraphy of the Micoene Hawthorn Group of northeastern Florida: Implications for eustacy and $\delta^{13}C$. Journal of Sedimentary Research 864:392–407.

Malloy, R. J., and R. J. Hurley. 1970. Geomorphology and geologic structure: Straits of Florida. Bulletin of the Geological Society of America 81:1947–72.

Mancini, E. A., and B. H. Tew. 1990. Relationships of Paleogene stage boundaries and unconformity-bounded depositional sequence contacts in Alabama and Mississippi. In Sequence stratigraphy as an exploration tool—concepts and practices in the Gulf Coast, 221–28. Program and Extended and Illustrated Abstracts, Gulf Coast Section, Society of Economic Paleontologists and Mineralogists, 11th Annual Research Conferences, Houston, Texas.

———. 1991. Paleogene sequence stratigraphy in Mississippi and Alabama. Geological Society of America Annual Meeting. Abstracts with Programs 23(1):62.

———. 1992. Paleogene stage and planktonic foraminiferal zone boundaries and unconformity-bounded depositional sequence contacts in southwestern Alabama (abstract). Bulletin of the American Association of Petroleum Geologists 75:628.

Mansfield, W. C. 1930. Miocene gastropods and scaphopods of the Choctawatchee formation of Florida. Florida Geological Survey Bulletin no. 3. 142 p.

———. 1931. Some Tertiary mollusks from southern Florida. Proceedings of the U.S. National Museum 79(21):1–12.

———. 1932. Miocene pelecypods of the Choctawhatchee Formation of Florida. Florida Geological Survey Bulletin no. 8. 240 p.

———. 1937. Mollusks of the Tampa and Suwannee limestones of Florida. Florida Geological Survey Bulletin no. 15. 334 p.

———. 1939. Notes on the upper Tertiary and Pleistocene mollusks of peninsular Florida. Florida Geological Survey Bulletin no. 18. 76 p.

Marella, R. L. 1992. Water withdrawals, use, and trends in Florida, 1990. U.S. Geological Survey Water-Resources Investigations Report no. 92-4140. 38 p.

Marino, J. N., and A. J. Mehta. 1986. Sediment volumes around Florida's east coast tidal inlets. Coastal and Ocean Engineering Laboratory Report no. 86/001. Gainesville: University of Florida. 71 p.

Marshall, L. G., R. F. Butler, R. E. Drake, G. H. Curtis, and R. H. Tedford. 1979. Calibration of the Great American Interchange. Science 204:272–79.

Marshall, L. G., R. Hoffstetter, and R. Pascual. 1983. Mammals and stratigraphy: Geochronology of the continental mammal-bearing Tertiary of South America. Palaeovertebrata (memoire extraordinaire). 93 p.

Martens, C. S., R. A. Berner, and J. K. Rosenfeld. 1978. Interstitial water chemistry of anoxic Long Island Sound sediments. 2. Nutrient regeneration and phosphate removal. Limnology and Oceanography 23:605–17.

Martens, C. S., J. P. Chanton, and C. K. Paull. 1991. Biogenic methane from abyssal brine seeps at the base of the Florida escarpment. Geology 19:851–54.

Martens, C. S., and R. C. Harris. 1970. Inhibition of apatite precipitation in the marine environment by Mg-ions. Geochimica et Cosmochimica Acta 34:621–25.

Martin, R. A. 1968. Late Pleistocene distribution of *Microtus pennsylvanicus*. Journal of Mammalogy 49:265–71.

———. 1969a. Line and grade in the extinct *medius* species group of *Sigmodon* [sic]. Science 167:1504–1506.

———. 1969b. Taxonomy of the giant Pleistocene beaver *Castoroides* from Florida. Journal of Paleontology 43:1033–41.

Martin, R. G. 1978. Northern and eastern Gulf of Mexico continental margin stratigraphic and structural framework. In Framework, facies, and oil-trapping characteristics of the upper continental margin, edited by A. H. Bouma, G. T. Moore, and J. M. Coleman, 21–42. Studies in Geology no. 7. American Association of Petroleum Geologists.

Masse, J. P., and J. Phillip. 1981. Cretaceous coral–rudist buildups of France. In European fossil reef models, edited by D. F. Toomey, 399–426. Society of Economic Paleontologists and Mineralogists Special Publication no. 30.

Matson, G. C., and F. G. Clapp. 1909. A preliminary report on the geology of Florida, with special reference to the stratigraphy. Florida Geological Survey 2nd Annual Report, 25–173.

Matson, G. C., and S. Sanford. 1913. Geology and ground waters of Florida. U.S. Geological Survey Water-Supply Paper no. 319. 445 p.

Mattraw, H. C., and B. J. Franks. 1986. Movement and fate of creosote waste in ground water, Pensacola, Florida—U.S. Geological Survey toxic-waste ground-water contamination program. U.S. Geological Survey Water-Supply Paper no. 2285. 63 p.

McArthur, J. M. 1978. Systematic variations in the contents of Na, Sr, CO_2, and SO_4 in marine carbonate fluorapatite and their relation to weathering. Chemical Geology 21:41–52.

———. 1985. Francolite geochemistry—compositional controls during formation, diagenesis, metamorphism and weathering. Geochimica et Cosmochimica Acta 49:23–35.

McArthur, J. M., R. A. Benmore, M. L. Coleman, C. Soldi, H. W. Yeh, and G. W. O'Brien. 1986. Stable isotopic characterization of francolite formation. Earth and Planetary Science Letters 77:20–34.

McArthur, J. M., A. R. Sahami, M. Thirwall, P. S> Hamilton, and A. O. Osborn. 1990. Dating phosphogenesis with strontium isotopes. Geochemica et Cosmochimica Acta 54:1343–51.

McBride, J. H. 1991. Constraints on the structure and tectonic development of the early Mesozoic South Georgia rift, southeastern United States—seismic reflection data processing and intepretation. Tectonics 10:1065–83.

McBride, J. H., and K. D. Nelson. 1988. Integration of CO-CORP deep reflection and magnetic anomaly analysis in the southeastern United States: Implications for origin of the Brunswick and East Coast magnetic anomalies. Bulletin of the Geological Society of America 100:436–45.

———. 1990. Integration of COCORP deep reflection and magnetic anomaly analysis in the southeastern United States: Implications for origin of the Brunswick and East Coast magnetic anomalies: Alternative interpretation and reply. Bulletin of the Geological Society of America 102:275–79.

McBride, J. H., K. D. Nelson, and L. D. Brown. 1989. Evidence and implications of an extensive early Mesozoic rift basin and basalt/diabase sequence beneath the southeast coastal plain. Bulletin of the Geological Society of America 101:512–20.

McBride, R. A., and T. F. Moslow. 1991. Origin, evolution, and distribution of shoreface sand ridges, Atlantic inner shelf, U.S.A. Marine Geology 97:57–85.

McClain, M. E., and E. Y. Karr. 1989. The Florida basement diary of a turbulent past. The Compass 66:52–58.

McClellan, G. H. 1980. Mineralogy of carbonate-fluorapatite. Journal of the Geological Society of London 137:675–81.

McClellan, G. H., and J. R. Lehr. 1969. Crystal chemical investigation of natural apatites. American Mineralogist 54:1379–91.

———. 1982. Phosphate rock impurities—good or bad? Proceedings of the 32nd Annual Meeting of the Fertilizer Industry Round Meeting, 238–47.

McClellan, G. H., and S. J. Van Kauwenbergh. 1990. Mineralogy of sedimentary apatites. In Phosphorite research and development, edited by A. J. G. Notholt and I. Jarvis, 23–31. Geological Society of London Special Publication no. 52.

McClellan, G. H., S. J. Van Kauwenbergh, and W. C. Isphording. 1985. An overview of the mineralogy of the phosphorite deposits of the southeastern U.S. In Guidebook, Eighth International Field Workshop and Symposium (Southeastern United States), edited by S. Snyder, 3–9. International Geological Congress Project no. 156—Phosphorites.

McCullough, L. M. 1969. The *Amusium ocalanum* biozone. Master's thesis, University of Florida, Gainesville. 53 p.

McFarlan, E. Jr., and L. S. Menes. 1991. Lower Cretaceous. In The Gulf of Mexico basin, edited by A. Salvador, 181–204. The geology of North America, vol. J. Geological Society of America.

McGinty, T. L. 1970. Mollusca of the 'Glades' unit of southern Florida: Part I—introduction and observations. Tulane Studies in Geology and Paleontology 8:53–56.

McGowan, B. 1990. Fifty million years ago. American Scientist 78:30–39

McIlreath, I. A., and N. P. James. 1984. Carbonate slopes. In Facies models, 2nd edition, edited by R. G. Walker, 245–257. Toronto: Geological Association of Canada.

McKenna, M.C. 1962. Collecting small fossils by washing and screening. Curator 5:221–35.

McKinney, M. L. 1984. Suwannee channel of the Paleogene coastal plain: Support for the 'carbonate suppression' model of basin formation. Geology 12:343–45.

McKinney, M. L., and D. S. Jones. 1983. Oligopygoid echinoids and the biostratigraphy of the Ocala Limestone in peninsular Florida. Southeastern Geology 24:21–29.

McKinney, M. L., and L. G. Zachos. 1986. Echinoids in biostratigraphy and paleoenvironmental reconstruction: A cluster analysis from the Eocene Gulf Coast. Palaios 1:420–23.

Meade, R. H. 1969. Landward transport of bottom sediments in the estuaries of the Atlantic coastal plain. Journal of Sedimentary Petrology 39:229–34.

Meeder, J. F. 1987. The paleoecology, petrology, and depositional model of the Pliocene Tamiami Formation, southwest Florida (with special reference to corals and reef development). Ph.D. diss., University of Miami (Florida). 748 p.

Mehta, A. J., and H. K. Brooks. 1973. Mosquito Lagoon barrier beach study. Shore and Beach 41:27–34.

Mehta, A. J., and C. P. Jones. 1977. Matanzas Inlet. Florida Sea Grant College Glossary of Inlets. Reports nos. 5 and 21. 79 p.

Meisburger, E. P., and D. B. Duane. 1971. Geomorphology and sediments of the inner continental shelf, Palm Beach to Cape Kennedy, Florida. U.S. Army Corps of Engineers Coastal Engineering Research Center Technical Memorandum no. 34. 82 p.

Meisburger, E. P., and M. E. Field. 1975. Geomorphology,

shallow structure, and sediments of the Florida inner continental shelf, Cape Canaveral to Georgia. U.S. Army Corps of Engineers Coastal Engineering Research Center Technical Memorandum no. 54. 119 p.

———. 1976. Neogene sediments of the Atlantic inner continental shelf off northern Florida. Bulletin of the American Association of Petroleum Geologists 60:2019–37.

Merkl, R. S. 1989. A sedimentological, mineralogical, and geochemical study of the fuller's earth deposits of the Miocene Hawthorne Group of south Georgia and north Florida. Ph.D. diss., Indiana University, Bloomington. 182 p.

Merriam, D. F. 1988. Some recent developments in the study of Florida Bay geology. The Compass 65:157–74.

Merriam, D. F., J. M. Fuhr, R. V. Jenkins, and P. J. Zimmerman. 1989. Pleistocene bedrock geology of Florida Bay, the Keys, and the Everglades (abstract). Bulletin of Marine Science 44:519–20.

Mertz, K. A. Jr. 1984. Diagenetic aspects, Sandholdt Member, Miocene Monterey Formation, Santa Lucia Mountains, California: Implications for depositional and burial environments. In Dolomites of the Monterey Formation and other organic-rich units, edited by R. E. Garrison, M. Kastner, and D. H. Zenger, 49–73. Society of Economic Paleontologists and Mineralogists, Pacific Section, no. 41.

Metcalf, S. J., and L. E. Hall. 1984. Sinkhole collapse induced by groundwater pumpage for freeze protection irrigation near Dover, Florida, January 1977. In Sinkholes: Their geology, engineering, and environmental impact, edited by B. F. Beck, 29–34. Rotterdam: A. A. Balkema.

Meyer, F. O. 1989. Siliciclastic influence on Mesozoic platform development: Baltimore Canyon trough, western Atlantic. In Controls on carbonate platform and basin development, edited by P. D. Crevello, J. L. Wilson, J. F. Sarg, and J. F. Reed, 213–32. Society of Economic Mineralogists and Paleontologists Special Publication no. 44.

Meyer, F. W. 1989. Hydrogeology, ground-water movement, and subsurface storage in the Floridan aquifer system in southern Florida. U.S. Geological Survey Professional Paper no. 1403-G. 59 p.

Michaud, D. 1988. Environmental developments. Mining Engineering 41:426–27.

Millan, A. A., and F. C. Townsend. 1981. Identification of expansive soils to minimize damage to residential structures. Engineering and Industrial Experiment Station Research Report no. 243*W07. Gainesville: University of Florida. 58 p.

Miller, J. A. 1982. Structural control of Jurassic sedimentation in Alabama and Florida. Bulletin of the American Association of Petroleum Geologists 66:1289–1301.

———. 1986. Hydrogeologic framework of the Floridan aquifer system in Florida and parts of Georgia, Alabama, and South Carolina. U.S. Geological Survey Professional Paper no. 1403-B. 91 p.

———. 1988. Coastal plain deposits. In Hydrogeology, edited by W. Back, J.S. R. Osenshein, and P.R. Seaber, 315–22. The Geology of North America, vol. O-2. Geological Society of America.

———. 1990. Ground water atlas of the United States—segment 6, Alabama, Florida, Georgia, and South Carolina. U.S. Geological Survey Hydrologic Investigations Atlas no. HA-730G. 28 p.

Miller, K. G., and R. G. Fairbanks. 1985. Oligocene to Miocene carbon isotope cycles and abyssal circulation changes. In The carbon cycle and atmospheric CO_2: Natural variations, Archean to present, edited by E. T. Sundquist and W. S. Broecker, 469–86. American Geophysical Union Monograph no. 32.

Miller, K. G., R. G. Fairbanks, and G. S. Mountain. 1987. Tertiary oxygen isotope synthesis, sea level history, and continental margin erosion. Paleoceanography 2:1–19.

Miller, K. G., J. D. Wright, and R. G. Fairbanks. 1991. Unlocking the ice house: Oligocene–Miocene oxygen isotopes, eustacy, and margin erosion. Journal of Geophysical Research 96:6829–48.

Miller, R. L., and H. Sutcliffe Jr. 1982. Water-quality and hydrogeologic data for three phosphate industry waste-disposal sites in central Florida, 1979 to 1980. U.S. Geological Survey, Water-Resources Investigations Report no. 81-84. 77 p.

———. 1984. Effects of three phosphate industrial sites on ground-water quality in central Florida, 1979 to 1980. U.S. Geological Survey Water-Resources Investigations Report no. 83-4256. 184 p.

Milliman, J. D., O. H. Pilkey, and D. A. Ross. 1972. Sediments of the continental margin off the eastern United States. Bulletin of the Geological Society of America 83:1315–34.

Milton, C. 1972. Igneous and metamorphic basement rocks of Florida. Florida Geological Survey Bulletin no. 55. 125 p.

Milton, C., and R. Grasty. 1969. Basement rocks of Florida and Georgia. Bulletin of the American Association of Petroleum Geologists 53:2483–93.

Missimer, T. M. 1992. Stratigraphic relationships of sediment facies within the Tamiami Formation of southwestern Florida: Proposed intraformational correlations. In Plio-Pleistocene stratigraphy and paleontology of southern Florida, edited by T. M. Scott and W. D. Allmon, 63–92. Florida Geological Survey Special Publication no. 36.

Missimer, T. M., and R. S. Banks. 1982. Miocene cyclic sedimentation in western Lee County, Florida. In Miocene of the southeastern United States, edited by T. M. Scott and S. B. Upchurch, 285–99. Florida Geological Survey Special Publication no. 25.

Missimer, T. M., and R. A. Gardner. 1976. High resolution seismic reflection profiling for mapping shallow water aquifers in Lee County, Florida. United States

Geological Survey Water-Resources Investigations Report no. 76-45. 30 p.

Mitchell-Tapping, H. J. 1980. Depositional history of the oolite of the Miami Limestone formation. Florida Scientist 43:116–25.

Mitchum, R. M. Jr. 1978. Seismic stratigraphic investigation of West Florida slope, Gulf of Mexico. In Framework, facies, and oil-trapping characteristics of the upper continental margin, edited by A. H. Bouma, G.T. Moore, and J.M. Coleman, 193–223. Studies in Geology no. 7. American Association of Petroleum Geologists.

Miyashiro, A. 1978. Nature of alkalic volcanic rocks. Contributions to Mineralogy and Petrology 66:91–104.

Moore, C. H. 1989. Carbonate Diagenesis and Porosity. Developments in Sedimentology no 46. Amsterdam: Elsevier. 338 p.

Moore, P. D., and D. J. Bellamy. 1974. Peatlands. New York: Springer-Verlag. 221 p.

Moores, E. M. 1991. Southwest U.S.–east Antarctic (SWEAT) connection: A hypothesis. Geology 19:425–28.

Morgan, G. S. 1989. New bats from the Oligocene and Miocene of Florida and the origins of the Neotropical chiropteran fauna. Journal of Vertebrate Paleontology 9:33A.

Morgan, G. S., O. J. Linares, and C. E. Ray. 1988. New species of vampire bats (Mammalia: Chiroptera: Desmodontidae) from Florida and Venezuela. Proceedings of the Biological Society of Washington 101:912–28.

Morgan, G. S., and R. B. Ridgway. 1987. Late Pliocene (late Blancan) vertebrates from the St. Petersburg Times site, Pinellas County, Florida, with a brief review of Florida Blancan faunas. Papers in Florida Paleontology no. 1. 22 p.

Mott, C. J. 1983. Earthquake history of Florida: 1727 to 1981. Florida Scientist 46:116–20.

Mueller, P. A., and J. Porch. 1983. Tectonic implications of Paleozoic and Mesozoic igneous rocks in the subsurface of peninsular Florida. Transactions of the Gulf Coast Association of Geological Societies 33:169–74.

Mueller, P. A., A. L. Heatherington, J. L. Wooden, R.D. Shuster, A.P. Nutman, and I.S. Williams. 1994. Precambrian zircons from the Florida basement: A Gondwanan connection. Geology 22:119–22.

Muhs, D. R., B. J. Szabo, L. McCartan, P.B. Maat, C.A. Bush and R.B. Halley. 1992. Uranium-series age estimates of corals from Quaternary marine sediments of southern Florida. In Plio-Pleistocene stratigraphy and paleontology of South Florida, edited by T. M. Scott and W. D. Allmon, 41–49. Florida Geological Survey Special Publication no. 36.

Mullins, H. T. 1983a. Structural controls of contemporary carbonate continental margins: Bahamas, Belize, Australia. In Platform margin and deepwater carbonates, edited by H. E. Cook, A.C. Hine, and H.T. Mullins, 2-1-2-57. Society of Economic Paleontologists and Mineralogists Short Course no. 12.

———. 1983b. Modern carbonate slopes and basins of the Bahamas. In Platform margin and deepwater carbonates, edited by H. E. Cook, A.C. Hine, and H.T. Mullins, 4-1-4-138. Society of Economic Paleontologists and Mineralogists Short Course no. 12.

Mullins, H. T., N. Breen, J. Dolan, R.W. Wellner, J.L. Petruccione, M. Gaylord, B. Anderson, A.J. Melillo, A.D. Jurgens, and D. Organe. 1992. Carbonate platforms along the southeast Bahamas–Hispaniola collision zone. Marine Geology 105:169–209.

Mullins, H. T., G. R. Dix, A. F. Gardulski, and L. S. Land. 1988c. Neogene deepwater dolomite from the Florida–Bahamas platform. In Sedimentology and geochemistry of dolostones, edited by V. Shukla and P. A. Baker, 235–43. Society of Economic Paleontologists and Mineralogists Special Publication no. 43.

Mullins, H. T., J. Dolan, N. Breen, et al. 1991. Retreat of carbonate platforms: Response to tectonic processes. Geology 19:1089–92.

Mullins, H. T., A. F. Gardulski, E. J. Hinchey, and A. C. Hine. 1988a. The modern carbonate ramp slope of central west Florida. Journal of Sedimentary Petrology 58:273–90.

Mullins, H. T., A. F. Gardulski, and A. C. Hine. 1986. Catastrophic collapse of the west Florida carbonate platform margin. Geology 14:167–70.

Mullins, H. T., A. F. Gardulski, A. C. Hine, A.J. Melillo, S.W. Wise, Jr., and J. Applegate. 1988b. Three-dimensional sedimentary framework of the carbonate ramp slope of central west Florida: A sequential stratigraphic perspective. Bulletin of the Geological Society of America 100:514–33.

Mullins, H. T., A. F. Gardulski, S. W. Wise, and J. Applegate. 1987. Middle Miocene oceanographic event in eastern Gulf of Mexico: Implications for seismic stratigraphic succession and loop current/Gulf Stream circulation. Bulletin of the Geological Society of America 98:702–13.

Mullins, H. T., and A. C. Hine. 1989. Scalloped bank margins—beginning of the end for carbonate platforms? Geology 17:30–33.

Mullins, H. T., and G. W. Lynts. 1977. Origin of the northwest Bahama Platform: Review and interpretation. Bulletin of the Seismological Society of America 88:1447–61.

Mullins, H. T., and A. C. Neumann. 1979. Geology of the Miami Terrace and its paleoceanographic implications. Marine Geology 30:205–32.

Multer, H. G. 1977. Field guide to some carbonate rock environments, Florida Keys and western Bahamas. Dubuque, Iowa: Kendall-Hunt. 415 p.

Multer, H. G., and J. E. Hoffmeister. 1968. Subaerial laminated crusts of the Florida Keys. Bulletin of the Geological Society of America 79:183–92.

Murphy, J. B., and R. D. Nance. 1989. Model for the evolution of the Avalonian–Cadomian belt. Geology 17:735–38.

Murray, G. E. 1961. Geology of the Atlantic and Gulf coastal province of North America. New York: Harper and Brothers. 692 p.

Nagda, N. L., M. D. Koontz, R. C. Fortman, W.A. Schoenborn and L.L. Mehegen. 1987. Florida statewide radiation study, Bartow. Florida Institute of Phosphate Research Report no. 05-029-057. 455 p.

Nance, R. D., J. Murphy, R. Strachan, R. D'Lemos and G. Taylor. 1991. Late Proterozoic tectonostratigraphic evolution of the Avalonian and Cadomian terranes. Precambrian Research 53:41–78.

Nathan, Y., and J. Lucas. 1976. Expériences sur la précipitation directe de l'apatite dans l'eau de mer—implications dans la genèse des phosphorites. Chemical Geology 18:181–86.

Nathan, Y., and H. Nielsen. 1980. Sulfur isotopes in phosphorites. In Marine phosphorites—geochemistry, occurrence, genesis, edited by Y. K. Bentor, 73–78. Society of Economic Paleontologists and Mineralogists Special Publication no. 29.

Neathery, T. L., and W. A. Thomas. 1975. Pre-Mesozoic basement rocks of the Alabama coastal plain. Transactions of the Gulf Coast Association of Geological Societies 25:86–99.

Nelsen, J. E., and R. N. Ginsburg. 1986. Calcium carbonate production by epibionts on *Thalassia* in Florida Bay. Journal of Sedimentary Petrology 56:622–28.

Nelson, K. D., J. A. Arnow, J. H. McBride, J.H. Eillemin, J. Huang, L. Zheng, J.E. Oliver, L.D. Brown, and S. Kaufman. 1985a. New COCORP profiling in the southeastern United States. Part 1: Late Paleozoic suture and Mesozoic rift basin. Geology 13:714–18.

Nelson, K. D., J. H. McBride, J. A. Arnow, J.E. Oliver, L. D. Brown, and S. Kaufman. 1985b. New COCORP profiling in the southeastern United States. Part 2: Brunswick and East Coast magnetic anomalies, opening of the north Atlantic. Geology 13:718–21.

Neumann, A. C., J. W. Kofoed, and G. H. Keller. 1977. Lithoherms in the Straits of Florida. Geology 5:4–10.

Neumann, A. C., and L. S. Land. 1975. Lime mud deposition and calcareous algae in the Bight of Abaco, Bahamas: A budget. Journal of Sedimentary Petrology 45:763–86.

Neumann, A. C., and I. G. Macintyre. 1985. Reef response to sea-level rise: Keep-up, catch-up, or give-up. Proceedings of the Fifth International Coral Reef Symposium, Tahiti, 3, 105–10.

Newton, C. R., H. T. Mullins, A. F. Gardulski, A. C. Hine, and G. R. Dix. 1987. Coral mounds on the west Florida slope: Unanswered questions regarding the development of deep-water banks. Palaios 2:359–67.

Nicol, D. 1991. Tethyan molluscs of the middle and late Eocene of peninsular Florida. Papers in Florida Paleontology no. 4. 6 p.

Nicol, D., and D. S. Jones. 1982. *Rotularia vernoni,* an annelid worm tube from the Eocene of peninsular Florida. Florida Scientist 45:139–42.

———. 1984. Mode of life of *Exputens ocalensis* (Malleidae), a Florida Eocene pelecypod. Florida Scientist 45:32–34.

Nicol, D., D. S. Jones, and J. W. Hoganson. 1989. *Anatipopecten* and the *Rotularia vernoni* Zone (Late Eocene) in peninsular Florida. Tulane Studies in Geology and Paleontology 22:55–59.

Nicol, D., G. D. Shaak, and J. W. Hoganson. 1976. The Crystal River Formation (Eocene) at Martin, Marion County, Florida. Tulane Studies in Geology and Paleontology 12:137–44.

NOAA (National Oceanographic and Atmospheric Administration). 1985. Gulf of Mexico and ocean zones strategic assessment: Data atlas. U.S. Department of Commerce 1.0-6.07.

Nummedal, D., G. F. Oertel, D. K. Hubbard, and A. C. Hine. 1977. Tidal inlet variability—Cape Hatteras to Cape Canaveral. In Coastal Sediments 1977: Fifth Symposium of the Waterway, Port, Coastal, and Ocean Division of the American Society of Civil Engineers, Charleston, S.C., 543–62.

O'Brien, G. W., and H. H. Veeh. 1980. Holocene phosphorite on the east Australian continental margin. Nature 288:690–92.

O'Connor, L. D. 1984. Gravity study of crustal structures in northeast peninsular Florida. Master's thesis, University of South Florida, Tampa. 62 p.

Odom, A. L., and J. F. Brown. 1976. Was Florida a part of North America in the lower Paleozoic? Geological Society of America, Northeast and Southeast Sections, Annual Meeting. Abstracts with Programs 8(2):237–38.

Officer, C. B., C. L. Drake, J. L. Pindell, and A. A. Meyerhoff. 1992. Cretaceous–Tertiary events and the Caribbean caper. GSA Today 2(4):69–75.

Oglesby, W. R., and M. M. Ball. 1971. Bouguer anomaly map of south Florida. Florida Geological Survey Map Series no. 41. 1 sheet.

Oglesby, W. R., M. M. Ball, and S. Chaki. 1973. Bouguer anomaly map of Florida and adjoining continental shelves. Florida Geological Survey Map Series no. 57. 1 sheet.

Oil and Gas Journal. 1989. Oil and Gas Journal 87:18.

Olsen, S. J. 1958. The fossil carnivore *Amphicyon intermedius* (sic) from the Thomas farm Miocene, Part I: Skull and dentition. Bulletin of the Museum of Comparative Zoology 118:157–77.

———. 1959. Fossil mammals of Florida. Florida Geological Survey Special Publication no. 6. 75 p.

———. 1960. The fossil carnivore *Amphicyon iongiramus* from the Thomas farm Miocene, Part II. Postcranial skeleton. Bulletin of the Museum of Comparative Zoology 123:3–45.

———. 1964a. Stratigraphic importance of a lower Miocene vertebrate fauna from north Florida. Journal of Paleontology 58:477–82.

———. 1964b. Vertebrate correlations and Miocene stratigraphy of north Florida fossil localities. Journal of Paleontology 58:600–604.

———. 1968. Miocene vertebrates and north Florida shorelines. Palaeogeography, Palaeoclimatology, Palaeoecology 5:127–34.

Olsson, A. A. 1961. Molluscs of the tropical eastern Pacific. Ithaca, N.Y.: Paleontological Research Institution. 574 p.

———. 1964. The geology and stratigraphy of south Florida. In Some Neogene mollusca from Florida and the Carolinas, edited by A. A. Olsson and R. E. Petit, 511–26. Bulletins of American Paleontology no. 47.

———. 1968. A review of late Cenozoic stratigraphy of southern Florida. In Late Cenozoic stratigraphy of southern Florida—a reappraisal, with additional notes on Sunoco-Felda and Sunniland oil fields, compiled by R. D. Perkins, 66–82. Miami Geological Society 2nd Annual Field Trip Guidebook.

———. 1972. Origin of the existing panamic biotas in terms of their geologic history and their separation by the isthmian land barrier. Bulletin of the Biological Society of Washington 2:117–23.

Olsson, A. A., and A. Harbison. 1953. Pliocene Mollusca of southern Florida, with special reference to those from north St. Petersburg. Philadelphia Academy of Natural Sciences Monograph no. 8. 457 p.

Opdyke, N. D., D. S. Jones, B. J. MacFadden, Smith, D. L., Mueller, P. A., and R. D. Schuster. 1987. Florida as an exotic terrane: Paleomagnetic and geochronologic investigation of lower Paleozoic rocks from the subsurface of Florida. Geology 15:900–903.

Opdyke, N. D., D. P. Spangler, D. L. Smith, Jones, D. S., and R. C. Lindquist. 1984. Origin of the epeirogenic uplift of Pliocene–Pleistocene beach ridges in Florida and development of the Florida karst. Geology 12:226–28.

Opyrchal, A. M., and K. L. Wang. 1981. Economic significance of the Florida phosphate industry: An input–output analysis. U.S. Bureau of Mines Information Circular no. 8850. 62 p.

Oslick, J. S., K. G. Miller, and M. D. Feigenson. 1992. Lower to Middle Miocene $^{87}Sr/^{86}Sr$ reference section: Ocean Drilling Program Hole 747A (abstract). EOS, Transactions of the American Geophysical Union 73 (supp.):171–72.

Osmond, J. K., J. R. Carpenter, and H. L. Windom. 1965. $^{230}Th/^{234}U$ age of the Pleistocene corals and oolites of Florida. Journal of Geophysical Research 70:1843–47.

Osmond, J. K., and J. B. Cowart. 1977. Uranium series isotopic anomalies in thermal ground waters from southwest Florida. In The geothermal nature of the Floridan Plateau, edited by D. L. Smith and G. M. Griffin, 131–48. Florida Geological Survey Special Publication no. 21.

Osmond, J. K., M. I. Kaufman, and J. B. Cowart. 1974. Mixing volume calculations, sources, and aging trends of Floridan aquifer water by uranium isotopic methods. Geochimica et Cosmochimica Acta 38:1083–1100.

Osmond, J. K., J. P. May, and W. F. Tanner. 1970. Age of the Cape Kennedy barrier and lagoon complex. Journal of Geophysical Research 75:469–79.

Ottman, R. D., P. L. Keyes, and M. A. Ziegler. 1973. Jay field—a Jurassic stratigraphic trap. Transactions of the Gulf Coast Association of Geologic Societies 22:146–75.

Otvos, E. G. 1981. Tectonic lineaments of Pliocene and Quaternary shorelines, northeast Gulf Coast. Geology 9:398–404.

———. 1985. Barrier platforms: Northern Gulf of Mexico. Marine Geology 63:285–306.

Oural, C. R., S. B. Upchurch, and H. R. Brooker. 1989. Radon progeny as sources of gross-alpha radioactivity anomalies. Health Physics 55:889–94.

Pardo, G. 1975. Geology of Cuba. In Ocean basins and margins, vol. 3, edited by A. E. M. Nairn and F. G. Stehli, 553–615. New York: Plenum Press.

Parker, G. G., and C. W. Cooke. 1944. Late Cenozoic geology of southern Florida with a discussion of the ground water. Florida Geological Survey Bulletin no. 27. 119 p.

Parker, G. G., G. E. Ferguson, S. K. Love, et al. 1955. Water resources of southeastern Florida. U.S. Geological Survey Water-Supply Paper no. 1255. 965 p.

Parkinson, R. W. 1989. Decelerating Holocene sea-level rise and its influence on southwest Florida coastal evolution: A transgressive/regressive stratigraphy. Journal of Sedimentary Petrology 59:960–72.

Patton, T. H. 1969. An Oligocene land vertebrate fauna from Florida. Journal of Paleontology 43:543–46.

Paull, C. K., J. P. Chanton, C. S. Martens, P. D. Fullagar, A. C. Neumann, and J. A. Coston. 1991b. Seawater circulation through the flank of the Florida Platform: Evidence and implications. Marine Geology 102:265–79.

Paull, C. K., R. P. Freeman-Lynde, T. J. Bralower, J. M. Gardemal, A. C. Neumann, B. D'Argenio, and E. Marsella. 1990a. Geology of the strata exposed on the Florida Escarpment. Marine Geology 91:177–94.

Paull, C. K., B. Hecker, R. Commeau, R. P. Freeman-Lynde, A. C. Neumann, W. P. Corso, S. Golubic, J. Hook, E. Sikes, and J. Curray. 1984. Biological communities at Florida Escarpment resemble hydrothermal vent communities. Science 226:965–67.

Paull, C. K., and A. C. Neumann. 1987. Continental margin brine seeps: Their geological consequences. Geology 15:545–48.

Paull, C. K., F. N. Speiss, J. R. Curray, and D. C. Twichell. 1990b. Origin of Florida Canyon and the role of spring sapping on the formation of submarine box canyons. Bulletin of the Geological Society of America 102:502–15.

Paull, C. K., D. C. Twichell, F. N. Speiss, and J. R. Curray. 1991a. Morphological development of the Florida Escarpment: Observations on the generation of time-transgressive unconformities in carbonate terrains. Marine Geology 101:181–201.

REFERENCES

Pegram, W. J. 1990. Development of continental lithospheric mantle as reflected in the chemistry of the Mesozoic Appalachian tholeiites, U.S.A. Earth and Planetary Science Letters 97:316–31.

Perkins, R. D. 1977. Depositional framework of Pleistocene rocks in south Florida. In Quaternary sedimentation in south Florida, edited by P. Enos and R. D. Perkins, 131–98. Geological Society of America Memoir no. 147.

Perkins, R. D., and P. Enos. 1968. Hurricane Betsy in the Florida–Bahamas area—geologic effects and comparison with Hurricane Donna. Journal of Geology 76:710–17.

Petuch, E. J. 1982a. Geographical heterochrony: contemporaneous coexistence of Neogene and Recent molluscan faunas in the Americas. Palaeogeography, Palaeoclimatology, Palaeoecology 37:277–312.

———. 1982b. Notes on the molluscan paleoecology of the Pinecrest beds at Sarasota, Florida, with a description of *Pyruella,* a stratigraphically important new genus (Gastropoda: Melongenidae). Proceedings of the Academy of Natural Sciences of Philadelphia 134:12–30.

———. 1986. The Pliocene reefs of Florida: Their geomorphological significance in the evolution of the Atlantic coastal ridge, southeastern Florida, U.S.A. Journal of Coastal Research 2:391–408.

———. 1988. Neogene history of tropical American mollusks. Charlottesville, Va.: Coastal Education and Research Foundation. 217 p.

———. 1989. Field guide to the Ecphoras. Charlottesville, Va.: Coastal Education and Research Foundation. 140 p.

———. 1990. New gastropods from the Bermont Formation (middle Pleistocene) of the Everglades basin. Nautilus 104:96–104.

———. 1991. New gastropods from the Plio-Pleistocene of southwestern Florida and the Everglades basin. Special W. H. Dall Paleontological Research Center Publication no. 1. 64 p.

———. 1992. The Pliocene pseudoatoll of southern Florida and its associated gastropod fauna. In Plio-Pleistocene stratigraphy and paleontology of southern Florida, edited by T. M. Scott and W. D. Allmon, 101–16. Florida Geological Survey Special Publication no. 36.

Pindell, J. 1985. Alleghenian reconstruction and subsequent evolution of the Gulf of Mexico, Bahamas, and proto-Caribbean. Tectonics 4:1–39.

Pindell, J., S. C. Cande, W. C. Pitman III, D. B. Rowley, J. F. Dewey, J. LaBreque, and W. Haxby. 1988. A plate-kinematic framework for models of Caribbean evolution. Tectonophysics 155:121–38.

Pindell, J., and J. F. Dewey. 1982. Permo-Triassic reconstruction of western Pangaea and the evolution of the Gulf of Mexico/Caribbean region. Tectonics 1:179–211.

Pinet, P. R., and P. Popenoe. 1985a. A scenario of Mesozoic–Cenozoic ocean circulation over the Blake Plateau and its environs. Bulletin of the Geological Society of America 96:618–26.

———. 1985b. Shallow seismic stratigraphy and post-Albian geologic history of the northern and central Blake Plateau. Bulletin of the Geological Society of America 96:627–38.

Pirkle, E. C. 1960. Kaolinitic sediments of peninsular Florida and origin of the kaolinite. Economic Geology 55:1382–1405.

Pirkle, E. C., W. A. Pirkle, and W. R. Yoho. 1974. The Green Cove Springs and Boulogne heavy mineral sand deposit on Trail Ridge in northern peninsular Florida. Economic Geology 69:1129–37.

———. 1977. The Highland heavy mineral sand deposit on Trail Ridge in northern peninsular Florida. Florida Geological Survey Report of Investigations no. 84. 50 p.

Pirkle, E. C., and W. R. Yoho. 1970. The heavy mineral ore body of Trail Ridge, Florida. Economic Geology 65:17–30.

Pirkle, E. C., W. R. Yoho, and A. T. Allen. 1964. Origin of the silica sand deposits of the Lake Wales Ridge area of Florida. Economic Geology 59:1107–39.

———. 1965. Hawthorn Bone Valley and Citronelle sediments in Florida. Quarterly Journal of the Florida Academy of Sciences 28:7–58.

Pirkle, E. C., W. H. Yoho, and C. W. Hendry Jr. 1970. Ancient sea level stands in Florida. Florida Geological Survey Bulletin no. 52. 61 p.

Pirkle, E. C., W. R. Yoho, and S. D. Webb. 1967. Sediments of the Bone Valley phosphate district of Florida. Economic Geology 6:237–61.

Pirkle, F. L., and L. J. Czel. 1983. Marine fossils from region of Trail Ridge, a Georgia–Florida landform. Southeastern Geology 24:31–38.

Pirkle, F. L., E. C. Pirkle, J. G. Reynolds, W. A. Pirkle, D. S. Jones, D. P. Spangler, and T. A. Goodman. 1991. Cabin Bluff heavy mineral deposits of southeastern Georgia. Economic Geology 86:436–43.

Pirkle, W. A. 1971. The offset course of the St. Johns River, Florida. Southeastern Geology 13:39–59.

Plummer, L. N. 1975. Mixing of sea water with calcium carbonate ground water. Geological Society of America Memoir no. 142, 219–36.

Poag, C. W. 1982. Biostratigraphy, sea level fluctuations, subsidence rates, and petroleum potential of the southeast Georgia embayment. In Proceedings: Second Symposium on the Geology of the Southeastern Coastal Plain, edited by D. D. Arden, B. F. Beck, and E. Morrow, 3–9. Georgia Geologic Survey Information Circular no. 53.

———. 1985. Geologic evolution of the U.S. Atlantic margin. In Quantitative studies in the geological sciences: A memoir to W. C. Krumbein, edited by E. H. T. Whitten, New York: Van Nostrand. 383 p.

———. 1991. Rise and demise of the Bahama–Grand Banks

Pojeta, J., J. Kriz, and J. M. Berdan. 1976. Silurian–Devonian pelecypods and Paleozoic stratigraphy of subsurface rocks in Florida and Georgia and related Silurian pelecypods from Bolivia and Turkey. U.S. Geological Survey Professional Paper no. 879. 32 p.

Pool, T. C. 1991. Uranium: You thought 1989 was bad. Engineering and Mining Journal 192:57–60.

———. 1992. Uranium: Is the worst over? Engineering and Mining Journal 193:45–47.

Popenoe, P. 1990. Paleoceanography and paleogeography of the Miocene of the southeastern United States. In Neogene to modern phosphates, edited by W. C. Burnett and S. R. Riggs, 352–80. Phosphate deposits of the World, vol. 3. Cambridge: Cambridge University Press.

Popenoe, P., V. J. Henry, and F. M. Idris. 1987. Gulf trough—the Atlantic connection. Geology 15:327–32.

Popenoe, P., F. A. Kohout, and F. T. Manheim. 1984. Seismic reflection studies of sinkholes and limestone dissolution features on the northeastern Florida shelf. In Sinkholes: Their geology, engineering, and environmental impact, edited by B. F. Beck, 43–57. Rotterdam: A. A. Balkema.

Portell, R. W. 1989. Fossil invertebrates from the banks of the Suwannee River at White Springs, Florida. In Miocene paleontology and stratigraphy of the Suwannee River basin of north Florida and south Georgia, edited by G. S. Morgan, 14–25. Guidebook for the 30th Annual Field Trip of the Southeastern Geological Society.

Power, P. 7 July 1992. Fertilizer firm to cut 75 jobs. Tampa Tribune, Business and Finance section, p. 1.

Prasad, S. 1985. Microsucrosic dolomite from the Hawthorn Formation (Miocene) of Florida: Distribution and development. Master's thesis, University of Miami (Florida). 124 p.

Pratt, A. E. 1990. Taphonomy of the large vertebrate fauna from the Thomas farm locality (Miocene, Hemingfordian), Gilchrist County, Florida. Bulletin of the Florida Museum of Natural History 35:35–130.

Pratt, A. E., and G. S. Morgan. 1989. New Sciuridae (Mammalia: Rodentia) from the early Miocene Thomas farm local fauna, Florida. Journal of Vertebrate Paleontology 9:98–100.

Price, W. A. 1954. Shorelines and coasts on the Gulf of Mexico. In Gulf of Mexico—its origin, waters, and marine life, edited by P. S. Galtsoff. U.S. Fish and Wildlife Bulletin 87:36–54.

Purdy, E. G. 1974. Reef configurations: Cause and effect. In Reefs in time and space, edited by L. F. LaPorte, 9–76. Society of Economic Paleontologists and Mineralogists Special Publication no. 18.

Puri, H. S. 1953. Contributions to the study of the Miocene of the Florida panhandle. Florida Geological Survey Bulletin no. 36. 345 p.

———. 1957. Stratigraphy and zonation of the Ocala Group. Florida Geological Survey Bulletin no. 38. 248 p.

Puri, H. S., and R. O. Vernon. 1964. Summary of the geology of Florida and guidebook to the classic exposures. Florida Geological Survey Special Publication no. 5 (revised). 312 p.

Puri, H. S., and G. Winston. 1974. Geologic framework of the high transmissivity zones in south Florida. Florida Geological Survey Special Publication no. 20. 120 p.

Randazzo, A. F. 1972. Petrography of the Suwannee Limestone. Florida Geological Survey Bulletin no. 54, 2:1–13.

———. 1976. The significance of the Williston and Inglis formations in stratigraphic nomenclature and classification. Guidebook for the 20th Annual Field Trip pf the Southeastern Geological Society, 49–62.

———. 1987. Tertiary sedimentology and evolution of the Florida Platform. Geological Society of America Annual Meeting. Abstracts with Programs 19(7):812.

Randazzo, A .F., and J. I. Bloom. 1985. Mineralogical changes along the freshwater/saltwater interface of a modern aquifer. Sedimentary Geology 43:219–39.

Randazzo, A. F., and D. J. Cook. 1987. Characterization of dolomite rocks from the coastal mixing zone of the Floridan aquifer, U.S.A. Sedimentary Geology 54:169–92.

Randazzo, A. F., and E. W. Hickey. 1978. Dolomitization in the Floridan aquifer. American Journal of Science 278:1177–84.

Randazzo, A. F., M. Kosters, D. S. Jones, and R. W. Portell. 1990. Paleoecology of shallow-marine carbonate environments, Middle Eocene of peninsular Florida. Sedimentary Geology 66:1–11.

Randazzo, A. F., and H. C. Saroop. 1976. Sedimentology and paleoecology of Middle and Upper Eocene carbonate shoreline sequences, Crystal River, Florida, U.S.A. Sedimentary Geology 15:259–91.

Randazzo, A. F., and L. G. Zachos. 1984. Classification and description of dolomitic fabric of rocks from the Floridan aqufier, U.S.A. Sedimentary Geology 37:151–62.

Rankin, D. W., A. A. Drake Jr., L. Glover III, R. Goldsmith, L. M. Hall, D. P. Murray, N. M. Ratcliffe, J. F. Read, D. T. Secor, and R. S. Stanley. 1989. Pre-orogenic terranes. In The Appalachian–Ouachita orogen in the United States, edited by R. D. Hatcher Jr., W. A. Thomas, and G. W. Viele, 7–100. The geology of North America, vol. F-2. Geological Society of America.

Rast, N., and J. W. Skehan. 1981. Possible correlation of Precambrian rocks of Newport, Rhode Island, with those of Anglesey, Wales. Geology 9:596–601.

Ray, C. E., and A. E. Saunders. 1984. Pleistocene tapirs in the eastern United States. In Contributions to Quaternary vertebrate paleoecology: A volume in memorial to John E. Guilday, edited by H. H. Genoways and M. R.

Dawson, 238–315. Pittsburgh: Carnegie Museum of Natural History.

Raymo, M. E., and W. F. Ruddiman. 1992. Tectonic forcing of the late Cenozoic climate. Nature 359:117–22.

Rea, R. A., and S. B. Upchurch. 1980. Influence of regolith properties on migration of septic tank effluent. Ground Water 18:118–25.

Read, J .F. 1985. Carbonate platform facies models. Bulletin of the American Association of Petroleum Geologists 64:1575–1612.

Reading, H. G. 1978. Sedimentary environments and facies. New York: Elsevier. 557 p.

Reagor, B. G., C. W. Stover, and S. T. Algermissen. 1987. Seismicity map of the State of Florida. U.S. Geological Survey Miscellaneous Field Studies Map no. MF-1056.

Reel, D. A. 1970. Geothermal gradients and heat flow in Florida. Master's thesis, University of South Florida, Tampa. 66 p.

Reel, D. A., and G. M. Griffin. 1971. Potentially petroliferous trends in Florida as defined by geothermal gradients. Transactions of the Gulf Coast Association of Geological Societies 21:31–36.

Reimers, C. E. 1982. Organic matter in anoxic sediments off central Peru: Relations of porosity, microbial decomposition, and deformation properites. Marine Geology 46:175–97.

Reimers, C. E., M. Kastner, and R. E. Garrison. 1990. The role of bacterial mats in phosphate mineralization, with particular reference to the Monterey Formation. In Neogene to modern phosphorites, edited by W. C. Burnett and S. R. Riggs, 300–311. Phosphate deposits of the world, vol. 3. Cambridge: Cambridge University Press.

Rich, T. H. V., and T. H. Patton. 1975. First record of a fossil hedgehog from Florida (Erinaceidae, Mammalia). Journal of Mammalogy 56:692–96.

Richards, H. G., and K. V. W. Palmer. 1953. Eocene mollusks from Citrus and Levy counties. Florida Geological Survey Bulletin no. 35. 95 p.

Riggs, S. R. 1979a. Petrology of the Tertiary phosphorite system of Florida. Economic Geology 74:195–220.

———. 1979b. Phosphorite sedimentation in Florida—a model phosphogenic system. Economic Geology 74:285–314.

———. 1984. Paleoceanographic model of Neogene phosphorite deposition, U.S. Atlantic continental margin. Science 223(4632):123–31.

———. 1986. Phosphogenesis and its relationship to exploration for Proterozoic and Cambrian phosphorites. In Proterozoic and Cambrian phosphorites, edited by P. J. Cook and J. H. Shergold, 352–369. Phosphate deposits of the world, vol. 1. Cambridge: Cambridge University Press.

Riggs, S. R., and D. F. Belknap. 1988. Upper Cenozoic processes and environments of continental margin sedimentation: eastern United States. In The Atlantic continental margin, U.S., edited by R. E. Sheridan and J. A. Grow, 131–76. The geology of North America, vol. I-2. Geological Society of America.

Riggs, S. R., and F. T. Manheim. 1988. Mineral resources of the U.S. Atlantic continental margin. In The Atlantic continental margin, U.S., edited by R. W. Sheridan and J. A. Grow, 365–85. The geology of North America, vol. I-2. Geological Society of America.

Robbin, D. M. 1984. A new Holocene sea level curve for the upper Florida Keys and Florida reef tract. In Environments of south Florida, past and present, part II, edited by P. J. Gleason, 437–58. Miami, Fla.: Miami Geological Society.

Robbin, D. M., and J. J. Stipp. 1979. Depositional rate of laminated soilstone crusts, Florida Keys. Journal of Sedimentary Petrology 49:175–80.

Roberts, H. H., L. J. Rouse, N. D. Walker, and J. H. Hudson. 1982. Coldwater stress in Florida Bay and northern Bahamas: A product of winter cold-air outbreaks. Journal of Sedimentary Petrology 52:145–55.

Roberts, H. H., R. Whelan, and W. G. Smith. 1977. Holocene sedimentation at Cape Sable, South Florida. Sedimentary Geology 18:25–60.

Robertson, J. S. 1974. Fossil bison of Florida. In Pleistocene mammals of Florida, edited by S. D. Webb, 214–46. Gainesville: University Press of Florida.

Robertson, J. S. 1976. Latest Pliocene mammals from Haile XV A, Alachua County, Florida. Bulletin of the Florida State Museum 20:111–86.

Rodgers, J. 1972. Latest Precambrian (post-Grenville) rocks of the Appalachian region. American Journal of Science 272:507–20.

Roof, S. R., H. T. Mullins, S. Gartner, T. C. Huang, E. Joyce, J. Printzman, and L. Tjalsma. 1991. Climatic forcing of cyclic carbonate sedimentation during the last 5.4 million years along the west Florida continental margin. Journal of Sedimentary Petrology 61:1070–88.

Rosenau, J. C., G. L. Faulkner, C. W. Hendry Jr. and R. W. Hull. 1977. Springs of Florida (revised). Florida Geological Survey Bulletin no. 31. 461 p.

Ross, M. I., and C. R. Scotese. 1988. A hierarchical tectonic model of the Gulf of Mexico and Caribbean region. Tectonophysics 115:139–68.

Roussel, J., and J. L. Liger. 1983. A review of deep structure and ocean–continent transition in the Senegal basin (West Africa). Tectonophysics 91:183–211.

Rowley, D. B. 1981. Accretionary collage of terrains assembled against eastern North America during the medial Ordovician Taconic orogeny. Geological Society of America Annual Meeting. Abstracts with Programs 13(7):542.

Rude, P. D., and R. C. Aller. 1991. Fluorine mobility during early diagenesis of carbonate sediment: An indicator of mineral transformation. Geochimica et Cosmochimica Acta 55:2491–2509.

Rukavina, M. 1989. Recycling: Grab the trend now. Rock Products 92:54–57.

Runnels, D. D. 1969. Diagenesis, chemical sediments, and the mixing of natural waters. Journal of Sedimentary Petrology 39:1188–1201.

Rupert, F. R. 1990. Geology of Gadsden County, Florida. Florida Geological Survey Bulletin no. 62. 61 p.

Safko, P. S., and J. J. Hickey. 1992. A preliminary approach to the use of borehole data, including television surveys, for characterizing secondary porosity of carbonate rocks in the Floridan aquifer system. U.S. Geological Survey, Water-Resources Investigations Report no. 91-4168. 70 p.

Salvador, A. 1987. Late Triassic–Jurassic paleogeography and origin of Gulf of Mexico basin. Bulletin of the American Association Petroleum Geologists 71:419–51.

———. 1991a. Introduction. In The Gulf of Mexico basin, edited by A. Salvador, 1–12. The geology of North America, vol. J. Geological Society of America.

———. 1991b. Origin and development of the Gulf of Mexico basin. In The Gulf of Mexico basin, edited by A. Salvador, 389–444. The geology of North America, vol. J. Geological Society of America.

Sanford, S. 1909. Topography and geology of southern Florida. Florida Geological Survey 2nd Annual Report, 177–231.

———. 1913. Geology and ground waters of Florida. U.S. Geological Survey Water-Supply Paper no. 319, 186–89.

Savage, R. J. G., and M. R. Long. 1986. Mammal evolution: An illustrated guide. New York: Facts on File. 259 p.

Sawyer, D. S., R. T. Buffler, and R. H. Pilger Jr. 1991. The crust under the Gulf of Mexico basin. In The Gulf of Mexico basin, edited by A. Salvador, 53–72. The Geology of North America, vol. J. Geological Society of America.

Schlager, W. 1981. The paradox of drowned reef and carbonate platforms. Bulletin of the Geological Society of America 92(1):197–211.

———. 1989. Drowning unconformities on carbonate platforms. In Controls on carbonate platform and basin development, edited by P. D. Crevello, J. L. Wilson, J. F. Sarg, and J. F. Reed, 15–25. Society of Economic Paleontologists and Mineralogists Special Publication no. 44.

———. 1992. Sedimentology and sequence stratigraphy of reefs and carbonate platforms. American Assocation of Petroleum Geologists Continuing Education Course Note Series no. 34. 71 p.

Schlager, W., R. T. Buffler, D. Angstadt, and R. Phair. 1984a. Geologic history of the southeastern Gulf of Mexico. Initial Reports of the Deep Sea Drilling Project 77:715–38.

———. 1984b. Deep Sea Drilling Project, Leg 77, southeastern Gulf of Mexico. Bulletin of the Geological Society of America 95:226–36.

Schlager, W., and O. Camber. 1986. Submarine slope angles, drowning unconformities, and self-erosion of limestone escarpments. Geology 14:762–65.

Schlager, W., and R. N. Ginsburg. 1981. Bahama carbonate platforms—the deep and the past. Marine Geology 44:1–24.

Schlanger, S. O., H. C. Jenkyns, and I. Premoli-Silva. 1981. Volcanism and vertical tectonics in Pacific Basin related to global Cretaceous transgressions. Earth and Planetary Science Letters 52:435–49.

Schlee, J. S., W. P. Dillon, and J. A. Grow. 1979. Structure of the continental slope off the eastern United States. In Geology of continental slopes, edited by L. J. Doyle and O. H. Pilkey, 95–117. Society of Economic Paleontologists and Mineralogists Special Publication no. 27.

Schlee, J. S., W. Manspeizer, and S. R. Riggs. 1988. Paleo-environments: Offshore Atlantic U.S. margin. In The Atlantic continental margin, edited by R. E. Sheridan and J. A. Grow, 365–85. The geology of North America, vol. I-2. Geological Society of America.

Schmidt, W. 1984. Neogene stratigraphy and geologic history of the Apalachicola embayment, Florida. Florida Geological Survey Bulletin no. 58. 146 p.

Schmidt, W., and M. Wiggs-Clark. 1980. Geology of Bay County, Florida. Florida Geological Survey Bulletin no. 57. 96 p.

Schnable, J. 1966. The evolution and development of part of the northwest Florida coast. Ph.D. diss., Florida State University, Tallhassee. 231 p.

Scholl, D. W., F. C. St. Craighead, and M. Stuiver. 1969. Florida submergence curve revised: Its relation to sedimentation rates. Science 163:562–64.

Scholl, D. W. and M. Stuiver. 1967. Recent submergence of southern Florida: A comparison with adjacent coasts and other eustatic data. Bulletin of the Geological Society of America 78:437–54.

Schopf, J. M. 1959. Sargassoid microfossil assemblage from black shale of early Paleozoic age in Florida and Georgia (abstract). Bulletin of the Geological Society of America 70:1671–72.

Schroeder, M. C., H. Klein, and N. D. May. 1958. Biscayne aquifer of Dade and Broward counties, Florida. Florida Geological Survey Report of Investigations no. 17. 56 p.

Scoffin, T. P. 1987. An introduction to carbonate sediments and rocks. New York: Chapman and Hall. 274 p.

Scotese, C. R., L. M. Gahagan, and R. L. Larson. 1988. Plate tectonic reconstructions of the Cretaceous and Cenozoic ocean basins. Tectonophysics 155:27–48.

Scott, T. M. 1976. Faults of Florida, or whose fault is this fault? (abstract). Florida Scientist 39 (supp. 1):14.

———. 1981. The paleoextent of the Miocene Hawthorn Formation in peninsular Florida (abstract). Florida Scientist 44 (supp. 1):42.

———. 1983. The Hawthorn Formation of northeastern Florida, Part I: The geology of the Hawthorn Formation of northeastern Florida. Florida Geological Survey Report of Investigations no. 94, 1–35.

———. 1988a. The Cypresshead Formation in northern

peninsular Florida. In Heavy mineral mining in northeast Florida and an examination of the Hawthorne Formation and post-Hawthorne clastic sediments, edited by F. L. Pirkle and J. G. Reynolds, 70–72. Southeastern Geological Society 29th Annual Field Trip Guidebook.

———. 1988b. The lithostratigraphy of the Hawthorn Group (Miocene) of Florida. Florida Geological Survey Bulletin no. 59. 148 p.

———. 1992. Coastal plains stratigraphy: The diochotomy of biostratigraphy and lithostratigraphy—a philosophical approach to an old problem. In The Plio-Pleistocene stratigraphy and paleontology of southern Florida, edited by T. M. Scott and W. D. Allmon, 21–25. Florida Geological Survey Special Publication no. 36.

Scott, T. M., and W. D. Allmon, eds. 1992. The Plio-Pleistocene stratigraphy and paleontology of southern Florida. Florida Geological Survey Special Publication no. 36. 194 p.

Scott, T. M., R. W. Hoenstine, M. S. Knapp, E. Lane, G. M. Ogden, R. Deuerling, and H. E. Neel. 1980. The sand and gravel resources of Florida. Florida Geological Survey Report of Investigations no. 90. 41 p.

Scott, T. M., J. M. Lloyd, and G. Maddox, eds. 1991. Florida's ground water quality monitoring program—hydrogeological framework. Florida Geological Survey Special Publication no. 32. 97 p.

Secor, D. T. Jr., A. W. Snoke, K. W. Bramlett, O. Costello, and O. P. Kimbrell. 1986. Character of the Alleghenian orogeny in the southern Appalachians: Part I, Alleghanian deformation in the eastern Piedmont of South Carolina. Bulletin of the Geological Society of America 97:1345–53.

Sellards, E. H. 1916. Fossil vertebrates from Florida: A new Miocene fauna; new Pliocene species; the Pleistocene fauna. Florida Geological Survey 8th Annual Report, 77–119.

———. 1919. Geologic section across the Everglades. Florida Geological Survey 12th Annual Report, 67–76.

SEUSSN Contributors. 1992. Seismicity of the southeastern United States. Bulletin of the Southeastern U.S. Seismic Network 26. 87 p.

Shackleton, N. J., M. A. Hall, and A. Boersma. 1984. Oxygen and carbon isotope data from Leg 74 foraminifers. Initial Reports of the Deep Sea Drilling Project 74:599–612.

Shackleton, N. J., and J. P. Kennett. 1975. Paleotemperature history of the Cenozoic and the initiation of Antarctic glaciation: Oxygen and carbon isotope analyses in DSDP sites 277, 279, and 281. Initial Reports of the Deep Sea Drilling Project 29:743–56.

Shaler, N. S. 1890. The topography of Florida. Bulletin of the Museum of Comparative Zoology 16:139–58.

Sheldon, R. P. 1980. Episodicity of phosphate deposition and deep ocean circulation—a hypothesis. In Marine phosphorites—geochemistry, occurrence, genesis, edited by Y. K. Bentor, 239–47. Society of Economic Paleontologists and Mineralogists Special Publication no. 29.

Sheridan, R. E., H. T. Mullins, J. A. Austin Jr., M. M. Ballard, and J. W. Ladd. 1988. Geology and geophysics in the Bahamas. In The Atlantic continental margin, U.S., edited by R. E. Sheridan and J. A. Grow, 329–64. The geology of North America, vol. I-2. Geological Society of America.

Shier, D. E. 1969. Vermetid reefs and coastal development in the Ten Thousand Islands, southwest Florida. Bulletin of the Geological Society of America 80:485–508.

Shinn, E. A. 1963. Spur and groove formation on the Florida reef tract. Journal of Sedimentary Petrology 33:291–303.

———. 1964. Recent dolomite, Sugarloaf Key. In South Florida carbonate sediments, edited by R. N. Ginsburg, 62–67. Guidebook for Annual Field Trip of the Geological Society of America.

———. 1966. Coral growth rate, an environmental indicator. Journal of Paleontology 40:233–40.

———1980. Geologic history of Grecian Rocks, Key Largo Coral Reef Marine Sanctuary. Bulletin of Marine Science 30:646–56.

———. 1988. The geology of the Florida Keys. Oceanus 31:46–53.

Shinn, E. A., J. H. Hudson, D. M. Robbin and B. Lidz. 1981. Spurs and grooves revisited—construction versus erosion, Looe Key Reef, Florida. In Proceedings of the Fourth International Coral Reef Symposium, Manila, 1:475–83.

Shinn, E. A., B. Lidz, J. L. Kindinger, J. H. Hudson, and R. B. Halley. 1989a. Reefs of Florida and the Dry Tortugas, a guide to the modern carbonate environments of the Florida Keys and the Dry Tortugas. American Geophysical Union International Geological Congress Field Trip no. T176. 54 p.

Shinn, E. A., R. P. Steinen, B. H. Lidz, and R. K. Swart. 1989b. Whitings, a sedimentologic dilemma. Journal of Sedimentary Petrology 59:147–61.

Siesser, W. B. 1983. Paleogene calcareous nannoplankton biostratigraphy: Mississippi, Alabama, and Tennessee. Mississippi Bureau of Geology Bulletin no. 125. 61 p.

Sigsby, R. J. 1976. Paleoenvironmental analysis of the Big Escambia Creek–Jay–Blackjack field area, Florida. Transactions of the Gulf Coast Association of Geological Societies 26:258–78.

Simo, J. A., R. W. Scott, and J. P. Masse. 1993. Cretaceous carbonate platforms: An overview. In Cretaceous carbonate platforms, edited by J. A. Sims, R. W. Scott, and P. Masse, 1–14. American Association of Petroleum Geologists Memoir no. 56.

Simpson, G. G. 1930. Tertiary land mammals of Florida. Bulletin of the American Museum of Natural History 59:149–211.

———. 1932. Miocene land mammals from Florida. Florida Geological Survey Bulletin no. 10, 11–41.

———. 1942. The beginnings of vertebrate paleontology in

North America. Proceedings of the American Philosophical Society 86:130–88.

Simpson, R. W., T. G. Hildenbrand, R. H. Godson, and M. F. Kane. 1987. Digital colored Bouguer gravity, free-air gravity, station location, and terrain maps for the conterminous United States. U.S. Geological Survey Map no. GP-953-B.

Sinclair, W. C., and J. W. Stewart. 1985. Sinkhole type, development, and distribution in Florida. Florida Geological Survey Map Series no. 110. 1 sheet.

Sinclair, W. C., J. W. Stewart, R. L. Knutilla, A. E. Gilboy, and R. L. Miller. 1985. Types, features, and occurrence of sinkholes in the karst of west-central Florida. U.S. Geological Survey Water-Resources Investigations Report no. 85-4126. 81 p.

Smith, D. L. 1978. Earthquake seismograph station at the University of Florida. Florida Scientist 41:35.

———. 1982. Review of the tectonic history of the Florida basement. Tectonophysics 88:1–22.

———. 1983. Basement model for the panhandle of Florida. Transactions of the Gulf Coast Association Geological Societies 33:203–208.

Smith, D. L., and W. T. Dees. 1982. Heat flow in the Gulf Coastal Plain. Journal of Geophysical Research 87:7687–93.

Smith, D. L., W. T. Dees, and D. W. Harrelson. 1981a. Geothermal conditions and their implications for basement tectonics in the Gulf coastal margin. Transactions of the Gulf Coast Association of Geological Societies 31:181–90.

Smith, D. L., and W. R. Fuller. 1977. Terrestrial heat flow values in Florida and the thermal effects of the aquifer system. In The geothermal nature of the Florida Plateau, edited by D. L. Smith and G. M. Griffin, 91–130. Florida Geological Survey Special Publication no. 21.

Smith, D. L., and M. A. Graves. 1986. A magnetic anomaly map of Polk County, Florida. Florida Scientist 49:141–47.

Smith, D. L., R. G. Gregory, and J. W. Emhof. 1981b. Geothermal measurements in the southern Appalachian Mountains and southeastern coastal plains. American Journal of Science 281:282–98.

Smith, D. L., A. L. Heatherington, and P. A. Mueller. 1992. Configuration and geochemistry of Triassic basins in Florida: basement controls. Geological Society of America, Southeastern Section, Annual Meeting. Abstracts with Programs 24(4):65.

Smith, D. L. and A. F. Randazzo. 1975. Detection of subsurface solution cavities in Florida using electrical resistivity measurements. *Southeastern Geology* 16:227–40.

———. 1986. Evaluation of electrical resistivity methods in the investigation of Karstic features, El Cajon Dam site, Honduras. *Engineering Geology* 22: 217–30.

———. 1989. History of seismological activity in Florida: Evidence of a uniquely stable basement. In Second Symposium on Current Issues Related to Nuclear Power Plant Structures, Equipment, and Piping, with Emphasis on Resolution of Seismic Issues in Low-Seismicity Regions, edited by J. C. Stepp, 2-37–2-58. Seismicity Owners Group and Electric Power Research Institute.

Smith, D. L., and G. J. Taylor. 1977. A Bouguer gravity anomaly map of Alachua County, Florida. Florida Scientist 40:149–54.

Smith, E. A. 1881. On the geology of Florida. American Journal of Science 121:292–309.

Snyder, S. W., M. W. Evans, A. C. Hine, and J. S. Compton. 1989. Seismic expression of solution collapse features from the Florida Platform. In Engineering and environmental impacts of sinkholes and karst, edited by B. F. Beck, 281–98. Rotterdam: A. A. Balkema.

Snyder, S. W., A. C. Hine, and S. R. Riggs. 1990. The seismic stratigraphic record of shifting Gulf Stream flow paths in response to Miocene glacioeustacy: Implications for phosphogenesis along the North Carolina continental margin. In Neogene to modern phosphorites, edited by W. C. Burnett and S. R. Riggs, 396–423. Phosphate deposits of the world, vol. 3. Cambridge: Cambridge University Press.

Society of Exploration Geophysicists. 1982. Gravity anomaly map of the United States. Society of Exploration Geophysicists. 2 sheets.

Sohl, N. F., E. Martinez, P. Salmeron-Urena, and F. Soto-Jaramillo. 1991. Upper Cretaceous. In The Gulf of Mexico basin, edited by A. Salvador, 205–44. The geology of North America, vol. J. Geological Society of America.

Spencer, R. V. 1948. Titanium minerals in Trail Ridge. U.S. Bureau of Mines Report of Investigations no. 4208. 21 p.

Spencer, S. M. 1989. The industrial minerals industry directory of Florida. Florida Geological Survey Information Circular no. 105. 51 p.

Sprinkle, C. L. 1989. Geochemistry of the Floridan aquifer system in Florida and in parts of Georgia, South Carolina, and Alabama. U.S. Geological Survey Professional Paper no. 1403-I. 105 p.

Sproul, C. R. 1977. Spatial distribution of ground water temperature in south Florida. In The geothermal nature of the Floridan Plateau, edited by D. L. Smith and G. M. Griffin, 65–89. Florida Geological Survey Special Publication no. 21. 30 p.

Sproul, C. R., D. H. Bogess, and H. J. Woodard. 1972. Saline-water intrusion from deep artesian sources in the McGregor Isles area of Lee County, Florida. Florida Geological Survey Information Circular no. 7.

Stanley, S. M. 1966. Paleoecology and diagenesis of Key Largo Limestone, Florida. Bulletin of the American Association of Petroleum Geologists 50:1927–47.

———. 1986. Anatomy of a regional mass extinction: Plio-Pleistocene decimation of the western Atlantic bivalve fauna. Palaios 1:17–36.

Stanley, S. M., and L. D. Campbell. 1981. Neogene mass

extinction of western Atlantic molluscs. Nature 293:457–59.

Stapor, F. W. 1973. Coastal sand budgets and Holocene beach ridge plain development, northwest Florida. Ph.D. diss., Florida State University, Tallahassee. 221 p.

Stapor, F. W., and T. D. Matthews. 1980. C-14 chronology of Holocene barrier islands, Lee County, Florida: A preliminary report. In Shorelines past and present, edited by W. F. Tanner, 47–67. Tallahassee: Department of Geology, Florida State University.

Stapor, F. W., T. D. Matthews, and F. E. Lindfors-Kearns. 1988. Episodic barrier island growth in southwest Florida: A response to fluctuating Holocene sea level? Miami Geological Society Memoir no. 3, 149–202.

Stapor, F. W., and J. P. May. 1983. The cellular nature of littoral drift along the northeast Florida coast. Marine Geology 51:217–37.

Stauble, D. K., and S. L. DaCosta. 1987. Evaluation of backshore protection techniques. Coastal Zone 1987, Waterways Division. American Society of Civil Engineering, 3233–47.

Stauble, D. K., and D. V. McNeill. 1985. Coastal geology and the occurrence of beachrock: central Florida Atlantic coast. Field Guide for the Annual Meeting of the Geological Society of America, part 1. 27 p.

Steinen, R. P., R. B. Halley and S. L. Videlock. 1977. Holocene dolomite locality in Florida Bay (abstract). Bulletin of the American Association of Petroleum Geologists 61:833.

Stephens, J. C. 1974. Subsidence of organic soils in the Florida Everglades—a review and update. In Environments of south Florida: Present and past, edited by P. J. Gleason, 191–237. Miami Geological Society Memoir no. 2.

Stockman, K. W., R. N. Ginsburg, and E. A. Shinn. 1967. The production of lime mud by algae in south Florida. Journal of Sedimentary Petrology 37:633–48.

Stover, C. W. 1986. Seismicity map of the conterminous United States and adjacent areas, 1975–1984. U.S. Geological Survey Map no. GP-984.

Stowasser, W. F. 1977. Phosphate. U.S. Bureau of Mines Mineral Commodity Profile no. 2. 46 p.

———. 1985. Phosphate rock. In Mineral facts and problems, edited by A. W. Keoerr, 114–15. U.S. Bureau of Mines Bulletin no. 675.

———. 1992. Phosphate: supply tightness in the mid-1990s? Engineering and Mining Journal 193:42–44.

Strom, R. N., and S. B. Upchurch. 1985. Palygorskite distribution and silicification in the phosphatic sediments of central Florida. In Guidebook for the 8th International Field Workshop and Symposium (Southeastern United States), 118–26. International Geological Congress Project 156—Phosphorites.

Stuckey, J. L. 1965. North Carolina: Its geology and mineral resources. North Carolina Department of Conservation and Natural Resources. 550 p.

Sussko, R. J., and R. A. Davis Jr. 1992. Siliciclastic-to-carbonate transition on the inner shelf embayment, southwest Florida. Marine Geology 107:51–60.

Swart, P. W., D. Berler, D. McNeill, M. Guzikowski, S. A. Harrison, and E. Dedick. 1989. Interstitial water geochemistry and carbonate diagenesis in the subsurface of a Holocene mud island in Florida Bay. Bulletin of Marine Science 44:490–514.

Swartz, F. M. 1949. Muscle marks, hinge, and overlap features and classification of some Leperditiidae. Journal of Paleontology 23:306–27.

Swedjemark, G. A. 1986. Swedish limitation schemes to decrease Rn daughters in indoor air. Health Physics 51:569–78.

Sweeney, J. W. 1979. Florida stakes its claim in the uranium market. Mining Engineering 31:1324–25.

Swift, D. J. P., J. W. Kofoed, F. P. Saulsbury, and P. Sears. 1972. Holocene evolution of the shelf surface, central and southern Atlantic coast of North America. In Shelf sediment transport: Process and pattern, edited by D. J. P. Swift, D. B. Duane, and O. H. Pilkey, 499–574. Stroudsburg, Penn.: Dowden, Hutchinson, and Ross.

Swihart, T., J. Hand, D. Barker, L. Bell, J. Carnes, C. Cosper, R. Deuerling, W. Hinkley, R. Leins, E. Livingston, and D. York. 1984. Water quality. In Water resources atlas of Florida, edited by E. A. Fernald and D. J. Patton, 68–91. Tallahassee: Florida State University.

Swinchatt, J. P. 1965. Significance of constituent composition, texture, and skeletal breakdown in some Recent carbonate sediments. Journal of Sedimentary Petrology 35:71–90.

Szabo, B. J., and R. B. Halley. 1988. ^{230}Th/^{234}U ages of aragonitic corals from the Key Largo Limestone of south Florida. American Quaternary Association, 10th Biennial Meeting, Amherst, Mass. Program and Abstracts 154.

Taft, W. H., and J. W. Harbaugh. 1964. Modern carbonate sediments of southern Florida, Bahamas, and Espiritu Santo Island, Baja California: A comparison of their mineralogy and chemistry. Stanford University Geological Sciences Publication no. 8. 133 p.

Tampa Tribune. 5 July 1992. Wave swamps cars, boats at Daytona. Tampa Tribune, Florida/Metro section, p. 6.

Tanner, W. F. 1960a. Florida coastal classification. Transactions of the Gulf Coast Association of Geological Societies 10:259–66.

———. 1960b. Bases of coastal classification. Southeastern Geology 2:13–22.

———. 1965. Recent faulting in south Florida (abstract). Geological Society of America, Southeastern Section, Annual Meeting, Nashville, Tenn. Program Notes 36.

Tappan, H. 1968. Primary production, isotopes, extinctions, and the atmosphere. Palaeogeography, Palaeoclimatology, Palaeoecology 4:187–210.

Tauvers, P. R., and W. R. Muehlberger. 1987. Is the Brunswick magnetic anomaly really the Alleghenian suture? Tectonics 6:331–42.

Tedford, R. H., and M. E. Hunter. 1984. Miocene

marine–nonmarine correlations, Atlantic and Gulf coastal plains, North America. Palaeogeography, Palaeoclimatology, and Palaeoecology 47:129–51.

Tedford, R. H., M. F. Skinner, R. W. Fields, et al. 1987. Faunal succession and biochronology of the Arikareean through Hemphillian interval (late Oligocene through earliest Pliocene epochs) in North America. In Mammals of North America: Geochronology and biostratigraphy, edited by M. O. Woodburne, 151–210. Berkeley: University of California Press.

Templeton, D. A. 1992. Zirconium and hafnium. U.S. Bureau of Mines Annual Report. 20 p.

Thomas, M. C. 1968. Fossil vertebrates: Beach and bank collecting for amateurs. Privately published. 72 p.

Thomas, W. A., T. M. Chowns, D. L. Daniels, T. L. Neatherly, L. Glover, and R. J. Gleason. 1989. The subsurface Appalachians beneath the Atlantic and Gulf coastal plains. In The Appalachian–Ouachita orogen in the United States, edited by R. D. Hatcher, W. A. Thomas, and G. W. Viele, 445–58. The Geology of North America, vol. F-2. Geological Society of America.

Tootle, C. H. 1991. Reserve, estimated production listed for Florida's 22 oil fields. Oil and Gas Journal 89:84–87.

Toscano, M., and A. C. Hine. 1992. Quaternary sea-level lowstand features on the southern Florida Keys upper continental slope: Seismic, submersible, and outcrop data. Geological Society of America Annual Meeting. Abstracts with Programs 24(7):A173.

Tóth, J. 1962. A theory of groundwater motion in small drainage basins in central Alberta. Journal of Geophysical Research 67:4375–87.

———. 1963. A theoretical analysis of groundwater flow in small drainage basins. Journal of Geophysical Research 68:4795–4812.

Toulmin, L. D. 1977. Stratigraphic distribution of Paleocene and Eocene fossils in the eastern Gulf Coast region. Geological Survey of Alabama Monograph no. 13(1). 602 p.

Trapp, H. Jr., C. A. Pascale, and J. B. Foster. 1977. Water resources of Okaloosa County and adjacent areas. U.S. Geological Survey Water-Resources Investigations Report no. 77-9. 83 p.

Tucker, H. I., and D. Wilson. 1932. Some new or otherwise interesting fossils from the Florida Tertiary. Bulletins of American Paleontology 18(65):41–62.

———. 1933. A second contribution to the Neogene paleontology of south Florida. Bulletins of American Paleontology 18(66):65–82.

Tucker, M. E., and V. P. Wright. 1990. Carbonate sedimentology. Oxford: Blackwell Scientific. 482 p.

Tuomey, M. 1851. Notice of the geology of the Florida Keys, and of the southern coast of Florida. American Journal of Science 61:390–94.

Turmel, R. J., and R. G. Swanson. 1976. The development of Rodriguez Bank, a Holocene mudbank in the Florida reef tract. Journal of Sedimentary Petrology 46:497–518.

Twichell, D. C., H. J. Knebel, and D. W. Folger. 1977. Delaware River: Evidence for its former extension to Wilmington submarine canyon. Science 195:483–84.

Twichell, D. C., C. K. Paull, and L. M. Parson. 1991. Terraces on the Florida escarpment: Implications for erosional process. Geology 19:897–900.

Tyler, A. N., and W. L. Erwin. 1975. Sunoco-Felda field, Hendry and Collier counties, Florida. In North American gas and oil fields, edited by J. Braunstein, 267–99. American Association of Petroleum Geologists Memoir no. 24.

Tyrell, W. W., and R. W. Scott. 1989. Early Cretaceous shelf margins, Vernon Parish, Louisiana. In Atlas of seismic stratigraphy, edited by A. W. Bally, 11–17. American Association of Petroleum Geologists Studies in Geology no. 3(27).

Upchurch, S. B. 1986. Use of water chemistry to identify interaquifer mixing—Lee County, Florida. Unpublished report, South Florida Water Management District, West Palm Beach. 96 p.

———. 1989. Hydrology and chemistry of ground waters in the central Florida phosphate district. In Florida phosphate deposits, compiled by T. Scott and J. Cathcart, 39–52. Field Trip Guidebook no. T178. 28th International Geological Congress. American Geophysical Union.

———. 1992. Quality of waters in Florida's aquifers. In Florida ground water quality monitoring program—background hydrogeochemistry, edited by G. L. Maddox, J. M. Lloyd, T. M. Scott, S. B. Upchurch, and R. Copeland, 12–52, 66–347. Florida Geological Survey Special Publication no. 34.

Upchurch, S. B., and F. W. Lawrence. 1984. Impact of ground-water chemistry on sinkhole development along a retreating scarp. In Sinkholes: Their geology, engineering, and environmental impact, edited by B. F. Beck, 189–195. Rotterdam: A. A. Balkema.

Upchurch, S. B., and J. R. Littlefield Jr. 1988. Evaluation of data for sinkhole-development risk models. Environmental Geology and Water Science 12:135–40.

Upchurch, S. B., P. Jewell IV, and E. DeHaven. 1992. Stratigraphy of Indian 'mounds' in the Charlotte Harbor area, Florida: Sea-level rise and paleoenvironments. In Culture and environment in the domain of the Calusa, edited by W. L. Marquardt, 59–103. Institute of Archaeology and Paleoenvironmental Studies, University of Florida Monograph no. 1.

Upchurch, S. B., C. R. Oural, D. W. Foss, and H. R. Brooker. 1991. Radiochemistry of uranium-series isotopes in ground water. Institute of Phosphate Research Publication no. 05-022-092. 199 p.

U.S. Department of Energy. 1979. Peat prospectus. Washington, D.C.: Division of Fuel Processing, U.S. Department of Energy. 79 p.

U.S. Environmental Protection Agency. 1978a. Draft areawide environmental impact statement—central Florida phosphate industry. U.S. Environmental Protection

Agency, Region IV, Publication no. 904/9–78-006. 180 p.

———. 1978b. Central Florida phosphate industry areawide asessment program—water, vol. V: Water quantity and quality. U.S. Environmental Protection Agency, Region IV, Publication no. 904/9–78-006(E). 170 p.

U.S. Geological Survey. 1974. Seismic risk map of the U.S. U.S. Geological Survey Earthquake Information Bulletin, November–December, 1974.

———. 1986a. Florida ground water quality. In National Water Summary 1986—, 205–214. U.S. Geological Water-Supply Paper no. 2325.

———. 1986b. National water summary 1985—hydrologic events and surface-water resources. U.S. Geological Survey Water-Supply Paper no. 2300. 506 p.

———. 1989. Florida floods and droughts. In National Water Summary 1988–89, 231–238. U.S. Geological Water-Supply Paper no. 2375.

———. 1990. National water summary 1987—hydrologic events and water supply and use. U.S. Geological Survey Water-Supply Paper no. 2350. 553 p.

Vail, P. R. 1987. Part I: Seismic stratigraphy interpretation procedure. In Atlas of seismic stratigraphy, edited by A. W. Bally, 1–10. Studies in Geology no. 27(1). American Association of Petroleum Geologists.

Vail, P. R., R. M. Mitchum Jr., R. G. Todd, J. M. Widmier, S. Thompson, III, J. B. Sangree, J. N. Bubb, and W. G. Hatlelid. 1977. Seismic stratigraphy and global changes of sea level. American Association of Petroleum Geologists Memoir no. 26, 49–212.

Van der Voo, R. 1988. Paleozoic paleogeography of North America, Gondwana, and intervening displaced terranes: Comparisons of paleomagnetism with paleoclimatology and biogeographical patterns. Bulletin of the Geological Society of America 100:311–24.

Van der Voo, R., F. J. Mauk, and R. B. French. 1976. Permian–Triassic continental configurations and the origin of the Gulf of Mexico. Geology 4:177–80.

Van Kauwenbergh, S. J., J. B. Cathcart, and G. H. McClellan. 1990. Mineralogy and alteration of the phosphate deposits of Florida. U.S. Geological Survey Bulletin no. 1914. 46 p.

Van Siclen, D. C. 1984. Early opening of initially closed Gulf of Mexico and central North Atlantic. Transactions of the Gulf Coast Association of Geological Societies 34:265–75.

———. 1990. Evolution of pre-Jurassic basement beneath northern Gulf of Mexico coastal plain. Transactions of the Gulf Coast Association of Geological Societies 40:839–50.

Van Wagoner, J. C., H. W. Posamentier, R. M. Mitchum, P. R. Vail, J. F. Sarg, T. S. Loutit, and J. Hardenbol. 1988. An overview of the fundamentals of sequence stratigraphy and key definitions. In Sea level changes: An integrated approach, edited by C. K. Wilgus, B. S. Hastings, C. G. St. C. Kendall, H. W. Posamentier, C. A. ross, and J. C. Van Wagoner, 39–45. Society of Economic Paleontologists and Mineralogists Special Publication no. 42.

Vaughan, T. W. 1900. The Eocene and lower Oligocene coral faunas of the United States, with description of a few doubtfully Cretaceous species. U.S. Geological Survey Monograph no. 39. 263 p.

———. 1910. A contribution to the geologic history of the Florida Plateau. In Papers from the Marine Biological Laboratory at Tortugas, 99–185. Carnegie Institute of Washington Publication no. 133(4).

———. 1919. Fossil corals from Central America, Cuba, and Puerto Rico, with an account of the American Tertiary, Pleistocene, and Recent coral reefs, 189–524. U.S. National Museum Bulletin no. 103.

Veeh, H. H., W. C. Burnett, and A. Soutar. 1973. Contemporary phosphorites on the continental margin of Peru. Science 181:844–45.

Venkatakrishan, R., and S. J. Culver. 1988. Plate boundaries in Africa and their implications for Pangean continental fit. Geology 16:322–25.

Vermeij, G. J. 1978. Biogeography and adaptation: Patterns of marine life. Cambridge, Mass.: Harvard University Press. 332 p.

Vermeij, G. J., and E. J. Petuch. 1986. Differential extinction in tropical American molluscs: Endemism, architecture, and the Panama land bridge. Malacologia 27:29–41.

Vernon, R. O. 1951. Geology of Citrus and Levy counties, Florida. Florida Geological Survey Bulletin no. 33. 256 p.

———. 1970. The beneficial uses of zones of high transmissivities in the Florida subsurface for water storage and waste disposal. Florida Geological Survey Information Circular no. 70. 70 p.

Vick, H. K., J. E. T. Channell, and N. D. Opdyke. 1987. Ordovician docking of the Carolina slate belt: Paleomagnetic data. Tectonics 6:573–83.

Vincent, E., and W. H. Berger. 1985. Carbon dioxide and polar cooling in the Miocene: The Monterey hypothesis. In The carbon cycle and atmospheric CO_2: Natural variations, Archean to present, edited by E. T. Sundquist and W. S. Broecker, 455–68. American Geophysical Union Monograph no. 32.

Vogt, P. R. 1972. Evidence for global synchronism in mantle plume convection, and possible significance for geology. Nature 240:338–42.

———. 1979. Global magmatic episodes: New evidence and implications for the steady-state mid-ocean ridge. Geology 7:93–98.

———. 1989. Volcanogenic upwelling of anoxic, nutrient-rich water: A possible factor in carbonate-bank/reef demise and benthic faunal extinctions? Bulletin of the Geological Society of America 101:1225–45.

Vokes, E. H. 1963. Cenozoic Muricidae of the western Atlantic region. Part I—*Murex, sensu stricto*. Tulane Studies in Geology and Paleontology 1:93–123.

———. 1989. An overview of the Chipola Formation, northwestern Florida. Tulane Studies in Geology and Paleontology 22:13–24.

Voorhies, M. R. 1981. Dwarfing the St. Helens eruption: An ancient ashfall creates Pompeii of prehistoric animals. National Geographic 159:66–75.

Vukovich, F. M., and G. A. Maul. 1985. Cyclonic eddies in the eastern Gulf of Mexico. Journal of Physical Oceanography 15:105–17.

Waldrop, J. S., and D. Wilson. 1990. Late Cenozoic stratigraphy of the Sarasota area. In Plio-Pleistocene stratigraphy and paleontology of outh Florida, compiled by W. D. Allmon and T. M. Scott, Guidebook for the 31st Annual Field Excursion of the Southeastern Geological Society.

Walker, E. P. 1975. Mammals of the world. 3rd edition. Baltimore: Johns Hopkins University Press. 1500 p.

Walker, K. R., G. Shanmugam, and S. C. Ruppel. 1983. A model for carbonate to terrigenous clastic sequences. Bulletin of the Geological Society of America 94:700–712.

Walker, S. T. 1984. Sediment structures on the west Florida slope and eastern Mississippi cone: Distribution and geologic implications. Master's thesis, University of South Florida St. Petersburg. 148 p.

Waller, T. R. 1969. The evolution of the *Argopecten gibbus* stock (Mollusca: Bivalvia), with emphasis on the Tertiary and Quaternary species of eastern North America. Journal of Paleontology 43 (supp.):1–124.

Walper, J. L., F. H. Henk Jr., E. J. Loudon, and S. N. Raschilla. 1979. Sedimentation on a trailing plate margin: The northern Gulf of Mexico. Transactions of the Gulf Coast Association of Geological Societies 29:188–201.

Walton, T. L. 1973. Littoral drift computation along the coast of Florida by means of ship wave observations. Coastal and Oceanographic Engineering Laboratory, University of Florida, Report no. 15. 97 p.

———. 1976. Littoral drift estimates along the coastline of Florida. Florida Sea Grant Program Report no. 13. 41 p.

Wanless, H. R. 1982. Editorial: Sea level is rising—so what? Journal of Sedimentary Petrology 52:1051–54.

———. 1989. The inundation of our coastlines: Past, present, and future, with a focus on South Florida. Sea Frontiers 35:264–71.

Wanless, H. R., and M. G. Tagett. 1989. Origin, growth, and evolution of carbonate mudbanks in Florida Bay. Bulletin of Marine Science 44:454–89.

Ward, L. W., and B. W. Blackwelder. 1987. Late Pliocene and early Pleistocene Mollusca from the James City and Chowan River formations at the Lee Creek mine. In Geology and paleontology of the Lee Creek mine, North Carolina, edited by C. E. Ray, 2:113–283. Smithsonian Contributions to Paleobiology no. 61.

Weaver, C. E., and K. C. Beck. 1977. Miocene of the southeastern United States: A model for chemical sedimentation in a peri-marine environment. Sedimentary Geology 17:1–234.

Webb, S. D. 1969. The Burge and Minnechaduza Clarendonian mammalian faunas of north-central Nebraska. University of California Publications in Geological Sciences no. 78. 191 p.

———. 1973. Pliocene pronghorns of Florida. Journal of Mammalogy 54:203–21.

———. 1974a. Chronology of Florida Pleistocene mammals. In Pleistocene mammals of Florida, edited by S. D. Webb, 5–31. Gainesville: University Presses of Florida.

———. 1974b. Pleistocene lamas of Florida, with a brief review of the Lamini. In Pleistocene Mammals of Florida, edited by S. D. Webb, 170–213. Gainesville: University Presses of Florida.

———. 1981a. *Kyptoceras amatorum,* new genus and species from the Pliocene of Florida, the last protoceratid artiodactyl. Journal of Vertebrate Paleontology 1:357–65.

———. 1981b. The Thomas farm fossil site. Plaster Jacket 37. 33 p.

———. 1989. Osteology and relationships of *Thinobadistes segnis,* the first mylodont sloth in North America. Advances in Neotropical Mammalogy 469–532.

Webb, S. D., and D. B. Crissinger. 1983. Stratigraphy and vertebrate paleontology of the central and southern phosphate districts of Florida. In The central Florida phosphate district—field trip guidebook, 28–72. Geological Society of America, Southeastern Section, Annual Meeting.

Webb, S. D., and R. C. Hulbert Jr. 1986. Systematics and evolution of *Pseudhipparion* (Mammalia, Equidae) from the late Neogene of the Gulf Coastal Plain and the Great Plains. In Vertebrates, phylogeny, and philosophy, edited by K. M. Flanagan and J. A. Lillegraven, 237–272. University of Wyoming Contributions to Geology Special Paper no. 3.

Webb, S. D., B. J. MacFadden, and J. A. Baskin. 1981. Geology and paleontology of the Love bone bed from the late Miocene of Florida. American Journal of Science 281:513–44.

Webb, S. D., G. S. Morgan, R. C. Hulbert Jr., D. S. Jones, B. J. MacFadden, and P. A. Mueller. 1989. Geochronology of a rich Pleistocene vertebrate fauna, Leisey shell pit, Tampa Bay, Florida. Quaternary Research 32:96–110.

Weedman, S. D., T. M. Scott, L. E. Edwards, G. L. Brewster-Wingard, and J. C. Libarkin. 1995. Preliminary analysis of integrated stratigraphic data from the Phred #1 corehole, Indian River County, Florida. U.S. Geological Survey Open-File Report no. 95-824. 63 p.

Weisbord, N. E. 1971. Corals from the Chipola and Jackson Bluff formations of Florida. Florida Geological Survey Bulletin no. 53. 100 p.

———. 1972. *Creusia neogenica,* a new species of coral-inhabiting barnacle from Florida. Tulane Studies in Geology and Paleontology 10:59–64.

———. 1973. New and little-known corals from the Tampa Formation of Florida. Florida Geological Survey Bulletin no. 56. 146 p.

REFERENCES

———. 1974. Late Cenozoic corals of south Florida. Bulletins of American Paleontology 66:259–511.

———. 1981. Two new balanid barnacles (Cirripedia) from the Pinecrest Sand of Sarasota, Florida. Tulane Studies in Geology and Paleontology 16: 97–104.

Wells, J. W. 1934. Some fossil corals from the West Indies. Proceedings of the U.S. National Museum 83:71–110.

Werdelin, L. 1985. Small Pleistocene felines of North America. Journal of Vertebrate Paleontology 5:194–210.

White, D. H. Jr., and W. Schmidt. 1991. Florida minerals yearbook—1989. Washington, D.C.: U.S. Bureau of Mines. 10 p.

———. 1993. Florida minerals yearbook—1991. Washington, D.C.: U.S. Bureau of Mines. 10 p.

White, J. A. 1987. The Archaeolaginae (Mammalia, Lagomorpha) of North America, excluding *Archaeolagus* and *Panolax*. Journal of Vertebrate Paleontology 7:425–50.

White, W. A. 1958. Some geomorphic features of central peninsular Florida. Florida Geological Survey Bulletin no. 41. 92 p.

———. 1970. The geomorphology of the Florida Peninsula. Florida Geological Survey Bulletin no. 51. 164 p.

White, W. A., R. O. Vernon, and H. S. Puri. 1964. Proposed physiographic divisions. Unpublished manuscript referred to in Puri, H. S. and Vernon, R. O. Summary of the Geology of Florida and guidebook to the classic exposures. Florida Geological Survey Special Publication no. 5 (revised). 312 p.

White, W. B. 1988. Geomorphology and hydrology of karst terrains. New York: Oxford University Press. 464 p.

Whiting, H. 1839. Cursory remarks upon east Florida, in 1838. American Journal of Science 35:47–64.

Whittington, H. B. 1953. A new Ordovician trilobite from Florida. Breviora of Harvard Museum of Comparative Zoology 17. 6 p.

———. 1992. Trilobites. Woodbridge, U.K.: Boydell. 145 p.

Whittington, H. B., and C. P. Hughes. 1972. Ordovician geography and faunal provinces deduced from trilobite distribution. Transactions of the Royal Philosophical Society of London 263(B):235–78.

Wicker, R. A., and D. L. Smith. 1978. Re-evaluating the Florida basement. Transactions of the Gulf Coast Association of Geologic Societies 28:681–87.

Wigley, T .M. L., and L. N. Plummer. 1976. Mixing of carbonate waters. Geochimica et Cosmochimica Acta 42:1117–39.

Wilkins, K. T. 1984. Evolutionary trends in Florida Pleistocene pocket gophers (genus *Geomys*), with description of a new species. Journal of Vertebrate Paleontology 3:166–81.

———. 1985. Pocket gophers of the genus *Thomomys* (Rodentia: Geomyidae) from the Pleistocene of Florida. Proceedings of the Biological Society of Florida 98:761–67.

Williams, D. F. 1988. Evidence for and against sea-level changes from the stable isotopic record of the Cenozoic. In Sea level changes: An integrated approach, edited by C. K. Wilgus, B. S. Hastings, C. G. St. C. Kendall, H. W. Posamentier, C. A. Ross, and J. C. Van Wagoner, 31–36. Society of Economic Paleontologists and Mineralogists Special Publication no. 42.

Williams, H., and R. D. Hatcher Jr. 1982. Suspect terranes and accretionary history of the Appalchian orogen. Geology 10:530–36.

———. 1983. Appalachian suspect terranes. In Contributions to the tectonics and geophysics of mountain chains, edited by R. D. Hatcher Jr., H. Williams, and I. Zietz, 33–53. Geological Society of America Memoir no. 158.

Williams, K. E., D. Nicol, and A. F. Randazzo. 1977. The geology of the western part of Alachua County, Florida. Florida Geological Survey Report of Investigations no. 85. 98 p.

Williams, S. J., K. Dodd, and K. K. Gohn. 1990. Coasts in crisis. U.S. Geological Survey Circular no. 1075. 32 p.

Willis, J. W. 1984. The shallow structure of Tampa Bay. Master's thesis, University of South Florida, St. Petersburg. 85 p.

Wilson, J. L. 1975. Carbonate facies in geologic history. New York: Springer-Verlag. 471 p.

Wilson, J. T. 1966. Did the Atlantic close and then re-open? Nature 211:676–81.

Wilson, P. A., and H. H. Roberts. 1992. Carbonate-periplatform sedimentation by density flows: A mechanism for rapid off-bank and vertical transport of shallow-water fines. Geology 20:713–16.

Wingard, G. L., P. J. Sugarman, L. E. Edwards, et al. 1993. Biostratigraphy and chronostratigraphy of the area between Sarasota and Lake Okeechobee, southern Florida: An intergrated approach (abstract). Geological Society of America, Southeastern Section, Annual Meeting. Abstracts with Programs 25(4):78.

Wingard, G. L., S. D. Weedman, T. M. Scott, L. E. Edwards, and R. C. Green. 1994. Preliminary analysis of integrated stratigraphic data from the South Venice corehole, Sarasota County, Florida. U.S. Geological Survey Open-File Report no. 95-3. 129 p.

Winker, C. D., and R. T. Buffler. 1988. Paleogeographic evolution of early deep-water Gulf of Mexico and margins, Jurassic to Middle Cretaceous (Comanchean). Bulletin of the American Association of Petroleum Geologists 72:318–46.

Winker, C. D., and J. D. Howard. 1977. Correlation of tectonically deformed shorelines on the southern Atlantic Coastal Plain. Geology 5:123–27.

Winston, G. O. 1976. Florida's Ocala Uplift is not an uplift. Bulletin of the American Association of Petroleum Geologists 60:992–94.

———. 1978. Rebecca Shoal reef complex (Upper Cretaceous and Paleocene) in south Florida. Bulletin of the American Association of Petroleum Geologists 62:121–27.

———. 1989. Rebecca Shoal barrier reef complex of Gulfian

and Paleocene age—onshore and offshore Florida (abstract). Bulletin of the American Assocation of Petroleum Geologists 73(3):426.

———. 1991. Atlas of structural evolution and facies development of the Florida–Bahama Platform—Triassic through Paleocene. Coral Gables, Fla.: Miami Geological Society. 39 p.

———. 1992. The Middle Jurassic Louann Formation in panhandle Florida. Coral Gables, Fla.: Miami Geological Society. 11 p.

———. 1993a. A regional analysis of the Oligocene–Eocene–Paleocene section of the panhandle using vertical lithologic suites. The Paleogene of Florida. Coral Gables, Fla.: Miami Geological Society. 19 p.

———. 1993b. A regional analysis of the Oligocene–Eocene section of the peninsula using vertical lithologic stacks. The Paleogene of Florida. Coral Gables, Fla.: Miami Geological Society. 33 p.

Wolansky, R. M., and D. S. Spangler. 1975. A gravity interpretation of the Ocala National Forest area, Florida. Geological Society of America, Southeastern Section, Annual Meeting. Abstracts with Programs 7:550–51.

Wolansky, R. M., and T. H. Thompson. 1987. Relation between ground water and surface water in the Hillsborough River Basin, west-central Florida. U.S. Geological Survey Water-Resources Investigations Report no. 87-4010. 58 p.

Woodburne, M. O. 1987. Cenozoic mammals of North America: Geochronology and biostratigraphy. Berkeley: University of California Press. 336 p.

Woodring, W. P. 1959. Tertiary Caribbean molluscan faunal province. International Oceanic Congress, American Association for the Advancement of Science, 299–300.

———. 1966. The Panama land bridge as a sea barrier. Transactions of the American Philosophical Society 110:425–33.

———. 1982. Geology and paleontology of Canal Zone and adjoining parts of Panama. Description of Tertiary mollusks (pelecypods: Proeamussiidae to Cuspidariidae; additions to families covered in P 306-E; additions to gastropods; cephalopods). U.S. Geological Survey Professional Paper no. 306-F, 542–845.

Woodruff, F., and S. M. Savin. 1985. $\delta^{13}C$ values of Miocene Pacific benthic foraminifera: Correlations with sea level and biological productivity. Geology 13:119–22.

———. 1989. Miocene deepwater oceanography. Paleoceanography 4:87–140.

———. 1991. Mid-Miocene Monterey isotope stratigraphy in the deep sea: High-resolution correlations, paleoclimatic cycles, and sediment preservation. Paleoceanography 6:755–806.

Wu, S., A. W. Bally, and C. Cramez. 1990a. Allochthonous salt, structure, and stratigraphy of the north-eastern Gulf of Mexico. Part II: Structure. Marine and Petroleum Geology 7:334–70.

Wu, S., P. R. Vail, and C. Cramez. 1990b. Allochthonous salt, structure, and stratigraphy of the north-eastern Gulf of Mexico. Part I: Stratigraphy. Marine and Petroleum Geology 7:318–33.

Wyrick, G. G. 1960. The ground water resources of Volusia County, Florida. Florida Geological Survey Report of Investigations no. 22. 65 p.

Yarnell, K. L. 1980. Systematics of late Miocene Tappiridae (Mammalia, Perissodactyla) from Florida and Nebraska. Master's thesis, University of Florida, Gainesville. 126 p.

Yurewicz, D. A., T. B. Marler, K. A. Meyerhotz, and E. X. Siroky. 1993. Early Cretaceous carbonate platform, north rim of the Gulf of Mexico, Mississippi and Louisiana. In Cretaceous carbonate platforms, edited by J. A. Simo, R. W. Scott, and J. P. Masse, 81–96. American Association of Petroleum Geologists Memoir no. 56.

Zachos, L. G., and G. D. Shaak. 1978. Stratigraphic significance of the Tertiary echinoid *Eupatagus ingens* Zachos. Journal of Paleontology 52:921–27.

Zartman, R. E. 1988. Three decades of geochronologic studies in the New England Appalachians. Bulletin of the Geological Society of America 100:1168–80.

Zeitz, I. 1982. Composite magnetic anomaly map of the United States. U.S. Geological Survey Map no. GP-954-A.

Zellars, M. E., and J. M. Williams. 1978. Evaluation of the phosphate deposits of Florida using the minerals availability system. U.S. Bureau of Mines Final Report, Contract no. J0377000. 196 p.

Zen, E.-An. 1983. Exotic terranes in the New England Appalachians—limits, candidates, and ages: A speculative essay. In Contributions to the tectonics and geophysics of mountain chains, edited by R. D. Hatcher, H. Williams, and I. Zietz, 55–82. Geological Society of America Memoir no. 158.

Zullo, V. A. 1992. Review of the Plio-Pleistocene barnacle fauna (Cirripedia) of Florida. In Plio-Pleistocene stratigraphy and paleontology of southern Florida, edited by T. M. Scott and W. D. Allmon, 117–32. Florida Geological Survey Special Publication no. 36.

Zullo, V. A., and W. B. Harris. 1992. Sequence stratigraphy of the marine Pliocene and lower Pleistocene deposits in southwestern Florida: Preliminary assessment. In Plio-Pleistocene stratigraphy and paleontology of southern Florida, edited by T. M. Scott and W. D. Allmon, 27–40. Florida Geological Survey Special Publication no. 36.

Contributors

JOHN S. COMPTON is senior lecturer of geological sciences at the University of Cape Town, South Africa, and former associate professor of marine science at the University of South Florida, St. Petersburg.

RICHARD A. DAVIS JR. is distinguished research professor of geology at the University of South Florida, Tampa.

JAMES L. EADES is associate professor of geology at the University of Florida, Gainesville.

ROBERT B. HALLEY is a marine geologist for the U.S. Geological Survey Center for Coastal and Regional Marine Studies in St. Petersburg.

ANN L. HEATHERINGTON is an assistant in geochemistry in the Department of Geology, University of Florida, Gainesville.

ALBERT C. HINE is professor of marine science at the University of South Florida, St. Petersburg.

DOUGLAS S. JONES is curator of fossil invertebrates, professor of geology and zoology, and director of the Florida Museum of Natural History, University of Florida, Gainesville.

KENNETH M. LORD is an independent geologist and attorney.

BRUCE J. MACFADDEN is curator of fossil vertebrates and professor of geology, Latin American studies, and zoology in the Department of Natural Sciences, Florida Museum of Natural History, University of Florida, Gainesville.

GUERRY H. MCCLELLAN is professor of geology at the University of Florida, Gainesville.

JAMES A. MILLER is a retired hydrogeologist for the U.S. Geological Survey in Norcross, Georgia.

PAUL A. MUELLER is professor of geology and chair of the Department of Geology at the University of Florida, Gainesville.

ANTHONY F. RANDAZZO is professor of geology at the University of Florida, Gainesville.

WALTER SCHMIDT is state geologist and chief of the Florida Geological Survey, Tallahassee.

THOMAS M. SCOTT is assistant state geologist for geologic investigations for the, Florida Geological Survey, Tallahassee.

DOUGLAS L. SMITH is professor of geology at the University of Florida, Gainesville.

SAM B. UPCHURCH is a principal for ERM South, Inc., in Tampa and courtesy professor of geology at the University of South Florida, Tampa.

Index

Page numbers in italics refer to figures and tables; numbers in bold refer to plates.

Abaco event, 179
Abbott, R. T., 112
Academy of Natural Sciences (Philadelphia), 90
acidity, 222, 226, 235
acid rain, 153, 222
Acropora: cervicornis, 258; *palmata,* **12,** 192, 254–59
Aepycamelus major, 133
Africa, hyenas in, 128. *See also* North Africa; South African shelf; West Africa
Agassiz, L., 89
aggregation, 232
agriculture: fertilizer for, 142, 195, *241,* 241–42; groundwater contamination by, 231, *243,* 243–44; irrigation in, 72; water supply for, 73, 75–76, 80, 82, 86–87
Agriotherium, 126
Agueguexitean subprovince, *107*
air pollution: and cement production, 146; and clay production, 150; and crushed-stone production, 145; hazards of, 247; and phosphate industry, 143–44; regulations on, 153–54
Alabama: aquifers in, 71, 82; Cretaceous/Tertiary boundary in, 48; horsts in, 25; and Panhandle coastal system, 166; sedimentary rocks in, 27–28
Alabama Promontory, 22, *23*
Alachua clays, 120
Alachua County: basement in, 14; fossils from, *90,* 91; lake disappearance in, 11; phosphate from, 141; zircon studies in, 31. *See also* Love Bone Bed
Alacran Reef, 186
Alafia River, 231
Alapaha River, 11
Alaska Sink, **9**
Albian, Platform during, *41*
alcyonarians, 192
Aldicarb (Temik), 244
algae: coralline, 101, 185–87; dasyclad, 92; in Florida Middle Ground, 185–86; green, 187; in Keys, 255–56; in shelf-slope system, 185–87, 189; Tethyan affinities of, 94
Alleghenian Orogeny, *23,* 31, 34
Allen, J. H., 103
Allmon, W. D., 112, 115–16
alpacas, 133
Altamaha anomaly, 35, 91
Altschuler, Z. S., 142
Alum Bluff, 64

Alum Bluff Group, 60–61, 64–65, 103, 105
Alum Bluff Stage, 59
aluminum oxyhydroxide, 235
Amblypgus americanus, 97
Amebelodontidae (family), 135, *135*
ammonites *(Dufrenoya taxana),* 92, *93*
Amphechinus, 125
amphibolites, 33
Amphicyonidae (family), 126
Amphicyon longiramus, 126, *126*
Amphistegina pinarensis cosdeni, 97
Ampian, S. G., 150
Amusium ocalanum, 98
Anadara: aequalitas, 117; notoflorida, 110; rustica, 114
Anastasia Formation, **4;** age of, 116; and aquifer system, 77; and coastal system, 158; and crushed-stone production, 145; lithology of, 59–61, 66; and sand and gravel production, 147; and shelf-slope system, 190, 193
Anastasia Island, 158–59, *159*
Anchitherium, 131
Anclote Key, 163, *163, 164*
anhydrite, 183, 235
anionic flotation, 142
annelids *(Rotularia vernoni), 98,* 101, 185
anomalies, magnetic and gravity, 16–18, *17, 27,* 35–36, 91
anoxia, in oceans, 178
Antarctica, 103, 105, 197
Antillan Orogeny, 172
Antilles Archipelago, 217
Antilocapra, 134
APAC shell pit, *111,* 112, 120, 123
Apalachicola: and subdivision of aquifer system, *83*
Apalachicola Basin, 16, *16,* 39, 64
Apalachicola Delta: coastal system of, 155–56, 166–68; geomorphology of, *167;* and sedimentation, 158
Apalachicola Embayment, 15, 46, 58, 65. *See also* South Georgia Rift
Apalachicola River, 5, 9. *See also* Apalachicola Delta
Aphelops, 130, *130*
aplodontids, 128
Appalachian basement, 19
Appalachian Mountains: as accreted terranes, 33; lithotectonic belts of, *34;* rejuvenation of, 46; sediments from, 7–8, 61, 101, 151, 166, 172
Applegate, A. V., 92

301

Applin, E. R., 92
Applin, P. L., 14, 15, 92
Aptian Stage, 50, 175, *175*
aquifer systems: and carbonate sediments, 82, 85; composition of, *238, 239;* contamination of, 76, 79, 229, 231; definition of, 71; extent of, *71;* flow path in, 235, 237; geothermal effects from, 20–21; hydrogeology of, 69–73; and injection wells, 82, 84, 87, 246; maturation patterns in, 238–39, *240;* mercury and lead in, *243,* 243–44; microbes in, 237; minerals in, 235, *236,* 237; nitrates in, *243,* 243–44; permeability of, 72, 88, 232; pesticides in, *243,* 243–44; pH in, 222; properties of, 72; quality of, 73, 88; recharge of, 69–71, 88; residence times in, 237; and saltwater intrusion, 237–38; and sinkhole development, 223–24; and storage and retrieval systems, 247; synthetic organics in, *243,* 243–44; water levels in, 70–71, *71;* withdrawal of freshwater from, *72, 72–73, 73,* 88. See also Biscayne aquifer system; Floridan aquifer system; intermediate aquifer system; sand and gravel aquifer system; surficial aquifer system
aragonite, *49;* in aquifer systems, 235; dissolution of, 207; inversion of, 51; in lime mud, 256; in ramp development, 180; in shelf-slope system, 187. See also corals
Arca: propearesta, 108; *wagneriana,* 114
Arcadia Formation: age of, 198; and aquifer system, 80, 82–83; composition of, 60–61; invertebrate fossils in, 101, 103, *104;* limestone and dolomite in, 198; reef corals in, 101
archaeocetes, *120,* 121, 136
Archaeohippus, 131; *blackbergi,* 131
Archaeolithothamnium parisiense, 94
archaic toothed whales, 136, *136*
Archer, 120, 122. See also Love Bone Bed
Arctonasua floridana, 126
Arden, D. D., 16
Arikareean, mammal faunas of, 122. See also Thomas Farm
Arkansas, fossils from, 93
armadillos, 123–25, *124*
Arrendondo site, mammal fossils at, 123
arsenic, 244
arthropod megafossils, 91
Arthur, J. D., 14, 28, 29
artiodactyls, 121–23, 132–35, *134*
Asterocyclina, 101
asteroids, *96*
Astrohippus stocki, 132
astronomical cycles, 45. See also Milankovitch cycles
Atkinson Formation, 50–51, 93
Atlantic Basin, 170
Atlantic Bight, 198
Atlantic Coastal system: Lowlands, *10,* 156; Plain, 9, 48, 111; Ridge, 66–67
Atlantic Ocean: and aquifer system, 76; and coastal systems, 156, 159–60; faunal provinces in, *103,* 115, *115;* formation of, 24–25, 29; interchanges with Gulf of Mexico, 258; molluscan diversity changes in, 112; and phosphorite formation, 215. See also Central American Isthmus

atmosphere: CO_2 in, 197–98, 221; radioactivity in, 248–49. See also air pollution; climate
Atrocoenia incrustans, 99
Aturia alabamensis, 98
Austin, J. A., Jr., 36
Austin beds, 93
Australia: Great Barrier Reef of, *170;* phosphorite formation in, 209
Australian National University, 31
Avalon-Carolina terrane, 22, *23,* 24, 33
Avalon complex, 15
Avalonian Delaware Piedmont, 34
Avalonian terrane, 33–35, *34*
Avon Park Formation: and aquifer systems, 82–84; and crushed-stone production, 145; description of, 50, 61, 94; dolomite in, *55;* invertebrate fossils in, *49,* 94–96, *95–96,* 99; oolitic carbonate rock in, **2,** 49, *49*

Babcock deep core site: diatoms and nannofossils from, 208; francolite from, 203–4, 206–8; location of, *199;* other sites compared to, 211–13, *212;* and phosphorite formation, 206, 210–13, *211;* samples from, *201, 202, 209;* $^{87}Sr/^{86}Sr$ ages from, 208
Back, W. B., 222–23, 238
bacterial mats, 203–5, *205*
Badiozomani, K., 223
Bahamas, mineral industries in, 154
Bahamas Bank, 94, 172, 176
Bahamas Fracture Zone (BFZ), 32, 174. See also Jay Fault
Bahamas Platform: development of, 25, 61, 169–76, *170;* effects of, 159; Florida Platform separation from, 172–74, *173;* and magnetic anomalies, 17; progradation of, 179; rudistid reefs along, 92; tectonic history of, 21, 24, 217
Baker County, heavy-mineral sands in, 152
Balaenoptera, 137
baleen whales, 137
Ball, M. M., 16, 255
Ballast Point, fossils from, 103
Banks, J. E., 13, 15
Banks, R. S., 64
Barbourofelis, 127, *127*
barnacles, 103, *110,* 185, 191
Barnett, R. S., 13–15
Barremian, Platform during, *41*
barrier islands: breaching of, 163; development of, 66, 158–59, 164, *164;* east-coast barrier system, 157–60, *159, 160;* in Holocene, *164,* 164–65; in marsh coastal system, 166–67; and sand ridge formation, 191; west-central barrier system, *162,* 162–65, *163.* See also inlets; *names of specific islands*
Barstovian, mammal faunas of, 122, 131, 135
basalts: and basement configuration, 14, 27–28, 30–31; in Carolina terranes, 28–29; island arc type of, 32; redbed type of, 13, 15, 28; systematics of, *29, 30;* tholeiite type of, 14–15, 28
basement: boundaries of, 171–75; components of, *14,* 15–16, *27;* description of, 13–15, 25–28; fossils from, 90–91;

geochemical characteristics of rocks in, 28–33; geothermal state of, 20–21; gravity and magnetic anomalies in, 16–18, *17;* origin of, 21–25; and seismicity, 16, 19–20, 25; structure of, 16, *16, 18,* 91–92, *171;* tectonic history of, 21–25, *22;* terranes in, 33–36. *See also* Gondwana basement

Basilosaurus, 121, 136, *136*

Bass, M. N., 14, 15, 32, 33

bats, 121, 125

Bay County, drainage system of, 5

bays, effects of, 163

beaches, 66, 161, 166–67

beach-ridge systems, 164, *165, 167, 168*

Bear Bluff Formation (N.C. and S.C.), 112

bear-dogs, 122, 126, *126*

bears, 126

beavers, 129

Beck, B. F., 223, 232–33

Beck, K. C., 149

Beggiatoa (bacterium), 203, 205, *205*

Belair slate belt, 34, *34*

bellerophontacean gastropod, 91

Belleview, gasoline-storage tanks in, 246

beneficiation, 142–43, 242

benthic foraminifers: age of, 206–8, *209;* in Avon Park Formation, *95;* composition of, 206–8; from core material, *93;* in limestones, 96, *97,* 101, *102;* and phosphogenesis, 209, 211, *212, 215;* in Platform margins, 178, 185, 189, 193; in Trinity rocks, 93

Berden, J. M., 14

Bering land bridge, 127, 134

Bermont Formation: age of, 116; description of, 120; and faunal-diversity issues, 115–16; fossils from, 116, *117;* structure of, 61, 66

Berriasian, Platform during, *41*

Berta, A., 127, 128

Big Badlands (S.D.), 133

Big Bend Coast (area), 4, 73, 156, 165–66, 168

Big Cypress Swamp, 161, 242

Big Pine Key, 253–54

biogenic carbonates, 94

bioherms, 178

biostratigraphy, age determined by, 197, 207

Birimian episode, 31–32

Biscayne aquifer system: components of, 71–72; contamination of, 79–80, 243, *243,* 246; description of, 76–80; extent of, *71;* freshwater/saltwater interface in, *80;* maintenance of, 246–47; permeability of, 76, 78–79; stratigraphic units of, 77, *77;* water movement in, 77–78, *78;* water quality of, 73; wells in, 77–78, *79;* withdrawal of freshwater from, 72, 80, *80*

Biscayne Bay, 76

bismuth, 248

bison, 134–35

bivalves: in Bermont Formation, *117;* boreholes with, 99, *99;* in Caloosahatchee Formation, 112, *114;* in Chipola Formation, *106;* fragments of, 93, *93;* in Jackson Bluff Formation, *108;* in limestones, 96, *102;* pectinid type of, 93; and phosphorite formation, 200; in Pinecrest Beds, 112; in shelf-slope system, 191–92; in Tamiami Formation, *110;* in Tampa Member, 103, *104*

Blackjack field, oil and gas from, 141

Blake, N. J., 186

Blake-Bahamas Escarpment, 172

Blake Plateau: chronostratigraphy of, *47,* deposition on, 172; phosphorite in, 198, 213; and sea-level rise, 177; structure of, 172, 181

Blarina, 125

bloating clay, 147–48, 150

Bloomberg, D., 228

BMA (Brunswick Magnetic Anomaly), 17, *27,* 35–36, 91

bobcats, 127–28

Boca Raton, location of Atlantic Coastal Ridge, 67, water supply for, 76

Bohemia, fossils from, 91

Bond, P., 9

Bone Valley Member: description of, 119–20, *120;* lithostratigraphic framework, 60, mammal fossils from, 122, 125–27, *128,* 130–34, *132, 134,* 136, *136;* phosphate from, 142

bony fish, 122

borophagine canids, 125

botany, 8

Boulder Zone: and aquifer system, *83,* 84–85, 223; dolomitic boulder from, *52;* formation of, 52–53; injection wells in, 87

boundstones, 49, 253

Bov Basin (Senegal and Guinea), and Platform origin, 22, 31–32

Bovidae (family), 134–35

Bowden Formation (Jamaica), 107

box canyons, in Escarpment, 182, *182,* 184, *184*

Boyle, J. R., 139

brachiopods *(Lingulepis floridaensis),* 31, *90,* 91, 93

Bradenton, cement production in, 146

Bradford County, heavy-mineral sands in, 151–52

Brandon, nitrate contamination in, 231

Brasiliano episode, 32

Brevard County: aquifer system in, 77, 84; rivers in, 12; water-yielding beds in, 80; wells in, 246

Brevard Platform, 58–59, 198

Bridgeboro Limestone, 101

brines, seepage of, 182–83, *183. See also* saltwater

Broecker, W. S., 254

Brooks, G. R., 185, 189

Brooks, H. K., 9, 12

Brooks Sink, 105

Brooksville, mammal fossils near, 122

Brooksville Ridge: formation of, 8, 66; and marsh coastal system, 165; naming of, 9; sinkholes on, 5; slope stability of, 233; and topographic inversion, 11, *11*

Broward County: aquifer system in, 76, 78; crushed-stone production in, 144; and flood-control system, *240;* gasoline-storage tanks in, 246;

Broward County (*continued*)
   limestone production in, 144–45; water supply for, 72–73
Brown, R. C., 120
Bruce Creek Limestone, 60
Brule Formation (S.D.), 121
Brunswick Magnetic Anomaly (BMA), 17, *27,* 35–36
Brunswick-Altamaha Magnetic Anomaly, 91
Bryant, J. D., 129
bryozoans: in Atkinson Formation, 93; in Bermont Formation, *117;* in Keys, 255; in Miami Limestone, 252–53; in Ocala Limestone, 96; in shelf-slope system, 185, 191, 193; in Suwannee Limestone, 101
Buckingham Limestone, 61, 111–12
Buckinghamian subprovince, 107, *107*
Buda local fauna, 122, 125, 127, 131, 132
Buffler, R. T., 48
Buliminella elegantissima, 208
Bunces Pass, **7,** 164
burrowing rodents (extinct), 128, *128*
Busycon: rapum, 113; *sicyoides,* 106

Caladesi Island, 163, *163*
calcite: age of, 208, *209;* in aquifer systems, 235; aragonite inversion to, 51; and dissolution, 222–23, 235; and phosphorite formation, 199–200, 207; in ramp development, 180; in shelf-slope system, 187
calcium-magnesium bicarbonate water, 237
calcium montmorillonites, 149
Calhoun County, drainage system of, 5, fossiliferous unit in, 105
California, phosphorite deposits in, 195, 213, 215, *215*
Calliostoma philantropus pontoni, 108
Callovian, Platform during, *40*
Caloosahatchee beds, 112; marl, 112
Caloosahatchee Formation: age of, 116; and aquifer system, 74; classification of, 112; and faunal-diversity issues, 115–16; formation of, 59, 61, 65–66; invertebrate fossils in, 105, 112, *113, 114,* 115–16; mammal fossils in, 120; and sand and gravel production, 147
Caloosahatchee River, 112, 239
Caloosahatchian Province, 103, *103,* 105, 107
Calophos wilsoni, 108, *109*
calyptogenid clams, 183
Camelidae (family), 122, 123, 133–34
camels, 122, 133–34
Campanian, Platform during, *41*
Campbell, K. M., 139
Campbell, L., 112
Campbell, L. D., 107
Campbell, R. B., 14, 217
Campeche Bank (Yucatan Peninsula), 94, 186
canals: construction of, 77, *78;* contaminants in, 79–80; drainage type of, 247; effects of, 78–79, *79;* salinity dams on, 237–38
Cancellaria propevenusta, 109

Canidae (family), 125–26
Canis, 126; *dirus,* 126
canyons, formation of, 182, *182,* 184, *184,* 193
Cape Canaveral: and Cenozoic sedimentation, 173; coastal system of, 155–56, 158, 160; seismic study near, *174;* and shelf-slope system, 191, *191*
Cape Orlando, 173
Cape Province (South Africa), phosphorite in, 204
cape-retreat massif, 191, *191*
Cape Romano, 4, *160,* 160–63
Cape Sable, 160–62, *161,* 256
Cape St. George, 167
Cape San Blas, 167–68, *168*
Capromeryx, 134
Captiva Island, **6,** 163
Captiva Pass, 221
capybaras, 129–30
carbon: and climate, 197–98; and global cooling events, 214–15; and phosphogenesis, 209–13; phosphorous linked to, 195
carbonate-fluorapatite, 235, 247–48
carbonate sediments: active areas for, 170–71; and aquifer systems, 82, 85; definition of, 49; deposition of, 39, 44–46, *46,* 57–58, 60–67; dissolution of, 51–53, 62; and dolomitization, 53–56; and eastern Platform boundary, 172; global distribution of, 44, *45;* lithification of, 53; permeability of, 72; sequence record of, 47–49; in shelf-slope system, 186–87, *187,* 189–93; stratigraphic units from, 50–51
Cardium: chipolanum, 106; *dalli,* 114
Card Sound Dolomite, 50
Caribbean Basin, 101
Caribbean Current, 180
Caribbean Plate, 56, 172, 217
Caribbean Sea: faunal age in, 105, 107; faunal provinces in, 115, *115;* and isthmus development, 180; reef composition in, 96; and sea-level rise, 177
Caricella obsoleta, 100
Carlsbad Caverns (N.M.), 53
carnivores, 121–23, 125–28, 136. *See also names of specific carnivores*
Carolinapecten eboreus, 110
Carolina slate belt, 34, *34*
Carolina terrane, 28–29, 33–35
Carpsfort Reef, coral species of, *253*
Carriacouan subprovince, *107*
Carroll, D., 14
Casey Key, 163
Cassis floridensis, 109
Castoridae (family), 129
Cathcart, J. B., 142
cats (true), 127–28
cattle, 134–35
Causaras, C. R., 77
caves/caverns: age of, 223; formation of, *51,* 51–53. *See also* sinkholes

Cayo Costa Island, **8,** 164, *165*
Cedar Keys Formation: and aquifer system, 82–84; Boulder Zone in, 52; formation of, 50, 94
cement: components of, 145–46, 150–51; industry for, 139, 145–47, *146, 147*
cementation: description of, 53, *53;* francolite in, 200, 202; geochemical conditions in, 53; reefs formed by, 193, *193;* and shelf-slope system, 189
Cenomanian, ramp development in, 177–78
Cenozoic: controversies over stratigraphy of, 107, 111; deposition in, 39, 57–59, 173; mammals in, 119–37; phosphogenesis in, 214–15, *215;* sea level in, *46, 47,* 61; southern Platform boundary in, 175, *176;* tectonic evidence from, 25; transition to, 94. See also Quaternary; Tertiary
Central America, submersion of, 103
Central American Isthmus: and biotic interchange, 123–25; development of, 115, 180, 258; and sediment cycles, *188*
Central and Southern Florida Flood Control District, 220, 239
Central Florida Phosphate District, *199*
Central Florida Platform, 198–99
Central Highlands, *10,* 147
cephalopods, 91
Cepolis crusta, 104
Cerithium: caloosaense, 113; *praecursor,* 104
Cervidae (family), 134
cetaceans, 122, 136–37
chain of bays, 161
Chalichotheridae (family), 132
Charleston (S.C.), earthquake at, 217
Charlotte County, aquifer system in, 80–81, 84
Charlotte Harbor: clay wastes spilled in, 242; and estuarine system, 184, hurricane activity in, 221; and shelf-slope system, 193; structure of, *185;* tides in, 157
Charlotte slate belt, 34
Charlton Clay Member, 149
Chasmaporthetes, 128
Chatahoochee Embayment, 46
Chatham Rise, 195
Chattahoochee anticline, 62
Chattahoochee Arch, 58–59, 65
Chattahoochee Formation, 59–60, 145
Chattahoochee River, 9
cheetahs, 127–28
Cheetham, A. H., 99
Chen, C. S., 99
chert, 200
Chesapecten septenarius, 110
Chevron Corporation, 140
Chicoreus chipolanus, 106
Chione: burnsii, 106; cancellata, 117; craspedonia, 102; ulocyma, 108
chipmunks, 129
Chipola Formation, 59–60, 103, 105, *106*

Chipola River valley, 9
chiropterans, 125
chitinozoans, 91
chloride: in aquifer systems, 76, 82; concentration of, 238, *238;* from septic tanks, *244,* 245
Choctawhatchee Bay, 167
Choctawhatchee Formation, 60
Choctawhatchee Stage, 59
Choffatella, 93; *decipiens, 93*
Chowan River Formation (Va. and N.C.), 112
Chowns, T. M., 14, 31, 34
cingulates, 124
Citronelle Formation: and aquifer system, 75; clay in, 150; formation of, 59–60, 65; and sand and gravel production, 147
Citrus County: dolomite production in, 144; limestones in, *165;* phosphate from, 141; rock formation in, 94
Claiborne Group, 51, 94
clams, calyptogenid, 183
Clapp, F. G., xvii, 8, 103
Clarendonian, mammal faunas of, 122–23, 131–32
Clavagella sp., *100*
Clay County: bloating clay in, 148–49; heavy-mineral sands in, 151–52; sand and gravel production in, 147; water-yielding beds in, 80
clay minerals: and groundwater quality, 235; in phosphorite, 199–200, 202, *202,* 205. See also kaolin, palygorskite; sepiolite
clays: and aquifer systems, 73, 75, 80, 235; in cement production, 145–46, 150; characteristics of, 148–50; in Hawthorn Group, 198; industry for, 139, *148,* 148–51, *149, 150;* and liners, 153, and phosphate industry, 144; in ramp development, 179; in shelf-slope system, 189, 190; shrinking and swelling of, 231–33, *233;* and sinkhole development, 223–25; as slimes in phosphate production, 143–44, *241,* 242; uses of, 150, 153–54. See also clay minerals; fullers earth clay
Clearwater: pumping center at, 86; water supply for, 82
climate: in depositional dynamics, 44; and diagenesis, 200; in Eocene/Oligocene boundary, 101; and faunal diversity, 115; and global cooling, 214–15; and mollusks, 105; and organic carbon, 197–98; and phosphogenesis, 214–15; and ramp development, 179; variations in, 56. See also atmosphere; greenhouse, hurricanes; icehouse, precipitation; temperature; tropical storms; waves; winds
clinoptilolite, 200
Clypeaster: rogersi, 101, 102; rosaceous, 117
Coastal Lowlands, 5, *6, 10*
Coastal Plains, 27, *199*
coastal systems: components of, 2–4; east-coast barrier, 157–60, *159, 160;* mangrove (southwest), **5,** 157, *160,* 160–62, *161, 162;* marsh coast of Big Bend area, 156, 165–66; mixed-energy type of, 158–59, 163, *163;* overview of, 155, *156;* of Panhandle, 156–57, 166–68, *167;* processes of, 155–57, *156, 157;* and sea levels, *157,* 157–58, 165; west-central barrier, *162,* 162–65, *163*

Cocoa Beach, aquifer system in, 84
COCORP (Consortium for Continental Reflection Profiling), 16, 35
Cody Escarpment, 7, 233
Coffee Mill Hammock Formation, 112, 115–16
Coharie Terrace, 2
Coleman site, 123
Collier County: aquifer system in, 74, 81; crushed-stone production in, 145; limestone production in, 144; marl production in, 144; oil in, 139, 242
Colpocoryphe exsul, 91
Colpophyllia, 255
Colton, R. C., 11
Columbia County: basement in, 14–15; hydrogeology of, *84;* phosphate from, 141
Comprehensive Environmental Response Compensation and Liability Act (1980), *245, 246.* See also superfund sites
Compton, J. S., 64, 210, 214
concrete aggregates, 145
Conepatus, 127
Coniacian-Santonian, Platform during, *41*
Coniglio, M., 253–54, 257
conodonts (*Drepanodus* sp.), 91
Conrad, T. A., 89–90, 103
Consortium for Continental Reflection Profiling (COCORP), 16, 35
construction: and cement production, 146; of radon-resistant structures, 249; sand, 148; and sinkholes, 226–29; slope-stability problems in, 233
contamination: and flow path length, 237; of groundwater, 229, 231, 242–49, *243;* hazards of, 243–46; sinkholes as pathways for, 229; of soils, 154
continental margin: and phosphorite formation, 213, *215;* reconstruction of, *196*
Contributions to the Tertiary Fauna of Florida (Dall), 90
conularids, 91
Conus: adversarius, 113; *designatus,* 104; *druidi, 109*
Cook, D. J., 223
Cooke, C. W., xvii, 8, 59, 147
Coosawhatchie Formation: clay in, 149; formation of, 60; invertebrate fossils in, 105; phosphorite in, 198
Copemys, 129
coquina, 66, 145
coral framestones, 187
coralline algae, 101, 185–87
corals: absence of, 96; age of, 105, 254; in Bermont Formation, *117;* boreholes in, 99, *99;* in Caloosahatchee Formation, *114;* growth of, 258–59; in Jackson Bluff Formation, 107, *108;* in Keys, 251–52, *253,* 254–59; living community of, **12;** in Oligocene, 101; origin of, 49; in Pinecrest Beds, 107; scleractinian, 101; in shelf-slope system, 187, 192–93; significance of, 254; stoloniferous, 183; in Tamiami Formation, *110;* in Tampa Member, 103, *104*
Cormohipparion, 132
Correlation of Stratigraphic Units of North America (COSUNA), 49–50, 60
Corso, W., 176, 181–82

Coskinolina, 94
Coskinoloides texanus, 93
COSUNA (Correlation of Stratigraphic Units of North America), 49–50, 60
Cotton Valley Group, 51, 92
Covington, J. M., 65
Cowart, J. B., 21, 53
Coya Granite (Senegal), and Platform origin, 22
crabs: galatheid, 183; *Ocalina floridana,* 98
Cramer, F. H., 91
crandallite, 142, 196, 206. See also phosphorite
Crane Key, 221
creosote, contamination by, 76
Cretaceous: carbonate-platform faunas in, 92–94; climate in, 56; deposition in, 39, 44, *45;* drowning in, 177–78; Escarpment in, 181–82, *182;* karst topography formed in, 223; Platform boundaries in, *41–43,* 171–76, *175, 176;* sea levels in, *46, 47,* 48, 172; sequence record of, 13, 27, 48, 92–93; stratigraphic units from, 50–51
Cretaceous/Tertiary boundary, 48, 94
Cribobulimina (Valvulina) cushmani, 95
crinoids, 91, 93
Crissinger, D. B., 122
Cronin, T. M., 115
crushed stone, 139, *144,* 144–45
crust, 17–18, 21. See also tectonics
Crystal Lake Quadrangle, drainage system of, *5*
Crystal River, 165, *166*
Cuba: formation of, 25; phosphorite deposits in, 195, 213; and southern Platform boundary, 175, *175*
Cuba-Bahamas Orogeny, 175
Cuban Orogeny, 172
Cuneolina, 93
Cutler site, 123
cyclic sedimentation, 44, *46,* 64
Cyclothyris, 93
Cymodocea, 94
Cynelos, 126
Cypraea cervus, 117
cypress dome, in Dade County, 4, *4;* over sinkhole, 226, *226*
Cypresshead Formation, 60–61, 65

Dade County: aquifer system in, 76–78; cement production in, 146; crushed-stone production in, 144; drainage system of, *4;* formations under, 66; gasoline-storage tanks in, 246; hydrogeology of, *84;* limestone production in, 144–45; oil extraction in, 242; water supply for, 72–73
Dall, W. H., 59, 90, 103, 105
Dallmeyer, R. D., 13, 15, 31–34
Dalziel, I. W. D., 36
dasyclad algae, 92
Dasypus: bellus, 124; *novemcinctus,* 124
Davis, J. H., 8, 152–53
Davis, R. A., Jr., 221
Daytona: beach at, 159; earthquake at, *19*
Daytona Beach: pumping center at, 86; rogue wave at, 218

DDT, contamination by, 244
Deep Sea Drilling Project Site 588: other sites compared to, 211–13, *212;* and phosphorite age, *206,* 206–8
deer, 131, 134
Dees, W. T., 20
Dell Limerock pit, *120*
deltas: burial by, 170; and coastal systems, 155–56, 166–68; geomorphology of, 158–59, *167;* wave refraction across, 163
deposition: active areas of, 170–71; and barrier island formation, 163–64; in basement, 16, 22; and coastal systems, 157–61, *159, 160,* 166; cycles of, 62, 64, *199;* and diagenetic processes, 51–56; dynamics of, *42–44,* 44–48, 156, 177; of evaporite rocks, 39, 44–45, 47, 92, 94; history of, 39, 57–59, 173; from hurricanes, 221; and phosphorite, 210; and shelf-slope systems, 184–89; structural framework for, 58–59. *See also* sedimentary rocks; sediments; *names of specific rock formations*
Dermomurex vaughani, 106
desalination, 247
Desmodus, 125
Desmophyllum willcoxi, 104
DeSoto Canyon, 176, 181
DeSoto County, phosphate from, 142
DeSoto site, 123
Destin, 167
Destin Dome, 140
Devils Millhopper State Geological Site, sinkholes at, 7, 224
dewatering, in phosphate mining, 144, 154, 241–42, 246
diabase, 14, 28
diagenesis: and climate, 200; effects of, 64, 94; and phosphorite origin, 202–6, *205;* processes of, 49, 51–56. *See also* cementation; dissolution; recrystallization
diagenetic minerals, 200, 202. *See also names of specific minerals*
diammonium phosphate, 142
diatoms, 105, 200, *201,* 208
Dicathais handgenae, 109
Dichocoenia caloosahatcheensis, 114
Dictyoconus, 93, 94; *americanus, 95; cookei, 102; floridanus, 93*
Didelphis virginiana, 124
Didymograptus: deflexus, 91; *protoindentus,* 91
Dinohippus mexicanus, 132
Dinohyus, 133
Diodora cattiliformis alumensis, 108
Dioplotherium, 136
diorite, 15
Diploria, 255
Discorinopsis gunteri, 102
disease, in bovid fossils, 135
dissolution: of anhydrite deposits, 183; and aquifer systems, 85, 88, 239; of carbonate sediments, 51–53, 62, *96,* 221; causes of, 184; and coastal systems, 166; of dolomite, 222–23; effects of, 56, 62; in Escarpment, 183–84; and permeability, 72; and phosphorite formation, 202, 207; and Platform, 51–53, 174; and sinkhole development, 224–26. *See also* caves/caverns

Dixie County: phosphate from, 141; wetlands in, 69
Dog Island, 166–67
dogs, 122, 125–26
Dogtown Clay Member, 149
Dolime Quarry, *96*
dolomites: age of, 197, 208, *209;* in aquifer systems, 80, 83, 88, 235; description of, **3**, 53, 54, *54;* dissolution of, 222–23; formation of, 54–56, 64, 210; in Keys, 55, 256; mining of, 142, 144–45; and phosphorite formation, 198–200, 202, *202,* 204–5, *205,* 207; in ramp development, 179, 180; in shelf-slope system, 187
dolomitization: and carbonate sediments, 53–56; description of, 53–56, *54;* effects of, 56, 64, 94; heterogeneous versus homogeneous, 54–55, *55;* in mixing zone, 54, 223; models of, *54;* and permeability, 72; in stratigraphic units, 50; time of, *55,* 55–56. *See also* Boulder Zone
dolosilt, 198, 205
dolphins, 136
Domning, D. P., 121, 135–36
Douglas, M. S., 4
Doyle, L. J., 185, 186
draglines, *120,* 142, 150, 153–54, 241, *241;* (dredges), 151
drainage density, 69, *70*
drainage systems: canals, 78, *78;* construction of, 239–40, *240;* description of, *4,* 4–5, *5,* 157–58; injection wells in, 82, 84, 87, 246; sinkholes as, 11, 229, 231, *231*
Drepanodus sp., 91
drilling, and sinkholes, 225, *225*
drought, 78, 239
drowning, processes of, 175–78
Droxler, A., 180
Dry Tortugas: carbonate-sand deposit in, 256; coral species of, *253;* geology of, 251
DSDP site. *See* Deep Sea Drilling Project Site 588
DSRV Alvin, 181
Duane, D. B., 67
DuBar, J. R., 65–66, 112, 116
Dufrenoya taxana, 92, 93
dugongs (sea cows), 94, 121, 135–36
Duncan Church facies, 101
Duncan, J. G., 49
Dunedin, structural damage in, 232, *233*
Dunedin Pass, 163, *163*
Duplin Formation, 107
Duplin Marl, 59
dust, regulations on, 146, 150–51
Duval County: aquifer system in, 84, 87–88; heavy-mineral sands in, 152

Eagle Mills Formation, 15
earthquakes: epicenters of, *19,* 19–20; intensities of, 217, *218;* occurrences of, 19, 217–18, *218*
east-coast barrier system, 157–60, *159, 160*
East Florida Flatwoods, 8
East Florida Shelf, *174,* 190–94
East Florida Slope, 193

East Pass, 167
ebb-tidal delta, *157,* 163–64, 191
echinoderms, 94, 255
echinoid-foraminifer grainstone, *53*
echinoids: in Atkinson Formation, 93; in Avon Park Formation, 94, *95,* 96; in Bermont Formation, *117;* in Caloosahatchee Formation, *114;* irregular, 101, *117;* in Ocala Limestone, *97;* in shelf-slope system, 189, 191; in Suwannee and Marianna limestones, *102;* in Tamiami Formation, *110*
Echinometra lucunrer, 114
economy: and cement production, 146; and clay production, 148; and crushed-stone production, 144–45; and foreign competition, 154; and heavy-minerals production, 151–52; and mineral industries, 139, *140,* 142–43; and paleontology, 90; and peat production, 153; and sand and gravel production, 148
Ecphora: bradleyae, 109; quadricostata umbilicata, 108
Ecuador, formations in, 107
EDB, contamination by, 244
edentates, 123–25
Edgar, 150
El Abra range (Mexico), rudistid-bioherm complex in, 178
electrical resistivity, 228
electricity, 141, 143, 152
Elephantidae (family), 135
Elephas maximus, 135
elk, 134
Enallopsammia sp., 193
Encope tamiamiensis, 110
endemism, of fauna, 105
Enhydritherium, 127
Enos, P., 221, 251, 255
Entelodontidae (family), 133
environmental geology: and airborne hazards, 247; and anthropogenic hazards, 240–46; definition of, 217; and natural hazards, 217–40; and water supply, 246–49
environmental issues: air pollution, 143–46, 150, 153–54; and cement production, 146–47; and clay production 150–51; crushed-stone production, 145; and heavy-minerals production, 152; and limestone mining, 240–41; noise pollution, 145; and oil and gas extraction, 242–43; and peat production, 153; and phosphate industry, 143–44, *241,* 241–42; regulations for, 153–54; and sand and gravel production, 148, 240; and waste disposal, 243–46. *See also* contamination; sinkholes; water pollution
Eocene: climate in, 56; configuration in, *99;* deposition in, *42–44,* 47–48, 58–59, 94; and dissolution of sediments, 52; diversity in, 50; invertebrate fossils in, 94–96, *95–100,* 99; mammal fossils in, 120–21; phosphorite formation in, 214–15; sea levels in, 46, 50; southern Platform boundary in, 175; stratigraphic units from, 50–51; Tethyan influence on, 99, *100. See also* Ocala Limestone
Eocene/Oligocene boundary, 101, 214–15

Eoceratoconcha weisbordi, 110
Eovasum vernoni, 100
epeirogenic uplift, 25
epibionts, 94, *96,* 256
epicontinental marine system. *See* marsh coast (Big Bend area)
Epicyon, 125
Eponides: jacksonensis, 97; mariannensis, 102
Equidae, 130–32
equines, 131–32, *132*
Equus, 119, 131–32
Eremotherium, 125
Erethizontidae (family), 129–30
erosion: of Canyon, *182;* of Escarpment, 179, 181–84, *182;* features created by, 4; in Pleistocene, 66; process of, 182–84; of ramp system, 179; and reef termination, 192; and research efforts, 3; and shelf-slope system, 186, 193; by storm action, 258. *See also* deposition; sediments
Escambia County: oil in, 140; sand and gravel production in, 147–48; water supply for, 75
Escarpia laminata, 183
Escarpment: box canyons in, 182, *182,* 184, *184;* and carbonate megabank, 172; and erosion, 179, 181–84, *182;* foraminifers in, 92; marine influences on, *2;* seismic profile of, *48;* structure of, *176;* unconformity in, 181–82, *182;* and western Platform boundary, 174
Esmeraldas Formation (Ecuador), 107
estuaries, development of, 158
euhedral crystals, 14
Eupatagus antillarum, 97
Eurasia: and fossil similarities, 91; hyenas in, 128
Europe: bears in, 126; graptolites in, 91; peat production in, 153
eurypterids, 91
eustacy, 47
Eutaw Formation, 51
eutrophication, of seawater, 170
Evans, C. C., 251
Evans, M., 164
Evans, M. W., 184
evaporite rocks: deposition of, 39, 44–45, 47, 92, 94; dissolution of, 53; permeability of, 72; sequence record of, 47–49; stratigraphic units from, 50–51
evapotranspiration, 69, *70,* 77, 235
Everglades: characteristics of, 69; and deposition from hurricanes, 221; drainage of, 4, *4,* 239; and oil and gas extraction, 242; and peat production, 153; Pleistocene sediments in, 67; sediments from, 161; seismograph station in, 19; water released to, 77, 247
extinctions: causes of, 135, *136;* and climate, 105; on Cretaceous/Tertiary boundary, 48, 94; evidence of, 123; implications of, 107; of mammals, 124–37; versus origination rates, 115

Fabiania (Pseudorbitolina) cubensis, 95
Fabularia vaughani, 95
False Cape, cape-retreat massif near, 191

false sabercats, 127, *127*
Falsilyria pycnopleura, 106
Fasciolaria: okeechobensis, 117; *scalarina, 113*
faults: in Cenozoic sediments, 59; and earthquake epicenters, *19;* mapping of, *22,* 217; in Mesozoic, 173; in northern peninsula area, 12; Pickens-Gilbertown, *18. See also* Jay Fault; tectonics
faunas: age of, 105, 107; in basement, 90–91; Buda local, 122, 125, 127, 131, 132; in Chipola Formation, 105; Cretaceous carbonate-platform type of, 92–94; demise of, 99, 101; diversity of, 89, 115–16, 123; in Eocene, 121; in Miocene, 122–23; in Neogene, 101, 103; in Ocala Limestone, 95–96; in Oligocene, 99, 101, 121; in Pleistocene, 123; in Pliocene, 123; SB-1A local, 122, 131; seaboard local, 122; and sequence record, 48; shelly, of southern part, 107, 111–12; White Springs local, 105, 122, 133; Willacoochee local, 131
Fayard, L. D., 246
Felda oil field, 140
feldspar, 199–200, 235
Felidae (family), 127–28
Fenneman, N. M., 8
Fernandina Beach: aquifer system in, 84; pumping center at, 86
Fernandina Permeable Zone, 84
Ficus eopapyratia, 106
Field, M. E., 67
finger ridges, 221
fish: bony, 122; zoarcid, 183
Fish, J. E., 77
Flagler County, age of rocks in, 32, rocks of, *29*
Flamingo, 221
Flatwoods, 8
Flint River, 9
flood-tidal delta, 167
flooding: areas prone to, *220;* control system for, 239, *240;* from hurricanes and tropical storms, 219–20; occurrences of, 239–40. *See also* drowning
Florida: coastal systems of, 2–4, *3,* 155–68; environmental geology of, 217–49; hydrogeology of, 69–73; invertebrate fossils of, 89–117; mammal fossils of, 119–37; minerals of, 139–54; origin of, 89–90; population of, 155; statutes of, 217; stratigraphy of, 49–51; structural features of, **1,** 1–8, *7,* 12, *58. See also* basement; Escarpment; Keys; Platform; South Florida; *names of specific cities and counties*
Florida-Bahamas Megabank, 170, *170,* 172, 175
Florida-Bahamas Platform: development of, 169; fluid flow in, 182–83, *183;* margins of, 175–76; separation of, 39, 172–74, *173;* structure of, 175, *175. See also* Bahamas Platform; Platform
Florida Bay: discharge to, 77; drowning of, 258; limestones under, 253; mud islands in, 251, 256; sediments in, 67; shelf division, 255
Florida Canyon, erosion of, 182, *182,* 184, *184*
Florida Caverns State Park, *51*

Floridaceras whitei, 130
Florida City, hurricane at, 220
Florida Current, *171,* 172, 175, 193, 256; *See also* Straits of Florida
Florida Department of Agriculture, 244
Florida Department of Environmental Regulation (now Protection), 243, 246
Florida Department of Transportation, 154
Florida Geological Survey: nomenclature of, 28, 49, 73; on peat resources, 152; reports published by, xvii, 8; on Suwannee River course, 11
Florida Ground-Water Quality Monitoring Program, *243*
Florida-Hatteras Slope, 193
Florida Institute for Phosphate Research, 242
Florida Keys. *See* Keys
Florida Legislature, 249; (Statute, 217)
Florida Magnetic Anomaly, *27,* 32, 36
Florida Middle Ground, sediments associated with, 185–86
Florida Museum of Natural History, 90, 120, 122, 125
Florida Phosphate Council, xviii
Florida Platform: Bahamas Platform separation from, 172–74, *173;* and carbonate loss, 9, 11; changing location of, 44–45; characteristics of, 12; from Cretaceous to Paleogene, 39, 44; and dissolution, 51–53, 174; drowning of, 175–78; fluid flow in, 182–83, *183;* formation of, 61–67, 169–76, *170;* karst processes on, 9; lithostratigraphic framework for, 59–61; and marine influences, 1–3, *2,* 25; northern boundary of, 171–72; organic-C burial on, 197, *197;* phosphogenic episodes on, 198, 209–15; post-depositional processes of, 51–56, 181; rudistid reefs along, 92; sediment patterns on, *62–63,* 101; stability of, 19, 25; stratigraphy of, 47–51, *174;* structure of, *2, 58,* 58–59, 170–71, *171, 199;* subsurface of, 173; suture zones in, 36; tectonic history of, 21–25; unconformities around, 180–81, *181. See also* basement; deposition; Platform margins; ramp system
Florida Sinkhole Research Institute, 229
Floridan aquifer system: and calcite dissolution, 222–23; complexity of, *84,* 84–85; components of, 72, 235, *238;* contamination of, *243,* 244, 249; description of, 71, 82–88; discharge from, 86; extent of, *71,* 82, *82;* flow paths in, 237; and geothermal gradients, 20; and limestone mining, 241; maturation patterns in, 238–39, *240;* permeability of, 82–84, 223; and potentiometric surface, *86,* 224, *224, 225, 226, 226, 227;* recharge for, 75, 81, 228; salt and freshwater in, 80, 82, 84, 87–88, 237; and sinkhole development, 223–24, 228; springs from, 85, *85,* 165; subdivisions of, 82–85, *83, 84;* thickness of, 84, *84, 85;* transmissivity of, 85, *86;* as waste repository, 82, 87; water injected in, 247; water movement in, *74,* 75, 85–86, 237; water quality of, 73, 87–88, *88;* wells in, 86–87, *87;* withdrawal of freshwater from, 72, 82, 86–87, *87*
flow path lengths, 85, 235, 237
fluorapatite, 200
fluorine, 143

fluorosilicic acid, 143
flyash, in cement production, 145–46, 153
flying squirrels, 129
foraminifers: age of, 197, *206*, 206–8; in Avon Park Formation, *49*, 94–96, *95–96*, 99; for biozonation classification, 101; composition of, 206–7; in Hawthorn Group, 199; in Jackson Bluff Formation, 107; in Keys, 255; in Neogene, 105; in Ocala Limestone, *97*; planktic, 93, 105, 178, 189, *206*; in ramp development, 178–79, 181; representative type of, *92*; and rock classification, 93, *93*; shallow-water type of, 179; in shelf-slope system, 185, 189, 191, 193; in Tampa Member, 103; in Wood River Formation, 92, *92*. *See also* benthic foraminifers
Fort Dallas Oolite, 252–54
Fort Lauderdale: aquifer system in, 78; geology of, *183*
Fort Meade, 141
Fort Pierce: buried reef near, 193; and imported cement, 146
Fort Pierce Formation, 92, *92*
Fort Preston Formation, 150
Fort Thompson Formation: age of, 116; and aquifer systems, 74, 77; classification of, 112; and crushed-stone production, 145; and faunal-diversity issues, 115–16; formation of, 59, 61, 66; and sand and gravel production, 147
Fort Walton Beach: aquifer system in, *83*; geothermal gradient at, 21; pumping center at, 86
fossils: in basement, 14; in caverns, 52; in Chipola Formation, *106*; collection of, 119–20; destruction of, 64; in Eocene, 94–96, *95–99*, 99; in Neogene, 101, 103, 105, 107; in Ocala Limestone, *97–99*; in Pleistocene sediments, 66; in Pliocene sediments, 65, *108*; sequences of, 91–93; in Suwannee Basin, 31; Tethyan affinities, 100. *See also* faunas; foraminifers; macrofossils; mammals (fossil); marine invertebrate fossils; microfossils; nannofossils
Fourmile Creek, drainage system of, 5
foxes, 125–26
Frailey, D., 132
France, cement clinkers imported from, 147
francolite: age of, 197, *206*, 206–8; alteration of, 200, 205–6; composition of, 207; crystals of, *201*; formation of, 195–96, 199; isotopic evidence on, *203*, 203–4, *204*. *See also* phosphorite
Frank, E. F., 232–33
Franklin County: aquifer system in, 84; heat flow values for, 20
Franklin phosphate pit no. 2, 122
Frazier, M. K., 129–30
Fredericksburg Group (Texas), 93
Fruitville Formation, 112
fuel-storage tanks, 245
Fuller, W. R., 20
fullers earth clay: industry for, 122, 139, 149–50, *150*; mammal fossils from, 122; resources of, 150
Fusinus: caloosaensis, 113; watermani, 117

Gadsden County: aquifer system in, 84, 88; clay in, 149; landslide in, 233
Gagaria mossomi, 102
Gainesville: aquifer system in, *83*; fossils in, 121; seismograph station at, 19; sinkholes in, 224; water supply for, 82
galatheid crabs, 183
Galiano, H., 127
gamma-ray log, on phosphorite abundance, 211–12, *212*
Gardner, J., 105
Gardulski, A. F., 179
gas. *See* natural gas; petroleum
gasoline storage, 245–46
Gasparilla Island, 164
gastropods: in Atkinson Formation, 93; in basement, 91; in Bermont Formation, *117*; in Caloosahatchee Formation, 112, *113*; in Chipola Formation, *106*; in Eocene/Oligocene boundary, 101; in Jackson Bluff Formation, *108*; *Orthaulax* type of, 101, 103, *106*; and phosphorite formation, 200, 204; in Pinecrest Beds, 112; in shelf-slope system, 183, 191; in Tamiami Formation, *109, 110*; in Tampa Member, 103, *104*
Gatun Formation (Panama), 107
Gatuncillo Formation (Panama), 96
Gatunian Province, 103, *103*, 105, 115
geoid, and depositional dynamics, 47
geologists, licensing of, 217
geology, dynamism of, 1. *See also* environmental geology; geomorphology; hydrogeology
geomorphology: of Apalachicola Delta, *167*; complexity of, 12; description of, 1–9, *7*; of east-coast barrier system, 158–59; of Escarpment, 181–82; of mangrove coastal system, 160–62; of Panhandle, 166–68; of west-central barrier system, 163–64
Geomyidae (family), 129
Geomys, 129
Georgia: aquifers in, 71, 82; boreholes in, 15; and Panhandle coastal system, 166; phosphorite deposits in, 195; reflection seismology in, 16; rocks in, 27, 28, 198; seismograph station in, 19
Georgia Basin, 28
Georgia Bight (Georgia Embayment), 155, 158
Georgia Channel System: abandonment of, 61, 179; reefs along, 101; as sediment barrier, 61, 94; structure of, 45–46, *46*, 57–58, 171
geothermal conditions: of basement, *20*, 20–21; and convection, 52–53; and sediment dissolution, 52–53
gibbsite, 235
gigaplatform, termination of, 169–70, *170*
Gilchrist County: phosphate from, 141; Thomas Farm at, 122. *See also* Thomas Farm
Ginsburg, R. N., 251, 255
giraffe-camels, 133
Gisortia harrisi, 100
Givens, C. R., 99
glacial episodes: cycles of, 115, 179; and organic carbon, 197–98; sediment production in, 187
Glades County: aquifer system in, 81; sand and gravel production in, 148
Glades Group, 50, 116.

Glaucomys, 128
glauconite, 181
Globigerina spp., 93
Globotruncana, 93
Glossotherium, 125
Glycymeris suwannensis, 101
glyptodonts, 123–25
Glyptotherium, 124–25
Goldstein, R. F., 91
Gondwana basement: and Avalonian terranes, 33; evidence for, 31–32; and marine invertebrate fossils, 91; and Platform formation, 21–22, *23,* 24; stability of, 19–20; and Tallahassee-Suwannee terrane affinity, 36
Gough, D. I., 16
graben hypothesis, 25, 172
Grace, S. R., 242
grains: color of, 206; phosphorite types of, 200, 202; and sediment reworking, 205; sizes of, 199–200
grainstones: echinoid-foraminifer, *53;* and marsh coastal system, 165; permeability of, *72;* in Platform margin, 178; in shelf-slope systems, 187, 189
Grand Banks Platform, *170*
Grandin sands, 150
granite, 15, 22, 27
graptolites (*Didymograptus deflexus* and *D. protoindentus*), 91
Grasty, R., 14, 28, 29
gravel. *See* sand and gravel
gravity, and patterns in basement, 16–18, *17, 18*
Great American Interchange, 123–25, 127
Great Bahama Bank, 179
Great Barrier Reef (Australia), *170*
Great Isaac borehole, *173*
Grecian Rocks, 258
green algae, 187
Green Cove Springs, heavy-mineral sands in, 151
greenhouse effect, 56, 179, 198, 214, 221
Green Swamp, aquifer system in, 222
Griffin, G. M., 20
Griscom plantation site, 122
groundwater: classification of, 87–88, *88;* composition of, *239;* contamination of, 229, 231, 242–49, *243;* and dissolution, 51–53; and dolomitization, 54; and escarpment domains, *6, 7;* flow path in, 85, 235, 237; importance of, *72,* 72–73; iron in, 73, 80; levels of, *70,* 70–71, *71;* maturation of, 238–39, *240;* pH of, 222; in Platform, 184, *184;* pumping impact on, 86, *86;* quality of, 73, 235–39; radioactivity in, 248–49; source of, 69, *70;* united system with surface water, 77; withdrawal of freshwater from, *72,* 72–73, *73,* 246–47. *See also* aquifer systems; springs
Groundwater Atlas of the United States (Miller), 71
guanacos, 133
Guinea, and Platform origin, 22, 31
Gulf Coast: coastal system of, 156, 166, 168; erosion of, 101; hurricane on, 220; marine transgression of, 92; stratigraphy of, 49–50
Gulf Coastal Lowlands: divisions of, *10;* and sand and gravel production, 147

Gulf Coastal Plain: heat flow values for, 20–21; Platform isolated from, 101; sediments from, 166; sequence record of, 48
Gulf County, aquifer system in, 84
Gulf Hammock, aquifer system in, *83*
Gulf of Mexico: basin of, 58, 64, 170, *170;* and Caribbean Current, 180; and coastal systems, 155–56; and depositional history, 39; discharge to, 77; drowning of, 177; formation of, 24–25, 170; interchanges with Atlantic, 258; sequence record of, 47–48; tectonism in, 25; temperature of, 257
Gulf Stream: changes in, 46, 62, 213; and faunal differences, 105; and isthmus development, 115; and phosphorite formation, *197,* 210, 213; sediment supply by, 89; temperature of, 257
Gulf Trough: formation of, 46, *46,* 58, 171, 199; reefs along, 101; sediments in, 65
Gulfian Series, 93
Gullivan Bay, 161
Gumbelina, 93
Gunter, H., 11
Gunteria floridana, 95
Guraban subprovince, *107*
Gut, H. J., 125
gypstacks, *241,* 242
gypsum: and aquifer systems, 84, 88, 235; deposition of, 47; radioactivity of, 242

Haile site, 123, 125
Halimeda, 187, 193, 255
Hall, D. J., 36
Halley, R. B., 251
Hallock, P., 178
Hamilton County: drainage system of, 11; phosphate from, 141
Hammock Lands, 8
Hanshaw, B. B., 222–23, 237, 238
Haq, B. U., 46–47, 61, 93, 210
Harbaugh, J. W., 251
Harbison, A., 112
Hardee County: basalts in, 30; classification of water in, 88; phosphate from, 142
hares, 130
Harmon, R. S., 254
Harper, R. M., 8
Harris, G. D., 59
Harrison, R. S., 253–54, 257
Harvard University, field crews from, 120
Hatcher, R. D., 22, 33
Haustellum: anniae, 117; gilli, 106
Hawk Channel, 255–256, 258
Hawthorn Group: age of, 198, 208; and aquifer systems, 80, 85, 235, 237; and coastal systems, 163; components of, 149, 198–200, 210, 232; and crushed-stone production, 145; deposition of, 64, 198–99, 202, 205; dolomitization of, 64; formations of, 59–61; fossils in, 64, 103, 105, 120, 136, 199; phosphate from, 60–61, 142;

Hawthorn Group (*continued*)
and radium, 248; and sea level, 64, 198; sediments from, 59, 62; and shelf-slope systems, 190; and sinkhole occurrence, 5, *6, 7*; in topographic inversion, 11, *11*
Hayes, M. O., 158
Haynesville-Cotton Valley sequences, 39
hazardous waste sites, *245,* 246
Hazel, J. E., 112
Hazlehurst Terrace, elevation of, 2–3
Healy, H. G., 2–3
heat flow values, *20,* 20–21
Heatherington, A. L., 28, 29, 32
heavy-mineral sands: industry for, 139, 151–52, *152;* mineralogy of, 151–52
hedgehogs, 125
Heilprin, A., 90, 105
Hemiauchenia, 134
Hemicyon johnhenryi, 126
Hemingfordian, mammal faunas of, 122. *See also* Thomas Farm
Hemphillian, mammal faunas of, 123. *See also* McGehee Farm; Mixsons Bone Bed; Moss Acres site
Hendry, C. W., 9
Hendry County: aquifer system in, 74; oil in, 139, 242; sand and gravel production in, 148
Henry, H. R., 20
herbivores, 123, 125, 130, 132. *See also names of specific herbivores*
Hermanson, J. W., 132
Hernando County: age of rocks in, 31; mineral industries in, 141, 144–46
Hesperocyonininae, 125
Heteromyidae (family), 129
Heterostegina ocalina, 97
Hexaplex hertweckorum, 109
Hexobelomeryx, 134; *simpsoni, 134*
Higgins, M., 36
high field strength elements (HFSE), 30, 32
highlands, 7–9, *10*
Highlands County: basalts in, 30–31; basement in, 15; sand and gravel production in, 147; wells in, 244
Highlands province, sinkholes in, 5, *6, 7*
Hillsborough County: cement production in, 146; flood-prone areas in, *220;* geochemical data from, *29;* mammal fauna from, 120, 123, *123;* phosphate from, 141; radium in, 249; radon in, 248; shell production in, 144; sinkholes in, **9,** 222, *225,* 227–28, *228;* water contamination in, 231, 244, *244;* water supply for, 73, 87, 247; wells in, 244. *See also* Leisy shell pit
Hillsborough River, *70,* 70–71, 246
Hine, A. C., 179, 184
hipparionines, 132, *132*
hipparions, 131
Hippoporidra sp., *117*
Hodell, D. A., 55
Hoffmeister, J. E., 251–53, 256
hogs (extinct), 133
Holarctica, carnivores in, 126

Holmes, C. W., 189
Holmes County, outcrops in, 145
Holmesina septentrionalis, 124, *124*
Holocene: and coastal systems, 162, *162,* 164–65; depositional dynamics in, 48, 67, 158, 166, 256; faunal diversity in, 116; francolite from, 207; hurricanes in, 221; reefs in, *192,* 192–93, 255–56; sea levels in, 67, *157,* 157–58, 165; sequence record of, 47–48; stability in, 19, 25
holothurians, 183
Homestead: hurricane at, 220; water supply for, 76
Homosassa River, 165
Homosassa Springs, 223
Homotherium, 127
Honeymoon Island, 164
Hoover Dike, construction of, 220
horned artiodactyls (extinct), 133
horses: description of, 130–32, *131, 132;* location of fossils of, 119; in Miocene, 122–23; in Oligocene, 121
horsts, formation of, 25
Howard, J. D., 9
Howell, B. F., 91
Howell Hook, reef structures in, 185–86
Hoyt, J. H., 9
Huddlestun, P. F., 45–46, 65
Hughes, C. P., 91
Hulbert, R. C., Jr., 122, 124, 131–32
Hull, R. W., 246
humic material, in aquifer systems, 235
Hunter, M. E., 111
Hurricane Andrew (1992), 220, *221*
Hurricane Donna (1960), 221, 239
Hurricane Dora (1969), 239
Hurricane Pass, 163
hurricanes: classification of, 219; effects of, 155–56, 163, 220–21; flooding from, 239–40; landfalls of, *156;* Saffir-Simpson scale for, 219, *219;* storm surges in, 219
Hutchinson, D. R., 36
Hyaenidae (family), 128
Hydrochaeridae (family), 129–30
Hydrochaeris hydrochaeris, 130
hydrofluoric acid, 143
hydrogen sulfide, 183
hydrogeology: nomenclature for, *71;* overview of, 69–73. *See also* aquifer systems; coastal systems; groundwater; sea levels; surface-water
hyenas, 128
hyolithids, 91
Hyotissa: haitensis, 110; podagrina, 98
Hyperaulax americana, 104
Hypohippus, 122, 131
Hypolagus, 130
hypsodonts, 131

Iapetus Ocean, 22, 171
icehouse effect, 56, 179
illite, 200

ilmenite, 151
Imperialian subprovince, *107*
incineration, 246
Indarctos, 126
India, elephants in, 135
Indian Ocean, drilling in, 208
Indian River, 159
Indian River County: aquifer system in, 73, *74,* 80; marl production in, 144; wells in, 75
industry, water supply for, 72–73, 75–76, 80, 82, 86–87, 143, 148, 150–52. *See also* air pollution; mineral industries; water pollution
infauna, in shelf-slope system, 187
Inglis site, 123, 125
injection wells, 82, 84, 87, 246
inlets: development of, 158–59, *163,* 163–64, 221; in marsh coastal system, 167–68; and storm activity, 221; in west-central barrier system, *162,* 162–63
Inoceramus, 93
insectivores, 121, 122, 125. *See also names of specific insectivores*
insurance, for building damage, 226, 228–29, 232
intermediate aquifer system: components of, 72, 235; contamination of, *243,* 244; description of, 57, 80–82; extent of, 71, *71,* 80, *80;* maturation in, 238–39; stratigraphic units in, 80, *81;* transmissivity of, 82; upper and lower aquifer in, 80–81; water movement in, 75, *81,* 81–82, 237; water quality of, 82; wells in, 81; withdrawal of freshwater from, 72, 82, *82*
Intermediate Coastal Lowlands, *10*
Intracoastal Formation, 60, 65
ion exchange, 232
iron, in groundwater, 73, 80
iron oxide, 75
iron oxyhydroxide, 235
iron redox cycle, 209
iron sulfides, 200
irrigation, 72
Irvingtonian sites, 127, 129, 133
Ischyrocyon, 126
I-75 local fauna, 124–25, 129–31, 133
island arc basalts, 32
islands: in marsh coastal system, 166; mud, 251, 256. *See also* barrier islands; *names of specific islands*
isopach maps, *43–44;* 58
Ivany, L. C., 94

Jackson, J. B. C., 115
Jackson Bluff, 64–65
Jackson Bluff Formation: formation of, 59–60, 65; invertebrate fossils in, 107, *108*
Jacksonbluffian subprovince, 107, *107*
Jackson County: crushed-stone production in, 145; dolomite in, 144–45; gravity anomaly in, 17; limestone in, 101, 145; wells in, 244
Jacksonville: aquifer system in, *83;* beach deposits at, 66; permeable (aquifer) zone in, 84; port of, 142; and river course, 12; shelf-slope system near, 190; water supply for, 82, 86
Jacksonville Basin, 58, 64, 199, 204, 213. *See also* Southeast Georgia Embayment
jaguars, 127–28
Jakes, P., 29
Jamaica, formations in, 107
James City Formation (Va. and N.C.), 112
Jay Fault: and basement configuration, *14,* 16, *16,* 18, *18;* and earthquake epicenters, 20; as lithospheric boundary, 36; location of, *19, 27;* and plate convergence, 22, *23,* 24–25
Jay field, oil and gas from, 140–41
Jefferson County: crushed-stone production in, 145; lake disappearance in, 11
Jenkyns, H. C., 177
Joeckel, R. M., 133
Johns Pass, 163
Jordan, G. F., 185
Juniper Creek Quadrangle, drainage system of, *5*
Jurassic: and basement configuration, 16, *16,* 91–92; deposition in, 91–92; Gulf of Mexico Basin in, 170, *170;* and Platform boundaries, 171–72, 174–75; Platform during, *40, 43, 45;* rifting during, 24–25, 39, 57, 170; sea levels in, *47;* sequence record of, 47–48; transform plate boundary in, 18

Kane, B. C., 150
kangaroo rats, 129
kaolin clay, industry for, 148, 150–51
kaolinite, 148, 150, 189, 200, 235
Karr, E. Y., 13
karst topography: characteristics of, 5, *6, 7;* and dissolution processes, 61–62; and eastern Platform boundary, 174; formation of, 51–52, 222–23; and marsh coastal system, 165–66, *167;* as natural hazard, 221–22; in Pleistocene, 66; and shelf-slope systems, 184–85; and springs formation, 85. *See also* sinkholes
Katz, B. G., 222, 238
Keen, A. M., 112
key, definition of, 251
Key Biscayne, 158
Key Largo: aquifer system at, *83;* borings on, 253–54, *254;* coral species of, *253;* fossils at, 93; structure of, *173, 255, 257*
Key Largo Limestone: age of, 116, 254–55; components of, 253–55, *254, 255;* and crushed-stone production, 145; deposition on, 251–52, 255; formation of, 59, 61, 67; and shelf-slope system, 190; thickness of, 253
Keys: age of, 116; borings on, *254;* coastal system of, 4, 156–57; coral species of, 251–52, *253,* 254–59; crushed-stone production in, 145; dolomite in, 55, 256; evolution of, 256–58; formation of, 251, *252;* future changes in, 258–59; geology of, **10, 11,** *173,* 251–59, *252;* marine influences on, *2;* and oil and gas extraction, 243; reefs in, 67, 189–90, *190,* 255–56; and sea levels, 157, 254, 256–59, *258;* siliciclastic sediments in, 65, 172; stratigraphic units of, 251–53;

Keys (*continued*)
    tropical storm at, 220; water supply for, 76, 247. *See also names of specific Keys*
Keystone Heights, sinkhole in, *225*
Key West: and hurricanes, 220; and sea level changes, *258*
Key West Oolite, 252
Kice Island, 163
Kimmeridgian, Platform during, *40*
Kimrey, J. O., 246
King, E. R., 16
King, W., 20
Kingena (brachiopod), 93
Kiokee slate belt, 34, *34*
Kissimmee River, 4, 239–40
Kissimmee Saddle, 199
Klein, H., 77
Klitgord, K. D., 13, 17
Knapp, M. S., 11
Knowles Limestone (Tex.), 175
Knowles, S. C., 221
Kohout, F. A., 20
Krantz, D. E., 115
Kuphus incrassatus, 102
Kurten, B., 126
Kyptoceras, 133; *amatorum, 134*

La Brea tar pits (Calif.), 127
La Costa Island, 221
Lafayette County, phosphate from, 141
lagomorphs, 130
Lake City: aquifer system at, *83;* drainage system of, 231; sinkholes in, 5, *6*
Lake County: classification of water in, 88; clay in, 149; Osceola Granite in, 32; sand and gravel production in, 148; wells in, 244
Lake Iamonia, 11
Lake Jackson, 11
Lakeland Ridge, 9
Lake Miccosukee, 11
Lake Okeechobee: Caloosahatchee River connected to, 239; drainage systems at, 4; as flood-control component, 247; hurricane damage to, 220; as lagoon in Pliocene, 65; nutrient loading in, 240; origin of, 9; significance of, 69; water level controlled for, 77
Lake Region, 8
Lake Upland, 66
Lake Wales Ridge: clay in, 149–50; naming of, 9; origins of, 8; and sand and gravel production, 147; sinkholes in, 5; slope stability in, 233
Lake Worth, 190, 192
Lama, 133–34
Lambert, W. D., 135
landfills, 153–54, 246. *See also* waste disposal
land-mammal ages, *121*
land reclamation, after mining, 148, 151, 152, *241,* 241–42
landslides, 218, 233, 235

land snails, 103, *104*
Lane, E., 5, 217, 223
large ion lithophile elements (LILE), 30, 32
Latirus maxwelli, 117
Laurentia, and plate convergence, 22, *23,* 24, 33, 36
Lawrence, F. W., 241
Lawson Limestone, 93–94
leached zone ores, 142
lead, *243,* 243–44, 248
LeConte, J., 89
Lee, F. W., 16
Lee County: aquifer systems in, 237, *239;* basement in, 15; coastal system of, **6,** 163, 221; mineral industries in, 144–45; oil in, 139, 242; water supply for, 80
Leg 101 Scientific Party, 172
Leidy, J., 119
Leisy shell pit, 120, 123, *123,* 132
Leon County: aquifer system in, 84; lake disappearance in, 11; sand and gravel production in, 147
Lepidocyclina: mantelli, 102; ocalana, 97; parvula, 102
Leptarctus, 127
Lepus, 130
Levy County: basalts in, 28–29; basement in, 14–15; fossils in, 105; phosphate from, 141; Platform development in, 61; and river meander pattern, 11; rock formation in, 94; wetlands in, 69. *See also* Mixsons Bone Bed
Liberia, basalts from, 28
Liberty County, 105
Lidz, B. H., 251, 255, 259
lightweight aggregate, 149
Lighty, R. G., 192
Lignumvitae Key, 251
LILE (large ion lithophile elements), 30, 32
lime mud, 72, 256
Lime Sink Region, 8
limestone: age of, 89–90; and aquifer systems, 73–74, 77, 80, 82–83; and coastal systems, 165–66; density changes in, 25; dissolution of, 11, 222; and drainage patterns, 5; invertebrate fossils in, 92, 101; karstic surface of, *165;* in Keys, 251–53, *252;* mining of, 144–45, 240–41; and permeability, 183, *183;* reserves of, 145; in scrubbers, 153; in shelf-slope system, 190; and sinkhole development, 223–25; stratigraphic units from, 50. *See also* cement; *names of specific limestones*
Limonian subprovince, *107*
Lingulepis floridaensis, 31, *90,* 91, 93
linguloid brachiopod, *90*
Liochlamys bulbosa, 113
lithification, 53
lithoclasts, 193
lithofacies, *40–42*
lithoherm, cementation formation of, 193, *193*
Lithophaga palmerae, 99, 100
Lithoplaision, 94, *95,* 96, *98*
Littlefield, J. R., 222, 229
Little Salt Springs, sinkholes at, 7

littoral drift, 159–60, *160*, 168
Lituonella, 94; *floridana, 95*
Live Oak: drainage wells in, 229, 231, *231;* elevation of, 11; limestone mining in, 241; local fauna near, 122
llamas, 133–34
Locker, S. D., 189
Long, L. T., 19, 21
Long Island Quadrangle, drainage system of, *4*
Long Key, coral species of, *253*
longshore transport, 156, 158
Looe Key Reef, coral-reef colonization on, **11,** 258
Loop Current: intensification of, *177, 179,* 179–82, *180;* precursor to, 179; and sediment cycles, 187, *188;* and shelf-slope system, 185–87, 189
Lophelia sp., 193
Louann Salt (Jurassic), 13, 35, 39, 92, 174
Louisiana: fossils from, 93; reef trends in, 176
Love Bone Bed: description of, 120, 122; diversity in, 123; mammal fossils from, 125–27, *126, 127, 130,* 130–33, 135
Lovers Key, 163
Lowell, R. P., 21
lowlands, divisions of, 8–9, *10*
Loxodonta africana, 135
Lutra, 127
Lynts, G. W., 172–74
Lyell, C., v
Lynx, 127–28; *rexroadensis, 128*
Lyons, W. G., 66, 111, 112, 115
Lytechinus variegatus plurituberculanus, 114

Maastrichtian: drowning in, 177; and Florida Platform, *42;* and foraminifer similarities, 93; ramp development in, 178
MacFadden, B. J., 122, 132
machairodonts, 127
Machairodus, 127
Macrocallista reposta, 108
macrofossils: in Bermont Formation, *117;* for biozonation classification, 101; in Caloosahatchee Formation, *113;* Cretaceous remains of, 93; location of, 91; in Tamiami Formation, *109, 110;* in Tampa Member, 103, *104*
Macurda, D. B., Jr., 184
Madison County: basement in, 14–15; trilobite in, 91
Madracis, 185
magnesian calcite, 187
magnesium, 142, 237
magnetic field, anomalies in, 16–18, 35–36, *171*
magnetostratigraphy, age determined by, 197, 207
malacofauna, 105
mammals (fossil): Blancan land, 123; geologic context of, 120–21, *121;* sequence of, 121–23; studies of, 119–20. *See also names of specific mammals*
mammoths, 135, *135*
Mammut americanum, 135, *135*
Mammutidae (family), 135, *135*

Manatee County: aquifer system in, 82; mineral industries in, 142, 144–46; water supply for, 246
Manatee River, 246
mangrove coast system (southwest): in Holocene, 162, *162;* overview of, **5,** *160,* 160–61; present morphology of, 161–62; sea level for, 157; and tides, 157, 161, *161*
Manicina areolata, 117
Mansfield, W. C., 101, 103, 105, 107, 111, 112
Marco Island, 163–64
Marianna Limestone, 101, *102,* 145
Marianna Lowlands, *10*
marine invertebrate fossils: from Cenozoic, 94–116; as evidence of uplift, 9, 11; influences on, 99, 101; from Mesozoic, 91–94; and origin of Florida, 89–90; from Paleozoic, 90–91
Marion County: clay in, 149; gasoline-storage tanks in, 246; geochemical data from, *29,* 31; mammal fossils in, 125; phosphate from, 141; sand and gravel production in, 148; sedimentary rocks in, 31; sinkholes in, *231*
Marks Head Formation, 60, 105, 198
marls, 144, 177
Marquesas Keys, 140, 251
Marquesas Supergroup, 50, 92, 93
marsh coast (Big Bend area), 4, 156, 165–66
marsupials, 124
Martin-Anthony Road Cut (Ocala), 122
Martin County: aquifer system in, 73; St. Lucie metamorphic complex in, 33; wells in, 244
masonry cement. *See* cement
mass wasting, 179, 182
mastodonts, 135, *135*
Matanzas Inlet, 159, *159,* 174
Matson, G. C., xvii, 8, 103
Mayo, mammal fossils from, *120*
McBride, J. H., 35–36
McBride, R. A., 191
McClain, M. E., 13
McGehee Farm, mammal fossils from, 120, 123–24, 130
MCSB (Mid-Cretaceous Sequence Boundary), 48, 50–51, 175, 178
MCU (Mid-Cretaceous Unconformity), 48
meander patterns, 11–12
megalonychids, 123–25
Megalonyx, 125
Megantereon, 127
megatheres, 125
Meigs Member, 149
Meisburger, E. P., 67
Mellita aclinensis, 110
Melongena consors, 109
Menoceras, 130
Mephitis, 127
Mercalli earthquake scale (modified), 217, *218*
Mercenaria tridacnoides, 110
mercury, 243, 243–44
Merritt Island, earthquake in, *19*

Merychippus, 131
Merycoidodontidae (family), 133
Mesohippus, 130
mesonychids, 136
Mesozoic: and basement formation, 13–16, *16,* 21–22, 27; Brunswick magnetic anomaly in, 36; deposition in, 39, 47; geochemical characteristics of rocks in, 28–31; northern platform boundary in, 171; sea levels in, *47;* Tallahassee-Suwannee terrane in, 35–36. See also Cretaceous; Jurassic; Triassic
metamorphic rocks, 27
Metaxytherium, 136; *floridanum, 136*
meteoric waters, phosphorite alteration of, 206
methane, 183
Mexico: cement clinkers imported from, 147; mineral industries in, 154; phosphorite deposits in, 215; rudistid-bioherm complex in, 178
Miami: aquifer system in, 78; earthquake offshore of, *19;* hurricane at, 220; injection wells in, 84; sediment of, *159,* 160; shelf-slope system near, 192
Miami-Dade Water and Sewer Department, 80
Miami Limestone: age of, 116, 254–55; and aquifer systems, 77; and coastal systems, 161; components of, 253–55, *254, 255;* and crushed-stone production, 145; deposition on, 251–52; description of, 223, 252–53; formation of, 61, 66–67; and shelf-slope system, 190, 193
Miami Oolite (Formation), 59, 252
Miami Terrace, 181, *183,* 193
mica, 200
Miccosukee Formation, 60, 65
mice, 129
microbes, in aquifers, 237
microfossils, 90–93
Microtus, 129
Mid-Cretaceous Sequence Boundary (MCSB), 48, 50–51, *174,* 175, 178
Mid-Cretaceous Unconformity (MCU), *45,* 48, 175, *176,* 182
Mid-Miocene Unconformity (MMU), *177, 179,* 182
Middle Ground Arch, 15–16, *170,* 174
Midnight Pass, 163
Midway fullers-earth mine, 122
Midway Group, 51, 94
Milankovitch cycles, 45; (frequency band), 187
miliolids, 93
Millepora, 185
Miller, J. A., 58
Miller, R. L., 242
Miltha: caloosaensis, 114; *carmenae,* 117
Milton, C., 14, 28, 29
mineral industries: cement, 139, 145–47, *146, 147;* clays, 139, *148,* 148–51, *149, 150;* crushed stone, 139, *144,* 144–45; foreign competition for, 154; fullers-earth, 122, 139, 149–50, *150;* hazards associated with, 240–43; heavy minerals, 139, 151–52, *152;* peat, 139, 141, *152,* 152–53, *153;* petroleum and gas, 139–41, *140;* phosphate rock, 139, *141,* 141–44, *143,* 195, *241,* 241–42; regulations and trends for, 153–54; sand and gravel, 139, *147,* 147–48, *148,* 240; value of, 139, *140,* 142–43; water for, 72–73, 75–76, 80, 82, 86–87, 143, 148, 150–52
minerals: clay, 199–200, 202, *202,* 205, 235; diagenetic, 200, 202; in soils and aquifers, 235, *236,* 237; weathering of, *236.* See also names of specific minerals
mining industries. See mineral industries
minks, 127
Miocene: climate in, 56; coastal plain in, *199;* and current intensification, 180; deposition in, *42, 44,* 48, 57, 58–60, 62, *62–63,* 64, *199;* distribution of rocks in, 25; divisions in, 59; and dolomitization, 55; fault movement in, 173; ice buildup in, 197; invertebrate fossils in, 103, *103,* 105, *106;* karst topography formed in, 223; major oceanographic event in, 181; mammal fossils in, 120, 122–23; ramp development in, 179–81; sea levels in, 52, 62, 64, 65, 210–11; seawater composition in, 207, 209–11; seawater temperatures in, 206. See also Love Bone Bed; phosphorite, formation of; Thomas Farm
Miogypsina intermedia, 102
Miracinonyx, 127–28
Missimer, T. M., 64
Mississippi, horsts in, 25
Mississippi River: oozes from, 187; sediments from, *176,* 181, 182, 184, 186, 189
Mitchell-Tapping, H. J., 254
Mitra lineolata, 113
Mixsons Bone Bed, 119, 123–24
Miyashiro, A., 29, 32
moles, 125
mollusks: age of, 103, 105, 197; boreholes from, 99, *99;* in Caloosahatchee Formation, 112, *113;* in Chipola Formation, 105, *106;* and climate, 105; extinctions of, 107; in Florida Middle Ground, 185; in Keys, 255–56; in Lawson Limestone, 93–94; in Marianna Limestone, *102;* in Miocene and Pliocene, 103; in Ocala Limestone, 95–96, *98–100,* 99; in Pinecrest Beds, 112; in Pliocene, 65, 103, 105, 107, *107, 108;* recrystallization of, 53; in seagrass community, 94; in shelf-slope system, 186–87, *187,* 189, 193; significance of, 64; in Suwannee Limestone, 101, *102;* in Tampa Member, 103, 105; Tethyan, in Florida, 99, *100;* in Wood River Formation, 92. See also bivalves
monazite, 151–52
Monroe County: aquifer system in, 76; crushed-stone production in, 145; formations under, 66; and offshore drilling, 140
Montastrea, 255; *canalis,* 104
Monterey Formation, 204, 213–14, *215*
Moore Haven, hurricane destruction of, 220
Moores, E. M., 36
moose, 134
Morgan, G. S., 125
Morocco: basalts from, 28; phosphorite offshore from, 204
Moropus, 132
Morris Bridge, water level at, *70,* 70–71

Morum chipolanum, 106
Moslow, T. F., 191
Moss Acres site, 120, 123, 132, 135
Mossom, S., xvii, 8
Mott, C. J., 19, 217
Mount Dora Ridge, 8
muds/mudstones: in Keys, 256; phosphorite in, 211; in Platform margin, 178; in shelf-slope system, 187
Muehlberger, W. R., 35
Mueller, P. A., 28, 29, 32
Muhs, D. R., 254
Mulinia sapotilla, 114
Mullins, H. T., 172–74, 177, 179–81
Multer, H. G., 251, 253, 256
municipal waste sites, 246
Muridae, 129
muscovites, 31
mussels (family *Mytilidae*), 183
Mustelidae (family), 127
Mylagaulidae (family), 128
Mylagaulus, 128, *128*
mylodontids, 124–25
Mylohyus, 133
mysticetes, 137
Mytilidae (family), 183

Namibian shelf, phosphorite in, 204, 209
Nannippus, 132; *minor, 132*
nannofossils, 64, 65, 208
Naples: geology of, *183;* hurricane at, 220
Naples Bay Group, 50
Nashua Formation, 60, 65–66
Nassau County: aquifer system in, 84; classification of water in, 88
natural gas: and dolomitization, 53; hazards of extracting, 242–43; industry for, 139–41, *140;* and porosity, 51; potential for, 20, 27; sulfur in, 140–41
Naval Oceanographic Office, 16
Nayadina (Exputens) ocalensis, 100
Nebraska, Arikareean sequence from, 133
Neithea, 93
Nelson, K. D., 16, 35–36
Neochoerus, 130
Neogene: deposition in, 58–62, *63,* 101, 187, *188,* 189; and dolomitization, 55; faulting in, 19; invertebrate fossils in, 101, 103, 105, 107; and marine environment, 25, 52; sea levels in, 61–62, 65; seawater composition in, 208; transition to, 101, 103. *See also* Miocene; Pliocene
Neohipparion trampasense, 131–32
Neolaganum: dalli, 94, *95; durhami, 97*
Neumann, A. C., 193
Newberry, mammal fossils from pit near, 122, 123. *See also* McGehee Farm
Newfoundland, lithologic assemblages of, 33
New Mexico, caverns in, 53
New Smyrna Beach, pumping center at, 86
Newton, C. R., 193

Nicaraguan Rise, 180
Nicol, D., 99
Nimravidae (family), 127
Nimravides, 127
Niokolo-Koba Group (Senegal), 32
nitrates, in groundwater, 231, *243,* 243–45, *244*
Nocatee Member, 60
noise pollution, and crushed-stone production, 145
Norphlet sands, 39, 140
North Africa: and fossil similarities, 91; phosphorite deposits in, 214, *215*
North America: plate of, 18; separated from South America, 24–25; Tallahassee-Suwannee terrane accretion to, 34–36. *See also* Central American Isthmus
North American Stratigraphic Code, 49
North Bunces Key, 164
North Captiva Island, **6,** 221
North Carolina: coastal systems of, 158; phosphorite deposits in, 195, 213
Northern Highlands: divisions of, *10;* drainage systems of, 4–5; formations underlying, 65
North Florida, crust of, 17–18
North Florida calc-alkaline volcanic suite, *27*
North Florida Mesozoic Volcanic suite, *30*
North Florida Paleozoic-Proterozoic Volcanic suite, *30,* 32
Nothrotheriops, 125
nuclear power, 141
Nummulites: (Camerina) vanderstoki, 97; (Operculinoides) dius, 102; (Operculinoides) ocalanus, 97

Oak Grove Formation, 59, 103, 105
Oak Grove Sand Formation, 60
obolids, 91
Ocala: fossils near, 119, 122, 123; tectonic displacement around, 9
Ocala Arch, 94, 142
Ocala High, 198, 200
Ocala Limestone: and aquifer systems, 82–84; and coastal systems, 165; and crushed-stone production, 145; description of, 50–51, 59, 95; invertebrate fossils in, 95–96, *97–100;* mammal fossils in, 121, 135–36; mollusks in, 99, *99;* outcrop area of, *96;* and post-deposition processes, *52, 53;* and shelf-slope system, 190; zones of, 96
Ocala Platform: formation of, 58–59, 61–62; and Pliocene sediments, 65; sinkhole damage around, 229; stress lines associated with, 11
Ocala Upland, 165
Ocala Uplift, 11, 25, 58
Ocalina floridana (crab), *98*
Ocean Drilling Project Site 747, 208
Ocean Reef Group, 50
Ochlockonee River, 4, 239
Ochopee Limestone, 61
Odessa, sinkhole in, *225*
Odobenidae (family), 126
Odobenus rosmarus, 126

Odocoileus virginianus, 131, 134
offshore drilling, 140, 242–43
oil. *See* petroleum
Okaloosa County, basement in, 15
Okeechobee Basin, 58, 64, 65
Okefenokee Swamp, 12, 165
Oklawaha River, 4
Oldsmar Formation, 50, 52, 82, 84, 94
Old Tampa Bay, 90
Oligocene: boundary of, 101; climate in, 56; deposition in, *42, 44,* 48, 58, 61, 199; and dolomitization, 55; invertebrate fossils in, 99, 101, *102;* mammal fossils in, 120–21; phosphorite formation in, 214–15; Platform boundary in, 172; ramp development in, 178–79; sea levels in, 46, 52, 57; stratigraphic units from, 50–51; Suwannee Channel destroyed in, 46
Oligocene/Miocene transition, 105
Oligopygus: haldemani, *97;* phelani, *97;* wetherbyi, *97*
Olsen, S. J., 122, 126
Olsson, A. A., 111, 112, 116
Onslow Bay, 213
ooids: in Keys, 251–53; in Miami Limestone, 252–53; in shelf-slope system, 187, 189, 193
oolite, **2,** 49, *49,* 59, 254
oozes, in shelf-slope system, 187
opal-CT, 200
Opdyke, N. D., 218
Ophiomorpha sp., 94, *95*
ophiuroids, *96*
opossums, 124
Orange County: Osceola Granite in, 32; water supply for, 87
Orange Island, 61
Orbitolina, 93, *93*
Ordovician, 14, 16, 22, 31
oreodonts, 122, 133, *133*
organic-carbon burial, 197, *197*
Orlando: aquifer system at, *83;* drainage system of, 231; injection wells in, 87; sinkholes near, 229, *229;* water supply for, 82
Orlando Ridge, 9
Orthaulax: gabbi, *106;* pugnax hernandoensis, 101, 103
orthoconic cephalopods, 91
Osceola complex, *14,* 15, 22, 27
Osceola County: batholith beneath, 15; Osceola Granite in, 32–33
Osceola Granite, *27,* 32–33
Osceola Low, 58
Osmond, J. K., 21, 66, 254
Osteoborus, 125
ostracodes, 91–93, 96, 101, 105, 107
Ostrea sp., 93
otters, 127
Otvos, E. G., 9
Ouachita Orogen, *23,* 24
Outer Banks (N.C.), 159
Oxfordian, Platform during, *40*
oxidation: of hydrogen sulfide, 183; of organic C, 214; of organic matter, 231–33; and phosphorite formation, 197–98, 200, 202–5, *205,* 209–10
oxygen, 214
oxyhydroxides, 235
oysters, 92, 165–66, *166*
Ozello, limestones in, *165*

Pacific Ocean: and isthmus development, 115; phosphorite formation in, 206, 211, *215;* volcanism in, 177–78
packstones: in Avon Park Formation, **2,** 49, *49;* in Escarpment, 181; in Key Largo, 253; in Platform margin, 178
Palaeolama, 134
Palatka: earthquake in, *19;* and river course, 12
paleoaltitude/longitude, 36
Paleocene: deposition in, 58, 94; phosphorite deposits in, 214; Platform during, *42, 43, 173;* sea levels in, 50; stratigraphic units from, 50–51; structure in, *171. See also* Ordovician
paleofelids, 127
Paleogene: deposition in, 46–47, 59; and dolomitization, 55; interoceanic circulation in, 99; Platform boundary in, 172; recrystallization in, 53; sea levels in, 61; sequence record of, 48–49; structural template in, 58; surface in, 61–62; transition from, 101, 103. *See also* Eocene; Oligocene; Paleocene
paleogeography, *40–42, 45, 99, 175*
paleokarst, 85, 120
Paleontological Research Institution (Ithaca), 90
paleontology: challenges to, 116; history of, 103, 119–20; role of, 89–90; techniques in, 125, 128
paleopathology, of bovids, 135
paleopole, 36
paleosols, 149
Paleozoic: and basement composition, 13–14, 16; deposition in, 16, 39; geochemical characteristics of rocks in, 31–32; invertebrate fossils in, 90–91; lithologic and faunal associations of, 33; Platform boundary in, 171; and Platform origin, 22, 24–25, 33, 91; terranes in, 33
Palm Beach County: aquifer system in, 73, 76; gasoline-storage tanks in, 246; heat flow values for, 20; limestone reserves in, 145; sediments in, 66; water supply for, 72
Palmer, K. V. W., 99
palygorskite, 64, 149, 200, 202
palynomorph spectra, 91
Pamlico Sand, 74
Pamlico Terrace, 2
pampatheres, 124
Pan-African episode, 31–32
Panama: capybaras in, 130; formations in, 107
Pangaea: assembly of, 33, 36; basalts related to, 28–29; and fossil evidence, 91–92; hot spot related to, 31; and Platform formation, 21–22, 24
Panhandle: coastal system of, 156–57, 166–68, *167;* continental influences on, 60; deposition in, 45–46, 57–59, 64; and diagenesis effects, 64; drainage systems of, 4–5; elevation of, 2; escarpments in, 9; geomorphology of,

318

166–68; heat flow values for, 20–21; highlands and ridges of, 7–8; Mesozoic rocks in, 28; sequence record of, 48; stratigraphic units of, 50–51; structural framework for, 58–59. *See also* Jackson Bluff Formation
Panopea floridana, 114
Panthera atrox, 127–28
Panther Camp Formation, 50
Parachucla Formation, 105
Parahippus leonensis, 131, 131–32
Paramylodon, 125
Pararotalia: (Rotalia) *byramensis, 102;* (Rotalia) *mexicana, 102*
Paris Basin, 94
Pasco County: crushed-stone production in, 145; part of Big Bend coast, 4; sinkhole in, *226;* water supply for, 247
Patterson, B., 120
Patton, T. H., 122
Paull, C. K., 182, 184
Paynes Prairie, 11
Peace River: clay wastes spilled in, 242; fossils along, 119; phosphate from, 141–42; water supply from, 247
Peace River Formation: age and composition of, 198; and aquifer system, 80; formation of, 60–61, 65
Peace River/Manasota Regional Water Supply Authority, 247
peat: formation of, 214; industry for, 139, 141, *152,* 152–53, *153;* types of, 152–53
peccaries, 133
Pecten perplanus, 102
pectinid bivalve *(Neithea),* 93
Pedro Bank (Caribbean), 177
Pegram, W. J., 28
pelecypods, 91
Pelican Shoal, 251
pellets, 193
peloids: age of, 208, *209;* and phosphorite formation, 200, *201,* 202, 204–6
Penholoway Terrace, 2
Penicillus, 187, 256
Peninsula: deposition in, 57–59, 64–65; development of, 89–90; division of sedimentation on, 39, 57; drainage of, 4–5, 157–58; faults in, 12; geochronologic cross-section of, *49;* highlands and ridges of, 7–8; lithostratigraphic framework for, 60–61; Mesozoic rocks in, 28; sequence record of, 48; stratigraphic units of, 50; structural framework for, 58–59; and wave and tidal energies, 156–57
Peninsular Arch: and aquifer systems, 84; carbonates over, 176; location of, 58, *58;* and northern Platform boundary, 171–72; role of, 39, 170, *170;* structure of, 16, *16, 171*
Penney Farms Formation, 60, 149, 198
Pensacola: aquifer system at, *83;* water supply for, 73, 75–76
Pensacola Bay, 167
Pensacola Clay, 60, 75
Peratherium, 124
Periarchus lyelli floridanus, 97
perissodactyls, 122, 123, 130–32
Perkins, R. D., 66–67, 221, 251, 253–55
Perna conradiana, 110

Peru, phosphorite deposits in, 195, 207, 209, 213
pesticides, *243,* 243–44
petroleum: and dolomitization, 53; hazards of extracting, 242–43; industry for, 139–41, *140;* and porosity, 51; potential for, 14, 20, 27; sulfur in, 140–41
Petuch, E. J., 65, 105, 107, 112, 115–16
Phalium inflatum, 117
Phanerozoic, 39, *196. See also* Cenozoic; Mesozoic; Paleozoic
Phenacocoelus luskensis, 122, 133
phosphate: deposition of, 62, 64–66, 173, 181; environmental problems with, 143–44, *241,* 241–42; fossils in deposits of, 127; geology of, 142; in groundwater, *244,* 245; in Hawthorn Group, 60–61; industry for, 139, *141,* 141–44, *143,* 195, *241,* 241–42; reserves of, 143; uranium in, 141–42, 242; uses of, 142–43. *See also* phosphogenesis; phosphorite
phosphogenesis: causes of, 204–5, *205;* and climate, 214–15; isotopic evidence of, *203,* 203–4, *204;* recorded in Babcock deep core, 210–11, *211;* timing of, 197, *197,* 211, 213–15, *215*
phosphoric acid, 143, 242
phosphorite: age of, 197, *206,* 206–8, *209,* 210–15, *215;* amount of, 64, 211–13; in aquifer systems, 235; carbonate banks capped by, 178; chemical alteration of, 205–6; comparison of, 211–12, *212;* formation of, 195–98, *196,* 200–206, *201, 202,* 204–6, *205,* 209–10, 214–15, *215;* grain types of, 200, *202;* in Hawthorn Group, 198–200; hazards of mining, 241–42; isotopic evidence on, 203–4; paleoceanographic significance of, 209–15; in ramp development, 181; reworking of, 197, 200, 204–5, *205,* 208, 210, *215;* ubiquitousness of, 195, *196. See also* phosphate
phosphorous, ubiquitousness of, 195, *196. See also* phosphogenesis; phosphorite
photosynthesis, 209
phyletic dwarfing, 131, 134
Phyllangia blakei, 108
phyllocarids, 91
Physeter, 136
physiography: definition of, 1; divisions of, 8–9; map of, *7*
Pickens-Gilbertown Fault, *18, 23*
Piedmont: basalts in, 28–29; metamorphic rocks in, *14;* sediments from, 166, 170, 172
pilosans, 124
Pinecrest Beds: age of, 107; classification of, 107, 111–12; and faunal-diversity issues, 116; formation of, 61; invertebrate fossils in, 103, 107, *109, 110,* 111–12; mammal fossils in, 120; shell accumulations in, *111,* 112
Pine Key Formation, 50, 93
Pinellas County: coastal system of, 163; flood-prone areas in, *220;* structural damage in, 232, *233;* water supply for, 247; wells in, 246
Pirkle, E. C., 142, 151
Pirkle, W. A., 12
Pitt landslide, 233
Placunanomia plicata, 110
Planorbella sp., *117*

Plant City, glass sand mined in, 147
Platform margins: bathymetric map of, *169;* drowning of, 175–78; eastern boundary of, 172–74; overview of, 169; southern boundary of, *175,* 175–76; western boundary of, 174–75. *See also* ramp system
Platygonus, 133
Pleistocene: conditions in, 115, *257;* controversies over stratigraphy of, 107, 111; deposition in, 57, 60, 66–67, 116; eastern Platform boundary in, 173; erosion of sediments from, 9; faunal diversity in, 115–16; formations placed in, 59; fossils from, 9, 11, *117,* 120–21, 123; Keys in, 251–55, *252;* Platform development in, *42,* 66–67; reefs formed in, 189–90, *190;* sediment patterns in, *63,* 66; and Suwannee River course, 12; uplift during, 9, 11, 25
Pleistocene overkill hypothesis, 135, *136*
Plesiogulo, 127
Plicatula hunterae, 110
Pliocene: conditions in, 115; controversies over stratigraphy of, 107, 111; deposition in, 60–61, *63,* 64–66, 112; erosion of sediments from, 9; faunal diversity in, 115; fossils from, 103, *109, 110,* 120, 123; isthmus development in, 115, 123; mollusks in, 65, 103, 105, 107, *107, 108;* Platform during, *42, 63,* 64–66; sea levels in, 65; and Suwannee River course, 12; and uplift, 25. *See also* Bone Valley Member
Pliocene/Pleistocene boundary, 112
Pliometanastes protistus, 124
Plio-Pleistocene, 9, 115–16
Plummer, L. N., 223
pocket gophers, 129, *129*
pocket mice, 129
Pojeta, J. J., 91
Poland, and fossil similarities, 91
Polk County: aquifer system in, 81; injection wells in, 87; magnetic studies in, 18; phosphate from, 141; sand and gravel production in, 148; water residence times in, 237; water supply for, 73, 87, 222; wells in, 244. *See also* Bone Valley Member
pollution. *See* air pollution; noise pollution; water pollution
polonium, 247–49
Polystira albidoides, 106
Pomatodelphis, 136
Popenoe, P., 174
Porch, J., 29, 32
porcupines, 123, 129–30
Porites, 255
Porpitella micra, 93
porpoises, 136
Port Charlotte: aquifer system in, 84; water treatment for, 247
portland cement. *See* cement
Port St. Joe, 4
potassium, 235
Pourtales Terrace, 181, *181, 183,* 189, *189,* 223
precipitation: acid rain in, 153, 222; and aquifer systems, 74; average, 69, *70;* chemistry of, *234,* 235; pH of, 222; retention facilities for, 77, *78;* and water level change, 70–71, *70–71;* water table response to, 78
proboscideans, 123, 135, *135*
Procyon, 126–27
Procyonidae (family), 126–27
Proheteromys, 129; *floridanus,* 129
pronghorns, 134, *134*
Proterozoic Eon, 27, 31–32
Protoceratidae (family), 133
Protosiren, 121, 135–36
Pseudhipparion simpsoni, 132
Pseudocyclammina, 93
Pseudomiltha megameris, 100
psychrosphere, 101
pteropods, 187, 189
Pulley Ridge, reef structures in, 185–86
Puma concolor, 128
pumping centers: effects of, 78, 81, 82; establishment of, 77, *78;* and groundwater movement changes, 86, *86;* and saltwater intrusion, 237–38, 247; and sinkhole rejuvenation, 227–28
Pungo River Formation (N.C.), 198, 213
Punta Gorda, 111
Punta Gorda Anhydrite, 50
Puntagavilanian subprovince, *107*
Puri, H. S., xvii, 9, 59, 101, 173
Putnam County: age of rocks in, 32; clay in, 150; geochemical data from, *29;* heavy-mineral sands in, 151; sand and gravel production in, 148
pyrite, 199–200, 202, 205, 210, 235

quartz sands: in barrier islands, 164; in basement rocks, 14; in cement production, 145–46; in coastal systems, 158, 161, 165, 167; industry for, 139, *147,* 147–48, *148,* 240; in Miami Limestone, 253; in phosphorite, 198–200; in ramp development, 181; in shelf-slope system, 186–87, *187,* 189–92, *191;* in surficial aquifer, 235; in Suwannee Basin Complex, 22. *See also* kaolin clay; sand and gravel
Quaternary: deposition in, 62, *63,* 187, *188,* 189; faunal diversity in, 89; reef development in, 186; sea levels in, 61–62; and shelf-slope systems, 184–94; uplift during, 11. *See also* Holocene; Pleistocene
quicksands, 255
Quincy, 122
Quincy fullers-earth mine, 122

rabbits, 130
raccoons, 126–27
radioactivity: in atmosphere, 247, 249; in groundwater, 248–49; origin of, 247–48, *248;* from phosphate industry, 241–42
radiolarians, 178
radium, 242, 247–49
radon, 143, 242, 247–49
rainfall. *See* precipitation
ramp system: development of, 178–81; and Loop Current in-

tensification, 179–82; and sediment facies, 186–87, 189; structure of, *179;* transition to, 175, 177–78. *See also* shelf-slope system
Rancholabrean sites, 129, 133
Randazzo, A. F., 94, 217, 223
rare earths: in gypsum, 242; in moazite, 151–52; radioactivity in, 247
rats, 129
raveling, 224, 226–28
Raysor Formation (Ga.), 65
Reagor, B. G., 19
Rebecca Shoal Dolomite, 50
reclamation, 143, 241, *241*
recrystallization: of benthic foraminifers, 208; of calcite, 51; of francolite, 207; of minerals, 200; of mollusks, 53
recycling, 154, 246
Reddick site, 123, 125
Redfish Pass, **6,** 163, 221
Red Snapper Sink, 174
reefs: age of, 89–90; in Chipola Formation, 105; composition of, 96, 177–78; coralgal, 101; destruction of, 178; development of, 185–86, 189–90, *190,* 255–58, *257;* in Eocene, 99; evolution of, 256–58; in Holocene, *192,* 192–93, 255–56; living community of, **12;** and mangrove coastal system, 162; and northern Platform boundary, 172; in Oligocene, 101; oyster-reef complexes, 165–66, *166;* and Platform margins, 176; repeated bathing of, 178; and sea levels, 254; sediments on, 255–56; in sequence record, 49; in shelf-slope system, 92, 185–86, *192,* 192–93; siliciclastics as foundation of, 65; in Smackover Formation, 92; in Straits of Florida, 193, *193;* zones for, 255, *256. See also* corals; rudists
Reel, D. A., 20
reserves, definition of, 139
Resource Conservation and Recovery Act, 153
resources, definition of, 139
reverse greenhouse effect, 214
reverse osmosis, 247
rhinoceroses, 122, 123, 130, *130*
Rhinocerotidae (family), 130
rhodolith rudstones, 187
Rhyncholampas: evergladensis, 110; *gouldii,* 101, *102*
rhynchonellid brachiopods, 91
rhyolitic rocks: in Africa and Florida, 22; age of, 36; alkali silica plot for, *29;* geochemical characteristics of, 29–31; in Mesozoic, 28; from South Florida Basin, 15
Richards, H. G., 91, 99
ridges, origins of, 7–8, 191–92
rifting: during Jurassic, 24–25, 39, 57, 170; during Triassic, 15–16, *16,* 21–22, 24, 39
Riggs, S. R., 62, 64
river of grass, 4. *See also* Everglades.
rivers: patterns of, 11–12; salinity dams on, 237–38; and shelf-slope systems, 193; water levels in, 70–71, *71. See also* deltas; *names of specific rivers*
road-base materials, 145

rock ridges, in shelf-slope system, 192
rodents, 121, 123, *128,* 128–30
Rodriguez Key, coral-reef colonization on, 258
rogue wave, 218
Rokelide Orogen: Guinea, 22; Sierra Leone, 33
Roof, S. R., 187
Rotularia vernoni, 98, 101, 185
rudists, 49, 92, 94, 178
Runnels, D. D., 223
runoff, 69, *70,* 82. *See also* drainage systems

sabellariid worms, 193
sabercats, 127
Saccocoma, 93
Saffir-Simpson hurricane scale, 219, *219*
St. Augustine, 19, 89, 158–59, 217
St. Helena Sound (S.C.), tidal range of, 158
St. Johns County: aquifer system in, 73; classification of water in, 88; crushed-stone production in, 145; wells in, 244
St. Johns Platform, 58–59, 198
St. Johns River, 4, 12, 88, 172, 217
St. Joseph Island, 166–67
St. Joseph Spit, 167, *167*
St. Lucie County: aquifer system in, 73; basement in, 15; metamorphic complex in, 33; Osceola Granite in, 32
St. Lucie metamorphic complex, 22, 27, *27,* 32–33
St. Lucie River estuary, 174
St. Marks Formation, 59–60, 145
St. Marys River, 12
St. Petersburg: aquifer system at, 83; flood-prone areas around, *220;* fossils at, 112; sea level at, 221; water quality in, 237; water supply for, 82
St. Vincent Island, 167
salinity dams, 237–38
saltwater: encroachment of, 73, 76, 78–79; and flooding, 219; intrusion of, 221, 237–38, 247; and radium presence, 248
saltwater/freshwater interface: configuration of, 78–79, *80;* effects of canals on, *79;* and karst formation, 222–23; and water quality, 237–38
Sambo Key, 251
sand: and aquifer systems, 73, 77, 80; heavy mineral, 139, 151–52, *152;* in Keys, 255–56; Norphlet, 140; in phosphate mining, 242; in reefs, 185; in shelf-slope system, 185–87, 189, 190; and sinkhole development, 223–25
sand and gravel: hazards of mining, 240; industry for, 139, *147,* 147–48, *148,* 240; reserves of, 148
sand and gravel aquifer system: components of, 71–72; contamination of, *243,* 244; extent of, *71,* 75; stratigraphic units in, 75, *75;* water movement in, 75–76; water quality of, 73, 76; wells in, 75–76; withdrawals of freshwater from, 72–73, 75–76, *76;* zones in, 75–76
sand-dune fields, 62
Sand Key, **10,** 163, 251
sand ridges, 191
sandstones, 14, 31, 91

Sandy Key, molluscan beaches at, 256
Sanford, and river course, 12
Sanford, S., 8, 252
Sanford High, 58–59, 173, 198, 200
San Gregorio Formation (Calif.), 215, *215*
Sanibel Island, 163–64, 247
Santa Fe River, 11, 123, 134
Santa Rosa County: oil in, 140, 242; sulfur removal facility in, 141; water supply for, 75
Santa Rosa Island, 156, 167
Sarasota: APAC shell pit at, *111*, 112, 120, 123; fossils from, *109, 110;* and sea-level changes, 158
Sarasota Arch, *16, 170*, 174
Sarasota Bay, 221
Sarasota County: aquifer system in, 81, 237; hurricanes in, 221; mineral industries in, 111–12, 144; seismograph station in, 19; water supply for, 80
SAR (sodium-absorption ratio), 232
SB–1A local fauna, 122, 131
Scalopus, 125
scarps, marine influences on, 2–3. *See also* terraces
Schizaster armiger, 97
Schizodelphis, 136
Schlager, W., 175, 178
Schmidt, W., 139
Schopf, J. M., 91
Sciuridae (family), 128–29
Sciurus, 128
scleractinian corals, 101
Scott, T. M., 49, 62, 64, 147
screen-washing, 125, 128, 130
seaboard local fauna, 122
sea cows, 94, 121, 135–36
sea-floor spreading, 47, 170, 175
seagrass, 94, *95, 96,* 187
sea levels: and anoxia, 178; and cape- and shole-retreat massifs, 191; in Cenozoic, *46, 47,* 61; changes in, 93, 170, *258;* and coastal systems, *157,* 157–58, 165; in Cretaceous, *46, 47,* 48, 172; and deposition, 44–48, *46, 47,* 61–67, 172, 187, *188,* 189; and diagenesis, 52–54; in Eocene, *46,* 50; in Eocene/Oligocene boundary, 101; in Holocene, 67, *157,* 157–58, 165; indicators of, 189–90; in Jurassic, *47;* and karst formation, 222–23; and Keys formation, 254, 256–59, *258;* and land elevation, 2, *2, 3;* in Mesozoic, *47;* in Miocene, 52, 62, 64, 65, 200, 210–11; in Neogene, 61–62, 65; in Oligocene, *46,* 52, 57; and organic carbon, *197,* 197–98, 214; in Paleocene, 50; in Paleogene, 61; and phosphogenic episodes, 205, *205,* 210–15, *211, 215;* in Pliocene, 65; in Quaternary, 61–62; and reef development, 186, 256–58, *258;* rise of, 221, 258–59; stratigraphy of, 47–51; in Tertiary, *47;* in Triassic, *47. See also* drowning
sea turtles, 94
seawater: composition of, 198, 206–12; and dolomitization, 55, *55;* and Platform permeability, 24, 183, *183;* and precipitation chemistry, 235; temperature of, 206. *See also* saltwater; saltwater/freshwater interface

sedimentary rocks: and basement configuration, 27; Cretaceous to Paleogene, 39, 44; depositional dynamics of, 44–47, *46,* 62, *62–63,* 64–66; distribution of marine, 25; geochemical characteristics of, 31–32; history of, 39, *40–43,* 57–58; mammal-fossil-bearing, *121;* Miocene to Holocene development of, 58–67; non-fossiliferous, 91; and sea level changes, 47–49; and shelf-slope systems, 186–89; stratigraphic units from, 50–51. *See also* deposition; diagenesis; sandstones; shales; siltstones
sediments: age of, 208, *209,* 210; comparisons of, 251; cycles for, 187, *188,* 189, 210; in Holocene, 67; from hurricanes, 221; and organic carbon, *197,* 197–98, 200, 202; and phosphorite formation, 195–200, 202–4, 209–10; on reefs, 255–56; reworking of, 204–5, *205,* 210–11, 214; and shelf-slope systems, 184–87, *187,* 189–93, 214; and weathering, 200. *See also* carbonate sediments; deposition; siliciclastic sediments
seismicity: description of, 19–20; and profile of West Florida Shelf, *48;* and reflection studies of basement, 16; and shelf-slope system, 189; and southern boundary studies, *176;* and Straits of Florida, 172–73, *173, 174*
Selenodontia, 132
Selenopeltis (province), 91
Sellards, E. H., 112, 115, 119
Selma Group, 51
Semele alumensis leonensis, 108
Seminole County, wells in, 244
Senegal: Coya Granite in, 22; and Platform origin, 22, 31–32
sepiolites, 64, 200, 202
Septastrea: crassa, 110; marylandica, 110
septic-tank systems, *244,* 245
sequence stratigraphy, 47, 65
Shaler, N. S., 8
shales: in basement, 14; carbonate banks capped by, 178; fossils from, 31, 91; in Suwannee Basin Complex, 22
sharks, 121, 122, 136, 208, *209*
shelf-slope systems: cape-retreat massif in, 191, *191;* corals in, 187, 192–93; in East Florida, 190–94; gastropods in, 183, 191; in Keys, 255–56; progradation of, 189, *189;* and sea-level changes, 214; sectors for, 190–92; sediment facies in, 186–87, *187,* 189, 193; in South Florida, 189–90; structure of, *185;* in West Florida, 184–89
shells: in Anastasia Formation, 60; aquifer underlain by, 73; in crushed-stone production, 144; in Hawthorn Group, 199–200; and hurricane activity, 221
Sheridan, R. E., 172
Shinn, E. A., 251, 255, 257–59
Shoal River Formation, 59–60, 103, 105
shole-retreat massif, 191
shovel-tuskers, 135, *135*
shrews, 125
SHRIMP ion microprobe, 31, 33
shrinking and swelling clays, 231–33. *See also* smectite and clays.
Siderastrea, 255
Sierra Leone, Rokelide Orogen in, 33

Siesta Key, 163–64
Sigmodon, 129
silex beds, 103
silica, 200
silicate gangue minerals, 151
siliciclastic sediments: and aquifer systems, 73, 83–84; deposition of, 39, 46, 57, 60–65, 101, 176, *177;* description of, 58–59; in Holocene, 67; and Keys formation, 254; in Miocene, 62, 64; origins of, 7–8; in Paleogene, 57, 61–62; permeability of, 72; and Platform boundaries, 172, 175; in Pleistocene, 66–67; in Pliocene, 64–66; sequence record of, 47–49; in shelf-slope system, 190–92; stratigraphic units from, 50–51; in topographic inversion, 11, *11*
silicon, 235
silt, 189
siltstones, *90,* 91
Silver Bluff Terrace, 2
Silver Springs, 223
Simmons, G., 20
Simpson, G. G., 119–20, 122
Sinclair, W. C., 223, 229
sinkholes: alluvial type of, *226,* 226–29, *227, 228;* and aquifer system, 85; classification and development of, 223–31; collapse type of, 223–25, *224;* cover-collapse type of, 5, 224, *225,* 229; description of, 221; differentiation of, 232–33, *233;* as drainage systems, 11, 229, 231, *231;* evaluation of risk for, 228–29; formation of, 51–52, 222–23, 229; insurance for, 228; and lake formation, 69–70; legal interpretations of, 226; other environmental impacts of, 229, 231; as pathways for contamination, 229; patterns of, 5, *6, 7;* and permeability, *183;* predicted types of, *230;* probability of, 229, *230;* raveling type of, 227–29; rock-collapse type of, 224; and shelf-slope systems, 184–85; solution type of, *225,* 225–26; springs converted to, 223; for waste disposal, 229, *231*
Sinum chipolanum, 106
Siphocypraea: carolinensis, 109; problematica, *113*
sirenians, 122, 135–36, *136*
Six Mile Creek, 103
Skolithos, 31, 91
skunks, 127
slate belts, 34, *34*
slimes, in phosphate production, 143–44, *241,* 242
slope, 193, 233, 235. *See also* shelf-slope systems
sloths, 123–25
slumps, 218, 233, 235
Smackover Limestone, 39, 92, 140
smectite: in Hawthorn Group, 149, 200; in shelf-slope system, 189; shrinking and swelling of, 231–33
Smilodon, 127; *fatalis,* 127; *populator,* 127
Smith, D. L., 13, 14, 17, 20, 22, 28, 217–18
snails, 103, *104*
sodium-absorption ratio (SAR), 232
sodium-bicarbonate water, 237
soil creep, 233, 235
soils: contamination of, 154; minerals in, 235, 237; permeability of, 232; pH of, 222; shrinking and swelling of, 231–33, *233;* in wetlands, 8
Solenastrea bournoni, 110
Sorex, 125
South Africa, phosphorite in, 204
South African shelf, phosphorite deposits in, 195, 204, 207
South America: basalts of, 28; bears in, 126; capybaras in, 130; and fossil similarities, 91; mammal immigration from, 123–25, 129–30; orogenic episodes in, 32; procyonids in, 126; separated from North America, 24–25
South Bay, storm surge at, 220
South Bunces Key, 164
South Carolina: aquifers in, 71, 82; coastal systems of, 158; earthquake in, 217; Hawthorn Group in, 198; phosphorite deposits in, 195
Southeastern Geological Society, *71*
Southeast Georgia Embayment, 46, 58, 65, 84, 173
South Florida: crust of, 17–18; shelf-slope systems in, 189–90; shelly faunas of, 107, 111–12; structural features of, *16;* volcanic rocks of, *14*
South Florida Basin, 16, *170,* 174, 176
South Florida Embayment, 58
South Florida Mesozoic Volcanic suite, *27,* 29–30, *30*
South Florida Shelf, 257
South Florida Water Management District, 76, 239, *240,* 246–47
South Georgia Rift: and earthquake epicenters, 20; formation of, 24–25; location of, *24;* and Platform formation, *14,* 15–16, *16, 18;* in Triassic, *24*
Southwest Florida Mesozoic Volcanic suite (SFMV), 29–31
Southwest Georgia Embayment, 15, 84. *See also* South Georgia Rift
Spain, cement clinkers imported from, 147
Spencer, R. V., 151
Spencer, S. M., 139
spillways, *240*
Spilogale, 127
Spirolina coyensis, 95
sponges, 185, 192
sponge spicules, 189
springs: contamination of, 231; converted to sinkholes, 223; location of, 85, *85;* and marsh coastal system, 165–66; and shelf-slope systems, 185
Sprinkle, C. L., 88, 222
squirrels, 122, 128–29
Stanley, S. M., 107, 112, 251
Statenville Formation, 60, 142, 198
staurolite, 146, 151
Steinen, R. P., 256
Stenella, 136
Stewart, J. W., 229
stoloniferous corals, 183
storm surges, 219–21, *220*
storm water, disposal of, 229, 231, 246
Stowasser, W. F., 141, 143
Straits of Florida: and Escarpment, 181; formation of, 172–73, 223; and Keys formation, 256;

Straits of Florida (*continued*)
and ramp development, 179; reefs in, 193, *193;* and shelf-slope system, 189; and southern Platform boundary, *175,* 175–76; structure of, *171, 173*
Stralopecten caloosaensis, 114
stratigraphy, controversies over, 49, 73–74, 107, 111
stream capture, 9
stromatoporoid, 92
Strombina aldrichi, 106
Strombus: aldrichi, 106; hertweckorum, 109; leidyi, 113; mayacensis, 117
strontium-isotope chronostratigraphy, 197
submerged lands, definition of, 8
Subpterynotus textilis, 109
subsidence: cause of, 232–33; damage from, 224, *225;* insurance disallowed for, 232; and sea-level rise, 177; versus sinkhole damage, *233.* See also sinkholes
sub-Zuni surface, 13
Sugarloaf Key, dolomites on, 256
sulfate-reduction zone, *203,* 203–5
sulfates, 88, 205, 235, 237, *238*
sulfide, 235
sulfur, in gas and oil, 140–41
sulfuric acid, 52–53, 143, 183, 242
sulfur isotopes, *203,* 203–4
Sumter County: crushed-stone production in, 144–45; phosphate from, 141
Sunderland (or Okefenokee) Terrace, 2
Sunniland, *83*
Sunniland Formation, 50, 139
Suoidea, 132
superfund sites, *245, 246*
surface lineaments, 9
surface mines, phosphorite from, 206
surface water: and aquifer systems, 76; contamination of, 229, 231, 248; and dissolution of sediments (erosion), 52; hazards associated with, 239–40; levels of, *70,* 70–71, *71;* major features of, *70;* pH of, 222; and shelf-slope systems, 184–85; source of, 69, *70;* united system with groundwater, 77; as water supply, 246–47. See also rivers
surficial aquifer system: components of, 71–72, 235; contamination of, *243;* extent of, *71;* maturation in, 238–39; permeability and transmissivity of, 75; and sinkholes, 224; stratigraphic units in, *74;* water movement in, *74,* 74–75, 237; water quality of, 73; wells in, 75; withdrawals of freshwater from, 72, 75, *75*
surficial rocks, geologic history of, 13–14
Sutcliffe, H., Jr., 242
Suwannee Basin complex: and basement configuration, 22, *23,* 24, 27; origin of, 31–32; and Platform formation, *14,* 15–16
Suwannee Channel (or Strait): and fossil patterns, 101; location of, 46, *46;* and ramp development, 178–79; as sediment barrier, 57, 94; structure of, *171,* 171–72
Suwannee County: dolomite production in, 144; drainage system of, 11; drainage wells in, 229, 231, *231;* limestone mining in, 241

Suwannee Current, *171,* 178–79
Suwannee Limestone: and aquifer system, 82–83; and crushed-stone production, 145; description of, 50–51; invertebrate fossils in, 101, *102;* in Platform, 61
Suwannee River, 11–12, 165–66, 222
Suwannee Saddle, *170,* 171
Suwannee terrane, 91
Swart, P. W., 256
SWEAT hypothesis, 36
Swedjemark, G. A., 249
Swift, D. J. P., 191
Swinchatt, J. P., 255
Sylvilagus, 130
synthetic organics, *243,* 243–44

Taft, W. H., 251
Talbot Terrace, 2
Tallahassee: fossils near, 122; water supply for, 82
Tallahassee Embayment, 46
Tallahassee Graben, 15. See also South Georgia Rift
Tallahassee Hills, 66, 233
Tallahassee-Suwannee terrane, 33–36, *35*
Tamiami Formation: age of, 107; and aquifer systems, 74, 80; classification of, 107, 111–12; and coastal system, 161; and crushed-stone production, 145; formation of, 59, 61, 65–66; invertebrate fossils in, 107, *109, 110;* and sand and gravel production, 147
Tamiami Trail, 111
Tamias, 128
Tampa: cement production in, 146; drainage system of, 231; flood-prone areas around, *220;* fossils from, 103; port of, 142; sea level at, 221; storm surge at, 219; water quality in, 237; water supply for, 82, 246
Tampa Basin, 16, *16*
Tampa Bay: fossils around, 103; limestones at, 89; precipitation at, 69; nitrate contamination in, 231; and shelf-slope system, 193; structure of, *185;* tides in, 157
Tampa Embayment, 39, *170*
Tampa Member: and aquifer system, 82–83; and coastal systems, 163; formation of, 59–60; macrofossils from, 103, *104,* 105
Tampa Stage, 59
Tanner, W. F., 9
Tapiridae (family), 132
tapirs, 123, 130, 132
Tapirus: simpsoni, 132; *veroensis,* 132
Tauvers, P. R., 35
Tavernier Key, coral-reef colonization on, 258
Tayassuidae (family), 133
Taylor County: mineral industries in, 141, 144–45; wetlands in, 69
tectonics: and Abaco event, 179; and deposition, 39, *40–43,* 47; development of, 39; and drowning processes, 178; and migration into colder climates, 170; in Neogene to Holocene, 58–59; and plate boundaries, 34–36; and plate convergence, 22, *23,* 24; and plate deformation, 9; and plate stability, 19, 25; and Platform boundaries,

172–76, *174;* regional history of, 21–22; and sea-floor spreading, 47, 170, 175; and sequence stratigraphy, 47–49. *See also* earthquakes; faults
Teleoceras, 130
Temik (Aldicarb), 244
temperature: and evapotranspiration, 69; of Gulf, 257; of seawater, 206. *See also* climate; glacial episodes
Tentaculites, 91
Ten Thousand Islands, **5,** 160–62, *162*
terraces: commercial sand of, 147; in Escarpment, 182; implications of, 9; location of, *3;* marine influences on, 2–3; reefs formed on, 189–90, *190;* and sea levels, 66; in shelf-slope system, 192. *See also* scarps
terranes: Avalonian-type, 33–35, *34;* boundaries of, 33–36, 91; and plate convergence, 22, *23, 24*
terrestrial vertebrate fauna, 101
Tertiary: deposition in, *45;* faunal diversity in, 89, 123; faunas in, 96, 105, 107; fossil-bearing strata in, 59; global cooling in, 214; sea level in, *47. See also* Bone Valley Formation; Neogene; Paleogene
Tethyan Province, 99
Tethys Current, 101
Tethys Seaway, 99, *99,* 103
Texas: fossils from, 93; reef trends in, 176
Thalassia, 256
Thalassodendron, 94, *96;* auricula-leporis, *95*
Thinobadistes, 124
tholeiites, characteristics of, 14–15, 28–30, *29*
Thomas, M. C., 120
Thomas, W. A., 15, 27
Thomas Farm: description of, 120, 122; mammal fossils at, 124–33, *126, 131;* techniques at, 125, 128
Thomomys orientalis, 129, *129*
thorium, 247–48
Three-Rooker Bar, 164
Thurber, D. L., 254
tidal prism, 157, *157,* 163, 167–68
tides: attenuation of, 159; and coastal systems, 156–57, *157,* 160–61, *161,* 163–64; and Keys formation, 254; and shelf-slope system, 190
titanium minerals, 151–52
Tithonian: deposition in, 92; Platform during, *40*
Tomarctus, 125
toothed whales, 136, *136*
Torreya Formation, 60, 105, 149
trace fossils, 94, *95, 96, 98*
Trail Ridge area: fossiliferous sediments behind, 11; minerals in, 151, 247; mines in, 146; origins of, 8; and river patterns, 12
Trans-Amazonian episode, 32
Transmarian molluscan province, *103*
transportation, and mineral industries, 144–45, 150
Tremarctos floridanus, 126
Triassic: deposition in, 91; land configuration in, *18;* Platform during, *40;* redbed rocks in, 13, 15; rifting during, 15–16, *16,* 21–22, 24, 39; sea levels in, *47;* South Georgia Rift in, *24*

Trichechus, 136
Trichecodon, 126
Trigonictus, 127
Trigonostoma hoerlei, 109
trilobite *(Colpocoryphe exsul),* 91
Trinity Group (Tex.), 93
Trocholina, 93
tropical storms: classification of, 219; and coastal systems, 155–56, 166; and shelf progradation, 189; storm surges in, 219, *219, 220. See also* hurricanes
tsunamis, 217–18
Tucker, H. I., 111, 112
Tuomey, M., 89–90
turbidites, 178–79
Turbinella: chipolanum, *106;* hoerlei, *117*
Turbo (Senectua) crenorugatus, 104
Turonian, Platform during, *41*
Turritella: etiwaensis, *108;* martinensis, 101; perattenuata, *113;* subgrundifera, *106;* tampae, *104*
turtle-grass *(Thalassia),* 256
Tuscaloosa Group, 51, 93
two-stage flotation, 142
Typhis obesus, 106

Udotea, 187
underground fuel-storage tanks, 245–46
ungulates: even-toed, 132–35; odd-toed, 130–32
uniformitarianism, 54, 251
United States, physiographic provinces of, *3*
U.S. Bureau of Mines, 141, 145, 153
U.S. Corps of Engineers, 240
U.S. Environmental Protection Agency, 146, 246
U.S. Geological Survey, 16, 49, 74, 229
U.S. National Museum of Natural History, 90
University of Florida, 120
University of South Florida, 228, *228*
Upchurch, S. B., 221, 229, 235, 241, 247
uranium, 141–43, 235, 247, *248,* 254–55
Urgonian faunal assemblages, 177
Urocyon, 125–26
Urosalpinx? inornata, 104
Ursidae (family), 126
Ursus speleaus, 126

Van der Voo, R., 34
Vasum: floridanum, *117;* horridum, *113;* locklini, *109*
Vaughan, T. W., 61
Velates floridanus, 100
Venericardia serricosta, 104
Venice Inlet, 163
Ventricolaria glyptoconcha, 104
Vernon, R. O., xvii, 8–9, 11, 59, 62, 101, 173, 217
Vero Beach site, 123, 132
vestimentiferan worms, 183
Vicksburg Group, 51
Vicugna, 134
vicuna, 133–34

Vogt, P. R., 177
Vokes, E., 105
Vokes, H., 105
volcanic rocks: age of, 32; and basement configuration, *14, 15*, 27; characteristics of, 29–32; significance of, 24
volcanism, 177–79
voles, 129
Volusia County, wells in, 244
Vulpes, 125–26

Wabasso beds, 65
Waccamaw Formation (N.C. and S.C.), 112
Waccasassa River, 222
wackestones: in Avon Park Formation, *49;* in Escarpment, 181; in Key Largo, 253; in Platform margin, 178
Wakulla County: crushed-stone production in, 145; sand and gravel production in, 147
Wakulla Springs: seismograph station at, 19; and sinkholes, 223
walruses, 126
Walton, T. L., 168
Walton County: elevation of, 2; nannofossils from, 64; outcrops in, 145
Wanless, H. R., 259
Washington County, outcrops in, 145
Washita Group (Texas), 93
waste disposal: hazards associated with, 243–46; sand and gravel pits for, 240; sinkholes used for, 229, *231*
water: budget of, *70;* conservation of, 143; and cover-collapse sinkholes, 224; for industry, 72–73, 75–76, 80, 82, 86–87, 143, 148, 150–52; regulations on, 154; supply of, 246–49. See also groundwater; surface water; water pollution; water quality
water conservation, 77, *78,* 150
water management districts, 239–40
water pollution: and cement production, 146; and clay production, 150; and crushed-stone production, 145; and groundwater, 229, 231, 242–49, *243;* and heavy-minerals production, 152; and limestone mining, 240–41; and peat production, 153; and phosphate industry, 143, 241–42; and sand and gravel mining, 240
water quality: assessment of, 243, *243;* of groundwater, 73; and maturation, 238–39, *240*
wavellite, 142, 196, 206. See also phosphorite
wave refraction, 163
waves: effects of, 155–60; and mangrove coastal system, 161, *161;* and marsh coastal system, 165–68; and Panhandle coastal system, 166; and shelf-slope system, 190; in tropical storms and hurricanes, 218–21, *219, 220;* and west-central barrier system, 162–64, *163*. See also tsunamis
Waycross (Ga.), 19
Wayne County, nannofossils in, 64
weasels, 127
weather, and coastal processes, 155–56
Weaver, C. E., 149
Webb, S. D., 122, 132

Weddell Sea, and phosphorite formation, *215*
Weisbord, N. E., 105
wells: in Biscayne aquifer system, 77–78, *79;* contamination of, 244, 246, 248; in Floridan aquifer system, 86–87, *87;* injection type of, 82, 84, 87, 246; in intermediate aquifer system, 81; Osceola Granite penetrated by, 32–33; in Paleozoic strata, 90–91; petroleum exploration, 27; St. Lucie metamorphic complex in, 33; in sand and gravel aquifer, 75–76; in surficial aquifer system, 75; water level in, *70,* 70–71
Werner evaporites, 13
West Africa: basalts of, 28; and fossil similarities, 91; orogenic episode of, 32; Platform rifted from, 21–22, 31, 32, 33; Tallahassee-Suwanne terrane correlations with, 34, 36
west-central barrier system, *162,* 162–65, *163*
West Coast Regional Water Supply Authority, 247
Western Bahamas Block, *16*
Western Highlands, 65
West Florida Basin, *170,* 174
West Florida Escarpment, *45,* 92, *169,* 172, 174, 176, 179, 181–84, 187
West Florida Shelf: description of, 169, 184–89; and foraminiferals in, 92; seismic profile of, *48;* structural features of, *16*
West Florida Slope, 169, *180,* 184, 186–87, 189
West Palm Beach: aquifer system in, *83;* beach deposits at, 66; injection wells in, 84
West Pass, 167
wetlands, 8, 69, *70,* 144, 237
whales, *120,* 121, *136,* 136–37
White, A. J. R., 29
White, D. H., Jr., 139
White, W. A., 5, 9, 11, 12
White, W. B., 222
White Bank, carbonate-sand deposit in, 256
White River Group, 121
White Springs, 11, 105
White Springs local fauna, 105, 122, 133
whitings, 256
Whittington, H. B., 91
Wicker, R. A., 17
Wicomico Terrace, 2
Wiggens Arch, 25
Wigley, T. M. L., 223
Wilcox Group, 51, 94
Willacoochee Creek, mammal fossils from, 122
Willacoochee local fauna, 131
Williams, C., *136*
Williams, C. T., 14, 31, 34
Williams, H., 33
Williston site, 119, 123. See also Mixsons Bone Bed
Wilson, D., 111, 112
winds: in hurricanes and tropical storms, 219–21; role of, 155, *156;* and shelf progradation, 189
Winker, C. D., 9
Winston, G. O., 50

Winter Haven Ridge, 9
Winter Park sinkhole, 229, *229*
Withlacoochee 4A site, 123, 126
Withlacoochee River, 4, 165
wolverines, 127
wolves, 125–26
Woodburne, M. O., 121
Woodring, W. P., 107
Wood River Formation, 92, *92*
worms: sabellariid, 193; vestimentiferan *(Escarpia laminata)*, 183
Wythella eldridgei, 97

Xenophora sp., 98
x-ray diffraction (XRD), 211–12, *212*

Yoho, W. R., 151
Yon, J. W., 9
Yorktown Formation, 107
Yorktownian subprovince, *107*
Young, E. J., 142
Yucatan Peninsula, 24–25, 186, 223
Yucatan Straits, 180
Yurewicz, M. C., 246

Zeitz, I., 16–17, 36
zeolite, 199–200
zircon, 31, 33, 151–52
zoarcid fish, 183
Zygorhiza, 121, 136, *136*

MESOZOIC																				P A	
JURASSIC				CRETACEOUS														PALEOCENE			
MIDDLE	UPPER				LOWER						UPPER						LOWER		UPPER		
BATHONIAN	CALLOVIAN	OXFORDIAN	KIMMERIDGIAN	TITHONIAN	BERRIASIAN	VALANGINIAN	HAUTRIVIAN	BARREMIAN	APTIAN	ALBIAN	CENOMANIAN	TURONIAN	CONIACIAN	SANTONIAN	CAMPANIAN	MAASTRICHTIAN	DANIAN		THANETIAN		
					COAHUILAN					COMANCHEAN			GULFIAN								
		ZULOAGAN		LA CASITAN	DURANGOAN			NUEVOLEONIAN		TRINITIAN	WASHITAN-FREDRICKSBURGIAN		WOODBINIAN	EAGLEFORDIAN	AUSTINIAN	TAYLORAN	NAVARROAN	MIDWAYAN			
		ZULOAGAN		LA CASITAN	DURANGOAN			NUEVOLEONIAN		TRINITIAN	WASHITAN-FREDRICKSBURGIAN		WOODBINIAN	EAGLEFORDIAN	AUSTINIAN	TAYLORAN	NAVARROAN	MIDWAYAN			

Norphlet | Louann/Werner | Smackover | Buckner | Haynesville | Cotton Valley | Hosston | Sligo | Pine Island | Rodessa | Ferry Lake | Paluxy | Mooringsport | Washitan-Fredricksburgian Undifferentiated | Tuscaloosa | Eutaw | Danzer | Mooreville | Demopolis | Prarie Bluff | Ripley | Selma | Clayton | Porters Creek | Naheola | Midway | Tuscahoma | Nanafalia | Wi

Norphlet | Louann | Smackover | Haynesville | Cotton Valley | Lower Cretaceous Undifferentiated | Tuscaloosa | Eutaw | Bluffown | Providence | Ripley | Clayton | Tuscahoma | Nanafalia | Wilc

Lower Cretaceous Undifferentiated | Atkinson | Pine Key | Lawson | Cedar Keys

Basalt | Wood River | Bone Island | Pumpkin Bay | Fort Pierce | Ocean Reef | Glades | Big Cypress | Naples Bay | Atkinson | Pine Key | Card Sound Dol. | Rebecca Shoal Dol. | Cedar Keys | Rebecca Shoal Dol.

← Marquesas Supergroup →